PROJECT MANAGEMENT

CASE STUDIES, THIRD EDITION

WITHDRAWN

DATE DUE

PROJECT MANAGEMENT

CASE STUDIES, THIRD EDITION

HAROLD KERZNER, Ph.D.

Senior Executive Director for Project Management
The International Institute for Learning
New York, New York

WILEY

John Wiley & Sons, Inc.

Library of Congress Cataloging-in-Publication Data:
Kerzner, Harold.
    Project management : case studies / Harold Kerzner.—3rd ed.
      p.   cm.
    Includes index.
    ISBN 978-0-470-27871-0 (pbk.)
      1. Project management—Case studies.   I. Title.
    HD69.P75K472   2009
    658.4′04—dc22
                                                              2008051312

Printed in the United States of America

10   9   8   7   6   5   4   3   2

Contents _____

Preface _____

Other than on-the-job training, case studies and situations are perhaps the best way to learn project management. Case studies allow the students to apply the knowledge learned in lectures. Case studies require that the students investigate what went right in the case, what went wrong, and what recommendations should be made to prevent these problems from reoccurring in the future. The use of cases studies is applicable both to undergraduate and graduate level project management courses, as well as to training programs in preparation to pass the exam to become a Certified Project Management Professional (PMP®) administered by the Project Management Institute.

Situations are smaller case studies and usually focus on one or two specific points that need to be addressed, whereas case studies focus on a multitude of problems. The table of contents identifies several broad categories for the cases and situations, but keep in mind that the larger case studies, such as Corwin Corporation and The Blue Spider Project, could have been listed under several topics. Several of the cases and situations have "seed" questions provided to assist the reader in the analysis of the case. An instructor's manual is available from John Wiley & Sons, Inc., to faculty members who adopt the book for classroom use.

Almost all of the case studies are factual. In most circumstances, the cases and situations have been taken from the author's consulting practice. Some educators prefer not to use case studies dated back to the 1970s and 1980s. It would be easy just to change the dates but inappropriate in the eyes of the author. The circumstances surrounding these cases and situations are the same today as they were thirty years ago. Unfortunately we seem to be repeating several of the mistakes made previously.

Harold Kerzner
The International Institute for Learning

PROJECT MANAGEMENT

CASE STUDIES, THIRD EDITION

Part 1

PROJECT MANAGEMENT METHODOLOGIES

As companies approach some degree of maturity in project management, it becomes readily apparent to all that some sort of standardization approach is necessary for the way that projects are managed. The ideal solution might be to have a singular methodology for all projects, whether they are for new product development, information systems, or client services. Some organizations may find it necessary to maintain more than one methodology, however, such as one methodology for information systems and a second methodology for new product development.

The implementation and acceptance of a project management methodology can be difficult if the organization's culture provides a great deal of resistance toward the change. Strong executive leadership may be necessary such that the barriers to change can be overcome quickly. These barriers can exist at all levels of management as well as at the worker level. The changes may require that workers give up their comfort zones and seek out new social groups.

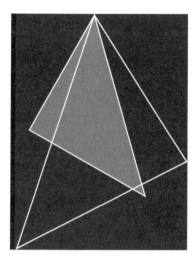

Lakes Automotive

Lakes Automotive is a Detroit-based tier-one supplier to the auto industry. Between 1995 and 1999, Lakes Automotive installed a project management methodology based on nine life-cycle phases. All 60,000 employees worldwide accepted the methodology and used it. Management was pleased with the results. Also, Lakes Automotive's customer base was pleased with the methodology and provided Lakes Automotive with quality award recognition that everyone believed was attributed to how well the project management methodology was executed.

In February 2000, Lakes Automotive decided to offer additional products to its customers. Lakes Automotive bought out another tier-one supplier, Pelex Automotive Products (PAP). PAP also had a good project management reputation and also provided quality products. Many of its products were similar to those provided by Lakes Automotive.

Because the employees from both companies would be working together closely, a singular project management methodology would be required that would be acceptable to both companies. PAP had a good methodology based on five life-cycle phases. Both methodologies had advantages and disadvantages, and both were well liked by their customers.

QUESTIONS

1. How do companies combine methodologies?
2. How do you get employees to change work habits that have proven to be successful?
3. What influence should a customer have in redesigning a methodology that has proven to be successful?
4. What if the customers want the existing methodologies left intact?
5. What if the customers are unhappy with the new combined methodology?

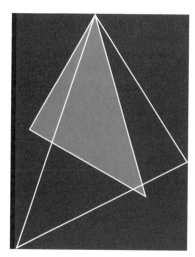

Ferris HealthCare, Inc.

In July of 1999, senior management at Ferris recognized that its future growth could very well be determined by how quickly and how well it implemented project management. For the past several years, line managers had been functioning as project managers while still managing their line groups. The projects came out with the short end of the stick, most often late and over budget, because managers focused on line activities rather than project work. Everyone recognized that project management needed to be an established career path position and that some structured process had to be implemented for project management.

A consultant was brought into Ferris to provide initial project management training for 50 out of the 300 employees targeted for eventual project management training. Several of the employees thus trained were then placed on a committee with senior management to design a project management stage-gate model for Ferris.

After two months of meetings, the committee identified the need for three different stage-gate models: one for information systems, one for new products/services provided, and one for bringing on board new corporate clients. There were several similarities among the three models. However, personal interests dictated the need for three methodologies, all based upon rigid policies and procedures.

After a year of using three models, the company recognized it had a problem deciding how to assign the right project manager to the right project. Project managers had to be familiar with all three methodologies. The alternative, considered impractical, was to assign only those project managers familiar with that specific methodology.

After six months of meetings, the company consolidated the three methodologies into a single methodology, focusing more upon guidelines than on policies and procedures. The entire organization appeared to support the new singular methodology. A consultant was brought in to conduct the first three days of a four-day training program for employees not yet trained in project management. The fourth day was taught by internal personnel with a focus on how to use the new methodology. The success to failure ratio on projects increased dramatically.

QUESTIONS

1. Why was it so difficult to develop a singular methodology from the start?
2. Why were all three initial methodologies based on policies and procedures?
3. Why do you believe the organization later was willing to accept a singular methodology?
4. Why was the singular methodology based on guidelines rather than policies and procedures?
5. Did it make sense to have the fourth day of the training program devoted to the methodology and immediately attached to the end of the three-day program?
6. Why was the consultant not allowed to teach the methodology?

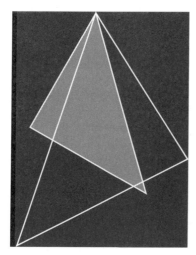

Clark Faucet Company

BACKGROUND

By 1999, Clark Faucet Company had grown into the third largest supplier of faucets for both commercial and home use. Competition was fierce. Consumers would evaluate faucets on artistic design and quality. Each faucet had to be available in at least twenty-five different colors. Commercial buyers seemed more interested in the cost than the average consumer, who viewed the faucet as an object of art, irrespective of price.

Clark Faucet Company did not spend a great deal of money advertising on the radio or on television. Some money was allocated for ads in professional journals. Most of Clark's advertising and marketing funds were allocated to the two semiannual home and garden trade shows and the annual builders trade show. One large builder could purchase more than 5,000 components for the furnishing of one newly constructed hotel or one apartment complex. Missing an opportunity to display the new products at these trade shows could easily result in a six- to twelve-month window of lost revenue.

CULTURE

Clark Faucet had a noncooperative culture. Marketing and engineering would never talk to one another. Engineering wanted the freedom to design new products, whereas marketing wanted final approval to make sure that what was designed could be sold.

The conflict between marketing and engineering became so fierce that early attempts to implement project management failed. Nobody wanted to be the project manager. Functional team members refused to attend team meetings and spent most of their time working on their own "pet" projects rather than the required work. Their line managers also showed little interest in supporting project management.

Project management became so disliked that the procurement manager refused to assign any of his employees to project teams. Instead, he mandated that all project work come through him. He eventually built up a large brick wall around his employees. He claimed that this would protect them from the continuous conflicts between engineering and marketing.

THE EXECUTIVE DECISION

The executive council mandated that another attempt to implement good project management practices must occur quickly. Project management would be needed not only for new product development but also for specialty products and enhancements. The vice presidents for marketing and engineering reluctantly agreed to try and patch up their differences, but did not appear confident that any changes would take place.

Strange as it may seem, nobody could identify the initial cause of the conflicts or how the trouble actually began. Senior management hired an external consultant to identify the problems, provide recommendations and alternatives, and act as a mediator. The consultant's process would have to begin with interviews.

ENGINEERING INTERVIEWS

The following comments were made during engineering interviews:

- "We are loaded down with work. If marketing would stay out of engineering, we could get our job done."
- "Marketing doesn't understand that there's more work for us to do other than just new product development."
- "Marketing personnel should spend their time at the country club and in bar rooms. This will allow us in engineering to finish our work uninterrupted!"

- "Marketing expects everyone in engineering to stop what they are doing in order to put out marketing fires. I believe that most of the time the problem is that marketing doesn't know what they want up front. This leads to change after change. Why can't we get a good definition at the beginning of each project?"

MARKETING INTERVIEWS

- "Our livelihood rests on income generated from trade shows. Since new product development is four to six months in duration, we have to beat up on engineering to make sure that our marketing schedules are met. Why can't engineering understand the importance of these trade shows?"
- "Because of the time required to develop new products [4–6 months], we sometimes have to rush into projects without having a good definition of what is required. When a customer at a trade show gives us an idea for a new product, we rush to get the project underway for introduction at the next trade show. We then go back to the customer and ask for more clarification and/or specifications. Sometimes we must work with the customer for months to get the information we need. I know that this is a problem for engineering, but it cannot be helped."

The consultant wrestled with the comments but was still somewhat perplexed. "Why doesn't engineering understand marketing's problems?" pondered the consultant. In a follow-up interview with an engineering manager, the following comment was made:

"We are currently working on 375 different projects in engineering, and that includes those which marketing requested. Why can't marketing understand our problems?"

QUESTIONS

1. What is the critical issue?
2. What can be done about it?
3. Can excellence in project management still be achieved and, if so, how? What steps would you recommend?
4. Given the current noncooperative culture, how long will it take to achieve a good cooperative project management culture, and even excellence?

5. What obstacles exist in getting marketing and engineering to agree to a singular methodology for project management?
6. What might happen if benchmarking studies indicate that either marketing or engineering are at fault?
7. Should a singular methodology for project management have a process for the prioritization of projects or should some committee external to the methodology accomplish this?

Part 2

IMPLEMENTATION OF PROJECT MANAGEMENT

The first step in the implementation of project management is to recognize the true benefits that can be achieved from using project management. These benefits can be recognized at all levels of the organization. However, each part of the organization can focus on a different benefit and want the project management methodology to be designed for their particular benefit.

Another critical issue is that the entire organization may not end up providing the same level of support for project management. This could delay the final implementation of project management. In addition, there may be some pockets within the organization that are primarily project-driven and will give immediate support to project management, whereas other pockets, which are primarily non–project-driven, may be slow in their acceptance.

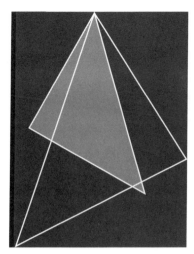

Kombs Engineering

In June 1993, Kombs Engineering had grown to a company with $25 million in sales. The business base consisted of two contracts with the U.S. Department of Energy (DOE), one for $15 million and one for $8 million. The remaining $2 million consisted of a variety of smaller jobs for $15,000 to $50,000 each.

The larger contract with DOE was a five-year contract for $15 million per year. The contract was awarded in 1988 and was up for renewal in 1993. DOE had made it clear that, although they were very pleased with the technical performance of Kombs, the follow-on contract must go through competitive bidding by law. Marketing intelligence indicated that DOE intended to spend $10 million per year for five years on the follow-on contract with a tentative award date of October 1993.

On June 21, 1993, the solicitation for proposal was received at Kombs. The technical requirements of the proposal request were not considered to be a problem for Kombs. There was no question in anyone's mind that on technical merit alone, Kombs would win the contract. The more serious problem was that DOE required a separate section in the proposal on how Kombs would manage the $10 million/year project as well as a complete description of how the project management system at Kombs functioned.

When Kombs won the original bid in 1988, there was no project management requirement. All projects at Kombs were accomplished through the traditional organizational structure. Line managers acted as project leaders.

In July 1993, Kombs hired a consultant to train the entire organization in project management. The consultant also worked closely with the proposal team in responding to the DOE project management requirements. The proposal was submitted to DOE during the second week of August. In September 1993, DOE provided Kombs with a list of questions concerning its proposal. More than 95 percent of the questions involved project management. Kombs responded to all questions.

In October 1993, Kombs received notification that it would not be granted the contract. During a post-award conference, DOE stated that they had no "faith" in the Kombs project management system. Kombs Engineering is no longer in business.

QUESTIONS

1. What was the reason for the loss of the contract?
2. Could it have been averted?
3. Does it seem realistic that proposal evaluation committees could consider project management expertise to be as important as technical ability?

Williams Machine Tool Company

For eighty-five years, the Williams Machine Tool Company had provided quality products to its clients, becoming the third largest U.S.-based machine tool company by 1990. The company was highly profitable and had an extremely low employee turnover rate. Pay and benefits were excellent.

Between 1980 and 1990, the company's profits soared to record levels. The company's success was due to one product line of standard manufacturing machine tools. Williams spent most of its time and effort looking for ways to improve its bread-and-butter product line rather than to develop new products. The product line was so successful that companies were willing to modify their production lines around these machine tools rather than asking Williams for major modifications to the machine tools.

By 1980, Williams Company was extremely complacent, expecting this phenomenal success with one product line to continue for twenty to twenty-five more years. The recession of the early 1990s forced management to realign their thinking. Cutbacks in production had decreased the demand for the standard machine tools. More and more customers were asking for either major modifications to the standard machine tools or a completely new product design.

The marketplace was changing and senior management recognized that a new strategic focus was necessary. However, lower-level management and the work force, especially engineering, were strongly resisting a change. The employees, many of them with over twenty years of employment at Williams Company, refused to recognize the need for this change in the belief that the glory days of yore would return at the end of the recession.

By 1995, the recession had been over for at least two years, yet Williams Company had no new product lines. Revenue was down, sales for the standard product (with and without modifications) were decreasing, and the employees were still resisting change. Layoffs were imminent.

In 1996, the company was sold to Crock Engineering. Crock had an experienced machine tool division of its own and understood the machine tool business. Williams Company was allowed to operate as a separate entity from 1995 to 1996. By 1996, red ink had appeared on the Williams Company balance sheet. Crock replaced all of the Williams senior managers with its own personnel. Crock then announced to all employees that Williams would become a specialty machine tool manufacturer and that the "good old days" would never return. Customer demand for specialty products had increased threefold in just the last twelve months alone. Crock made it clear that employees who would not support this new direction would be replaced.

The new senior management at Williams Company recognized that eighty-five years of traditional management had come to an end for a company now committed to specialty products. The company culture was about to change, spearheaded by project management, concurrent engineering, and total quality management.

Senior management's commitment to product management was apparent by the time and money spent in educating the employees. Unfortunately, the seasoned twenty-year-plus veterans still would not support the new culture. Recognizing the problems, management provided continuous and visible support for project management, in addition to hiring a project management consultant to work with the people. The consultant worked with Williams from 1996 to 2001.

From 1996 to 2001, the Williams Division of Crock Engineering experienced losses in twenty-four consecutive quarters. The quarter ending March 31, 2002, was the first profitable quarter in over six years. Much of the credit was given to the performance and maturity of the project management system. In May 2002, the Williams Division was sold. More than 80 percent of the employees lost their jobs when the company was relocated over 1,500 miles away.

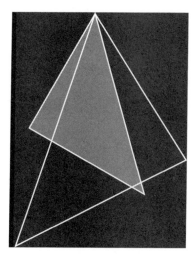

Wynn Computer Equipment (WCE)

In 1965, Joseph Wynn began building computer equipment in a small garage behind his house. By 2002, WCE was a $1 billion a year manufacturing organization employing 900 people. The major success found by WCE has been attributed to the nondegreed workers who have stayed with WCE over the past fifteen years. The nondegreed personnel account for 80 percent of the organization. Both the salary structure and fringe benefit packages are well above the industry average.

CEO PRESENTATION

In February 2002, the new vice president and general manager made a presentation to his executive staff outlining the strategies he wished to see implemented to improve productivity:

> Our objective for the next twelve months is to initiate a planning system with the focus on strategic, developmental, and operational plans that will assure continued success of WCE and support for our broad objectives. Our strategy is a four-step process:
>
> - To better clarify expectations and responsibility
> - To establish cross-functional goals and objectives

- To provide feedback and performance results to all employees in each level of management
- To develop participation through teamwork

The senior staff will merely act as a catalyst in developing long- and short-term objectives. Furthermore, the senior staff will participate and provide direction and leadership in formulating an integrated manufacturing strategy that is both technology- and human-resources-driven. The final result should be an integrated project plan that will:

- Push decision making down
- Trust the decision of peers and people in each organization
- Eliminate committee decisions

Emphasis should be on communications that will build and convey ownership in the organization and a *we* approach to surfacing issues and solving problems.

In April 2002, a team of consultants interviewed a cross section of Wynn personnel to determine the "pulse" of the organization. The following information was provided:

- "We have a terrible problem in telling our personnel (both project and functional) exactly what is expected on the project. It is embarrassing to say that we are a computer manufacturer and we do not have any computerized planning and control tools."
- "Our functional groups are very poor planners. We, in the project office, must do the planning for them. They appear to have more confidence in and pay more attention to our project office schedules than to their own."
- "We have recently purchased a $65,000 computerized package for planning and controlling. It is going to take us quite a while to educate our people. In order to interface with the computer package, we must use a work breakdown structure. This is an entirely new concept for our people."
- "We have a lack of team spirit in the organization. I'm not sure if it is simply the result of poor communications. I think it goes further than that. Our priorities get shifted on a weekly basis, and this produces a demoralizing effect. As a result, we cannot get our people to live up to either their old or new commitments."
- "We have a very strong mix of degreed and nondegreed personnel. All new, degreed personnel must 'prove' themselves before being officially accepted by the nondegreed personnel. We seem to be splitting the organization down the middle. Technology has become more important than loyalty and tradition and, as a result, the nondegreed personnel, who believe themselves to be the backbone of the organization, now feel cheated. What is a proper balance between experience and new blood?"

- "The emphasis on education shifts with each new executive. Our nondegreed personnel obviously are paying the price. I wish I knew what direction the storm is coming from."

- "My department does not have a database to use for estimating. Therefore, we have to rely heavily on the project office for good estimating. Anyway, the project office never gives us sufficient time for good estimating so we have to ask other groups to do our scheduling for us."

- "As line manager, I am caught between the rock and the hard spot. Quite often, I have to act as the project manager and line manager at the same time. When I act as the project manager I have trouble spending enough time with my people. In addition, my duties also include supervising outside vendors at the same time."

- "My departmental personnel have a continuous time management problem because they are never full-time on any one project, and all of our projects never have 100 percent of the resources they need. How can our people ever claim ownership?"

- "We have trouble in conducting up-front feasibility studies to see if we have a viable product. Our manufacturing personnel have poor interfacing with advanced design."

- "If we accept full project management, I'm not sure where the project managers should report. Should we have one group of project managers for new processes/products and a second group for continuous (or old) processes/products? Can both groups report to the same person?"

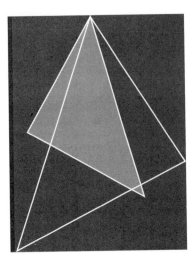

The Reluctant Workers

Tim Aston had changed employers three months ago. His new position was project manager. At first he had stars in his eyes about becoming the best project manager that his company had ever seen. Now, he wasn't sure if project management was worth the effort. He made an appointment to see Phil Davies, director of project management.

Tim Aston: "Phil, I'm a little unhappy about the way things are going. I just can't seem to motivate my people. Every day, at 4:30 P.M., all of my people clean off their desks and go home. I've had people walk out of late afternoon team meetings because they were afraid that they'd miss their car pool. I have to schedule morning team meetings."

Phil Davies: "Look, Tim. You're going to have to realize that in a project environment, people think that they come first and that the project is second. This is a way of life in our organizational form."

Tim Aston: "I've continually asked my people to come to me if they have problems. I find that the people do not think that they need help and, therefore, do not want it. I just can't get my people to communicate more."

Phil Davies: "The average age of our employees is about forty-six. Most of our people have been here for twenty years. They're set in their ways. You're the first person that we've hired in the past three years. Some of our people may just resent seeing a thirty-year-old project manager."

Tim Aston: "I found one guy in the accounting department who has an excellent head on his shoulders. He's very interested in project management. I asked his boss if he'd release him for a position in project management, and his boss just laughed at me, saying something to the effect that as long as that guy is doing a good job for him, he'll never be released for an assignment elsewhere in the company. His boss seems more worried about his personal empire than he does in what's best for the company.

"We had a test scheduled for last week. The customer's top management was planning on flying in for firsthand observations. Two of my people said that they had programmed vacation days coming, and that they would not change, under any conditions. One guy was going fishing and the other guy was planning to spend a few days working with fatherless children in our community. Surely, these guys could change their plans for the test."

Phil Davies: "Many of our people have social responsibilities and outside interests. We encourage social responsibilities and only hope that the outside interests do not interfere with their jobs.

"There's one thing you should understand about our people. With an average age of forty-six, many of our people are at the top of their pay grades and have no place to go. They must look elsewhere for interests. These are the people you have to work with and motivate. Perhaps you should do some reading on human behavior."

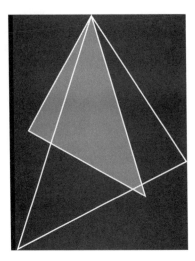

Hyten Corporation

On June 5, 1998, a meeting was held at Hyten Corporation, between Bill Knapp, director of sales, and John Rich, director of engineering. The purpose of the meeting was to discuss the development of a new product for a special customer application. The requirements included a very difficult, tight-time schedule. The key to the success of the project would depend on timely completion of individual tasks by various departments.

Bill Knapp: "The Business Development Department was established to provide coordination between departments, but they have not really helped. They just stick their nose in when things are going good and mess everything up. They have been out to see several customers, giving them information and delivery dates that we can't possibly meet."

John Rich: "I have several engineers who have MBA degrees and are pushing hard for better positions within engineering or management. They keep talking that formal project management is what we should have at Hyten. The informal approach we use just doesn't work all the time. But I'm not sure that just any type of project management will work in our division."

Knapp: "Well, I wonder who Business Development will tap to coordinate this project? It would be better to get the manager from inside the organization instead of hiring someone from outside."

Exhibit I. Organizational chart of the automotive division, Hyten Corporation

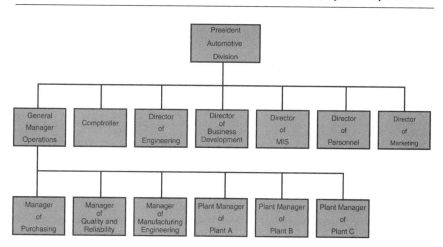

COMPANY BACKGROUND

Hyten Company was founded in 1982 as a manufacturer of automotive components. During the Gulf War, the company began manufacturing electronic components for the military. After the war, Hyten continued to prosper.

Hyten became one of the major component suppliers for the Space Program, but did not allow itself to become specialized. When the Space Program declined, Hyten developed other product lines, including energy management, building products, and machine tools, to complement their automotive components and electronics fields.

Hyten has been a leader in the development of new products and processes. Annual sales are in excess of $600 million. The Automotive Components Division is one of Hyten's rapidly expanding business areas (see the organizational chart in Exhibit I).

THE AUTOMOTIVE COMPONENTS DIVISION

The management of both the Automotive Components Division and the Corporation itself is young and involved. Hyten has enjoyed a period of continuous growth over the past fifteen years as a result of careful planning and having the right people in the right positions at the right time. This is emphasized by the fact that within five years of joining Hyten, every major manager and division head has been promoted to more responsibility within the corporation. The management staff of the Automotive Components Division has an average age of forty and no one is over fifty. Most of the middle managers have MBA degrees and a few have Ph.D.s. Currently, the Automotive Components Division has three manufacturing plants at various locations throughout the country. Central offices and most of the nonproduction functions are located at the main plant. There has been some effort by past presidents to give each separate plant some minimal level of purchasing, quality, manufacturing engineering and personnel functions.

INFORMAL PROJECT MANAGEMENT AT HYTEN CORPORATION

The Automotive Components Division of Hyten Corporation has an informal system of project management. It revolves around each department handling their own functional area of a given product development or project. Projects have been frequent enough that a sequence of operations has been developed to take a new product from concept to market. Each department knows its responsibilities and what it must contribute to a project.

A manager within the Business Development Department assumes informal project coordination responsibility and calls periodic meetings of the department heads involved. These meetings keep everyone advised of work status, changes to the project, and any problem areas. Budgeting of the project is based on the cost analysis developed after the initial design, while funding is allocated to each functional department based on the degree of its involvement. Funding for the initial design phase is controlled through business development. The customer has very little control over the funding, manpower, or work to be done. The customer, however, dictates when the new product design must be available for integration into the vehicle design, and when the product must be available in production quantities.

THE BUSINESS DEVELOPMENT DEPARTMENT

The Business Development Department, separate from Marketing/Sales, functions as a steering group for deciding which new products or customer requests are to be pursued and which are to be dropped. Factors which they consider in making these decisions are: (1) the company's long- and short-term business plans, (2) current sales forecasts, (3) economic and industry indicators, (4) profit potential, (5) internal capabilities (both volume and technology), and (6) what the customer is willing to pay versus estimated cost.

The duties of Business Development also include the coordination of a project or new product from initial design through market availability. In this capacity, they have no formal authority over either functional managers or functional employees. They act strictly on an informal basis to keep the project moving, give status reports, and report on potential problems. They are also responsible for the selection of the plant that will be used to manufacture the product.

The functions of Business Development were formerly handled as a joint staff function where all the directors would periodically meet to formulate short-range plans and solve problems associated with new products. The department was formally organized three years ago by the then-38-year-old president as a recognition of the need for project management within the Automotive Components Division.

Manpower for the Business Development Department was taken from both outside the company and from within the division. This was done to honor the Corporation's commitment to hire people from the outside only after it was determined that there were no qualified people internally (an area that for years has been a sore spot to the younger managers and engineers).

When the Business Development Department was organized, its level of authority and responsibility was limited. However, the Department's authority and responsibility have subsequently expanded, though at a slow rate. This was done so as not to alienate the functional managers who were concerned that project management would undermine their "empire."

INTRODUCTION OF FORMAL PROJECT MANAGEMENT AT HYTEN CORPORATION

On July 10, 1998, Wilbur Donley was hired into the Business Development Department to direct new product development efforts. Prior to joining Hyten, he worked as project manager with a company that supplied aircraft hardware to the government. He had worked both as an assistant project manager and as a project manager for five years prior to joining Hyten.

Shortly after his arrival, he convinced upper management to examine the idea of expanding the Business Development group and giving them responsibility for formal project management. An outside consulting firm was hired to give an in-depth seminar on project management to all management and supervisor employees in the Division.

Prior to the seminar, Donley talked to Frank Harrel, manager of quality and reliability, and George Hub, manager of manufacturing engineering, about their problems and what they thought of project management.

Frank Harrel is thirty-seven years old, has an MBA degree, and has been with Hyten for five years. He was hired as an industrial engineer and three years ago was promoted to manager of quality and reliability. George Hub is forty-five years old and has been with Hyten for twelve years as manager of manufacturing engineering.

Wilbur Donley: "Well, Frank, what do you see as potential problems to the timely completion of projects within the Automotive Components Division?"

Frank Harrel: "The usual material movement problems we always have. We monitor all incoming materials in samples and production quantities, as well as in-process checking of production and finished goods on a sampling basis. We then move to 100 percent inspection if any discrepancies are found. Marketing and Manufacturing people don't realize how much time is required to inspect for either internal or customer deviations. Our current manpower requires that schedules be juggled to accommodate 100 percent inspection levels on 'hot items.' We seem to be getting more and more items at the last minute that must be done on overtime."

Donley: "What are you suggesting? A coordination of effort with marketing, purchasing, production scheduling, and the manufacturing function to allow your department to perform their routine work and still be able to accommodate a limited amount of high-level work on 'hot' jobs?"

Harrel: "Precisely, but we have no formal contact with these people. More open lines of communication would be of benefit to everyone."

Donley: "We are going to introduce a more formal type of project management than has been used in the past so that all departments who are involved will actively participate in the planning cycle of the project. That way they will remain aware of how they affect the function of other departments and prevent overlapping of work. We should be able to stay on schedule and get better cooperation."

Harrel: "Good, I'll be looking forward to the departure from the usual method of handling a new project. Hopefully, it will work much better and result in fewer problems."

Donley: "How do you feel, George, about improving the coordination of work among various departments through a formal project manager?"

George Hub: "Frankly, if it improves communication between departments, I'm all in favor of the change. Under our present system, I am asked to make estimates of cost and lead times to implement a new product. When the project begins, the Product Design group starts making changes that require new cost figures and lead times. These changes result in cost overruns and in not meeting schedule dates. Typically, these changes continue right up to the production start date. Manufacturing appears to be the bad guy for not meeting the scheduled start date. We need someone to coordinate the work of various departments to prevent this continuous redoing of various jobs. We will at least have a chance at meeting the schedule, reducing cost, and improving the attitude of my people."

PERSONNEL DEPARTMENT'S VIEW OF PROJECT MANAGEMENT

After the seminar on project management, a discussion was held between Sue Lyons, director of personnel, and Jason Finney, assistant director of personnel. The discussion was about changing the organization structure from informal project management to formal project management.

Sue Lyons: "Changing over would not be an easy road. There are several matters to be taken under consideration."

Jason Finney: "I think we should stop going to outside sources for competent people to manage new projects that are established within Business Development. There are several competent people at Hyten who have MBA's in Systems/Project Management. With that background and their familiarity with company operations, it would be to the company's advantage if we selected personnel from within our organization."

Lyons: "Problems will develop whether we choose someone form inside the company or from an outside source."

Finney: "However, if the company continues to hire outsiders into Business Development to head new projects, competent people at Hyten are going to start filtering to places of new employment."

Lyons: "You are right about the filtration. Whoever is chosen to be a project manager must have qualifications that will get the job done. He or she should not only know the technical aspect behind the project, but should also be able to work with people and understand their needs. Project managers have to show concern for team members and provide them with work challenge. Project managers must work in a dynamic environment. This often requires the implementation of change. Project managers must be able to live with change and provide necessary leadership to implement the change. It is the project manager's responsibility to develop an atmosphere to allow people to adapt to the changing work environment.

"In our department alone, the changes to be made will be very crucial to the happiness of the employees and the success of projects. They must feel they are being given a square deal, especially in the evaluation procedure. Who will do the evaluation? Will the functional manager be solely responsible for the evaluation when, in fact, he or she might never see the functional employee for the duration of a project? A functional manager cannot possibly keep tabs on all the functional employees who are working on different projects."

Finney: "Then the functional manager will have to ask the project managers for evaluation information."

Lyons: "I can see how that could result in many unwanted situations. To begin with, say the project manager and the functional manager don't see eye to eye on things. Granted, both should be at the same grade level and neither one has authority over the other, but let's say there is a situation where the two of them disagree as to either direction or quality of work. That puts the functional employee in an awkward position. Any employee will have the tendency of bending toward the individual who signs his or her promotion and evaluation form. This can influence the project manager into recommending an evaluation below par regardless of how the functional employee performs. There is also the situation where the employee is on the project for only a couple of weeks, and spends most of his or her time working alone, never getting a chance to know the project manager. The project manager will probably give the functional employee an average rating, even though the employee has done an excellent job. This results from very little contact. Then what do you do when the project manager allows personal feelings to influence his or her evaluation of a functional employee? A project manager who knows the functional employee personally might be tempted to give a strong or weak recommendation, regardless of performance."

Finney: "You seem to be aware of many difficulties that project management might bring."

Lyons: "Not really, but I've been doing a lot of homework since I attended that seminar on project management. It was a good seminar, and since there is not much written on the topic, I've been making a few phone calls to other colleagues for their opinions on project management."

Finney: "What have you learned from these phone calls?"

Lyons: "That there are more personnel problems involved. What do you do in this situation? The project manager makes an excellent recommendation to the functional manager. The functional employee is aware of the appraisal and feels he or she should be given an above average pay increase to match the excellent job appraisal, but the functional manager fails to do so. One personnel manager from another company incorporating project management ran into problems when the project manager gave an employee of one grade level responsibilities of a higher grade level. The employee did an outstanding job taking on the responsibilities of a higher grade level and expected a large salary increase or a promotion."

Finney: "Well, that's fair, isn't it?"

Lyons: "Yes, it seems fair enough, but that's not what happened. The functional manager gave an average evaluation and argued that the project manager had no business giving the functional employee added responsibility without first checking with him. So, then what you have is a disgruntled employee ready to seek employment elsewhere. Also, there are some functional managers who will only give above-average pay increases to those employees who stay in the functional department and make that manager look good."

Lyons: "Right now I can see several changes that would need to take place. The first major change would have to be attitudes toward formal project management and hiring procedures. We do have project management here at Hyten but on an informal basis. If we could administer it formally, I feel we could do the company a great service. If we seek project managers from within, we could save on time and money. I could devote more time and effort on wage and salary grades and job descriptions. We would need to revise our evaluation forms—presently they are not adequate. Maybe we should develop more than one evaluation form: one for the project manager to fill out and give to the functional manager, and a second form to be completed by the functional manager for submission to Personnel."

Finney: "That might cause new problems. Should the project manager fill out his or her evaluation during or after project completion?"

Lyons: "It would have go be after project completion. That way an employee who felt unfairly evaluated would not feel tempted to screw up the project. If an employee felt the work wasn't justly evaluated, that employee might decide not to show up for a few days—these few days of absence could be most crucial for timely project completion."

Finney: "How will you handle evaluation of employees who work on several projects at the same time? This could be a problem if employees are really enthusiastic about one project over another. They could do a terrific job on the project they are interested in and slack off on other projects. You could also have functional people working on departmental jobs but charging their time to the project overhead. Don't we have exempt and nonexempt people charging to projects?"

Lyons: "See what I mean? We can't just jump into project management and expect a bed of roses. There will have to be changes. We can't put the cart before the horse."

Finney: "I realize that, Sue, but we do have several MBA people working here at Hyten who have been exposed to project management. I think that if we start putting our heads together and take a systematic approach to this matter, we will be able to pull this project together nicely."

Lyons: "Well, Jason, I'm glad to see that you are for formal project management. We will have to approach top management on the topic. I would like you to help coordinate an equitable way of evaluating our people and to help develop the appropriate evaluation forms."

PROJECT MANAGEMENT AS SEEN BY THE VARIOUS DEPARTMENTS

The general manager arranged through the personnel department to interview various managers on a confidential basis. The purpose of the interview was to evaluate the overall acceptance of the concept of formal project management. The answers to the question, "How will project management affect your department?" were as follows:

Frank Harrel, quality and reliability manager

> Project management is the actual coordination of the resources of functional departments to achieve the time, cost, and performance goals of the project. As a consequence, personnel interfacing is an important component toward the success of the project. In terms of quality control, it means less of the attitude of the structured workplace where quality is viewed as having the function of finding defects and, as a result, is looked upon as a hindrance to production. It means that the attitude toward quality control will change to one of interacting with other departments to minimize manufacturing problems. Project management reduces suboptimization among functional areas and induces cooperation. Both company and department goals can be achieved. It puts an end to the "can't see the forest for the trees" syndrome.

Harold Grimes, plant manager

> I think that formal project management will give us more work than long-term benefits. History indicates that we hire more outside people for new positions than we promote from within. Who will be hired into these new project management jobs? We are experiencing a lot of backlash from people who are required to teach new people the ropes. In my opinion, we should assign inside MBA graduates with project management training to head up projects and not hire an outsider as a formal project manager. Our present system would work fine if inside people were made the new managers in the Business Development Department.

Herman Hall, director of MIS

> I have no objections to the implementation of formal project management in our company. I do not believe, however, that it will be possible to provide the reports needed by this management structure for several years. This is

due to the fact that most of my staff are deeply involved in current projects. We are currently working on the installation of minicomputers and on-line terminals throughout the plant. These projects have been delayed by the late arrival of new equipment, employee sabotage, and various start-up problems. As a result of these problems, one group admits to being six months behind schedule and the other group, although on schedule, is 18 months from their scheduled completion date. The rest of the staff currently assigned to maintenance projects consists of two systems analysts who are nearing retirement and two relatively inexperienced programmers. So, as you can readily see, unless we break up the current project teams and let those projects fall further behind schedule, it will be difficult at this time to put together another project team

The second problem is that even if I could put together a staff for the project, it might take up to two years to complete an adequate information system. Problems arise from the fact that it will take time to design a system that will draw data from all the functional areas. This design work will have to be done before the actual programming and testing could be accomplished. Finally, there would be a debugging period when we receive feedback from the user on any flaws in the system or enhancements that might be needed. We could not provide computer support to an "overnight" change to project management.

Bob Gustwell, scheduling manager

I am happy with the idea of formal project management, but I do see some problems implementing it. Some people around here like the way we do things now. It is a natural reaction for employees to fight against any changes in management style.

But don't worry about the scheduling department. My people will like the change to formal project management. I see this form of management as a way to minimize, of not eliminate, schedule changes. Better planning on the part of both department and project managers will be required, and the priorities will be set at corporate level. You can count on our support because I'm tired of being caught between production and sales.

John Rich, director of engineering

It seems to me that project management will only mess things up. We now have a good flowing chain of command in our organization. This new matrix will only create problems. The engineering department, being very technical, just can't take direction from anyone outside the department. The project office will start to skimp on specifications just to save time and dollars. Our products are too technical to allow schedules and project costs to affect engineering results.

Bringing in someone from the outside to be the project manager will make things worse. I feel that formal project management should not be implemented at Hyten. Engineering has always directed the projects, and we should keep it that way. We shouldn't change a winning combination.

Fred Kuncl, plant engineering

I've thought about the trade-offs involved in implementing formal project management at Hyten and feel that plant engineering cannot live with them. Our departmental activities are centered around highly unpredictable circumstances, which sometimes involve rapidly changing priorities related to the production function. We in plant engineering must be able to respond quickly and appropriately to maintenance activities directly related to manufacturing activities. Plant engineering is also responsible for carrying out critical preventive maintenance and plant construction projects.

Project management would hinder our activities because project management responsibilities would burden our manpower with additional tasks. I am against project management because I feel that it is not in the best interest of Hyten. Project management would weaken our department's functional specialization because it would require cross-utilization of resources, manpower, and negotiation for the services critical to plant engineering.

Bill Knapp, director of marketing

I feel that the seminar on formal project management was a good one. Formal project management could benefit Hyten. Our organization needs to focus in more than one direction at all times. In order to be successful in today's market, we must concentrate on giving all our products sharp focus. Formal project management could be a good way of placing individual emphasis on each of the products of our company. Project management would be especially advantageous to us because of our highly diversified product lines. The organization needs to efficiently allocate resources to projects, products, and markets. We cannot afford to have expensive resources sitting idle. Cross-utilization and the consequent need for negotiation ensures that resources are used efficiently and in the organization's best overall interest.

We can't afford to continue to carry on informal project management in our business. We are so diversified that all of our products can't be treated alike. Each product has different needs. Besides, the nature of a team effort would strengthen our organization.

Stanley Grant, comptroller

In my opinion, formal project management can be profitably applied in our organization. Management should not, however, expect that project management would gain instant acceptance by the functional managers and functional employees, including the finance department personnel.

The implementation of formal project management in our organization would have an impact on our cost control system and internal control system, as well.

In the area of cost control, project cost control techniques have to be formalized and installed. This would require the accounting staff to: (1) break comprehensive cost summaries into work packages, (2) prepare commitment reports for "technical decision makers," (3) approximate report data and (4) concentrate talent on major problems and opportunities. In project management, cost commitments on a project are made when various functional departments, such as engineering, manufacturing and marketing, make technical decisions to take some kind of action. Conventional accounting reports do not show the cost effects of these technical decisions until it is too late to reconsider. We would need to provide the project manager with cost commitment reports at each decision state to enable him or her to judge when costs are getting out of control. Only by receiving such timely cost commitment reports, could the project manager take needed corrective actions and be able to approximate the cost effect of each technical decision. Providing all these reports, however, would require additional personnel and expertise in our department.

In addition, I feel that the implementation of formal project management would increase our responsibilities in finance department. We would need to conduct project audits, prepare periodic comparisons of actual versus projected costs and actual versus programmed manpower allocation, update projection reports and funding schedules, and sponsor cost improvement programs.

In the area of internal control, we will need to review and modify our existing internal control system to effectively meet our organization's goals related to project management. A careful and proper study and evaluation of existing internal control procedures should be conducted to determine the extent of the tests to which our internal auditing procedures are to be restricted. A thorough understanding of each project we undertake must be required at all times.

I'm all in favor of formal project management, provided management would allocate more resources to our department so we could maintain the personnel necessary to perform the added duties, responsibilities, and expertise required.

After the interviews, Sue Lyons talked to Wilbur Donley about the possibility of adopting formal project management. As she put it,

You realize that regardless of how much support there is for formal project management, the general manager will probably not allow us to implement it for fear it will affect the performance of the Automotive Components Division.

QUESTIONS

1. What are some of the major problems facing the management of Hyten in accepting formalized project management? (Include attitude problems/personality problems.)
2. Do any of the managers appear to have valid arguments for their beliefs as to why formal project management should not be considered?

3. Are there any good reasons why Hyten should go to formal project management?
4. Has Hyten taken a reasonable approach toward implementing formal project management?
5. Has Hyten done anything wrong?
6. Should formal project management give employees more room for personal growth?
7. Will formalized project management make it appear as though business development has taken power away from other groups?
8. Were the MBAs exposed to project management?
9. Were the organizational personnel focusing more on the problems (disadvantages) or advantages of project management?
10. What basic fears do employees have in considering organizational change to formal project management?
11. Must management be sold on project management prior to implementation?
12. Is it possible that some of the support groups cannot give immediate attention to such an organizational change?
13. Do functional managers risk a loss of employee loyalty with the new change?
14. What recommendations would you make to Hyten Corporation?
15. Is it easier or more difficult to implement a singular methodology for project management after the company has adopted formal project management rather than informal project management?
16. Is strategic planning for project management easier or more difficult to perform with formal project management in place?

Macon, Inc.

Macon was a fifty-year-old company in the business of developing test equipment for the tire industry. The company had a history of segregated departments with very focused functional line managers. The company had two major technical departments: mechanical engineering and electrical engineering. Both departments reported to a vice president for engineering, whose background was always mechanical engineering. For this reason, the company focused all projects from a mechanical engineering perspective. The significance of the test equipment's electrical control system was often minimized when, in reality, the electrical control systems were what made Macon's equipment outperform that of the competition.

Because of the strong autonomy of the departments, internal competition existed. Line managers were frequently competing with one another rather than focusing on the best interest of Macon. Each would hope the other would be the cause for project delays instead of working together to avoid project delays altogether. Once dates slipped, fingers were pointed and the problem would worsen over time.

One of Macon's customers had a service department that always blamed engineering for all of their problems. If the machine was not assembled correctly, it was engineering's fault for not documenting it clearly enough. If a component failed, it was engineering's fault for not designing it correctly. No matter what problem occurred in the field, customer service would always put the blame on engineering.

As might be expected, engineering would blame most problems on production claiming that production did not assemble the equipment correctly and did not maintain the proper level of quality. Engineering would design a product and then throw it over the fence to production without ever going down to the manufacturing floor to help with its assembly. Errors or suggestions reported from production to engineering were being ignored. Engineers often perceived the assemblers as incapable of improving the design.

Production ultimately assembled the product and shipped it out to the customer. Oftentimes during assembly the production people would change the design as they saw fit without involving engineering. This would cause severe problems with documentation. Customer service would later inform engineering that the documentation was incorrect, once again causing conflict among all departments.

The president of Macon was a strong believer in project management. Unfortunately, his preaching fell upon deaf ears. The culture was just too strong. Projects were failing miserably. Some failures were attributed to the lack of sponsorship or commitment from line managers. One project failed as the result of a project leader who failed to control scope. Each day the project would fall further behind because work was being added with very little regard for the project's completion date. Project estimates were based upon a "gut feel" rather than upon sound quantitative data.

The delay in shipping dates was creating more and more frustration for the customers. The customers began assigning their own project managers as "watchdogs" to look out for their companies' best interests. The primary function of these "watchdog" project managers was to ensure that the equipment purchased would be delivered on time and complete. This involvement by the customers was becoming more prominent than ever before.

The president decided that action was needed to achieve some degree of excellence in project management. The question was what action to take, and when.

QUESTIONS

1. Where will the greatest resistance for excellence in project management come from?
2. What plan should be developed for achieving excellence in project management?
3. How long will it take to achieve some degree of excellence?
4. Explain the potential risks to Macon if the customer's experience with project management increases while Macon's knowledge remains stagnant.

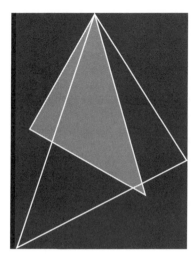

Continental
Computer
Corporation

"We have a unique situation here at Continental," remarked Ed White, Vice President for Engineering.

> We have three divisions within throwing distance of one another, and each one operates differently. This poses a problem for us at corporate headquarters because career opportunities and administrative policies are different in each division. Now that we are looking at project management as a profession, how do we establish uniform career path opportunities across all divisions?

Continental Computer Corporation (CCC) was a $9 billion a year corporation with worldwide operations encompassing just about every aspect of the computer field. The growth rate of CCC had exceeded 13 percent per year for the last eight years, primarily due to the advanced technology developed by their Eton Division, which produces disk drives. Continental is considered one of the "giants" in computer technology development, and supplies equipment to other computer manufacturers.

World headquarters for CCC is in Concord, Illinois, a large suburb northwest of Chicago in the heart of Illinois's technology center. In addition to corporate headquarters, there are three other divisions: the Eton Division, which manufactures disk drives, the Lampco Division, which is responsible for Department of Defense (DoD) contracts such as for military application, satellites, and so on, and the Ridge Division, which is the primary research center for peripherals and terminals.

According to Ed White:

Our major problems first began to surface during the early nineties. When we restructured our organization, we assumed that each division would operate as a separate entity (i.e., strategic business unit) without having to communicate with one another except through corporate headquarters. Therefore, we permitted each of our division vice presidents and general managers to set up whatever organizational structure they so desired in order to get the work accomplished. Unfortunately, we hadn't considered the problem of coordinating efforts between sister divisions because some of our large projects demanded this.

The Lampco Division is by far the oldest, having been formed in 1989. The Lampco Division produces about $2 billion worth of revenue each year from DoD funding. Lampco utilizes a pure matrix structure. Our reason for permitting the divisions to operate independently was cost reporting. In the Lampco Division, we must keep two sets of books: one for government usage and one for internal control. This was a necessity because of DoD's requirement for earned value reporting on our large, cost-reimbursable contracts. It has taken us about five years or so to get used to this idea of multiple information systems, but now we have it well under control.

We have never had to lay people off in the Lampco Division. Yet, our computer engineers still feel that a reduction in DoD spending may cause massive layoffs here. Personally, I'm not worried. We've been through lean and fat times without having to terminate people.

The big problem with the Lampco Division is that because of the technology developed in some of our other divisions, Lampco must subcontract out a good portion of the work (to our other divisions). Not that Lampco can't do it themselves, but we do have outstanding R&D specialists in our other divisions.

We have been somewhat limited in the salary structure that we can provide to our engineers. Our computer engineers in the Lampco Division used to consider themselves as aerospace engineers, not computer engineers, and were thankful for employment and reasonable salaries. But now the Lampco engineers are communicating more readily with our other divisions and think that the grass is greener in these other divisions. Frankly, they're right. We've tried to institute the same wage and salary program corporate-wide, but came up with problems. Our engineers, especially the younger ones who have been with us five or six years, are looking for management positions. Almost all of our management positions in engineering are filled with people between thirty-five and forty years of age. This poses a problem in that there is no place for these younger engineers to go. So, they seek employment elsewhere.

We've recently developed a technical performance ladder that is compatible to our management ladder. At the top of the technical ladder we have our consultant grade. Here our engineers can earn just about any salary based, of course, on their performance. The consultant position came about because of a problem in our Eton Division. I would venture to say that in the entire computer world, the most difficult job is designing disk drives. These people are specialists in a world of their own. There are probably only twenty-five people in the world who possess this expertise. We have five of them here at Continental. If one of our competitors would come in here and lure away just two of these guys, we would literally have to close down the Eton Division. So we've developed a consultant category. Now the word has spread and all of our engineers are applying for transfer to the Eton Division so as to become eligible for this new pay grade. In the Lampco Division alone I have had over fifty requests for transfer from engineers who now consider themselves as computer engineers. To make matters worse, the job market in computer technology is so good today that these people could easily leave us for more money elsewhere.

We've been lucky in the Lampco Division. Most of our contracts are large, and we can afford to maintain a project office staffed with three or four project engineers. These project engineers consider themselves as managers, not engineers. Actually they're right in doing so because theoretically they are engineering managers, not doers. Many of our people in Lamco are title-oriented and would prefer to be a project engineer as opposed to any other position. Good project engineers have been promoted, or laterally transferred, to project management so that we can pay them more. Actually, they do the same work.

In our Eton Division, we have a somewhat weird project management structure. We're organized on a product form rather than a project form of management. The engineers are considered to be strictly support for the business development function, and are not permitted to speak to the customers except under special circumstances. Business development manages both the product lines and R&D projects going on at one time. The project leader is selected by the director of engineering and can be a functional manager or just a functional employee. The project leader reports to his normal supervisor. The project leader must also report informally to one of the business development managers who is also tracking this project. This poses a problem in that when a conflict occurs, we sometimes have to take it up two or three levels before it can be resolved. Some conflicts have been so intense that they've had to be resolved at the corporate level.

The Eton Division happens to be our biggest money maker. We're turning out disk drives at an incredible rate and are backlogged with orders for at least six months. Many of our top R&D engineers are working in production support capacities because we cannot get qualified people fast enough. Furthermore, we have a yearly turnover rate in excess of 10 percent among our engineers below thirty years of age. We have several engineers who are earning more than their department managers. We also have five consultant engineers who are earning more than the department managers. We also have four consultant engineers who are earning as much as division managers.

We've had the greatest amount of problems in this division. Conflicts continuously arise due to interdependencies and misunderstandings. Our product line managers are the only people permitted to see the customers. This often alienates our engineering and manufacturing people, who are often called upon to respond to customer requests.

Planning is another major problem that we're trying to improve upon. We have trouble getting our functional mangers to make commitments. Perhaps this is a result of our inability to develop a uniform procedure for starting up a program. We always argue about when to anchor down the work. Our new, younger employees want to anchor everything down at once, whereas the poor project managers say not to anchor down anything. We, therefore, operate at all levels of the spectrum.

We can carry this problem one step further. How do we get an adequate set of objectives defined initially? We failed several times before because we couldn't get corporate agreement or understanding. We're trying to establish a policy for development of an architectural design document that will give good front-end definition.

Generally we're O.K. if we're simply modifying an existing product line. But with new product lines we have a problem in convincing people, especially our old customers.

The Ridge Division was originally developed to handle all corporate R&D activities. Unfortunately, our growth rate became so large and diversified that this became impractical. We, therefore, had to decentralize the R&D activities. This meant that each division could do their own R&D work. Corporate then had the responsibility for resolving conflicts, establishing priorities, and ensuring that all division are well-informed of the total R&D picture. Corporate must develop good communication channels between the divisions so that duplication of effort does not occur.

Almost all of our technical specialists have advanced degrees in engineering disciplines. This poses a severe problem for us, especially since we have a pure traditional structure. When a new project comes up, the project is assigned to the functional department that has the majority of the responsibility. One of the functional employees is then designated as the project manager. We realize that the new project manager has no authority to control resources that are assigned to other departments. Fortunately, our department managers realize this also, and usually put forth a concerted effort to provide whatever resources are needed. Most of the conflicts that do occur are resolved at the department manager level.

When a project is completed, the project manager returns to his or her former position as an engineering member of a functional organization. We've been quite concerned about these people that continuously go back and forth between project management and functional project engineering. This type of relationship is a must in our environment because our project managers must have a command of technology. We continuously hold in-house seminars on project management so as to provide our people with training in management skills, cost control, planning, and scheduling. We feel that we've been successful in this regard. We are always afraid that if we continue to grow, we'll have to change our structure and throw the company into chaos. Last time when we began to grow, corporate reassigned some of our R&D activities to other divisions. I often wonder what would have happened if this had not been done.

For R&D projects that are funded out of house, we generally have no major management problems for our project managers or project engineers. For corporate funded projects, however, life becomes more complex mainly because we have a tough time distinguishing when to kill a project or to pour money into it. Our project managers always argue that with just a little more corporate funding they can solve the world's greatest problems.

From the point of view of R&D, our biggest problems are in "grass roots projects." Let me explain what I mean by this. An engineer comes up with an idea and wants some money to pursue it. Unfortunately, our division managers are not budgeted for "seed monies" whenever an employee comes up with an idea for research or new product development. Each person must have a charge number to bill his time against. I know of virtually no project manager who would out-and-out permit someone to do independent research on a budgeted project.

So the engineer comes to us at corporate looking for seed money. Occasionally, we at corporate provide up to $50,000 for short-term seed money. That $50,000 might last for three to four months if the engineer is lucky. Unfortunately, obtaining the money is the lesser of the guy's problems. If the engineer needs support from another department, he's not going to get it because his project is just an informal "grass roots" effort, whereas everything else is a clearly definable, well-established project. People are reluctant to attach themselves to a "grass roots" effort because history has shown that the majority of them will be failures.

The researcher now has the difficult job of trying to convince people to give him support while continuously competing with other projects that are clearly defined and have established priorities. If the guy is persistent, however, he has a good chance to succeed. If he succeeds, he gets a good evaluation. But if he fails, he's at the mercy of his functional manager. If the functional manager felt that this guy could have been of more value to the company on a project basis, the he's liable to grade him down. But even with these risks, we still have several "seed money" requests each month by employees looking for glory.

Everyone sat around the gable listening to Ed White' comments. What had started out as a meeting to professionalize project management as a career path position, uniformly applied across all divisions seemed to have turned into a complaint session. The problems identified by Ed White now left people with the notion that there may be more pressing problems.

QUESTIONS

1. Is it common for companies to maintain two or more sets of books for cost accounting?

2. Is the matrix structure well suited for the solution to the above question?
3. Why do most project management structures find the necessity for a dual ladder system?
4. Should companies with several different types of projects have a uniform procedure for planning projects?
5. Is it beneficial to have to take conflicts up two or three levels for resolution?
6. Should project managers be permitted to talk to the customer even if the project is in support of a product line?
7. Should corporate R&D be decentralized?
8. What is meant by seed money?
9. How does control of seed money differ in a decentralized versus a centralized R&D environment?
10. Should the failure of a "grass roots" project affect an employee's opportunity for promotion?
11. If you were the vice president of either engineering or R&D, would you prefer centralized or decentralized control?
12. In either case, how would you handle each of the previously defined problems?

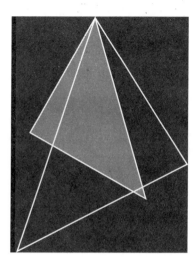

Goshe Corporation

"I've called this meeting to try to find out why we're having a difficult time upgrading our EDP [Electronic Data Processing] Department to an MIS [Managment Information Systems] Division," remarked Herb Banyon, executive vice president of Goshe Corporation.

> Last year we decided to give the EDP Department a chance to show that it could contribute to corporate profits by removing the department from under the control of the Finance Division and establishing an MIS Division. The MIS Division should be a project-driven division using a project management methodology. I expected great results. I continuously get reports stating that we're having major conflicts and personality clashes among the departments involved in these MIS projects and that we're between one month to three months behind on almost all projects. If we don't resolve this problem right now, the MIS Division will be demoted to a department and once again find itself under the jurisdiction of the finance director.

BACKGROUND

In June 1997, Herb Banyon announced that Goshe Corporation would be giving salary increases amounting to an average of 7 percent companywide, with the percent distribution as shown in Exhibit I. The EDP Department, especially the scientific programmers, were furious because this was the third straight year they had received below-average salary increases. The scientific programmers felt that they were performing engineering-type work and, therefore, should be paid according to the engineering pay scale. In addition, the software that was developed by the scientific programs was shortening schedules and lowering manufacturing costs. The scientific programmers were contributing to corporate profitability.

The year before, the scientific programmers had tried to convince management that engineering needed its own computer and that there should be established a separate engineering computer programming department within the Engineering Division. This suggestion had strong support form the engineering community because they would benefit by having complete control of their own computer. Unfortunately, management rejected the idea, fearing that competition and conflict would develop by having two data processing units, and that one centralized unit was the only viable solution.

As a result of management's decision to keep the EDP Department intact and not give them a chance to demonstrate that they can and do contribute to profits, the EDP personnel created a closed shop environment and developed a very hostile attitude toward all other departments, even those within their own Finance Division.

THE MEETING OF THE MINDS

In January 1998, Banyon announced the organizational restructuring that would upgrade the EDP Department. Al Grandy, the EDP Department manager, was given a promotion to division manager, provided that he could adequately manage the MIS project activities. By December 1988, it became apparent that something must be done to remedy the deteriorating relationship between the functional departments and the MIS personnel. Banyon called a meeting of all functional and divisional managers in hopes that some of the problems could be identified and worked out.

Herb Banyon: "For the past ten months I've watched you people continuously arguing back and forth about the MIS problems, with both sides always giving me the BS about how we'll work it out. Now, before it's too late, let's try to get at the root cause of the problem. Anyone want to start the ball rolling?"

Cost accounting manager: "The major problem, as I see it, is the lack of interpersonal skills employed by the MIS people. Our MIS personnel have received only on-the-job training. The Human Resources Department has never provided us with any project management training, especially in the behavioral areas of project management. Our organization here is, or should I say has been up to now, purely traditional, with each person reporting to and working for and with one manager. Now we have horizontal projects in which the MIS project leaders must work with several functional managers, all of whom have different management styles, different personalities, and different dispositions. The MIS group just can't turn around in one or two weeks and develop these necessary skills. It takes time and training."

Training manager: "I agree with your comments. There are two types of situations that literally demand immediate personnel development training. The first situation is when personnel are required to perform in an organizational structure that has gone from the relatively simple, pure structure to a complex, partial matrix structure. This is what has happened to us. The second situation is when the task changes from simple to complex.

"With either situation by itself, there is usually some slack time. But when both occur almost instantaneously, as is our case, immediate training should be undertaken. I told this to Grandy several times, but it was like talking to deaf ears. All he kept saying was that we don't have time now because we're loaded down with priority projects."

Al Grandy: "I can see from the start that we're headed for a rake-Grandy-over-the-coals meeting. So let me defend each accusation as it comes up. The day Banyon announced the organizational change, I was handed a list of fifteen MIS projects that had to be completed within unrealistic time schedules. I performed a manpower requirements projection and found that we were understaffed by 35 percent. Now I'm not stupid. I understand the importance of training my people. But how am I supposed to release my people for these training sessions when I have been given specific instructions that each of these fifteen projects had a high priority? I can just see myself walking into your office, Herb, telling you that I want to utilize my people only half-time so that they can undergo professional development training."

Banyon: "Somehow I feel that the buck just got passed back to me. Those schedules that I gave you appeared totally realistic to me. I just can't imagine any simple computer program requiring more time than my original estimates. And had you come to me with a request for training, I would have checked with personnel and then probably would give you the time to train your people."

Engineering manager: "I wish to make a comment or two about schedules. I'm not happy when an MIS guy walks into my office and tells me, or should I say demands, that certain resources be given to him so that he can meet a schedule or milestone date that I've had no input into establishing. My people are just not going to become pawns in the power struggle for MIS supremacy. My people become very defensive if they're not permitted to participate in the planning activities, and I have to agree with them."

Manufacturing manager: "The Manufacturing Division has a project with the MIS group for purchasing a hardware system that will satisfy our scheduling and material handling system requirements. My people wanted to be involved in the hardware selection process. Instead, the MIS group came to us with proposal in hand identifying a system that was not a practical extension of the state of the art and that did not fall within our cost and time constraints.

"We in manufacturing, being nice guys, modified our schedules to be compatible with the MIS project leaders' proposal. We then tried to provide more detailed information for the MIS team so that . . ."

Grandy: "Just a minute here! Your use of the word *we* is somewhat misleading. Project management is designed and structured so that sufficient definition of work to be performed can be obtained in order that a more uniform implementation can result. My people requested a lot of detailed information from your staff and were told to do the work ourselves and find our own information. After all, as one of the functional employees put it, if we're going to pass all of the responsibility over to you guys in project management; you people can just do it all.

"Therefore, because my people had insufficient data, between us we ended up creating a problem, which was further intensified by a lack of formal communication between the MIS group and the functional departments, as well as between the functional departments themselves. I hold functional management responsible for this problem because some of the managers did not seem to have understood that they are responsible for the project work under their cognizance. Furthermore, I consider you, the manufacturing manager, as being remiss in your duties by not reviewing the performance of our personnel assigned to the project."

Manufacturing manager: "Your people designed a system that was way too complex for our needs. Your people consider this project as a chance for glory. It is going to take us ten years to grow into this complex system you've created."

Grandy: "Let me make a few comments about our delays in the schedule. One of our projects was a six-month effort. After the third month, there was a new department manager assigned in the department that was to be the prime user of this project. We were then given a change in user requirements and incurred additional delays in waiting for new user authorization.

"Of course, people problems always affect schedules. One of my most experienced people became sick and had to be replaced by a rookie. In addition, I've tried to be a 'good guy' by letting my people help out some of the functional managers when non-MIS problems occur. This other work ended up encroaching on staff time to a degree where it impacted the schedules.

"Even though the MIS group regulates computer activities, we have no control over computer downtime or slow turnabout time. Turnabout time is directly proportional to our priority lists, and we all know that these lists are established from above.

*Exhibit I.   **Goshe organizational chart. Note: Percentages indicate 1997 salary increases***

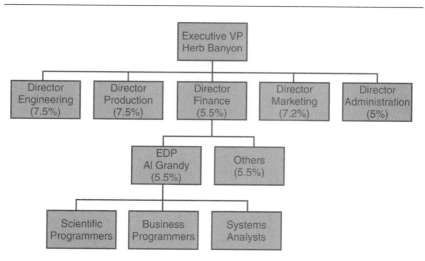

"And last, we have to consider both company and project politics. All the MIS group wanted to do was to show that we can contribute to company profits. Top management consistently tries to give us unwanted direction and functional management tries to sabotage our projects for fear that if we're successful, then it will be less money for their departments during promotion time."

Banyon: "Well, I guess we've identified the major problem areas. The question remaining is: What are we going to do about it?"

QUESTIONS

1. What are the major problems in the case study?
2. What are the user group's perceptions of the problem?
3. Was the company committed to project management?
4. Was project management forced upon the organization?
5. Did Goshe jump blindly into project management, or was there a gradual introduction?
6. Did the company consider the problems that could manifest themselves with the implementation of change (i.e., morale)?
7. Did the company have a good definition of project management?

8. Should there have been a new set of company policies and procedures when the MIS group was developed?
9. How were project deadlines established?
10. Who established responsibilities for resource management?
11. Was there an integrated planning and control system?
12. Was there any training for division or project managers
13. Should Grandy have been promoted to his current position, or should someone have been brought in from outside?
14. Can Grandy function effectively as both a project manager and a division manager?
15. Do you feel that Banyon understands computer programming?
16. Did anyone consider employee performance evaluations?
17. Did the company have good vertical communications?
18. Can a company without good vertical communications still have (or develop) good horizontal communications?
19. With the development of the MIS group, should each division be given 7 percent in the future?
20. What are the alternatives that are available?
21. What additional recommendations would you make?

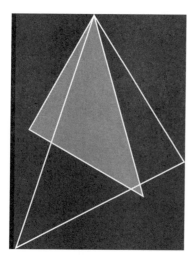

Acorn Industries

Acorn Industries, prior to July of 1996, was a relatively small midwestern corporation dealing with a single product line. The company dealt solely with commercial contracts and rarely, if ever, considered submitting proposals for government contracts. The corporation at that time functioned under a traditional form of organizational structure, although it did possess a somewhat decentralized managerial philosophy within each division. In 1993, upper management decided that the direction of the company must change. To compete with other manufacturers, the company initiated a strong acquisition program whereby smaller firms were bought out and brought into the organization. The company believed that an intensive acquisition program would solidify future growth and development. Furthermore, due to their reputation for possessing a superior technical product and strong marketing department, the acquisition of other companies would allow them to diversify into other fields, especially within the area of government contracts. However, the company did acknowledge one shortcoming that possibly could hurt their efforts—it had never fully adopted, nor implemented, any form of project management.

In July of 1996, the company was awarded a major defense contract after four years of research and development and intensive competition from a major defense organization. The company once again relied on their superior technological capabilities, combined with strong marketing efforts, to obtain the contract. According to Chris Banks, the current marketing manager at Acorn Industries, the successful proposal for the government contract was submitted

solely through the efforts of the marketing division. Acorn's successful marketing strategy relied on three factors when submitting a proposal:

1. Know exactly what the customer wants.
2. Know exactly what the market will bear.
3. Know exactly what the competition is doing and where they are going.

The contract awarded in July 1996 led to subsequent successful government
contracts and, in fact, eight more were awarded amounting to $80 million each.
These contracts were to last anywhere from seven to ten years, taking the company
into early 2009 before expiration would occur. Due to their extensive growth, especially with the area of government contracts as they pertained to weapon systems,
the company was forced in 1997 to change general managers. The company
brought in an individual who had an extensive background in program management and who previously had been heavily involved in research and development.

PROBLEMS FACING THE GENERAL MANAGER

The problems facing the new general manager were numerous. Prior to his arrival,
the company was virtually a decentralized manufacturing organization. Each division within the company was somewhat autonomous, and the functional managers
operated under a Key Management Incentive Program (KMIP). The prior general
manager had left it up to each division manager to do what was required.
Performance had been measured against attainment of goals. If the annual objective was met under the KMIP program, each division manager could expect to
receive a year-end bonus. These bonuses were computed on a percentage of the
manager's base pay, and were directly correlated to the ability to exceed the annual
objective. Accordingly, future planning within each division was somewhat stagnant, and most managers did not concern themselves with any aspect of organizational growth other than what was required by the annual objective.

Because the company had previously dealt with a single product line and
interacted solely with commercial contractors, little, if any, production planning
had occurred. Interactions between research and development and the production
engineering departments were virtually nonexistent. Research and Development
was either way behind or way ahead of the other departments at any particular
time. Due to the effects of the KMIP program, this aspect was likely to continue.

CHANGE WITHIN THE ORGANIZATIONAL STRUCTURE

To compound the aforementioned problems, the general manager faced the
unique task of changing corporate philosophy. Previously, corporate management
was concerned with a single product with a short term production cycle. Now,
however, the corporation was faced with long-term government contracts, long

cycles, and diversified products. Add to this the fact that the company was almost void of any individuals who had operated under any aspect of program management, and the tasks appeared insurmountable.

The prime motivating factor for the new general manager during the period from 1997 to 1999 was to retain profitability and maximize return on investment. In order to do this, the general manager decided to maintain the company's commercial product line, operating it at full capacity. This decision was made because the company was based in solid financial management and the commercial product line had been extremely profitable. According to the general manager, Ken Hawks,

> The concept of keeping both commercial and government contracts separate was a necessity. The commercial product line was highly competitive and maintained a good market share. If the adventure into weaponry failed, the company could always fall back on the commercial products. At any rate, the company at this time could not solely rely on the success of government contracts, which were due to expire.

In 1996, Acorn reorganized its organizational structure and created a project management office under the direct auspices of the general manager (see Exhibit I).

EXPANSION AND GROWTH

In late 1996, Acorn initiated a major expansion and reorganization within its various divisions. In fact, during the period between 1996 and 1997, the government contracts resulted in the acquiring of three new companies and possibly the acquisition of a fourth. As before, the expertise of the marketing department was heavily relied upon. Growth objectives for each division were set by corporate headquarters with the advice and feedback of the division managers. Up to 1996, Acorn's divisions had not had a program director. The program management functions for all divisions were performed by one program manager whose expertise was entirely within the commercial field. This particular program manager was concerned only with profitability and did not closely interact with the various customers. According to Chris Banks,

> The program manager's philosophy was to meet the minimum level of performance required by the contract. To attain this, he required only adequate performance. As Acorn began to become more involved with government contracts, the position remained that given a choice between high technology and low reliability, the company would always select an acquisition with low technology and high reliability. If we remain somewhere in between, future government contracts should be assured.

At the same time, Acorn established a Chicago office headed by a group executive. The office was mainly for monitoring for government contracts. Concurrently, an office was established in Washington to monitor the trends within the Department of Defense and to further act as a lobbyist for government contracts. A position of director of marketing was established to interact with the program office on contract proposals. Prior to the establishment of a director of program management position in 1997, the marketing division had been responsible for contract proposals. Acorn believed that marketing would always, as in the past, set the tone for the company. However, in 1997, and then again in 1998 (see Exhibits II and III), Acorn underwent further organizational changes. A full-time director of project management was appointed, and a program management office was set up, with further subdivisions of project managers responsible for the various government contracts. It was at this time that Acorn realized the necessity of involving the program manager more extensively in contract proposals. One faction within corporate management wanted to keep marketing responsible for contract proposals. Another decided that a combination between the marketing input and the expertise of the program director must be utilized. According to Chris Banks,

> We began to realize that marketing no longer could exclude other factors within the organization when preparing contract proposals. As project management became a reality, we realized that the project manager must be included in all phases of contract proposals.

Prior to 1996, the marketing department controlled most aspects of contract proposals. With the establishment of the program office, interface between the marketing department and the program office began to increase.

RESPONSIBILITIES OF THE PROJECT MANAGER

In 1997, Acorn, for the first time, identified a director of project management. This individual reported directly to the general manager and had under his control:

1. The project managers
2. The operations group
3. The contracts group

Under this reorganization, the director of project management, along with the project managers, possessed greater responsibility relative to contract proposals. These new responsibilities included:

1. Research and development
2. Preparation of contract proposals
3. Interaction with marketing on submittal of proposals
4. Responsibility for all government contracts
   a. Trade-off analysis
   b. Cost analysis
5. Interface with engineering department to insure satisfaction of customer's desires

Due to the expansion of government contracts, Acorn was now faced with the problem of bringing in new talent to direct ongoing projects. The previous project manager had had virtual autonomy over operations and maintained a singular philosophy. Under his tenure, many bright individuals left Acorn because future growth and career patterns were questionable. Now that the company is diversifying into other product lines, the need for young talent is crucial. Project management is still in the infancy stage.

Acorn's approach to selecting a project manager was dependent upon the size of the contract. If the particular contract was between $2 and $3 billion, the company would go with the most experienced individual. Smaller contracts would be assigned to whoever was available.

INTERACTION WITH FUNCTIONAL DEPARTMENTS

Due to the relative newness of project management, little data was available to the company to fully assess whether operations were successful. The project managers were required to negotiate with the functional departments for talent. This aspect has presented some problems due to the long-term cycle of most government contracts. Young talent within the organization saw involvement with projects as an opportunity to move up within the organization. Functional managers, on the other hand, apparently did not want to let go of young talent and were extremely reluctant to lose any form of autonomy.

Performance of individuals assigned to projects was mutually discussed between the project manager and the functional manager. Problems arose, however, due to length of projects. In some instances, if an individual had been assigned longer to the project manager than to the functional manager, the final evaluation of performance rested with the project manager. Further problems thus occurred when performance evaluations were submitted. In some instances, adequate performance was rated high in order to maintain an individual within the project scheme. According to some project managers, this aspect was a reality that must be faced, due to the shortage of abundant talent.

CURRENT STATUS

In early 1998, Acorn began to realize that a production shortage relative to government contracts would possibly occur in late 2001 or early 2003. Acorn initiated a three-pronged attack to fill an apparent void:

1. Do what you do best.
2. Look for similar product lines.
3. Look for products that do not require extensive R&D.

To facilitate these objectives, each division within the corporation established its own separate marketing department. The prime objective was to seek more federal funds through successful contract proposals and utilize these funds to increase investment into R&D. The company had finally realized that the success of the corporation was primarily attributed to the selection of the proper general manager. However, this had been accomplished at the exclusion of proper control over R&D efforts. A more lasting problem still existed, however. Program management was still less developed than in most other corporations.

Exhibit I.   1996 organizational structure

Exhibit II.   1997 organizational structure

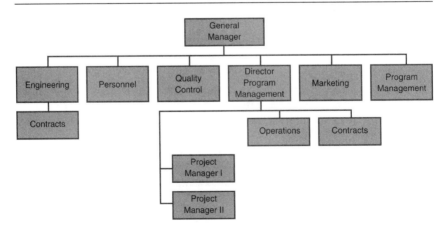

Exhibit III.   1998 organizational structure (10/1/98)

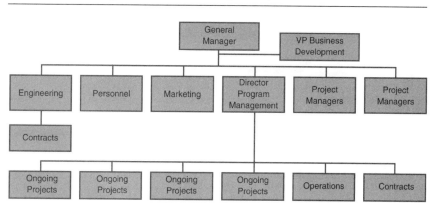

QUESTIONS

1. What are the strengths of Acorn?
2. What are the weaknesses of Acorn?
3. What are your recommendations?
4. Additional questions:
   A. Why was project management so slow in getting off the ground?
   B. Can marketing continue to prepare proposals without functional input?
   C. What should be the working relationship between the product manager and the proposal?
   D. Does KMIP benefit project management?
   E. Should KMIP be eliminated?

MIS Project
Management at
First National Bank

During the last five years, First National Bank (FNB) has been one of the fastest-growing banks in the Midwest. The holding company of the bank has been actively involved in purchasing small banks thoughout the state of Ohio. This expansion and the resulting increase of operations had been attended by considerable growth in numbers of employees and in the complexity of the organizational structure. In five years the staff of the bank has increased by 35 percent, and total assets have grown by 70 percent. FNB management is eagerly looking forward to a change in the Ohio banking laws that will allow statewide branch banking.

INFORMATION SERVICES DIVISION (ISD) HISTORY

Data processing at FNB has grown at a much faster pace than the rest of the bank. The systems and programming staff grew from twelve in 1980 to more than seventy-five during the first part of 1987. Because of several future projects, the staff was expected to increase by 50 percent during the next two years.

Prior to 1982, the Information Services Department reported to the executive vice president of the Consumer Banking and Operations Division. As a result, the first banking applications to be computerized were in the demand deposit, savings, and consumer credit banking areas. The computer was seen as a tool to

Exhibit I. *Information Services Division organizational chart*

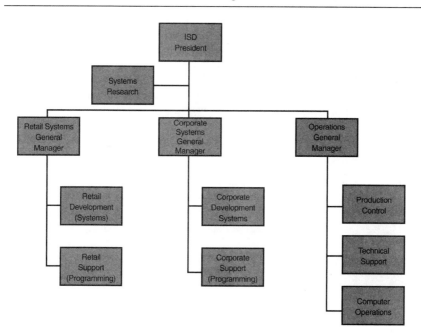

speed up the processing of consumer transactions. Little effort was expended to meet the informational requirements of the rest of the bank. This caused a high-level conflict, since each major operating organization of the bank did not have equal access to systems and programming resources. The management of FNB became increasingly aware of the benefits that could accrue from a realignment of the bank's organization into one that would be better attuned to the total information requirements of the corporation.

In 1992 the Information Services Division (ISD) was created. ISD was removed from the Consumer Banking and Operations Division to become a separate division reporting directly to the president. An organizational chart depicting the Information Services Division is shown in Exhibit I.

PRIORITIES COMMITTEE

During 1992 the Priorities Committee was formed. It consists of the chief executive officer of each of the major operating organizations whose activities are directly affected by the need for new or revised information systems.

Exhibit II. **First National Bank organizational chart**

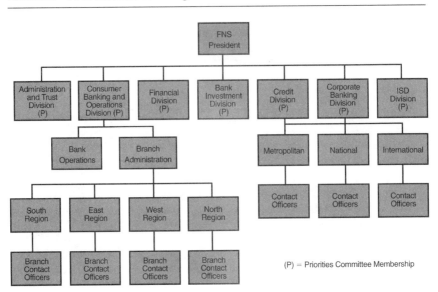

(P) = Priorities Committee Membership

The Priorities Committee was established to ensure that the resources of systems and programming personnel and computer hardware would be used only on those information systems that can best be cost justified. Divisions represented on the committee are included in Exhibit II.

The Priorities Committee meets monthly to reaffirm previously set priorities and rank new projects introduced since the last meeting. Bank policy states that the only way to obtain funds for an information development project is to submit a request to the Priorities Committee and have it approved and ranked in overall priority order for the bank. Placing potential projects in ranked sequence is done by the senior executives. The primary document used for Priorities Committee review is called the project proposal.

THE PROJECT PROPOSAL LIFE CYCLE

When a user department determines a need for the development or enhancement of an information system, it is required to prepare a draft containing a statement of the problem from its functional perspective. The problem statement is sent[jy[bnto the president of ISD, who authorizes Systems Research (see Exhibit I) to prepare an impact statement. This impact statement will include a general overview from ISD's perspective of:

- Project feasibility
- Project complexity
- Conformity with long-range ISD plans
- Estimated ISD resource commitment
- Review of similar requests
- Unique characteristics/problems
- Broad estimate of total costs

The problem and impact statements are then presented to the members of the Priorities Committee for their review. The proposals are preliminary in nature, but they permit the broad concept (with a very approximate cost attached to it) to be reviewed by the executive group to see if there is serious interest in pursuing the idea. If the interest level of the committee is low, then the idea is rejected. However, if the Priorities Committee members feel the concept has merit, they authorize the Systems Research Group of ISD to prepare a full-scale project proposal that contains:

- A detailed statement of the problem
- Identification of alternative solutions
- Impact of request on:
  - User division
  - ISD
  - Other operating divisions
- Estimated costs of solutions
- Schedule of approximate task duration
- Cost-benefit analysis of solutions
- Long-range implications
- Recommended course of action

After the project proposal is prepared by systems research, the user sponsor must review the proposal and appear at the next Priorities Committee meeting to speak in favor of the approval and priority level of the proposed work. The project proposal is evaluated by the committee and either dropped, tabled for further review, or assigned a priority relative to ongoing projects and available resources.

The final output of a Priorities Committee meeting is an updated list of project proposals in priority order with an accompanying milestone schedule that indicates the approximate time span required to implement each of the proposed projects.

The net result of this process is that the priority-setting for systems development is done by a cross section of executive management; it does not revert by default to data processing management. Priority-setting, if done by data processing, can lead to misunderstanding and dissatisfaction by sponsors of the projects that did not get ranked high enough to be funded in the near future. The project proposal cycle at FNB is diagrammed in Exhibit III. Once a project has risen to the

Exhibit III.   *The project proposal cycle*

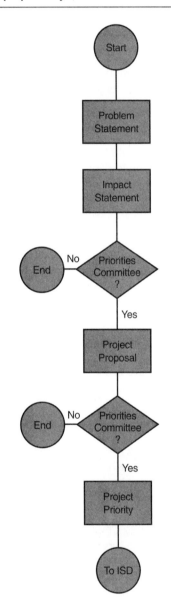

top of the ranked priority list, it is assigned to the appropriate systems group for systems definition, system design and development, and system implementation.

The time spent by systems research in producing impact statements and project proposals is considered to be overhead by ISD. No systems research time is directly charged to the development of information systems.

PROJECT LIFE CYCLE

As noted before, the systems and programming staff of ISD has increased in size rapidly and was expected to expand by another 50 percent over the next two years. As a rule, most new employees have previous data processing experience and training in various systems methodologies. ISD management recently implemented a project management system dedicated to providing a uniform step-by-step methodology for the development of management information systems. All project work is covered by tasks that make up the information project development life cycle at FNB. The subphases used by ISD in the project life cycle are:

1. Systems definition
   a. Project plan
   b. User requirements
   c. Systems definition
   d. Advisability study
2. Systems design and development
   a. Preliminary systems design
   b. Subsystems design
   c. Program design
   d. Programming and testing
3. System implementation
   a. System implementation
   b. System test
   c. Production control turnover
   d. User training
   e. System acceptance

PROJECT ESTIMATING

The project management system contains a list of all normal tasks and subtasks (over 400) to be performed during the life cycle of a development project. The project manager must examine all the tasks to determine if they apply to a given project. The manager must insert additional tasks if required and delete tasks that do not apply. The project manager next estimates the amount of time (in hours) to complete each task of each subphase of the project life cycle.

The estimating process of the project management system uses a "moving window" concept. ISD management feels that detailed cost estimating and time schedules are only meaningful for the next subphase of a project, where the visibility of the tasks to be performed is quite clear. Beyond that subphase, a more summary method of estimating is relied on. As the project progresses, new segments of the project gain visibility. Detailed estimates are made for the next major portion of the project, and summary estimates are done beyond that until the end of the project.

Estimates are performed at five intervals during the project life cycle. When the project is first initiated, the funding is based on the original estimates, which are derived from the list of normal tasks and subtasks. At this time, the subphases through the advisability study are estimated in detail, and summary estimates are prepared for the rest of the tasks in the project. Once the project has progressed through the advisability study, the preliminary systems design is estimated in detail, and the balance of the project is estimated in a more summary fashion. Estimates are conducted in this manner until the systems implementation plan is completed and the scope of the remaining subphases of the project is known. This multiple estimating process is used because it is almost impossible at the beginning of many projects to be certain of what the magnitude of effort will be later on in the project life cycle.

FUNDING OF PROJECTS

The project plan is the official document for securing funding from the sponsor in the user organization. The project plan must be completed and approved by the project manager before activity can begin on the user requirements subphase (1b). An initial stage in developing a project plan includes the drawing of a network that identifies each of the tasks to be done in the appropriate sequence for their execution. The project plan must include a milestone schedule, a cost estimate, and a budget request. It is submitted to the appropriate general manager of systems and programming for review so that an understanding can be reached of how the estimates were prepared and why the costs and schedules are as shown. At this time the general manager can get an idea of the quantity of systems and programming resources required by the project. The general manager next sets up a meeting with the project manager and the user sponsor to review the project plan and obtain funding from the user organization.

The initial project funding is based on an estimate that includes a number of assumptions concerning the scope of the project. Once certain key milestones in the project have been achieved, the visibility on the balance of the project becomes much clearer, and reestimates are performed. The reestimates may result in refunding if there has been a significant change in the project. The normal milestone refunding points are as follows:

1. After the advisability study (1d)
2. After the preliminary systems design (2a)
3. After the program design (2c)
4. After system implementation (3a)

The refunding process is similar to the initial funding with the exception that progress information is presented on the status of the work and reasons are given to explain deviations from project expenditure projections. A revised project plan is prepared for each milestone refunding meeting.

During the systems design and development stage, design freezes are issued by the project manager to users announcing that no additional changes will be accepted to the project beyond that point. The presence of these design freezes is outlined at the beginning of the project. Following the design freeze, no additional changes will be accepted unless the project is reestimated at a new level and approved by the user sponsor.

SYSTEM QUALITY REVIEWS

The key element in ensuring user involvement in the new system is the conducting of quality reviews. In the normal system cycles at FNB, there are ten quality reviews, seven of which are participated in jointly by users and data processing personnel, and three of which are technical reviews by data processing (DP) personnel only. An important side benefit of this review process is that users of a new system are forced to become involved in and are permitted to make a contribution to the systems design.

Each of the quality review points coincides with the end of a subphase in the project life cycle. The review must be held at the completion of one subphase to obtain authorization to begin work on the tasks of the next subphase of the project.

All tasks and subtasks assigned to members of the project team should end in some "deliverable" for the project documentation. The first step in conducting a quality review is to assemble the documentation produced during the subphase for distribution to the Quality Review Board. The Quality Review Board consists of between two and eight people who are appointed by the project manager with the approval of the project sponsor and the general manager of systems and programming. The minutes of the quality review meeting are written either to express "concurrence" with the subsystem quality or to recommend changes to the system that must be completed before the next subphase can be started. By this process the system is fine-tuned to the requirements of the members of the review group at the end of each subphase in the system. The members of the Quality Review Board charge their time to the project budget.

Quality review points and review board makeup are as follows:

Review	Review Board
User requirements	User oriented
Systems definition	User oriented
Advisability study	User oriented
Preliminary systems design	User oriented
Subsystems design	Users and DP
Program design	DP
Programming and testing	DP
System implementation	User oriented
System test	User oriented
Production control turnover	DP

To summarize, the quality review evaluates the quality of project subphase results, including design adequacy and proof of accomplishment in meeting project objectives. The review board authorizes work to progress based on their detailed knowledge that all required tasks and subtasks of each subphase have been successfully completed and documented.

PROJECT TEAM STAFFING

Once a project has risen to the top of the priority list, the appropriate manager of systems development appoints a project manager from his or her staff of analysts. The project manager has a short time to review the project proposal created by systems research before developing a project plan. The project plan must be approved by the general manager of systems and programming and the user sponsor before the project can be funded and work started on the user requirements subphase.

The project manager is "free" to spend as much time as required in reviewing the project proposal and creating the project plan; however, this time is "charged" to the project at a rate of $76 per hour. The project manager must negotiate with a "supervisor," the manager of systems development, to obtain the required systems analysts for the project, starting with the user requirements subphase. The project manager must obtain programming resources from the manager of systems support. Schedule delays caused by a lack of systems or programming resources are to be communicated to the general manager by the project manager. All ISD personnel working on a project charge their time at a rate of $76 per hour. All computer time is billed at a rate of $164 per hour.

There are no user personnel on the project team; all team members are from ISD.

CORPORATE DATABASE

John Hart had for several years seen the need to use the computer to support the corporate marketing effort of the bank. Despite the fact that the majority of the bank's profits were from corporate customers, most information systems effort was directed at speeding up transactions handling for small unprofitable accounts.

Mr. Hart had extensive experience in the Corporate Banking Division of the bank. He realized the need to consolidate information about corporate customers from many areas of the bank into one corporate database. From this information corporate banking services could be developed not only to better serve the corporate customers, but also to contribute heavily to the profit structure of the bank through repricing of services.

The absence of a corporate database meant that no one individual knew what total banking services a corporate customer was using, because corporate services were provided by many banking departments. It was also impossible to determine how profitable a corporate customer was to the bank. Contact officers did not have regularly scheduled calls. They serviced corporate customers almost on a hit-or-miss basis. Unfortunately, many customers were "sold" on a service because they walked in the door and requested it. Mr. Hart felt that there was a vast market of untapped corporate customers in Ohio who would purchase services from the bank if they were contacted and "sold" in a professional manner. A corporate database could be used to develop corporate profiles to help contact officers sell likely services to corporations.

Mr. Hart knew that data about corporate customers was being processed in many departments of the bank, but mainly in the following divisions:

- Corporate banking
- Corporate trust
- Consumer banking

He also realized that much of the information was processed in manual systems, some was processed by time-sharing at various vendors, and other information was computerized in many internal information systems.

The upper management of FNB must have agreed with Mr. Hart because in December of 1996 the Corporate Marketing Division was formed with John Hart as its executive vice president. Mr. Hart was due to retire within the year but was honored to be selected for the new position. He agreed to stay with the bank until "his" new system was "off the ground." He immediately composed a problem statement and sent it to the ISD. Systems Research compiled a preliminary impact statement. At the next Priorities Committee meeting, a project proposal was authorized to be done by Systems Research.

The project proposal was completed by Systems Research in record time. Most information was obtained from Mr. Hart. He had been thinking about the systems requirements for years and possessed vast experience in almost all areas of the bank. Other user divisions and departments were often "too busy" when approached for information. A common reply to a request for information was, "That project is John's baby; he knows what we need."

The project proposal as prepared by Systems Research recommended the following:

- Interfaces should be designed to extract information from existing computerized systems for the corporate database (CDB).
- Time-sharing systems should be brought in-house to be interfaced with the CDB.
- Information should be collected from manual systems to be integrated into the CDB on a temporary basis.
- Manual systems should be consolidated and computerized, potentially causing a reorganization of some departments.
- Information analysis and flow for all departments and divisions having contact with corporate customers should be coordinated by the Corporate Marketing Division.
- All corporate database analysis should be done by the Corporate Marketing Division staff, using either a user-controlled report writer or interactive inquiry.

The project proposal was presented at the next Priorities Committee meeting where it was approved and rated as the highest priority MIS development project in the bank. Mr. Hart became the user sponsor for the CDB project.

The project proposal was sent to the manager of corporate development, who appointed Jim Gunn as project manager from the staff of analysts in corporate development. Jim Gunn was the most experienced project manager available. His prior experience consisted of successful projects in the Financial Division of the bank.

Jim reviewed the project proposal and started to work on his project plan. He was aware that the corporate analyst group was presently understaffed but was assured by his manager, the manager of corporate development, that resources would be available for the user requirements subphase. He had many questions concerning the scope of the project and the interrelationship between the

Corporate Marketing Division and the other users of corporate marketing data. But each meeting with Mr. Hart ended with the same comment: "This is a waste of time. I've already been over this with Systems Research. Let's get moving." Jim also was receiving pressure from the general manager to "hurry up" with the project plan. Jim therefore quickly prepared his project plan, which included a general milestone schedule for subphase completion, a general cost estimate, and a request for funding. The project plan was reviewed by the general manager and signed by Mr. Hart.

Jim Gunn anticipated the need to have four analysts assigned to the project and went to his manager to see who was available. He was told that two junior analysts were available now and another analyst should be free next week. No senior analysts were available. Jim notified the general manager that the CDB schedule would probably be delayed because of a lack of resources, but received no response.

Jim assigned tasks to the members of the team and explained the assignments and the schedule. Since the project was understaffed, Jim assigned a heavy load of tasks to himself.

During the next two weeks the majority of the meetings set up to document user requirements were canceled by the user departments. Jim notified Mr. Hart of the problem and was assured that steps would be taken to correct the problem. Future meetings with the users in the Consumer Banking and Corporate Banking Divisions became very hostile. Jim soon discovered that many individuals in these divisions did not see the need for the corporate database. They resented spending their time in meetings documenting the CDB requirements. They were afraid that the CDB project would lead to a shift of many of their responsibilities and functions to the Corporate Marketing Division.

Mr. Hart was also unhappy. The CDB team was spending more time than was budgeted in documenting user requirements. If this trend continued, a revised budget would have to be submitted to the Priorities Committee for approval. He was also growing tired of ordering individuals in the user departments to keep appointments with the CDB team. Mr. Hart could not understand the resistance to his project.

Jim Gunn kept trying to obtain analysts for his project but was told by his manager that none were available. Jim explained that the quality of work done by the junior analysts was not "up to par" because of lack of experience. Jim complained that he could not adequately supervise the work quality because he was forced to complete many of the analysis tasks himself. He also noted that the quality review of the user requirements subphase was scheduled for next month, making it extremely critical that experienced analysts be assigned to the project. No new personnel were assigned to the project. Jim thought about contacting the general manager again to explain his need for more experienced analysts, but did not. He was due for a semiyearly evaluation from his manager in two weeks.

Even though he knew the quality of the work was below standards, Jim was determined to get the project done on schedule with the resources available to him. He drove both himself and the team very hard during the next few weeks. The quality review of the user requirement subphase was held on schedule. Over 90 percent of the assigned tasks had to be redone before the Quality Review Board would sign-off on the review. Jim Gunn was removed as project manager.

Three senior analysts and a new project manager were assigned to the CDB project. The project received additional funding from the Priorities Committee. The user requirements subphase was completely redone despite vigorous protests from the Consumer Banking and Corporate Banking divisions.

Within the next three months the following events happened:

- The new project manager resigned to accept a position with another firm.
- John Hart took early retirement.
- The CDB project was tabled.

SYNOPSIS

All projects at First National Bank (FNB) have project managers assigned and are handled through the Information Services Division (ISD). The organizational structure is not a matrix, although some people think that it is. The case describes one particular project, the development of a corporate database, and the resulting failure. The problem at hand is to investigate why the project failed.

QUESTIONS

1. What are the strengths of FNB?
2. What are the major weaknesses?
3. What is the major problem mentioned above? Defend your answer.
4. How many people did the project manager have to report to?
5. Did the PM remain within vertical structure of the organization?
6. Is there anything wrong if a PM is a previous co-worker of some team members before the team is formed?
7. Who made up the project team?
8. Was there any resistance to the project by company management?
9. Was there an unnecessary duplication of work?
10. Was there an increased resistance to change?
11. Was the communication process slow or fast?

12. Was there an increased amount of paperwork?
13. What are reasonable recommendations?
14. Does the company have any type of project management methodology?
15. Could the existence of a methodology have alleviated any of the above problems?
16. Did the bank perform strategic planning for project management or did it simply rush into the project?
17. Why do organizations rush into project management without first performing strategic planning for project management or, at least, some form of benchmarking against other organizations?

Cordova Research Group

Cordova Research Group spent more than thirty years conducting pure and applied research for a variety of external customers. With the reduction, however, in R&D funding, Cordova decided that the survival of the firm would be based upon becoming a manufacturing firm as well as performing R&D. The R&D culture was close to informal project management with the majority of the personnel holding advanced degrees in technical disciplines. To enter the manufacturing arena would require hiring hundreds of new employees, mostly nondegreed.

QUESTIONS

1. What strategic problems must be solved?
2. What project management problems must be solved?
3. What time frame is reasonable?
4. If excellence can be achieved, would it occur most likely using formal or informal project management?

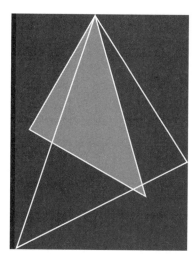

Cortez Plastics

Cortez Plastics was having growing pains. As the business base of the company began to increase, more and more paperwork began to flow through the organization. The "informal" project management culture that had worked so well in the past was beginning to deteriorate and was being replaced by a more formal project management approach. Recognizing the cost implications of a more formal project management approach, senior management at Cortez Plastics decided to take some action.

QUESTIONS

1. How can a company maintain informal project management during periods of corporate growth?
2. If the organization persists in creeping toward formal project management, what can be done to return to a more informal approach?
3. How would you handle a situation where only a few managers or employees are promoting the more formal approach?

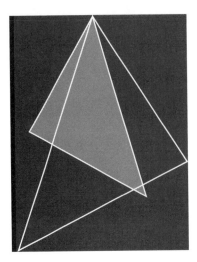

L. P. Manning Corporation

In March 2001, the Marketing Division of the L. P. Manning Corporation performed a national survey to test the public's reaction to a new type of toaster. Manning had achieved success in the past and established itself as a leader in the home appliance industry.

Although the new toaster was just an idea, the public responded favorably. In April of the same year, the vice presidents for planning, marketing, engineering, and manufacturing all met to formulate plans for the development and ultimately the production of the new toaster. Marketing asserted that the manufacturing cost must remain below $70 per unit or else Manning Corporation would not be competitive. Based on the specifications drawn up in the meeting, manufacturing assured marketing that this cost could be met.

The engineering division was given six months to develop the product. Manning's executives were eager to introduce the product for the Christmas rush. This might give them an early foothold on a strong market share.

During the R&D phase, marketing continually "pestered" engineering with new designs and changes in specifications that would make the new product easier to market. The ultimate result was a one-month slip in the schedule.

Pushing the schedule to the right greatly displeased manufacturing personnel. According to the vice president for manufacturing, speaking to the marketing manager: "I've just received the final specifications and designs from engineering. This is not what we had agreed on last March. These changes will cause us to lose at least one additional month to change our manufacturing planning. And because we're already one month behind, I don't see any way that we could reschedule our Christmas production facilities to accommodate this new product. Our established lines must come first. Furthermore, our estimating department says that these changes will increase the cost of the product by at least 25 to 35 percent. And, of course, we must include the quality control section, which has some questions as to whether we can actually live with these specifications. Why don't we just cancel this project or at least postpone it until next year?"

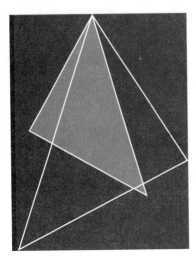

Project
Firecracker

"Don, project management is the only way to handle this type of project. With $40 million at stake we can't afford not to use this approach."

"Listen, Jeff, your problem is you take seminars given by these ivory tower professors and you think you're an expert. I've been in this business for forty years and I know how to handle this job—and it isn't through project management."

HISTORY AND BACKGROUND

Jeff Pankoff, a registered professional engineer, came to work for National Corporation after receiving a mechanical engineering degree. After he arrived at National, he was assigned to the engineering department. Soon thereafter, Jeff realized that he needed to know more about statistics, and he enrolled in the graduate school of a local university. When he was near completion of his master of science degree, National transferred Jeff to one of its subsidiaries in Ireland to set up an engineering department. After a successful three years, Jeff returned to National's home office and was promoted to chief engineer. Jeff's department increased to eighty engineers and technicians. Spending a considerable time in administration, Jeff decided an MBA would be useful, so he enrolled in a program at a nearby university. At the time when this project began, Jeff was near the end of the MBA program.

National Corporation, a large international corporation with annual sales of about $600 million, employs 8,000 people worldwide and is a specialty machine, component, and tool producer catering to automotive and aircraft manufacturers. The company is over a hundred years old and has a successful and profitable record.

National is organized in divisions according to machine, component, and tool production facilities. Each division is operated as a profit center (see Exhibit I). Jeff was assigned to the Tool Division.

National's Tool Division produces a broad line of regular tools as well as specials. Specials amounted to only about 10 percent of the regular business, but over the last five years had increased from 5 percent to the current 10 percent. Only specials that were similar to the regular tools were accepted as orders.

National sells all its products through about 3,000 industrial distributors located throughout the United States. In addition, National employs 200 sales representatives who work with the various distributors to provide product seminars.

The traditional approach to project assignments is used. The engineering department, headed by Jeff, is basically responsible for the purchase of capital equipment and the selection of production methods used in the manufacture of the product. Project assignments to evaluate and purchase a new machine tool or to determine the production routing for a new product are assigned to the engineering department. Jeff assigns the project to the appropriate section, and, under the direction of a project engineer, the project is completed.

Exhibit I.   The Tool Division of National Corporation

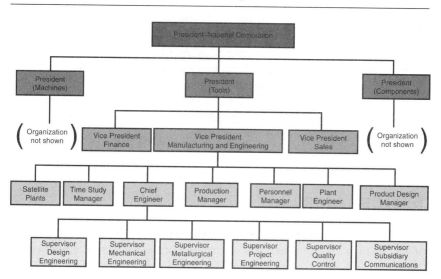

The project engineer works with all the departments reporting to the vice president, including production, personnel, plant engineering, product design (the project engineer's link to sales), and time study. As an example of the working relationship, the project engineer selects the location of the new machine and devises instructions for its operation with production. With personnel the engineer establishes the job descriptions for the new jobs as well as for the selection of people to work on the new machine. The project engineer works with plant engineering on the moving of the machine to the proper location and instructs plant engineering on the installation and services required (air, water, electricity, gas, etc.). It is very important that the project engineer work very closely with the product design department, which develops the design of the product to be sold. Many times the product designed is too ambitious an undertaking or cannot be economically produced. Interaction between departments is essential in working out such problems.

After the new machine is installed, an operator is selected and the machine is ready for production. Time study, with the project engineer's help, then establishes the incentive system for the job.

Often a customer requests certain tolerances that cannot be adhered to by manufacturing. In such a case, the project engineer contacts the product design department, which contacts the sales department, which in turn contacts the customer. The communication process is then reversed, and the project engineer gets an answer. Based on the number of questions, the total process may take four to five weeks.

As the company is set up, the engineering department has no authority over time study, production, product design, or other areas. The only way that the project engineer can get these departments to make commitments is through persuasion or through the chief engineer, who could go to the vice president of manufacturing and engineering. If the engineer is convincing, the vice president will dictate to the appropriate manager what must be done.

Salaries in all departments of the company are a closely guarded secret. Only the vice president, the appropriate department manager, and the individual know the exact salary. Don Wolinski, the vice president of manufacturing and engineering, pointed out that this approach was the "professional way" and an essential aspect of smooth business operations.

THE ILL-FATED PROJECT

Jeff Pankoff, the chief engineer for National, flew to Southern California to one of National's (tool) plants. Ben Ehlke, manager of the Southern California plant, wanted to purchase a computer numerical controlled (CNC) machining center for

$250,000. When the request came to Jeff for approval, he had many questions and wanted some face-to-face communication.

The Southern California plant supplied the aircraft industry, and one airplane company provided 90 percent of the Southern California plant's sales. Jeff was mainly concerned about the sales projections used by Ehlke in justifying the machining center. Ehkle pointed out that his projections were based on what the airplane company had told him they expected to buy out the next five years. Since this estimate was crucial to the justification, Jeff suggested that a meeting be arranged with the appropriate people at the airplane company to explore these projections. Since the local National sales representative was ill, the distributor salesman, Jack White, accompanied Jeff and Ben. While at the airplane company (APC), the chief tool buyer of APC, Tom Kelly, was informed that Jeff was there. Jeff received a message from the receptionist that Tom Kelly wanted to see him before he left the building. After the sales projections were reviewed and Jeff was convinced that they were as accurate and as reliable as they possibly could be, he asked the receptionist to set up an appointment with Tom Kelly.

When Jeff walked into Kelly's office the fireworks began. He was greeted with, "What's wrong with National? They refused to quote on this special part. We sent them a print and asked National for their price and delivery, indicating it could turn into a sizable order. They turned me down flat saying that they were not tooled up for this business. Now I know that National is tops in the field and that National can provide this part. What's wrong with your sales department?"

All this came as a complete surprise to Jeff. The distributor salesman knew about it but never thought to mention it to him. Jeff looked at the part print and asked, "What kind of business are you talking about?" Kelly said, without batting an eye, "$40 million per year."

Jeff realized that National had the expertise to produce the part and would require only one added machine (a special press costing $20,000) to have the total manufacturing capability. Jeff also realized he was in an awkward situation. The National sales representative was not there, and he certainly could not speak for sales. However, a $40 million order could not be passed over lightly. Kelly indicated that he would like to see National get 90 percent of the order if they would only quote on the job. Jeff told Kelly that he would take the information back and discuss it with the vice presidents of sales, manufacturing, and engineering and that most likely the sales vice president would contact him next.

On the return flight, Jeff reviewed in his mind his meeting with Kelly. Why did Bob Jones, National's sales vice president, refuse to quote? Did he know about the possible $40 million order? Although Jeff wasn't in sales, he decided that he would do whatever possible to land this order for National. That evening Jack White called from California. Jack said he had talked to Kelly after Jeff left and told Kelly that if anybody could make this project work, it would be Jeff Pankoff. Jeff suggested that Jack White call Bob Jones with future reports concerning this project.

The next morning, before Jeff had a chance to review his mail, Bob Jones came storming into his office. "Who do you think you are committing National to accept an order on your own without even a sales representative present? You know that all communication with a customer is through sales."

Jeff replied, "Let me explain what happened."

After Jeff's explanation, Jones said, "Jeff, I hear what you're saying, but no matter what the circumstances, all communications with any customer must go through proper channels."

Following the meeting with Jones, Jeff went to see Wolinski, his boss. He filled Wolinski in on what had happened. Then he said, "Don, I've given this project considerable thought. Jones is agreeable to quoting this job. However, if we follow our normal channels, we will experience too many time delays and problems. Through the various stages of this project, the customer will have many questions and changes and will require continuous updating. Our current system will not allow this to happen. It will take work from all departments to implement this project, and unless all departments work under the same priority system, we won't have a chance. What we need, Don, is project management. Without this approach where one man heads the project with authority from the top, we just can't make it work."

Wolinski looked out the window and said, "We have been successful for many years using our conventional approach to project work. I grant you that we have not had an order of this magnitude to worry about, but I see no reason why we should change even if the order were for $100 million."

"Don, project management is the only way to handle this type of project. With $40 million at stake we can't afford not to use this approach."

"Listen Jeff, your problem is you take seminars given by these ivory tower professors and you think you're an expert. I've been in this business for forty years and I know how to handle this job—and it isn't through project management. I'll call a meeting of all concerned department managers so we can get started on quoting this job."

That afternoon, Jeff and the other five department managers were summoned to a meeting in Wolinski's office. Wolinski summarized the situation and informed the assembled group that Jeff would be responsible for the determination of the methods of manufacture and the associated manufacturing costs that would be used in the quotation. The method of manufacture, of course, would be based on the design of the part provided by product design. Wolinski appointed Jeff and Waldo Novak, manager of product design, as coheads of the project. He further advised that the normal channels of communication with sales through the product design manager would continue as usual on this project.

The project began. Jeff spent considerable time requesting clarification of the drawings submitted by the customer. All these communications went through Waldo. Before the manufacturing routing could be established for quotation purposes, questions concerning the drawings had to be answered. The customer was getting anxious to receive the quotation because its management had to select a supplier within eight weeks. One week was already lost owing to communication delay. Wolinski decided that to speed up the quoting process he would send Jeff and Waldo along with Jones, the sales vice president, to see the customer. This meeting at APC helped clarify many questions. After Jeff returned, he began laying out the alternative routing for the parts. He assigned two of his most creative technicians and an engineer to run isolated tests on the various methods of manufacturing. From the results he would then finalize the routing that would be used for quoting. Two weeks of the eight were gone, but Jeff was generally pleased until the phone rang. It was Waldo.

"Say, Jeff, I think if we change the design on the back side of the part, it will add to its strength. In fact, I've assigned one of my men to review this and make this change, and it looks good."

While this conversation was going on, Wolinski popped into Jeff's office and said that sales had promised that National would ship APC a test order of 100 pieces in two weeks. Jeff was irate. Product design was changing the product. Sales was promising delivery of a test order that no one could even describe yet.

Needless to say, the next few days were long and difficult. It took three days for Jeff and Waldo to resolve the design routing problem. Wolinski stayed in the background and would not make any position statement except that he wanted everything "yesterday." By the end of the third week the design problem was resolved, and the quotation was prepared and sent out to the customer. The quotation was acceptable to APC pending the performance of the 100 test parts.

At the start of the fourth week, Jeff, with the routing in hand, went to Charlie Henry, the production manager, and said he needed 100 parts by Friday. Charlie looked at the routing and said, "The best I can do is a two-week delivery."

After discussing the subject for an hour, the two men agreed to see Wolinski. Wolinski said he'd check with sales and attempt to get an extension of one week. Sales asked the distributor salesman to request an extension. Jack White was sure it would be okay so he replied to Bob Jones without checking that the added week was in fact acceptable.

The 100 pieces went out in three weeks rather than two. That meant the project was at the end of the sixth week and only two remained. Inspection received the test pieces on Monday of the seventh week and immediately reported them not to be in specification. Kelly was upset. He was counting heavily on National to provide these parts. Kelly had received four other quotations and test orders from National's competitors. The prices were similar, and the test parts were to specification. However, National's parts, although out of specification, looked better than their competitors'. Kelly reminded Jones that the customer now had only nine days left before the contract would be let. That meant the 100 test parts had to be made in nine days. Jones immediately called Wolinski, who agreed to talk to his people to try to accomplish this.

The tools were shipped in eleven days, two days after the customer had awarded orders to three of National's competitors. Kelly was disappointed in

National's performance but told Jones that National would be considered for next year's contract, at least a part of it.

Jeff, hearing from Waldo that National lost the order, returned to his office, shut the door, and thought of the hours, nearly round the clock, that were spent on this job. Hours were wasted because of poor communications, nonuniform priorities, and the fact that there was no project manager. "I wonder if Wolinski learned his lesson; probably not. This one cost the company at least $6 million in profits, all because project management was not used." Jeff concluded that his work was really cut out for him. He decided that he must convince Wolinski and others of the advantages of using project management. Although Wolinski had attended a one-day seminar on project management two years ago, Jeff decided that one of his objectives during the coming year would be to get Wolinski to the point where he would, on his own, suggest becoming more knowledgeable concerning project management. Jeff's thought was that if the company was to continue to be profitable it must use project management.

The phone rang, it was Wolinski. He said, "Jeff, do you have a moment to come down to my office? I'd like to talk about the possibility of using, on a trial basis, this project management concept you mentioned to me a few months ago."

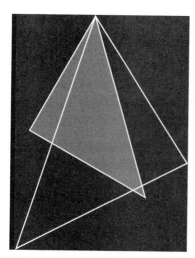

Philip Condit and the Boeing 777: From Design and Development to Producton and Sales*

Following his promotion to Boeing CEO in 1988, Frank Shrontz looked for ways to stretch and upgrade the Boeing 767—an eight-year-old wide-body twin jet—in order to meet Airbus competition. Airbus had just launched two new 300-seat wide-body models, the two-engine A330 and the four-engine A340. Boeing had no 300-seat jetliner in service, nor did the company plan to develop such a jet.

To find out whether Boeing's customers were interested in a double-decker 767, Philip Condit, Boeing Executive Vice President and future CEO (1996) met with United Airlines Vice President Jim Guyette. Guyette rejected the idea outright, claiming that an upgraded 767 was no match to Airbus's new model transports. Instead, Guyette urged Boeing to develop a brand new commercial jet, the most advanced airplane of its generation.[1] Shrontz had heard similar suggestions from other airline carriers. He reconsidered Boeing's options, and decided to abandon the 767 idea in favor of a new aircraft program. In December 1989, accordingly, he announced the 777 project and put Philip Condit in charge of its management. Boeing had launched the 777 in 1990, delivered the first jet in 1995, and by February 2001, 325 B-777s were flying in the services of the major international and U.S. airlines.[2]

---

*This case was presented by Isaac Cohen, San Jose State University, at the 2000 North American Case Research Association (NACRA) workshop. Reprinted by permission from the *Case Research Journal*. Copyright 2000 by Isaac Cohen and the North American Case Research Association.

Condit faced a significant challenge in managing the 777 project. He wanted to create an airplane that was preferred by the airlines at a price that was truly competitive. He sought to attract airline customers as well as cut production costs, and he did so by introducing several innovations—both technological and managerial—in aircraft design, manufacturing, and assembly. He looked for ways to revitalize Boeing's outmoded engineering production system, and update Boeing's manufacturing strategies. And to achieve these goals, Condit made continual efforts to spread the 777 program-innovations companywide.

Looking back at the 777 program, this case focuses on Condit's efforts. Was the 777 project successful, and was it cost effective? Would the development of the 777 allow Boeing to diffuse the innovations in airplane design and production beyond the 777 program? Would the development of the 777's permit Boeing to revamp and modernize its aircraft manufacturing system? Would the making and selling of the 777 enhance Boeing competitive position relative to Airbus, its only remaining rival?

THE AIRCRAFT INDUSTRY

Commercial aircraft manufacturing was an industry of enormous risks where failure was the norm, not the exception. The number of large commercial jet makers had been reduced from four in the early 1980s—Boeing, McDonnell Douglas, Airbus, and Lockheed—to two in late 1990s, turning the industry into a duopoly, and pitting the two survivors—Boeing and Airbus—one against the other. One reason why aircraft manufacturers so often failed was the huge cost of product development.

Developing a new jetliner required an up-front investment of up to $15 billion (2001 dollars), a lead time of five to six years from launch to first delivery, and the ability to sustain a negative cash flow throughout the development phase. Typically, to break even on an entirely new jetliner, aircraft manufacturers needed to sell a minimum of 300 to 400 planes and at least 50 planes per year.[3] Only a few commercial airplane programs had ever made money.

The price of an aircraft reflected its high development costs. New model prices were based on the average cost of producing 300 to 400 planes, not a single plane. Aircraft pricing embodied the principle of learning by doing, the so called *learning curve*[4]: workers steadily improved their skills during the assembly process, and as a result, labor cost fell as the number of planes produced rose.

The high and increasing cost of product development prompted aircraft manufacturers to utilize subcontracting as a risk-sharing strategy. For the 747, the 767, and the 777, the Boeing Company required subcontractors to share a substantial part of the airplane's development costs. Airbus did the same with its own latest models. Risk sharing subcontractors performed detailed design work and assembled major subsections of the new plane while airframe integrators (i.e., aircraft manufacturers) designed the aircraft, integrated its systems and equipment, assembled the entire plane, marketed it, and provided customer support for twenty to thirty years. Both the airframe integrators and their subcontractors were supplied by thousands of domestic and foreign aircraft components manufacturers.[5]

Neither Boeing, nor Airbus, nor any other post-war commercial aircraft manufacturer produced jet engines. A risky and costly venture, engine building had become a highly specialized business. Aircraft manufacturers worked closely with engine makers—General Electric, Pratt and Whitney, and Rolls Royce—to set engine performance standards. In most cases, new airplanes were offered with a choice of engines. Over time, the technology of engine building had become so complex and demanding that it took longer to develop an engine than an aircraft. During the life of a jetliner, the price of the engines and their replacement parts was equal to the entire price of the airplane.[6]

A new model aircraft was normally designed around an engine, not the other way around. As engine performance improved, airframes were redesigned to exploit the engine's new capabilities. The most practical way to do so was to stretch the fuselage and add more seats in the cabin. Aircraft manufacturers deliberately designed flexibility into the airplane so that future engine improvements could facilitate later stretching. Hence the importance of the "family concept" in aircraft design, and hence the reason why aircraft manufacturers introduced families of planes made up of derivative jetliners built around a basic model, not single, standardized models.[7]

The commercial aircraft industry, finally, gained from technological innovations in two other industries. More than any other manufacturing industry, aircraft construction benefited from advances in material applications and electronics. The development of metallic and nonmetallic composite materials played a key role in improving airframe and engine performance. On the one hand, composite materials that combined light weight and great strength were utilized by aircraft manufacturers; on the other, heat-resisting alloys that could tolerate temperatures of up to 3,000 degrees were used by engine makers. Similarly, advances in electronics revolutionized avionics. The increasing use of semiconductors by aircraft manufacturers facilitated the miniaturization of cockpit instruments, and more important, it enhanced the use of computers for aircraft communication, navigation, instrumentation, and testing.[8] The use of computers contributed, in addition, to the design, manufacture, and assembly of new model aircraft.

THE BOEING COMPANY

The history of the Boeing company may be divided into two distinct periods: the piston era and the jet age. Throughout the piston era, Boeing was essentially a military contractor producing fighter aircraft in the 1920s and 1930s, and bombers during World War II. During the jet age, beginning in the 1950s, Boeing had become the world's largest manufacturer of commercial aircraft, deriving most of its revenues from selling jetliners.

Boeing's first jet was the 707. The introduction of the 707 in 1958 represented a major breakthrough in the history of commercial aviation; it allowed Boeing to gain a critical technological lead over the Douglas Aircraft Company, its closer competitor. To benefit from government assistance in developing the 707, Boeing produced the first jet in two versions: a military tanker for the Air Force (k-135) and a commercial aircraft for the airlines (707-120). The company, however, did not recoup its own investment until 1964, six years after it delivered the first 707, and twelve years after it had launched the program. In the end, the 707 was quite profitable, selling 25 percent above its average cost.[9] Boeing retained the essential design of the 707 for all its subsequent narrow-body single-aisle models (the 727, 737, and 757), introducing incremental design improvements, one at a time.[10] One reason why Boeing used shared design for future models was the constant pressure experienced by the company to move down the learning curve and reduce overall development costs.

Boeing introduced the 747 in 1970. The development of the 747 represented another breakthrough; the 747 wide body design was one of a kind; it had no real competition anywhere in the industry. Boeing bet the entire company on the success of the 747, spending on the project almost as much as the company's total net worth in 1965, the year the project started.[11] In the short-run, the outcome was disastrous. As Boeing began delivering its 747s, the company was struggling to avoid bankruptcy. Cutbacks in orders as a result of a deep recession, coupled with production inefficiencies and escalating costs, created a severe cash shortage that pushed the company to the brink. As sales dropped, the 747's break-even point moved further and further into the future.

Yet, in the long run, the 747 program was a triumph. The Jumbo Jet had become Boeing's most profitable aircraft and the industry's most efficient jetliner. The plane helped Boeing solidify its position as the industry leader for years to come, leaving McDonnell Douglas far behind, and forcing the Lockheed Corporation to exit the market. The new plane, furthermore, contributed to Boeing's manufacturing strategy in two ways. First, as Boeing increased its reliance on outsourcing, six major subcontractors fabricated 70 percent of the value of the 747 airplane,[12] thereby helping Boeing reduce the project's risks. Second, for the first time, Boeing applied the family concept in aircraft design to a wide-body jet, building the 747 with wings large enough to support a stretched fuselage with bigger engines, and offering a variety of other modifications in the 747's basic design. The 747-400 (1989) is a case in point. In 1997, Boeing sold the stretched and upgraded 747-400 in three versions, a standard jet, a freighter, and a "combi" (a jetliner whose main cabin was divided between passenger and cargo compartments).[13]

Boeing developed other successful models. In 1969, Boeing introduced the 737, the company's narrow-body flagship, and in 1982 Boeing put into service two additional jetliners, the 757 (narrow-body) and the 767 (wide-body). By the early 1990s, the 737, 757, and 767 were all selling profitably. Following the introduction of the 777 in 1995, Boeing's families of planes included the 737 for short-range travel, the 757 and 767 for medium-range travel, and the 747 and 777 for medium- to long-range travel (Exhibit I).

In addition to building jetliners, Boeing also expanded its defense, space, and information businesses. In 1997, the Boeing Company took a strategic gamble, buying the McDonnell Douglas Company in a $14 billion stock deal. As a result of the merger, Boeing had become the world's largest manufacturer of military aircraft, NASA'S largest supplier, and the Pentagon's second largest contractor (after Lockheed). Nevertheless, despite the growth in its defense and space businesses, Boeing still derived most of its revenues from selling jetliners. Commercial aircraft revenues accounted for 59 percent of Boeing's $49 billion sales in 1997 and 63 percent of Boeing's $56 billion sales in 1998.[14]

Following its merger with McDonnell, Boeing had one remaining rival: Airbus Industrie.[15] In 1997, Airbus booked 45 percent of the worldwide orders for commercial jetliners[16] and delivered close to 1/3 of the worldwide industry output. In 2000, Airbus shipped nearly 2/5 of the worldwide industry output (Exhibit II).

Airbus's success was based on a strategy that combined cost leadership with technological leadership. First, Airbus distinguished itself from Boeing by incorporating the most advanced technologies into its planes. Second, Airbus managed to cut costs by utilizing a flexible, lean production manufacturing system that stood in a stark contrast to Boeing's mass production system.[17]

Exhibit I. *Total number of commercial jetliners delivered by the Boeing Company, 1958–2/2001[a]*

Model	No. Delivered	First Delivery
B-707	1,010 (retired)	1958
B-727	1,831 (retired)	1963
B-737	3,901	1967
B-747	1,264	1970
B-757	953	1982
B-767	825	1982
B-777	325	1995
B-717	49	2000
Total:	**10,158**	

[a]McDonnell Douglas commercial jetliners (the MD-11, MD-80, and MD-90) are excluded.
Sources: Boeing Commercial Airplane Group, *Announced Orders and Deliveries as of 12/31/97; The Boeing Company 1998 Annual Report*, p. 35.
"Commercial Airplanes: Order and Delivery Summary," *http://www.Boeing com/commercial/orders/index.html.* Retrieved from Web, March 20, 2001.

Exhibit II.   Market share of shipments of commercial aircraft, Boeing, McDonnell Douglas (MD), Airbus, 1992–2000

	1992	1993	1994	1995	1996	1997	1998	1999	2000
Boeing	61%	61%	63%	54%	55%	67%	71%	68%	61%
MD	17	14	9	13	13				
Airbus	22	25	28	33	32	33	29	32	39

Source: Aerospace Facts and Figures, 1997–98, p. 34; *Wall Street Journal* (December 3, 1998, and January 12, 1999); *The Boeing Company 1997 Annual Report,* p. 19; data supplied by Mark Luginbill, Airbus Communication Director (November 16, 1998, February 1, 2000, and March 20, 2001).

As Airbus prospered, the Boeing company was struggling with rising costs, declining productivity, delays in deliveries, and production inefficiencies. Boeing Commercial Aircraft Group lost $1.8 billion in 1997 and barely generated any profits in 1998.[18] All through the 1990s, the Boeing Company looked for ways to revitalize its outdated production manufacturing system on the one hand, and to introduce leading edge technologies into its jetliners on the other. The development and production of the 777, first conceived of in 1989, was an early step undertaken by Boeing managers to address both problems.

THE 777 PROGRAM

The 777 program was Boeing's single largest project since the completion of the 747. The total development cost of the 777 was estimated at $6.3 billion and the total number of employees assigned to the project peaked at nearly 10,000. The 777's twin-engines were the largest and most powerful ever built (the diameter of the 777's engine equaled the 737's fuselage), the 777's construction required 132,000 uniquely engineered parts (compared to 70,000 for the 767), the 777's seat capacity was identical to that of the first 747 that had gone into service in 1970, and its manufacturer empty weight was 57 percent greater than the 767's. Building the 777 alongside the 747 and 767 at its Everett plant near Seattle, Washington, Boeing enlarged the plant to cover an area of seventy-six football fields.[19]

Boeing's financial position in 1990 was unusually strong. With a 21 percent rate of return on stockholder equity, a long-term debt of just 15 percent of capitalization, and a cash surplus of $3.6 billion, Boeing could gamble comfortably.[20] There was no need to bet the company on the new project as had been the case with the 747, or to borrow heavily, as had been the case with the 767. Still, the decision to develop the 777 was definitely risky; a failure of the new jet might have triggered an irreversible decline of the Boeing Company and threatened its future survival.

The decision to develop the 777 was based on market assessment—the estimated future needs of the airlines. During the fourteen-year period, 1991–2005, Boeing market analysts forecasted a +100 percent increase in the number of passenger miles traveled worldwide, and a need for about 9,000 new commercial jets. Of the total value of the jetliners needed in 1991–2005, Boeing analysts forecasted a \$260 billion market for wide body jets smaller than the 747. An increasing number of these wide-body jets were expected to be larger than the 767.[21]

A CONSUMER-DRIVEN PRODUCT

To manage the risk of developing a new jetliner, aircraft manufacturers had first sought to obtain a minimum number of firm orders from interested carriers, and only then commit to the project. Boeing CEO Frank Shrontz had expected to obtain one hundred initial orders of the 777 before asking the Boeing board to launch the project, but as a result of Boeing's financial strength on the one hand, and the increasing competitiveness of Airbus on the other, Schrontz decided to seek the board's approval earlier. He did so after securing only one customer: United Airlines. On October 12, 1990, United had placed an order for thirty-four 777s and an option for an additional thirty-four aircraft, and two weeks later, Boeing's board of directors approved the project.[22] Negotiating the sale, Boeing and United drafted a handwritten agreement (signed by Philip Condit and Richard Albrecht, Boeing's executive vice presidents, and Jim Guyette, United's executive vice president) that granted United a larger role in designing the 777 than the role played by any airline before. The two companies pledged to cooperate closely in developing an aircraft with the "best dispatch reliability in the industry" and the "greatest customer appeal in the industry." "We will endeavor to do it right the first time with the highest degree of professionalism" and with "candor, honesty, and respect" [the agreement read]. Asked to comment on the agreement, Philip Condit, said: "We are going to listen to our customers and understand what they want. Everybody on the program has that attitude."[23] Gordon McKinzie, United's 777 program director agreed: "In the past we'd get brochures on a new airplane and its options. . . wait four years for delivery, and hope we'd get what we ordered. This time Boeing really listened to us."[24]

Condit invited other airline carriers to participate in the design and development phase of the 777. Altogether, eight carriers from around the world (United, Delta, American, British Airways, Qantas, Japan Airlines, All Nippon Airways, and Japan Air System) sent full-time representatives to Seattle; British Airways alone assigned seventy-five people at one time. To facilitate interaction between its design engineers and representatives of the eight carriers, Boeing introduced an initiative called "Working Together." "If we have a problem," a British Airways production manager explained, "we go to the source—design engineers on the IPT [Integrated Product Teams], not service engineer(s). One of the frustrations on the 747 was that we rarely got to talk to the engineers who were doing the work."[25]

"We have definitely influenced the design of the aircraft," a United 777 manager said, mentioning changes in the design of the wing panels that made it easier for airline mechanics to access the slats (slats, like flaps, increased lift on takeoffs and landings), and new features in the cabin that made the plane more attractive to passengers.[26] Of the 1,500 design features examined by representatives of the airlines, Boeing engineers modified 300 (see Exhibit III). Among changes made by Boeing was a redesigned overhead bin that left more stand-up headroom for passengers (allowing a six-foot-three tall passenger to walk from aisle to aisle), "flattened" side walls that provided the occupant of the window seat with more room, overhead bin doors that opened down and made it possible for shorter passengers to lift baggage into the overhead compartment, a redesigned reading lamp that enabled flight attendants to replace light bulbs, a task formerly performed by mechanics, and a computerized flight deck management

Exhibit III. *The 777: Selected design features proposed by Boeing airline customers and adapted by the Boeing Company*

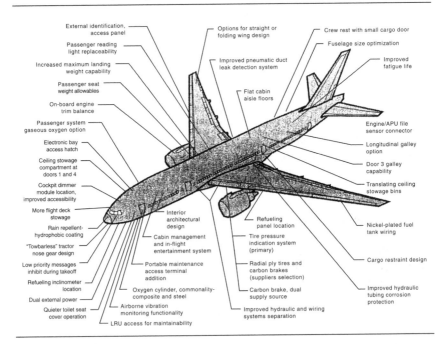

Source: The Boeing Company.

system that adjusted cabin temperature, controlled the volume of the public address system, and monitored food and drink inventories.[27]

More important were changes in the interior configuration (layout plan) of the aircraft. To be able to reconfigure the plane quickly for different markets of varying travel ranges and passenger loads, Boeing's customers sought a flexible plan of the interior. On a standard commercial jet, kitchen galleys, closets, lavatories, and bars were all removable in the past, but were limited to fixed positions where the interior floor structure was reinforced to accommodate the "wet" load. On the 777, by contrast, such components as galleys and lavatories could be positioned anywhere within several "flexible zones" designed into the cabin by the joint efforts of Boeing engineers and representatives of the eight airlines. Similarly, the flexible design of the 777's seat tracks made it possible for carriers to increase the number of seat combinations as well as reconfigure the seating arrangement quickly. Flexible configuration resulted, in turn, in significant cost savings; airlines no longer needed to take the aircraft out of service for an extended period of time in order to reconfigure the interior.[28]

The airline carriers also influenced the way in which Boeing designed the 777 cockpit. During the program definition phase, representatives of United Airlines, British Airways, and Qantas—three of Boeing's clients whose fleets included a large number of 747-400s—asked Boeing engineers to model the 777 cockpit on the 747-400s. In response to these requests, Boeing introduced a shared 747/777 cockpit design that enabled its airline customers to use a single pool of pilots for both aircraft types at a significant cost savings.[29]

Additionally, the airline carriers urged Boeing to increase its use of avionics for in-flight entertainment. The 777, as a consequence, was equipped with a fully computerized cabin. Facing each seat on the 777, and placed on the back of the seat in front, was a combined computer and video monitor that featured movies, video programs, and interactive computer games. Passengers were also provided with a digital sound system comparable to the most advanced home stereo available, and a telephone. About 40 percent of the 777's total computer capacity was reserved for passengers in the cabin.[30]

The 777 was Boeing's first fly by wire (FBW) aircraft, an aircraft controlled by a pilot transmitting commands to the moveable surfaces (rudder, flaps, etc.) electrically, not mechanically. Boeing installed a state of the art FBW system on the 777 partly to satisfy its airline customers, and partly to challenge Airbus' leadership in flight control technology, a position Airbus had held since it introduced the world's first FBW aircraft, the A-320, in 1988.

Lastly, Boeing customers were invited to contribute to the design of the 777's engine. Both United Airlines and All Nippon Airlines assigned service engineers to work with representatives of Pratt and Whitney (P&W) on problems associated with engine maintenance. P&W held three specially scheduled "airline conferences." At each conference, some forty airline representatives clustered around a

Exhibit IV. 777 supplier contracts

U.S. Suppliers of Structural Components

Astech/MCI	Santa Ana, CA	Primary exhaust cowl assembly (plug and nozzle)
Grumman Aerospace	Bethpage, NY	Spoilers, inboard flaps
Kaman	Bloomfield, CT	Fixed training edge
Rockwell	Tulsa, OK	Floor beams, wing leading edge slats

International Suppliers of Structural Components

AeroSpace Technologies of Australia	Australia	Rudder
Alenia	Italy	Wing outboard flaps, radome
Embrace-Empresa Brasiera de Aeronautica	Brazil	Dorsal fin, wingtip assembly
Hawker de Havilland	Australia	Elevators
Korean Air	Korea	Flap support fairings, wingtip assembly
Menasco Aerospace/ Messier-Bugatti	Canada/France	Main and nose landing gears
Mitsubishi Heavy Industries, Kawasaki Heavy Industries, and Fuji Heavy Industries[a]	Japan	Fuselage panels and doors, wing center section wing-to-body fairing, and wing in-spar ribs
Short Brothers	Ireland	Nose landing gear doors
Singapore Aerospace Manufacturing	Singapore	Nose landing gear doors

U.S. Suppliers of Systems and Equipment

AlliedSignal Aerospace Company, AiResearch Divisions	Torrance, CA	Cabin pressure control system, air supply control system, integrated system controller, ram air turbine
Bendix Wheels and	South Bend, IN	Wheel and brakes
Garrett Divisions	Phoenix/Tempe, AZ	Auxiliary power unit (APU), air-driven unit
BFGoodrich	Troy, OH	Wheel and brakes
Dowly Aerospace	Los Angeles, CA	Thrust reverser actuator system
Eldec	Lynnwood, WA	Power supply electronics
E-Systems, Montek Division	Salt Lake City, UT	Stabilizer trim control module, secondary hydraulic brake, optional folding wingtip system
Honeywell	Phoenix, AZ Coon Rapid, MN	Airplane information management system (AIMS), air data/inertial reference system (ADIRS)
Rockwell, Collins Division	Cedar Rapids, IA	Autopilot flight director system, electronic library system (ELS) displays
Sundstrand Corporation	Rockford, IL	Primary and backup electrical power systems
Teijin Seiki America	Redmond, WA	Power control units, actuator control electronics
United Technologies, Hamilton Standard Division	Windsor Lock, CT	Cabin air-conditioning and temperature control systems, ice protection system

International Suppliers of Systems and Equipment

General Electric Company (GEC) Avionics	United Kingdom	Primary flight computers
Smiths Industries	United Kingdom	Integrated electrical management system (ELMS), throttle control system actuator, fuel quantityindicating system (FQIS)

[a]Program partners
Source: James Woolsey, "777, Boeing's New Large Twinjet," *Air Transport World* (April 1994), p. 24.

full scale mock-up of the 777 engine and showed Pratt and Whitney engineers gaps in the design, hard-to-reach points, visible but inaccessible parts, and accessible but invisible components. At the initial conference, Pratt and Whitney picked up 150 airline suggestions, at the second, fifty, and at the third, ten more suggestions.[31]

A GLOBALLY MANUFACTURED PRODUCT

Twelve international companies located in ten countries, and eighteen more U.S. companies located in twelve states, were contracted by Boeing to help manufacture the 777. Together, they supplied structural components as well as systems and equipment. Among the foreign suppliers were companies based in Japan, Britain, Australia, Italy, Korea, Brazil, Singapore, and Ireland; among the major U.S. subcontractors were the Grumman Corporation, Rockwell (later merged with Boeing), Honeywell, United Technologies, Bendix, and the Sunstrand Corporation (Exhibits IV and V). Of all foreign participants, the Japanese played the largest role. A consortium made up of Fuji Heavy Industries, Kawasaki Heavy Industries, and Mitsubishi Heavy Industries had worked with Boeing on its wide-body models since the early days of the 747. Together, the three Japanese subcontractors produced 20 percent of the value of the 777's airframe (up from 15 percent of the 767s). A group of 250 Japanese engineers had spent a year in Seattle working on the 777 alongside Boeing engineers before most of its members went back home to begin production. The fuselage was built in sections in Japan and then shipped to Boeing's huge plant at Everett, Washington for assembly.[32]

Boeing used global subcontracting as a marketing tool as well. Sharing design work and production with overseas firms, Boeing required overseas carriers to buy the new aircraft. Again, Japan is a case in point. In return for the contract signed with the Mitsubishi, Fuji, and Kawasaki consortium—which was heavily subsidized by the Japanese government—Boeing sold forty-six 777 jetliners to three Japanese air carriers: All Nippon Airways, Japan Airlines, and Japan Air System.[33]

A FAMILY OF PLANES

From the outset, the design of the 777 was flexible enough to accommodate derivative jetliners. Because all derivatives of a given model shared maintenance, training, and operating procedures, as well as replacement parts and components, and because such derivatives enabled carriers to serve different markets at lower costs, Boeing's clients were seeking a family of planes built around a basic model, not a single 777. Condit and his management team, accordingly, urged Boeing's engineers to incorporate the maximum flexibility into the design of the 777.

Exhibit V.   The builders of the Boeing 777

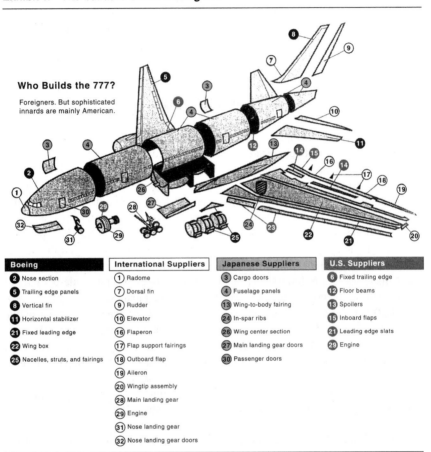

Who Builds the 777?

Foreigners. But sophisticated innards are mainly American.

Boeing	International Suppliers	Japanese Suppliers	U.S. Suppliers
2 Nose section	**1** Radome	**3** Cargo doors	**6** Fixed trailing edge
5 Trailing edge panels	**7** Dorsal fin	**4** Fuselage panels	**12** Floor beams
8 Vertical fin	**9** Rudder	**13** Wing-to-body fairing	**13** Spoilers
11 Horizontal stabilizer	**10** Elevator	**24** In-spar ribs	**15** Inboard flaps
21 Fixed leading edge	**16** Flaperon	**26** Wing center section	**21** Leading edge slats
22 Wing box	**17** Flap support fairings	**27** Main landing gear doors	**29** Engine
25 Nacelles, struts, and fairings	**18** Outboard flap	**30** Passenger doors	
	19 Aileron		
	20 Wingtip assembly		
	28 Main landing gear		
	29 Engine		
	31 Nose landing gear		
	32 Nose landing gear doors		

Source: Jeremy Main, "Corporate Performance: Betting on the 21st Century Jet," *Fortune* (April 20, 1992), p. 104.

The 777's design flexibility helped Boeing manage the project's risks. Offering a family of planes based on a single design to accommodate future changes in customers' preferences, Boeing spread the 777 project's risks among a number of models all belonging to the same family.

The key to the 777's design efficiency was the wing. The 777 wings, exceptionally long and thin, were strong enough to support vastly enlarged models. The first model to go into service, the 777-200, had a 209-foot-long fuselage, was designed to carry 305 passengers in three class configurations, and had a travel range of 5,900 miles in its original version (1995), and up to 8,900 miles in its extended version (1997). The second model to be introduced (1998), the 777-300, had a stretched fuselage of 242 feet (ten feet longer than the 747), was configured for 379 passengers (three-class), and flew to destinations of up to 6,800 miles away. In all-tourist class configuration, the stretched 777-300 could carry as many as 550 passengers.[34]

DIGITAL DESIGN

The 777 was the first Boeing jetliner designed entirely by computers. Historically, Boeing had designed new planes in two ways: paper drawings and full-size models called mock-ups. Paper drawings were two dimensional and therefore insufficient to account for the complex construction of the three dimensional airplane. Full-scale mock-ups served as a backup to drawings.

Boeing engineers used three classes of mock-ups. Made up of plywood or foam, class 1 mock-ups were used to construct the plane's large components in three dimensions, refine the design of these components by carving into the wood or foam, and feed the results back into the drawings. Made partly of metal, class 2 mock-ups addressed more complex problems such as the wiring and tubing of the airframe, and the design of the machine tools necessary to cut and shape the large components. Class 3 mock-ups gave the engineers one final opportunity to refine the model and thereby reduce the need to keep on changing the design during the actual assembly process or after delivery.[35]

Despite the engineers' efforts, many parts and components did not fit together on the final assembly line but rather "interfered" with each other, that is, overlapped in space. The problem was both pervasive and costly, Boeing engineers needed to rework and realign all overlapping parts in order to join them together.

A partial solution to the problem was provided by the computer. In the last quarter of the twentieth century, computer aided design was used successfully in car manufacture, building construction, machine production, and several other industries; its application to commercial aircraft manufacturing came later, both in the United States and in Europe. Speaking of the 777, Dick Johnson, Boeing chief engineer for digital design, noted the "tremendous advantage" of computer application:

> With mock-ups, the . . . engineer had three opportunities at three levels of detail to check his parts, and nothing in between. With Catia [Computer aided three dimensional, interactive application] he can do it day in and day out over the whole development of the airplane.[36]

Catia was a sophisticated computer program that Boeing bought from Dassault Aviation, a French fighter planes builder. IBM enhanced the program to improve image manipulation, supplied Boeing with eight of its largest mainframe computers, and connected the mainframes to 2,200 computer terminals that Boeing distributed among its 777 design teams. The software program showed on a screen exactly how parts and components fit together before the actual manufacturing process took place.[37]

A digital design system, Catia had five distinctive advantages. First, it provided the engineers with 100 percent visualization, allowing them to rotate, zoom, and "interrogate" parts geometrically in order to spotlight interferences. Second, Catia assigned a numerical value to each drawing on the screen and thereby helped engineers locate related drawings of parts and components, merge them together, and check for incompatibilities. Third, to help Boeing's customers service the 777, the digital design system created a computer simulated human— a Catia figure playing the role of the service mechanic—who climbed into the three dimensional images and showed the engineers whether parts were serviceable and entry accessible. Fourth, the use of Catia by all 777 design teams in the United States, Japan, Europe, and elsewhere facilitated instantaneous communication between Boeing and its subcontractors and ensured the frequent updating of the design. And fifth, Catia provided the 777 assembly line workers with graphics that enhanced the narrative work instructions they received, showing explicitly on a screen how a given task should be performed.[38]

DESIGN-BUILD TEAMS (DBTs)

Teaming was another feature of the 777 program. About thirty integrated-level teams at the top and more than 230 design-build teams at the bottom worked together on the 777.[39] All team members were connected by Catia. The integrated-level teams were organized around large sections of the aircraft; the DBTs around small parts and components. In both cases, teams were cross-functional, as Philip Condit observed:

> If you go back . . . to earlier planes that Boeing built, the factory was on the bottom floor, and Engineering was on the upper floor. Both Manufacturing and Engineering went back and forth. When there was a problem in the factory, the engineer went down and looked at it. . . .
> With 10,000 people [working on the 777], that turns out to be really hard. So you start devising other tools to allow you to achieve that—the design-build team. You break the airplane down and bring Manufacturing, Tooling, Planning, Engineering, Finance, and Materials all together [in small teams].[40]

Under the design-build approach, many of the design decisions were driven by manufacturing concerns. As manufacturing specialists worked alongside engineers, engineers were less likely to design parts that were difficult to produce and needed to be redesigned. Similarly, under the design-build approach, customers' expectations as well as safety and weight considerations were all incorporated into the design of the aircraft; engineers no longer needed to "chain saw"[41] structural components and systems in order to replace parts that did not meet customers expectations, were unsafe, or were too heavy.

The design of the 777's wing provides an example. The wing was divided into two integration-level teams, the *leading-edge* (the forward part of the wing) and the *trailing-edge* (the back of the wing) team. Next, the trailing-edge team was further divided into ten design-build teams, each named after a piece of the wing's trailing edge (Exhibit VI). Membership in these DBTs extended to two groups of outsiders: representatives of the customer airlines and engineers employed by the foreign subcontractors. Made up of up to twenty members, each DBT decided its own mix of insiders and outsiders, and each was led by a team leader. Each DBT included representatives from six functional disciplines: engineering, manufacturing, materials, customer support, finance, and quality assurance. The DBTs met twice a week for two hours to hear reports from team members, discuss immediate goals and plans, divide responsibilities, set time lines, and take specific notes of all decisions taken.[42] Described by a Boeing official as *little companies,* the DBTs enjoyed a high degree of autonomy from management supervision; team members designed their own tools, developed their own manufacturing plans, and wrote their own contracts with the program management, specifying deliverables, resources, and schedules. John Monroe, a Boeing 777 senior project manager remarked:

> The team is totally responsible. We give them a lump of money to go and do th[eir] job. They decide whether to hire a lot of inexpensive people or to trade numbers for resources. It's unprecedented. We have some $100 million plus activities led by non-managers.[43]

Exhibit VI. *The ten DBTs ("little companies") responsible for the wing's trailing edge*

- Flap Supports Team
- Inboard Flap Team
- Outboard Flap Team
- Flaperon[a] Team
- Aileron[a] Team
- Inboard Fixed Wing and Gear Support Team
- Main Landing Gear Doors Team
- Spoilers[b] Team
- Fairings[c] Team

---

[a]The flaperon and aileron were movable hinged sections of the trailing edge that helped the plane roll in flight. The flaperon was used at high speed, the aileron at low speed.
[b]The spoilers were the flat surfaces that lay on top of the trailing edge and extended during landing to slow down the plane.
[c]The fairing were the smooth parts attached to the outline of the wing's trailing edge. They helped reduce drag.
Source: Karl Sabbagh, *21st Century Jet: The Making and Marketing of the Boeing 777* (New York: Scribner, 1996), p. 73.

EMPLOYEES' EMPOWERMENT AND CULTURE

An additional aspect of the 777 program was the empowering of assembly line workers. Boeing managers encouraged factory workers at all levels to speak up, offer suggestions, and participate in decision making. Boeing managers also paid attention to a variety of "human relations" problems faced by workers, problems ranging from childcare and parking to occupational hazards and safety concerns.[44]

All employees entering the 777 program—managers, engineers, assembly line workers, and others—were expected to attend a special orientation session devoted to the themes of team work and quality control. Once a quarter, the entire "777 team" of up to 10,000 employees met offsite to hear briefings on the aircraft status. Dressed casually, the employees were urged to raise questions, voice complaints, and propose improvements. Under the 777 program, managers met frequently to discuss ways to promote communication with workers. Managers, for example, "fire fought" problems by bringing workers together and empowering them to offer solutions. In a typical *firefight* session, Boeing 777 project managers learned from assembly line workers how to improve the process of wiring and tubing the airframe's interior: "staffing" fuselage sections with wires, ducts, tubs, and insulation materials before joining the sections together was easier than installing the interior parts all at once in a preassembled fuselage.[45]

Under the 777 program, in addition, Boeing assembly line workers also were empowered to appeal management decisions. In a case involving middle managers, a group of Boeing machinists sought to replace a nonretractable jig (a large device used to hold parts) with a retractable one in order to ease and simplify their jobs. Otherwise they had to carry heavy equipment loads up and down stairs. Again and again, their supervisors refused to implement the change. When the machinists eventually approached a factory manager, he inspected the jig personally, and immediately ordered the change.[46]

Under the 777 program, work on the shop floor was ruled by the *Bar Chart*. A large display panel placed at different work areas, the Bar Chart listed the name of each worker, his or her daily job description, and the time available to complete specific tasks. Boeing had utilized the Bar Chart system as a "management visibility system" in the past, but only under the 777 program was the system fully computerized. The chart showed whether assembly line workers were meeting or missing their production goals. Boeing industrial engineers estimated the time it took to complete a given task and fed the information back to the system's computer. Workers ran a scanner across their ID badges and supplied the computer with the data necessary to log their job progress. Each employee "sold" his/her completed job to an inspector, and no job was declared acceptable unless "bought" by an inspector.[47]

LEADERSHIP AND MANAGEMENT STYLE

The team in charge of the 777 program was led by a group of five vice presidents, headed by Philip Condit, a gifted engineer who was described by one Wall Street analyst as "a cross between a grizzly bear and a teddy bear. Good people skills, but furious in the marketplace."[48] Each of the five vice presidents rose through the ranks, and each had a twenty-five to thirty years experience with Boeing. All were men.[49]

During the 777 design phase, the five VPs met regularly every Tuesday morning in a small conference room at Boeing's headquarters in Seattle in what was called the "Muffin Meeting." There were no agendas drafted, no minutes drawn, no overhead projectors used, and no votes taken. The homemade muffins served during the meeting symbolized the informal tone of the forum. Few people outside the circle of five had ever attended these weekly sessions. Acting as an informal chair, Condit led a freewheeling discussion of the 777 project, asking each VP to say anything he had on his mind.[50]

The weekly session reflected Boeing's sweeping new approach to management. Traditionally, Boeing had been a highly structured company governed by engineers. Its culture was secretive, formal, and stiff. Managers seldom interacted, sharing was rare, divisions kept to themselves, and engineers competed with each other. Under the 777 program, Boeing made serious efforts to abandon its secretive management style. Condit firmly believed that open communication among top executives, middle managers, and assembly line workers was indispensable for improving morale and raising productivity. He urged employees to talk to each other and share information, and he used a variety of management tools to do so: information sheets, orientation sessions, question and answer sessions, leadership meetings, regular workers as well as middle managers, Condit introduced a three-way performance review procedure whereby managers were evaluated by their supervisors, their peers, and their subordinates.[51] Most important, Condit made teamwork the hallmark of the 777 project. In an address titled "Working Together: The 777 Story" and delivered in December 1992 to members of the Royal Aeronautics Society in London,[52] Condit summed up his team approach:

> [T]eam building is . . . very difficult to do well but when it works the results are dramatic. Teaming fosters the excitement of a shared endeavor and creates an atmosphere that stimulates creativity and problem solving. But building team[s] . . . is hard work. It doesn't come naturally. Most of us are taught from an early age to compete and excel as individuals. Performance in school and performance on the job are usually measured by individual achievement. Sharing your ideas with others, or helping others to enhance their performance, is often viewed as contrary to one's self interest.
>
> This individualistic mentality has its place, but . . . it is no longer the most useful attitude for a workplace to possess in today's world. To create a high performance organization, you need employees who can work together in a way that promotes continual learning and the free flow of ideas and information.

THE RESULTS OF THE 777 PROJECT

The 777 entered revenue service in June 1995. Since many of the features incorporated into the 777's design reflected suggestions made by the airline carriers, pilots, mechanics, and flight attendants were quite enthusiastic about the new jet. Three achievements of the program, in airplane interior, aircraft design, and aircraft manufacturing, stood out.

Configuration Flexibility
The 777 offered carriers enhanced configuration flexibility. A typical configuration change took only seventy-two hours on the 777 compared to three weeks in competing aircraft. In 1992, the Industrial Design Society of America granted Boeing its Excellence Award for building the 777 passenger cabin, honoring an airplane interior for the first time.[53]

Digital Design
The original goal of the program was to reduce "change, error, and rework" by 50 percent, but engineers building the first three 777s managed to reduce such modification by 60 percent to 90 percent. Catia helped engineers identify more than 10,000 interferences that would have otherwise remained undetected until assembly, or until after delivery. The first 777 was only 0.023 inch short of perfect alignment, compared to as much as 0.5 inch on previous programs.[54] Assembly line workers confirmed the beneficial effects of the digital design system. "The parts snap together like Lego blocks," said one mechanics.[55] Reducing the need for reengineering, replanning, retooling, and retrofitting, Boeing's innovative efforts were recognized yet again. In 1993, the Smithsonian Institution honored the Boeing 777 division with its Annual Computerworld Award for the manufacturing category.[56]

Empowerment
Boeing 777 assembly line workers expressed a high level of job satisfaction under the new program. "It's a whole new world," a fourteen-year Boeing veteran mechanic said, "I even like going to work. It's bubbly. It's clean. Everyone has confidence."[57] "We never used to speak up," said another employee, "didn't dare. Now factory workers are treated better and are encouraged to offer ideas."[58] Although the Bar Chart system required Boeing 777 mechanics to work harder and faster as they moved down the learning curve, their principal union organization, the International Association of Machinists, was pleased with Boeing's new approach to labor–management relations. A union spokesman reported that under the 777 program, managers were more likely to treat problems as

opportunities from which to learn rather than mistakes for which to blame. Under the 777 program, the union representative added, managers were more respectful of workers' rights under the collective bargaining agreement.[59]

UNRESOLVED PROBLEMS AND LESSONS LEARNED

Notwithstanding Boeing's success with the 777 project, the cost of the program was very high. Boeing did not publish figures pertaining to the total cost of Catia. But a company official reported that under the 777 program, the 3D digital design process required 60 percent more engineering resources than the older, 2D drawing-based design process. One reason for the high cost of using digital design was slow computing tools: Catia's response time often lasted minutes. Another was the need to update the design software repeatedly. Boeing revised Catia's design software four times between 1990 and 1996, making the system easier to learn and use. Still, Catia continued to experience frequent software problems. Moreover, several of Boeing's outside suppliers were unable to utilize Catia's digital data in their manufacturing process.[60]

Boeing faced training problems as well. One challenging problem, according to Ron Ostrowski, director of 777 engineering, was "to convert people's thinking from 2D to 3D. It took more time than we thought it would. I came from a paper world and now I am managing a digital program."[61] Converting people's thinking required what another manager called an "unending communication" coupled with training and retraining. Under the 777 program, Ostrowski recalled, "engineers had to learn to interact. Some couldn't, and they left. The young ones caught on" and stayed.[62]

Learning to work together was a challenge to managers, too. Some managers were reluctant to embrace Condit's open management style, fearing a decline in their authority. Others were reluctant to share their mistakes with their superiors, fearing reprisals. Some other managers, realizing that the new approach would end many managerial jobs, resisted change when they could, and did not pursue it wholeheartedly when they could not. Even top executives were sometimes uncomfortable with Boeing's open management style, believing that sharing information with employees was likely to help Boeing's competitors obtain confidential 777 data.[63]

Teamwork was another problem area. Working under pressure, some team members did not function well within teams and had to be moved. Others took advantage of their newborn freedom to offer suggestions, but were disillusioned and frustrated when management either ignored these suggestions, or did not act upon them. Managers experienced different team-related problems. In several cases, managers kept on meeting with their team members repeatedly until they arrived at a solution desired by their bosses. They were unwilling to challenge senior executives, nor did they trust Boeing's new approach to teaming. In other cases, managers distrusted the new digital technology. One engineering manager instructed his team members to draft paper drawings alongside Catia's digital designs. When Catia experienced a problem, he followed the drawing, ignoring the computerized design, and causing unnecessary and costly delays in his team's part of the project.[64]

Extending the 777 Revolution

Boeing's learning pains played a key role in the company's decision not to implement the 777 program companywide. Boeing officials recognized the importance of team work and Catia in reducing change, error, and rework, but they also realized that teaming required frequent training, continuous reinforcement, and ongoing monitoring, and that the use of Catia was still too expensive, though its cost was going down (in 1997, Catia's "penalty" was down to 10 percent). Three of Boeing's derivative programs, the 737 Next Generation, the 757-300, and the 767-400, had the option of implementing the 777's program innovations, and only one, the 737, did so, adopting a modified version of the 777's cross-functional teams.[65]

Yet the 777's culture was spreading in other ways. Senior executives took broader roles as the 777 entered service, and their impact was felt through the company. Larry Olson, director of information systems for the 747/767/777 division, was a former 777 manager who believed that Boeing 777 employees "won't tolerate going back to the old ways." He expected to fill new positions on Boeing's next program—the 747X—with former 777 employees in their forties.[66] Philip Condit, Boeing CEO, implemented several of his own 777's innovations, intensifying the use of meeting among Boeing's managers, and promoting the free flow of ideas throughout the company. Under Condit's leadership, all mid-level managers assigned to Boeing Commercial Airplane Group, about sixty people, met once a week to discuss costs, revenues, and production schedules, product by product. By the end of the meeting—which sometimes ran into the evening—each manager had to draft a detailed plan of action dealing with problems in his/her department.[67] Under Condit's leadership, more important, Boeing developed a new "vision" that grew out of the 777 project. Articulating the company's vision for the next two decades (1996–2016), Condit singled out "Customer satisfaction," "Team leadership," and "A participatory workplace," as Boeing's core corporate values.[68]

CONCLUSION: BOEING, AIRBUS, AND THE 777

Looking back at the 777 program twelve years after the launch and seven years after first delivery, it is now (2002) clear that Boeing produced the most successful commercial jetliner of its kind. Airbus launched the A330 and A340 in 1987,

Exhibit VII. Total number of MD11, A330, A340, and 777 airplanes delivered during 1995–2001

	1995	1996	1997	1998	1999	2000	2001
McDonnell Douglas/ Boeing MD11	18	15	12	12	8	4	2
Airbus A330	30	10	14	23	44	43	35
Airbus A340	19	28	33	24	20	19	20
Boeing 777	13	32	59	74	83	55	61

Source: For Airbus, Mark Luginbill Airbus Communication Director, February 1, 2000, and March 11, 2002. For Boeing, *The Boeing Company Annual Report,* 1997, p. 35, 1998, p. 35; "Commerical Airplanes: Order and Delivery, Summary," *http//www.boeing.com/commercial/order/index.html.* Retreived from Web, February 2, 2000, and March 9, 2002.

and McDonnell Douglas launched a new 300-seat wide body jet in the mid 1980s, the three-engine MD11. Coming late to market, the Boeing 777 soon outsold both models. The 777 had entered service in 1995, and within a year Boeing delivered more than twice as many 777s as the number of MD11s delivered by McDonnell Douglas. In 1997, 1998, 1999, and 2001, Boeing delivered a larger number of 777s than the combined number of A330s and A340s delivered by Airbus (Exhibit VII). A survey of nearly 6,000 European airline passengers who had flown both the 777 and the A330/A340 found that the 777 was preferred by more than three out of four passengers.[69] In the end, a key element in the 777's triumph was its popularity with the traveling public.

NOTES

1. Rodgers, Eugene. *Flying High: The Story of Boeing* (New York: Atlantic Monthly Press, 1996), 415–416; Michael Dornheim, "777 Twinjet Will Grow to Replace 747-200," *Aviation Week and Space Technology* (June 3, 1991): 43.
2. "Commercial Airplanes: Order and Delivery, Summary," http/www. boeing.com/commercial/orders/index.html. Retrieved from Web, February 2, 2000.
3. Donlon, P. "Boeing's Big Bet" (an interview with CEO Frank Shrontz), *Chief Executive* (November/December 1994): 42; Dertouzos, Michael, Richard Lester, and Robert Solow, *Made in America: Regaining the Productive Edge* (New York: Harper Perennial, 1990), 203.
4. John Newhouse, *The Sporty Game* (New York: Alfred Knopf, 1982), 21, but see also 10–20.
5. Mowery, David C., and Nathan Rosenberg. "The Commercial Aircraft Industry," in Richard R. Nelson, ed., *Government and Technological Progress:*

A Cross Industry Analysis (New York: Pergamon Press, 1982), 116; Dertouzos et al., *Made in America,* 200.

6. Dertouzos et al., *Made in America,* 200.
7. Newhouse, *Sporty Game,* 188. Mowery and Rosenberg, "The Commercial Aircraft Industry," 124–125.
8. Mowery and Rosenberg, "The Commercial Aircraft Industry," 102–103, 126–128.
9. Rae, John B. *Climb to Greatness: The American Aircraft Industry, 1920–1960* (Cambridge, Mass.: MIT Press, 1968), 206–207; Rodgers, *Flying High,* 197–198.
10. Spadaro, Frank. "A Transatlantic Perspective," *Design Quarterly* (Winter 1992): 23.
11. Rodgers, *Flying High,* 279; Newhouse, *Sporty Game,* Ch. 7.
12. Hochmuth, M. S. "Aerospace," in Raymond Vernon, ed., *Big Business and the State* (Cambridge: Harvard University Press, 1974), 149.
13. Boeing Commercial Airplane Group, *Announced Orders and Deliveries as of 12/31/97,* Section A 1.
14. *The Boeing Company 1998 Annual Report,* 76.
15. Formed in 1970 by several European aerospacc firms, the Airbus Consortium had received generous assistance from the French, British, German, and Spanish governments for a period of over two decades. In 1992, Airbus had signed an agreement with Boeing that limited the amount of government funds each aircraft manufacturer could receive, and in 1995, at long last, Airbus had become profitable. "Airbus 25 Years Old," *Le Figaro,* October 1997 (reprinted in English by Airbus Industrie); Rodgers, *Flying High,* Ch. 12; *Business Week* (30 December 1996): 40.
16. Charles Goldsmith, "Re-engineering, After Trailing Boeing for Years, Airbus Aims for 50% of the Market," *Wall Street Journal* (March 16, 1998).
17. "Hubris at Airbus, Boeing Rebuild," *Economist,* 28 (November 1998).
18. *The Boeing Company 1997 Annual Report,* 19; *The Boeing Company 1998 Annual Report,* 51.
19. Donlon, "Boeing's Big Bet," 40; John Mintz, "Betting It All on 777" *Washington Post* (March 26, 1995); James Woolsey, "777: A Program of New Concepts," *Air Transport World* (April 1991): 62; Jeremy Main, "Corporate Performance: Betting on the 21st Century Jet," *Fortune* (April 20, 1992), 104; James Woolsey, "Crossing New Transport Frontiers," *Air Transport World* (March 1991): 21; James Woolsey, "777: Boeing's New Large Twinjet," *Air Transport World* (April 1994): 23; Michael Dornheim, "Computerized Design System Allows Boeing to Skip Building 777 Mockup," *Aviation Week and Space Technology* (June 3, 1991): 51; Richard O'Lone, "Final Assembly of 777 Nears," *Aviation Week and Space Technology* (October 2, 1992): 48.

20. Rodgers, *Flying High,* 42.
21. *Air Transport World* (March 1991): 20; *Fortune* (April 20, 1992), 102–103.
22. Rodgers, *Flying High,* 416, 420–424.
23. Richard O'Lone and James McKenna, "Quality Assurance Role was Factor in United's 777 Launch Order," *Aviation Week and Space Technology* (October 29, 1990): 28–29; *Air Transport World* (March 1991): 20.
24. Quoted in the *Washington Post* (March 25, 1995).
25. Quoted in Bill Swectman, "As Smooth as Silk: 777 Customers Applaud the Aircraft's First 12 Months in Service," *Air Transport World* (August 1996): 71, but see also *Air Transport World* (April 1994): 24, 27.
26. Quoted in *Fortune* (April 20, 1992), 112.
27. Rodgers, *Flying High,* 426; *Design Quarterly* (Winter 1992): 22; Polly Lane, "Boeing Used 777 to Make Production Changes," *Seattle Times* (May 7, 1995).
28. *Design Quarterly* (Winter 1992): 22; The Boeing Company, *Backgrounder: Pace Setting Design Value-Added Features Boost Boeing 777 Family* (May 15, 1998).
29. Boeing, *Backgrounder,* (May 15, 1998); Sabbagh, *21st Century Jet,* p. 49.
30. Karl Sabbagh, *21st Century Jet: The Making and Marketing of the Boeing 777* (New York: Scribner, 1996), 264, 266.
31. Sabbagh, *21st Century Jet,* 131–132
32. *Air Transport World* (April 1994): 23; *Fortune* (April 20, 1992), 116.
33. *Washington Post* (March 26, 1995); Boeing Commercial Airplane Group, 777 Announced Order and Delivery Summary...As of 9/30/99.
34. Rodgers, *Flying High,* 420–426; *Air Transport World* (April 1994): 27, 31; "Leading Families of Passenger Jet Airplanes," Boeing Commercial Airplane Group, 1998.
35. Sabbagh, *21st Century Jet,* 58.
36. Quoted in Sabbagh, *21st Century Jet,* 63.
37. *Aviation Week and Space Technology* (June 3, 1991): 50, (October 12, 1992), p. 49; Sabbagh *21st Century Jet,* p. 62.
38. George Taninecz, "Blue Sky Meets Blue Sky," *Industry Week* (December 18, 1995); 49–52; Paul Proctor, "Boeing Rolls Out 777 to Tentative Market," *Aviation Week and Space Technology* (October 12, 1992): 49.
39. *Aviation Week and Space Technology* (April 11, 1994): 37; *Aviation Week and Space Technology* (June 3, 1991): 35.
40. Quoted in Sabbagh, *21st Century Jet,* 68–69.
41. This was the phrase used by Boeing project managers working on the 777. See Sabbagh, *21st Century Jet,* Ch. 4.
42. *Fortune* (April 20, 1992), 116; Sabbagh, *21st Century Jet,* 69–73; Wolf L. Glende, "The Boeing 777: A Look Back," The Boeing Company, 1997, 4.
43. Quoted in *Air Transport World* (August 1996): 78.

44. Richard O'Lone, "777 Revolutionizes Boeing Aircraft Development Process," *Aviation Week and Space Technology* (June 3, 1992): 34.
45. O. Casey Corr. "Boeing's Future on the Line: Company's Betting its Fortunes Not Just on a New Jet, But on a New Way of Making Jets," *Seattle Times* (August 29, 1993); Polly Lane, "Boeing Used 777 to Make Production Changes, Meet Desires of Its Customers," *Seattle Times* (May 7, 1995); *Aviation Week and Space Technology* (June 3, 1991): 34.
46. *Seattle Times* (August 29, 1993).
47. *Seattle Times* (May 7, 1995, and August 29, 1993).
48. Quoted in Rodgers, *Flying High,* 419–420.
49. Sabbagh, *21st Century Jet,* 33.
50. Sabbagh, *21st Century Jet,* 99.
51. Dori Jones Young, "When the Going Gets Tough, Boeing Gets Touchy-Feely, *Business Week* (January 17, 1994): 65–67; *Fortune* (April 20, 1992), 117.
52. Reprinted by The Boeing Company, Executive Communications, 1992.
53. Boeing, *Backgrounder* (May 15, 1998).
54. *Industry Week* (December 18, 1995): 50–51; *Air Transport World* (April 1994).
55. *Aviation Week and Space Technology* (April 11, 1994): 37.
56. Boeing, *Backgrounder,* "Computing & Design/Build Process Help Develop the 777." Undated.
57. *Seattle Times* (August 29, 1993).
58. *Seattle Times* (May 7, 1995).
59. *Seattle Times* (August 29, 1993).
60. Glende, "The Boeing 777: A Look Back," 1997, 10; *Air Transport World* (August 1996): 78.
61. *Air Transport World* (April 1994): 23.
62. *Washington Post* (March 26, 1995).
63. *Seattle Times* (May 7, 1995); Rodgers, *Flying High,* 441.
64. *Seattle Times* (May 7, 1995); Rodgers, *Flying High,* 441–442.
65. Glende, "The Boeing 777: A Look Back," 1997, 10.
66. *Air Transport World* (August 1996), 78.
67. "A New Kind of Boeing," *Economist* (January 22, 2000), 63.
68. "Vision 2016," The Boeing Company 1997.
69. "Study: Passengers Voice Overwhelming Preference for Boeing 777, http/www.boeing.com/news/releases/1999. Retrieved from Web 11/23/99.

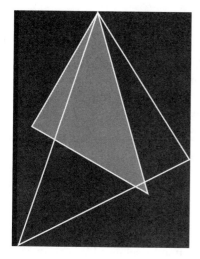

IVEY

Richard Ivey School of Business
The University of Western Ontario

AMP of Canada (A)

Stéphane Marchak prepared this case under the supervision of Professor E.F. Peter Newson solely to provide material for class discussion. The authors do not intend to illustrate either effective or ineffective handling of a managerial situation. The authors may have disguised certain names and other identifying information to protect confidentiality.

Version: (A) 1999-12-10

Perplexed, Doris Puddington pondered the most important decision she had to make in the year since she had become controller at AMP of Canada. In April 1996 she had joined AMP with assurances that the existing transactional processing system JBA did not require major rework. However, in the summer of 1997, she had learned that JBA was not year 2000 compliant. To solve this problem, she had to choose from three alternatives: upgrading JBA, implementing

AMPICS, or implementing SAP. She knew that Canadian management preferred to implement the very popular SAP system, but her IS manager did not think that users were ready for SAP and preferred an option involving JBA. AMP's headquarters management preferred the software package that many AMP companies already used, called AMPICS. Doris now wondered which solution to choose and how to persuade Canadian management, the Canadian IS department, and headquarters management to support this decision.

AMP INCORPORATED

AMP was founded in 1941 to support the war effort by selling electrical connectors to the U.S. government. AMP commercialized the technique of crimping, whereby an electrical contact is quickly attached to a wire using a hand tool instead of being slowly soldered. After WWII, AMP enjoyed rapid growth from the expanding computer electronics industry. In 1955, AMP changed its name from Aero-Marine Products to AMP. Since then, AMP had become the world's leading manufacturer of electrical and electronic connectors (see Exhibit I for examples of connectors). Global revenue in 1996 was U.S.$5.5 billion (see Exhibits II to III). AMP had 48,000 employees in 53 countries in North America, Europe, and Asia. Nearly half of those employees worked in the world headquarters of Harrisburg, Pennsylvania.

AMP served more than 200,000 customers around the world, including Siemens, Sony, Intel, Apple, Motorola, Boeing, and Ford (see Exhibit IV). AMP offered more than 800,000 part numbers in more than 470 product lines. AMP's principal products were still electrical and electronic connectors, but AMP now sold cable and cable assemblies, printed wiring boards, panel assemblies, networking and premise wiring systems, optical fiber and electro-optical products, and components for wireless communications systems (see Exhibit V). AMP had many competitors who offered smaller product lines than AMP and operated in far fewer industries. Companies such as Molex, Thomas and Betts, Berg, Robinson Nugent, Panduit, Amphenol, and Foxconn were all major competitors to AMP. AMP earned about 20 percent of market share in the approximately $27 billion electrical and electronic connection devices market.

AMP OF CANADA

AMP of Canada was a wholly owned subsidiary of AMP Incorporated. AMP of Canada was responsible for sales and manufacturing in Canada. AMP first opened a sales office in Toronto, Ontario, in 1952. A manufacturing facility was

Exhibit I. Sample AMP products

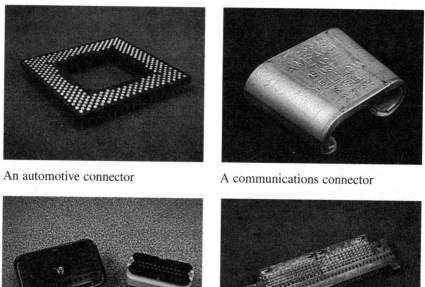

An automotive connector   A communications connector

An integrated circuit (PC) connector   A utility connector (tap)

Exhibit II. AMP Incorporated consolidated statement of income year ended December 31

	1994	1995	1996
Net Sales	4,369,067	5,227,226	5,468,028
Cost of Sales	2,884,185	3,539,715	3,902,733
Gross Income	1,484,882	1,687,511	1,565,295
Selling, General and Administrative Expenses	824,945	969,512	964,589
Restructuring and One-time Charges	—	—	98,000
Income from Operations	659,937	717,999	502,706
Interest Expense	(29,153)	(36,847)	(31,156)
Other Deductions, net	(31,972)	(13,418)	(33,242)
Income before Income Taxes	598,812	667,734	438,308
Income Taxes	225,022	240,400	151,324
Net Income	373,790	427,334	286,984
Net Income per Share	1.72	1.96	1.31

Note: All figures except per share data are in thousands of U.S. dollars.
Source: FreeEdgar.com, March 1999

Exhibit III. **AMP Incorporated consolidated balance sheets, year ended December 31**

	1994	1995	1996
Assets			
Current Assets:			
Cash and cash equivalents	244,568	212,538	223,779
Securities available for sale	156,708	58,197	27,971
Receivables	908,390	1,011,460	1,025,850
Inventories	641,953	762,803	786,623
Deferred income taxes	135,498	137,043	184,273
Other current assets	87,183	95,867	107,684
Total current assets	2,174,300	2,277,908	2,356,180
Property, Plant and Equipment	3,713,660	4,352,026	4,690,819
Less—Accumulated depreciation	2,138,978	2,413,760	2,663,211
Property, plant and equipment, net	1,574,682	1,938,266	2,027,608
Investments and other assets	343,564	288,565	301,917
Total Assets	4,092,546	4,504,739	4,685,705
Liabilities and Shareholders' Equity			
Current Liabilities:			
Short-term debt	182,338	318,169	419,411
Payables, trade and other	403,947	460,892	463,261
Accrued payrolls and employee benefits	156,322	168,667	164,842
Accrued income taxes	247,997	196,417	201,169
Other accrued liabilities	116,318	121,948	196,212
Total current liabilities	1,106,922	1,266,093	1,444,895
Long-term debt	278,843	212,485	181,599
Deferred income taxes	34,249	45,768	48,037
Other liabilities	176,777	212,365	221,276
Total liabilities	1,596,791	1,736,711	1,895,807
Shareholders' Equity:			
Common stock, without par value—Authorized 700,000,000 shares	70,135	79,580	80,866
Other capital	80,105	83,454	85,325
Deferred compensation	−4,568	−2,489	−6,896
Cumulative translation adjustments	129,612	156,837	112,179
Net unrealized investment gains	21,585	19,423	6134
Retained earnings	2,442,317	2,667,755	2,695,990
Treasury stock, at cost	−243,431	−236,532	−183,700
Total shareholders' equity	2,495,755	2,768,028	2,789,898
Total Liabilities and Shareholders' Equity	4,092,546	4,504,739	4,685,705

Note: All figures except per share data are in thousands of U.S. dollars.
Source: FreeEdgar.com, March 1999

Exhibit IV. AMP business segments

The operations of AMP are worldwide and can be grouped into several geographic segments. Operations outside the United States are conducted through wholly owned subsidiary companies that function within assigned, principally national, markets. The subsidiaries manufacture locally where required by market conditions and/or customer demands, where cost beneficial, and where permitted by economies of scale. Most are also self-financed. However, while they operate fairly autonomously, there are substantial intersegment and intrasegment sales.

Pertinent financial data are given by major geographic segments for the year ended December 31:

	1994	1995	1996
Net sales to trade customers:			
United States	1,955,329	2,238,594	2,414,652
Europe	1,308,604	1,698,407	1,725,377
Asia/Pacific	892,085	1,059,095	1,054,027
Americas	213,049	231,130	273,972
Total	4,369,067	5,227,226	5,468,028
Intersegment sales:			
United States	399,968	482,962	559,361
Europe	49,274	64,688	79,265
Asia/Pacific	73,706	95,443	94,254
Americas	15,301	11,222	13,160
Eliminations	(538,249)	(654,315)	(746,040)
Total	—	—	—
Pretax income:			
United States	300,173	398,826	264,715
Europe	184,666	192,807	141,884
Asia/Pacific	112,302	74,305	21,308
Americas	21,797	11,096	15,401
Eliminations	(20,126)	(9,300)	(5,000)
Total	598,812	667,734	438,308
Identifiable assets:			
United States	2,495,379	2,676,394	2,778,391
Europe	956,351	1,135,606	1,221,972
Asia/Pacific	905,289	1,022,667	1,081,335
Americas	107,874	121,489	170,755
Eliminations	(372,347)	(451,417)	(566,748)
Total	4,092,546	4,504,739	4,685,705

Note: All figures are in thousands of U.S. dollars.
Source: AMP 1996 Annual Report from FreeEdgar.com, March 1999.

opened in 1969. By 1997 AMP of Canada had 326 full-time employees, many of whom had worked at AMP for years. The average length of service was about ten years. Everyone operated on a first-name basis, from the employee on the factory floor to the general manager. AMP of Canada still purchased most of the product it sold in the Canadian market from AMP Incorporated in the United States, but had a strategy to grow the manufacturing portion of its business to 50 percent of sales by 2000. AMP of Canada also operated manufacturing facilities in Ottawa and Montreal to broaden the product offering made from the Markham plant.

Exhibit V. **AMP's percentage of sales by market categories**

	1986	1996
Transportation	28%	28%
Communications	11%	23%
Computer & office	28%	20%
Industrial & commercial	11%	13%
Consumer goods	11%	8%
Aerospace & military	5%	5%
Power & utility, construction, misc.	6%	3%

Source: AMP Web site, March 1999.

The Markham manufacturing plant was located just outside Toronto, Ontario. Most of AMP of Canada's manufacturing and engineering was done at Markham. Purchasing, accounting, information systems, and other functions were centralized in Markham. AMP of Canada opened a manufacturing facility near Ottawa, Ontario, in 1991. Ottawa manufactured fiber-optic connectors used in communications applications for customers like Northern Telecom. AMP of Canada inherited the manufacturing plant in Montreal, Quebec, from AMP's 1994 purchase of the French company, Simel. The Montreal plant manufactured specialized utility connectors, mostly for the Hydro Quebec utility company. To maintain AMP's global strategy of country-level operations, after the purchase of Simel the Montreal plant was assigned to AMP of Canada.

AMP of Canada also had sales offices throughout the country. Sales offices were used to take orders for local customers and to support the local field sales-people. Offices were located in Dartmouth, Montreal, Ottawa, Markham, Windsor, Guelph, Winnipeg, Calgary, and Burnaby.

AMP OF CANADA'S ORDER PROCESS

Sales orders were taken from all locations and entered into AMP of Canada's transactional processing system JBA. A few large customers transmitted sales orders directly to AMP of Canada using electronic data interchange (EDI). Most smaller customers phoned or faxed AMP of Canada's main order-entry department, called *inside sales*. An inside sales representative (ISR) would enter the customer's requested part numbers and quantities, and quote pricing and delivery information. This information included the unit price at the quantity the customer wanted (AMP's pricing used quantity scales), the minimum order quantity, the minimum packaging quantity (applicable for orders with multiple shipment dates), the lead time, and so on. Most of this information was loaded from AMP Inc.'s SS40 data transmissions. See Exhibit VI for a list of terms.

AMP collected and redistributed data around the world using a series of transmissions called SS40s, which stood for Systems Services specification # 40. Their two main functions were supporting business analysis and transactional processing. Certain SS40s were used to collect information globally for analysis in Harrisburg. For example, the SS40-05 collected sales information at the customer/part level, and the SS40-10 collected standard cost information. The SS40s were also used to supply local countries with information necessary to interact with related AMP companies. For example, the weekly SS40-01 data transmission collected part number information (including intercompany pricing, lead times, minimum order quantities, etc.) from all AMP companies around the world. This information was then retransmitted to those companies, who loaded

Exhibit VI. **List of terms**

AMP	Used to stand for Aero-Marine Products, but was changed in 1955 to one word pronounced the same way as the electrical current measurement. Although AMP is a word instead of an acronym, it is always spelled with capitalized letters.
AMPICS	Version of BPCS software modified for AMP.
ASAP	An implementation methodology for SAP called accelerated SAP (ASAP).
BPCS	An ERP software system sold by IBM.
BPP	Business process procedure.
CO	Controlling module of SAP.
ERP	Enterprise Resource Planning software. It integrates major organizational functions like accounting, manufacturing, sales, and human resources.
FI	Financial Accounting module of SAP.
ISR	An inside sales representative at AMP who places customer orders into the transactional processing system.
JBA	An ERP vendor with headquarters in Studley, UK.
MM	Materials Management module of SAP.
MRP	The Material Requirements Planning algorithm used to drive manufacturing and purchasing activity. The program takes customer and manufacturing part number quantity requirements and generates purchase orders and / or manufacturing orders to fulfill these requirements.
PM	Plant maintenance module of SAP.
PP	Production planning module of SAP.
PS	Project systems module of SAP.
QM	Quality management module of SAP.
SAP	German software company standing for systems, applications, and products (SAP) in data processing.
SD	Sales and distribution module of SAP.
SS40	AMP's Systems Services specification number 40, which defines how data were transmitted between related-AMP companies.
WF	Workflow module of SAP.
WINS	AMP's worldwide inventory system used to track availability of inventory from each related-AMP organization around the world.
WM	Warehouse management module of SAP.

the data into their local systems. In total, there were more than twenty SS40 data feeds, and the number was growing. During 1996, at least one of AMP's nine programmers was 100 percent dedicated to programming new SS40 interfaces. Although customers did not use these interfaces, they saw much of the data they supplied.

If the customer was satisfied with the pricing and delivery information produced by JBA, the ISR would complete the order. Every night an MRP run was carried out in JBA, where new customer requirements were turned into purchase requisitions. AMP of Canada's Material Control department analyzed new requirements and turned the purchase requisitions into purchase orders. Purchase orders were transmitted to related-AMP companies using EDI. Most purchase orders were sent to AMP Inc. in the United States. Every day AMP Inc. transmitted the list of parts it was shipping on the daily truck. AMP of Canada used that transmission to load inventory information into the warehouse module of JBA. After the truck was loaded, packing slips were run to tell the shippers in the warehouse what part numbers and quantities to pick for customer shipments. For certain customers, AMP of Canada transmitted Advance Shipping Notices (ASNs) to alert them to the pending shipment. The warehouse picked, packed, and shipped parts to customers.

This process took longer if the customer was not satisfied with the pricing and delivery information. For example, if a customer wanted a shorter lead time, the ISR would check AMP's global inventory system, called WINS. The WINS system was updated daily to display on-hand quantities by plant, but did not allow orders to be placed. If there was stock in another plant, the ISR would create a special sales order that also created a purchase order. The ISR then expedited the purchase order to make sure the related-AMP company shipped the stock already on hand instead of creating a new production order. This process had more manual intervention and took longer, but not as long as other situations.

If a customer wanted inventory of frequently purchased parts to be continuously available, AMP of Canada would create a stocking program for the customer. Stocking programs required AMP of Canada to buy and warehouse a customer-specified inventory level. AMP of Canada had stocking programs with several large customers for hundreds of parts. Because parts in customer stocking programs were in high demand, these parts were sometimes shipped to other customers. If the original customer wanted to buy a part that AMP stocked for it, it would contact inside sales to place an order. The ISR would check JBA for inventory in the Markham plant, and sometimes would not find enough to meet the requirement. The ISR would contact material control to ask them to find inventory and expedite the shipment. Material control would contact planners around the world to find stock. Sometimes a planner would have to contact a local vendor to make sure the raw material would be available for a production order. Eventually, material control found stock, received a delivery date, and relayed the information to the ISR. The ISR contacted the customer to ask if the lead time was acceptable. If the customer was satisfied, the ISR would place the order in JBA.

THE JBA IMPLEMENTATION AND EXPERIENCE

AMP of Canada started implementing JBA version 2.0 in late 1992 to replace MAPICS as the transactional processing system. MAPICS was a basic transactional processing system sold by IBM. JBA was selected to replace MAPICS over software from J. D. Edwards and AMPICS because of its strong distribution functionality. Most of JBA's customers were located in Great Britain and operated in retail industries.

A part-time and cross-functional project team of ten people was employed for system design, testing, and training. Consultants were also hired to assist with the design and testing. Because the project team still worked at their old jobs, the design and testing of the system were late, and users spent little or no time in training. After five months, the financial modules were the first to be implemented in late 1992. The sales and materials modules took eight months to implement, and went live in April 1993.

After JBA was implemented, there were many problems that took months to resolve. Employees who remembered the JBA implementation were determined not to duplicate the mistakes made in that project. Apart from go-live issues, users also began requesting that the IS department change the coding of the base software package to accommodate new requirements, such as the customer *price and delivery screen.* The sales organization asked IS to reprogram JBA to display pricing and delivery information on one screen because they were unhappy with the number of screens they had to access to find all the information that customers wanted. In addition to the price and delivery screen, the entire order entry system was rewritten to accommodate sales requirements.

Because users in manufacturing had expressed interest in putting manufacturing transactions in JBA, in 1995 release 2.6 of JBA was implemented. This version was selected instead of release 3 because release 3 did not work with release 2 modules. Release 2.6 was a special version of JBA written for a few customers that supported manufacturing and worked with release 2 modules. It was also less expensive than release 3. There were many coding problems with the manufacturing module of release 2.6, such as the lack of integration between manufacturing and finance. In 1995, AMP of Canada rewrote the product-costing module of JBA to accommodate corporate costing standards. In 1996, AMP spent $100,000 to rewrite JBA's invoice-matching program to accommodate three-way invoice matching.

By 1997, IS manager Richard Stoveld estimated there were one million lines of custom code in JBA. To upgrade JBA, each custom program would have to be examined to determine whether the program's functionality was included in the new version of the software, or whether the program would have to be completely rewritten. At a cost of $1 per line of code, Richard estimated that it would cost $1 million to upgrade JBA.

PROBLEMS WITH JBA AND LESSONS LEARNED

AMP of Canada had many problems with both the JBA software and the implementation. For example, in manufacturing planning, manual forecasts could be entered for manufactured parts, but were always overwritten by the MRP program. This MRP feature caused all manufacturing forecasting and planning to be done manually without the aid of MRP.

Finance also had many problems with JBA. For example, all inventory, cost of sales, and manufacturing general ledger postings were done manually at month-end. Although JBA supported integrated financials, there were so many problems in testing that this functionality was never activated. The sheer volume of manual postings Finance made required thousands of pages of reports which IS had to write and maintain.

There were many lessons learned from the JBA experience:

- The most important lesson AMP learned was that there should be no modifications made to the base system. AMP of Canada had learned the hard way that modifying the system made support much more difficult to obtain from the vendor. Software upgrades also became nearly impossible. Modifying the system also put design knowledge in the hands of contracted consultants, who could leave for other projects more easily than full-time AMP employees.
- *The project team should be full-time instead of part-time.* AMP had learned that it was impossible for one person to do both a functional job and a systems job, especially the implementation of a complex and integrated ERP system. It was also important for the project to be led by users instead of IS, which was how the JBA project was run.
- *There should be more and better end-user training.* Too many users did not learn basic functions, such as entering a customer order, until AMP was already live with JBA.
- *The data would have to be cleaned up before conversion.* Not enough time was spent fixing data problems before the data were loaded into JBA. This caused massive problems after go-live trying to correct both the data and transactions produced with those data.

By early 1997, the manufacturing and costing modules had been stabilized. Doris had declared that 1997 would be a quiet year for the IS department after the major work of 1996. It was ironic for Doris to learn in the summer that JBA was not year 2000 compliant. She now faced a difficult decision about how to provide a system to support AMP of Canada's business requirements beyond 2000.

OPTION 1: UPGRADE JBA

The IS manager, Richard Stoveld, had informed Doris that there were two ways the existing system could be changed to meet the year 2000 problem. First, the source code of JBA could be rewritten to accommodate the extra two digits for the year. The technological risk of this option was relatively low, because the IS department and user community already had a great deal of experience rewriting JBA. However, this option would cause AMP of Canada to lose all support from the vendor. Alternatively, AMP of Canada could implement the new version of JBA that was year 2000 compliant. The new version also offered an attractive Windows-based interface, instead of the mainframe terminal displays that the current version featured.

Richard thought AMP of Canada should implement "vanilla" JBA (i.e., the new version without extensive modifications). He was hired as AS/400 team leader a few months earlier because of his knowledge of the IBM mainframe. Richard had experience with BPCS, and like many IT professionals had heard horror stories about SAP implementations being late and over budget. Richard was not strongly opposed to implementing SAP, but didn't think users were ready for that system because of all the modifications they had come to expect.

There were several disadvantages to upgrading JBA. The most important problem was that the user community was very dissatisfied with the software. Users across all functions routinely complained about the system and its problems. There was a running joke that JBA stood for "Just Bloody Awful." JBA was also hinting that the new version with Canadian tax settings would only be available in the summer of 1998. Assuming a year for implementation, that left very little time before the year 2000.

OPTION 2: IMPLEMENT AMPICS

AMPICS was a version of IBM's BPCS software modified for AMP's business processes. AMPICS was a corporate standard targeted for AMP's small- to medium-sized businesses. Harrisburg strongly favored this option because there was a great deal of AMPICS experience in the United States that could be drawn upon to implement this system. These existing resources made AMPICS the least expensive alternative to AMP as a whole. Harrisburg also thought that AMPICS could be used as a transition to meet the year 2000 problem until the global SAP project was rolled out. Harrisburg committed that if AMP of Canada implemented AMPICS, AMP of Canada would be one of the early units that would be migrated to SAP. Harrisburg's target for AMP of Canada to switch to SAP was 2001.

The users' perception of AMPICS was even worse than their perception of JBA. Users believed AMPICS would be a step backward from JBA, since JBA was selected over J. D. Edwards and AMPICS. Furthermore, because AMPICS was used mostly in small countries with sales offices only, the manufacturing capabilities of AMPICS were very limited. This functionality was critical to support the Canadian strategy of growing manufacturing. Some important functionality developed for JBA would also have to be rewritten for AMPICS, including the customer price and delivery screen.

OPTION 3: IMPLEMENT SAP

SAP was the largest ERP vendor in the world and fourth-largest software company, with 1997 revenues of approximately U.S.$3.4 billion (6.0 billion DM), up 60 percent from the prior year. By comparison, JBA's 1997 revenue was approximately U.S.$370 million (£221 million). An ERP system supports enterprise resource planning, where all major functions such as sales, manufacturing, purchasing, and accounting are integrated in the same system. This integration saves time and money by reducing or eliminating data transfers between functional systems and allowing enterprise-level planning of resources such as inventory, fixed assets, personnel, etc.

Canadian management strongly favored implementing SAP. Harrisburg had evaluated several ERP packages and selected SAP as the standard for larger manufacturing business units. AMP was in the process of designing one global information systems architecture, with SAP as its core ERP software. AMP Inc. and several countries in Europe were implementing the financial modules of SAP (FI and CO) effective January 1, 1998. The global design of the sales (SD), materials (MM), and manufacturing (PP) modules would proceed after the financial modules were implemented. Doris thought that Canada could "piggy-back" on AMP's decision to use SAP and implement all modules of the system immediately to satisfy the year 2000 problem. Once all major AMP companies were using SAP, inventory could be planned and tracked worldwide. This promised significant improvements in customer service and cost savings from inventory reduction. In addition, many existing SS40 data transmissions would no longer be required, because the information would all be in the same system.

Unfortunately, AMP of Canada did not meet Harrisburg's size criteria for an immediate implementation. Doris thought it would be very difficult to convince Harrisburg that Canada should be allowed to implement a software application that was recommended only for AMP's larger businesses. Other hurdles to overcome were cost and time. Most SAP implementations cost between $5 and $10 million and took well over a year to implement. Because year 2000 problems were already appearing in JBA, Doris thought Canada had roughly one year to complete the implementation. Because of AMP of Canada's size, Doris believed an acceptable implementation cost target would be between $1 and $2 million.

There were other disadvantages to SAP. First, Harrisburg's recommendation was to use consultants from SAP itself instead of local business partners, which would significantly increase the cost. Second, Harrisburg wanted the system box to be located in the United States instead of Canada, which represented a significant loss of system control. Third, Harrisburg wanted the project team to consist of Americans from the global SAP team. These people would have to learn about how the Canadian business worked, which would take extra time, and then design a system to meet those needs. Fourth, retaining people from the project team would be difficult because market demand for SAP consultants was very high. AMP had already started to lose people to consulting because of market demand.

Only a week earlier, Doris had accepted an invitation to attend a special SAP sales presentation. At the meeting, a local SAP business partner named Optimum presented the story of how they helped a Montreal company implement SAP in one year at a cost of under $1 million using SAP's new ASAP rapid-implementation methodology. The inexpensive project had taken a short time because the project did not try to reengineer business processes, but rather implemented only "vanilla" functionality, which would be customized and improved later. Doris was excited by this possibility of a low cost and rapid SAP implementation. As she prepared for several conference calls with IS management in Harrisburg, she wondered which option to recommend and how to argue the case.

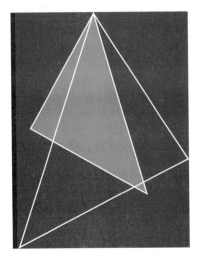

IVEY

Richard Ivey School of Business
The University of Western Ontario

Lipton Canada

INTRODUCTION

August 1997: Five months into a ten-month timeline, and Tim Pallant knew the January 1, 1998, project deadline was in jeopardy. The team had been unable to find a practical way to configure SAP R/3[1] to handle trade spending in the flexible manner the company required. As project leader of Lipton Canada's SAP R/3 implementation team, he had to make a decision very soon on what to do next. The basic functionality required to support this process had been identified during the system evaluation, but prototyping had revealed that, in general usage, it would become a maintenance nightmare. Now, despite the many possibilities he and his team had explored, he could see no easy way to make it work.

At this point he had identified three alternatives, and he didn't particularly like any of them. One option would be to rewrite a section of the SAP R/3 software—without the appropriate functionality, the company wasn't going to get the benefits it had anticipated. Alternatively, the company could give up on the way trade spending was handled. The final alternative would be to keep searching for a configured solution within SAP R/3 for a little while longer. Given SAP's complexity, Tim wasn't fully convinced that all possibilities had been exhausted. He didn't even want to think about the ultimate alternative— abandon SAP R/3 altogether and quickly tackle year 2000 compliance some other way.

BACKGROUND

Lipton Canada was a division within Unilever Canada that manufactured and sold packaged food products, predominantly to the retail trade. Becel, Red Rose, Lipton Sidedishes, Soupworks, and Ragu were a few of its main brands. Its head office was located in Toronto, and it operated four manufacturing plants and three sales offices located across the country.

The business case to implement SAP R/3 for handling order management had been developed in the fall of 1996. Up to then, Lipton had been using a highly tailored order management system (OMAR) that had been developed in-house fifteen years earlier. By 1996 the company knew it had to either update the existing system to make it year 2000 compliant, or replace it. While the straight cost of updating OMAR was expected to be less than purchasing and implementing SAP R/3, other considerations made the SAP proposal attractive.

OMAR had been continuously enhanced over the years and was finely tuned to provide excellent support for the business needs of the day, but it was based on older technology and would need considerable modification to support future

---

[1]SAP are the initials of the German corporate name. The company has chosen to be known throughout the world as SAP (like IBM). An English equivalent is Standard Application Packages. R/3 stands for Release 3 and uses client server architecture. R/2 ran on mainframes.

needs. As the system aged, integration with other, more recently acquired systems became increasingly difficult. In addition, support and maintenance depended on the knowledge of a small number of Lipton employees, making the company vulnerable to normal turnover of personnel.

Unilever as a whole had also adopted a policy of moving to common open systems across the company. Not only was this expected to alleviate IT management problems such as those mentioned, but it would also support the parent company's desire to integrate its global operations. SAP R/3 software had been recommended by Unilever and it was already in use in over thirty Unilever operating companies worldwide, including Lipton United States.

Even so, Lipton Canada wanted to ensure that SAP R/3 would meet the company's specific needs in the extended order management process. During the fall of 1996 an assessment was conducted. (Only order management was under consideration, although other SAP R/3 modules, particularly finance, were candidates for future implementation.) The first step was to ensure that SAP R/3 would support current practices in order management. A high level analysis of existing processes was conducted, and compared with SAP R/3 functionality. Several gaps were identified, specifically for certain payment processes related to distribution, and for many EDI transactions, but by and large SAP R/3 appeared to be a good substitute for OMAR, and offered significant additional functionality in some key areas. It was judged that the gaps could be handled by add-on modules, implemented at designated "user exits" in the software.

The second step was to identify strategic initiatives in order management that were planned for the near future, and that would require either new systems or enhancements to OMAR. If these initiatives were already supported by SAP R/3, the cost to develop the capabilities could be avoided. The assessment determined that in addition to providing year 2000 compliance, SAP R/3 offered flexibility in pricing, apparent improvements to the planning and execution of trade spending, and the ability to track contribution by customer. Further, it would enable migration from the current AS/400 platform to a UNIX environment.

THE IMPLEMENTATION

By the end of 1996 the assessment was completed. SAP R/3 appeared to match about 80 to 90 percent of Lipton's requirements, and a plan was in place to address the gaps. A budget and timeline were established, based both on some consulting advice and on the experience of other organizations. Ten months, while an aggressive target, was considered feasible. In February 1997, the business case was presented to the executive committee, and the proposal was approved. Eleven people, six from the business and five from IT, were selected to be on the project team, with Tim Pallant as the leader.

The users chosen for the team were selected on the basis of their competence and knowledge of their particular functional area. In Tim's words:

> The argument I used with the various departments to get the right people is that these are the people who are going to design how your department works for the next five years. You need to release your best people to work full-time on the project, and back-fill their positions. By and large I got the people I wanted—it was a very strong team. But that's what you need, the senior people—a team of leaders.

Tim himself, coming from the sales and trade marketing area, was well versed in the business processes being affected, and he also had considerable experience working with IT on various aspects of the company's systems.

To support the team members from Lipton and add specific SAP R/3 knowledge, a well-established systems implementation consulting practice was hired. The consulting firm assigned six consultants to the Lipton project, although this number would vary over the life of the project. The consultants were expected to lead the implementation, but also to provide sufficient knowledge transfer for Lipton to be self-sufficient by the end of the project. Tim expected three things from the consultants:

1. *Leadership/mentoring.* Lipton had no experience with SAP R/3, and needed the expertise of the consultants.
2. *Depth of knowledge.* Whether specific individuals on the team personally knew all the details or not, a big consulting practice had the capacity to leverage both its special knowledge of the product and corporate experience from previous implementations.
3. *Workmanship.* In addition to providing knowledge and leadership, the individual consultants were expected to be able to execute the mechanics of implementation—i.e., configure the software.

The project team reported to the executive committee, which had overall project oversight, and met biweekly with the Business Review Group (BRG), senior managers who had line responsibility for the business processes being affected.

In accordance with the SAP R/3 implementation methodology, the first six weeks were spent developing a blueprint of existing organizational processes. Unlike the high-level analysis of processes conducted during the assessment phase, this analysis was detailed, and provided not only a documented starting point, but was also intended to help educate the consulting team on Lipton's way of doing business. As the Lipton users worked on developing the blueprint, the consultants explained how SAP R/3 handled the same processes, and indicated where Lipton might have to alter the way it handled certain operations. In fact, to the frustration of the team members, the consultants seemed to respond to most issues with the phrase "SAP doesn't work that way." Before long, Lipton team members found they weren't willing to accept this response without pushing back, demanding a fuller explanation and more research.

Once the blueprint was complete, configuration began. For most areas, SAP R/3 processes, even where different from existing practices, were acceptable. Because the users on the team were fairly senior, they had the mandate to make process changes in their areas. If changes had a significant business impact (such as altering the level of service to customers, or requiring additional resources), they were taken to the BRG for discussion. In accordance with recommended practice, this group was expected to provide rapid (forty-eight hour) turnaround on issues brought to them by the project team.

In general, issues were presented to the BRG with the team's proposed solution, and for the most part they were resolved in a timely fashion. On those occasions where the BRG delayed making a decision or changed its mind, the team would have to reconfigure the software, which put pressure on the completion date.

TRADE SPENDING

By August decisions had been made on how to handle most business processes. The major exception turned out to be trade spending, where the team had run into a brick wall. From the beginning they had known that existing processes would have to change, but they had expected that SAP's built-in functionality would be an improvement. For trade spending, the SAP R/3 solution appeared to be an unacceptable step backward in terms of both the information provided and the ease of use.

Trade spending was the process through which Lipton supported cooperative promotional work with its customers. Lipton agreed to set aside a certain amount of money; when the customer promoted a product through flyers or special displays, Lipton covered some of the cost. Apart from managing the mechanics of the approval process and payment to customers, the software was expected to support analysis of return on trade spending, both by promotional program and by customer.

Ironically, the initial assessment of the software had suggested that SAP R/3 would address inefficiencies in the existing process. Lipton wanted to plan and execute trade spending by individual customer. The original system had been designed on the basis of regional planning. Numerous work-arounds plus great flexibility in allowance type and payment had resulted in a system that provided the desired customer-specific functionality, but not efficiently. SAP R/3 offered customer-specific planning and execution as part of the base system, and provided the opportunity to simplify the overall trade spending processes. With improved process efficiency and more effective tracking of trade spending, the expectation was that SAP R/3 would be an improvement.

That said, there was general recognition that the approach in SAP R/3 was quite different from what the company had been used to, and would have a significant effect on the way that customer agreements were set up and the flow of funds tracked. For example, decisions would have to be made on whether account managers would now be responsible for setting up and maintaining their own agreements, and how this would affect overall control. However, because of the overall potential that SAP R/3 offered for the customer management process, these changes were considered worth implementing.

Unfortunately, when the team encountered the details of configuration, they realized that instead of an improvement over the existing system, they might end up with something worse. The original expectation was that the customer rebate agreement process in SAP R/3 could easily handle the functionality required for trade spending. Unfortunately, to do what Lipton required increased the clerical workload dramatically. Rather than improving efficiency, following this route would slow the process down and require increased staff to handle the process. Part of the increase in workload came from the required complicated, multiscreen data entry, which not only meant the process took longer, but invited errors and update difficulties. This was particularly problematic because the new system would operate in real-time, so errors could have widespread ripple effects before—and after—being corrected.

After trying many different ways to configure the software without success, a proposal was made to create custom screens to make data entry easier. However, the implications for long-term maintenance were not acceptable. Members of the BRG were also unwilling to forgo some of the functionality that they felt added value both for Lipton and for customers, but which also added to the complexity of the process.

By August, Tim felt he was facing a brick wall. The team had been working flat out, desperately trying to come up with a solution, and stress was mounting. Deep down, he couldn't believe that SAP R/3 was unable to handle the process in the way the company wanted to operate. Trade marketing was not, after all, unique to Lipton, and with all SAP R/3's built-in functionality, he felt sure there was a way to make it work. Hadn't other companies faced the same problem? He was irritated that the consultants had provided little help on this. While they were competent enough at straight configuration, they hadn't been able to provide much leadership on helping him assess his options, nor had they been able to find any answers within their own organization or through SAP. Their pat response was to change the process to fit with the software, but that meant SAP R/3 would reduce rather than add value—not what Lipton expected from its investment.

Unfortunately, time was growing short. The team was already behind schedule for going live on January 1, and he didn't know how much longer he should ask them to keep trying to solve the problem. Since the fiscal year ended in December, missing the January deadline would require a much more complicated cut-over. Tim had to sort out his options, and take a proposal to the Business Review Group and the executive committee within the next few days.

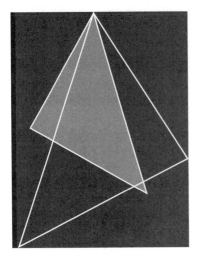

IVEY

Richard Ivey School of Business

The University of Western Ontario

Riverview
Children's Hospital

Bradley J. Dixon prepared this case under the supervision of Professors James A. Erskine and Duncan G. Copeland solely to provide material for class discussion. The authors do not intend to illustrate either effective or ineffective handling of a managerial situation. The authors may have disguised certain names and other identifying information to protect confidentiality.

On Thursday, February 15, 1990, Louis Bernard, the assistant executive director of finance at Riverview Children's Hospital in Toronto, reviewed the latest financial statements from the new computerized financial system. His fears of a slower-than-expected implementation were confirmed. The fiscal year-end of March 30 was fast approaching, and the new system was not ready for the external auditors who would begin their audit in mid-April. Even though the implementation was already eight months late, Louis was tempted to delay the system implementation until after the audit.

RIVERVIEW HOSPITAL BACKGROUND

Founded in 1899 as a Home for Incurable Children, Riverview Children's Hospital had developed over the years into a modern eighty-seven-bed children's facility providing long-term care and rehabilitation for infants, children, and young adults. Riverview patients were chronically ill, physically handicapped children who were educable. The most common afflictions were cerebral palsy, spina bifida, and muscular dystrophy. Officially classified as a chronic care hospital, it had become one of the most respected pediatric facilities in Canada. Riverview currently enjoyed a three-year accreditation status, the highest award granted by the Canadian Council on Hospital Accreditation. (See Mission Statement—Exhibit I.)

Since his arrival in 1987, Mark Thompson, the executive director, had guided Riverview toward enhancing its leadership role in providing services to its target population.

Recently, Riverview had expanded into providing long-term acute care for eight ventilator-dependent children. This program required special approval from the Ministry of Health to fund the additional staffing and specialized equipment requirements. Additionally, many other programs had been expanded or

---

Exhibit I. A statement of mission

Philosophy
Riverview Children's Hospital is committed to providing high-quality inpatient and outpatient services for physically disabled children and young adults through ongoing programs of rehabilitation, health care, education, and research. This care involves the family or guardian, and is provided in an environment serving the whole person to promote optimum individual growth, development and integration into the community.

Structure and Role
Riverview Children's Hospital, an eighty-seven-bed Chronic Care facility, shall operate within the requirements of the Ontario Public Hospitals Act and strive to:

- Assess and meet each patient's physical, mental, spiritual, social, recreational and educational needs.
- Promote an atmosphere of caring support to patients, their families, staff and volunteers.
- Liaise with other health services to fulfil its role in providing a continuum of care to the community.
- Encourage research and scholarly works to enhance the quality of life for the disabled.
- Provide education and training for health care personnel and the public.
- Exclude service for the management of those conditions, which primarily require ongoing critical and/or diagnostic services of an Acute Care Hospital.

enhanced since Mark Thompson's arrival at Riverview to replace the previous executive director, who was removed by the board of trustees.

More than 95 percent of the operating budget of $10 million came from the Ontario Ministry of Health. The 1989 fiscal operating deficit of $200,000 was funded by the Riverview foundation, which had grown into a sizable ($20 million in assets) foundation that supported disabled children through grants to Riverview and other institutions.

Riverview, like all public hospitals, was run by a board of trustees. The board had always consisted of a large majority of women, as women had started Riverview when they were driven from the board of a major children's hospital more than ninety years ago. The board took an active role in the administration of the hospital and met regularly with the hospital executive management group.

The board had several committees that also met regularly to set policy and review management decisions: the executive committee, the joint planning committee, and the finance and audit committee. The finance and audit committee met every month and comprised nine board members, three of whom were Chartered Accountants.

In April 1987, Riverview had been given the responsibility for managing the eventual closing down of another chronic-care children's hospital twenty kilometers north of Riverview in Thornhill, Ontario. The Thornhill Heights Hospital had been privately owned, and the physical condition was deemed too inadequate by the Ministry of Health to warrant continued operation. The Ministry purchased the facility and gave the management team at Riverview the responsibility for managing the Thornhill Heights facility until it was closed. The phase-out period was estimated to be at least five to seven years.

Louis Bernard

Louis obtained an undergraduate degree in business administration in 1982, joined a major accounting firm, and received his Chartered Accountant designation in 1985. At the accounting firm, Louis had the opportunity to learn about healthcare accounting as an external auditor of hospitals and medical supply companies. Through his accounting firm's consultancy practice, he was given the opportunity to become the interim finance director at Riverview in April 1987. The opening at Riverview had arisen from the recent dismissal, by the hospital board, of the previous assistant executive director of finance, who had held the position for less than nine months. In August 1987, Louis was offered the position of assistant executive director (AED) of finance.

The assistant executive director of finance was responsible for all facets of the finance function at Riverview: treasury, accounting, auditing, and office management (see Exhibit II). When Thornhill Heights was acquired in April 1987, a part-time AED of Finance was hired. In the fall of 1987, the part-time contract

Exhibit II.   Assistant executive director of finance job responsibilities

The assistant executive director of finance is responsible for all facets of the finance function at Riverview and Thornhill: Treasury, Accounting and Auditing, and Office Management.
Treasury responsibilities include:

- Negotiate revenue from the Ministry of Health.
- Manage the cash and investments of the Riverview Hospital and Foundation.
- Prepare capital assets budgeting.
- Advise the board of the financial implication of decisions.
- Oversee all donations, bequests, and estate matters.

Accounting and audit responsibilities include:

- Submit financial statements of board of trustees.
- Produce quarterly reports to the Ministry of Health.
- Ensure the accounting system is current and accurate.
- Establish policies and procedures to prevent errors and fraud.

Office management responsibilities include:

- Ensure smooth functioning of all financial procedures.
- Respond to questions and requests from the departments.
- Manage the telephone system and photocopiers.
- Supervise office staff.

was not renewed and Louis was given the finance responsibility for both facilities. Louis spent between one and two days per week at Thornhill.

Finance and Computer Departments

As the assistant executive director of finance, Louis Bernard was responsible for all aspects of financial management at Riverview and Thornhill. At Riverview, his staff consisted of seven people organized into two departments, accounting and materials management (see Exhibit III). Three Thornhill staff reported to Louis; the accounting clerk, the payroll clerk, and the receptionist. Job responsibilities, educational background, employment history, and Louis' comments on the staff are detailed in Exhibit IV.

The computer and communications department was formed in January 1989 at the same time that a computer room was being constructed to house the new computer hardware. Wilma Lo was promoted to computer coordinator, reporting to Mark Thompson, who was overseeing the new system implementation. Previously, Wilma was the word-processing coordinator and reported to Louis Bernard. The computer vendor's technical staff were favorably impressed with Wilma Lo's enthusiasm and felt that she could manage the computer operations. As the computer coordinator, Wilma was responsible for the operations and the

Exhibit III. Organizational chart

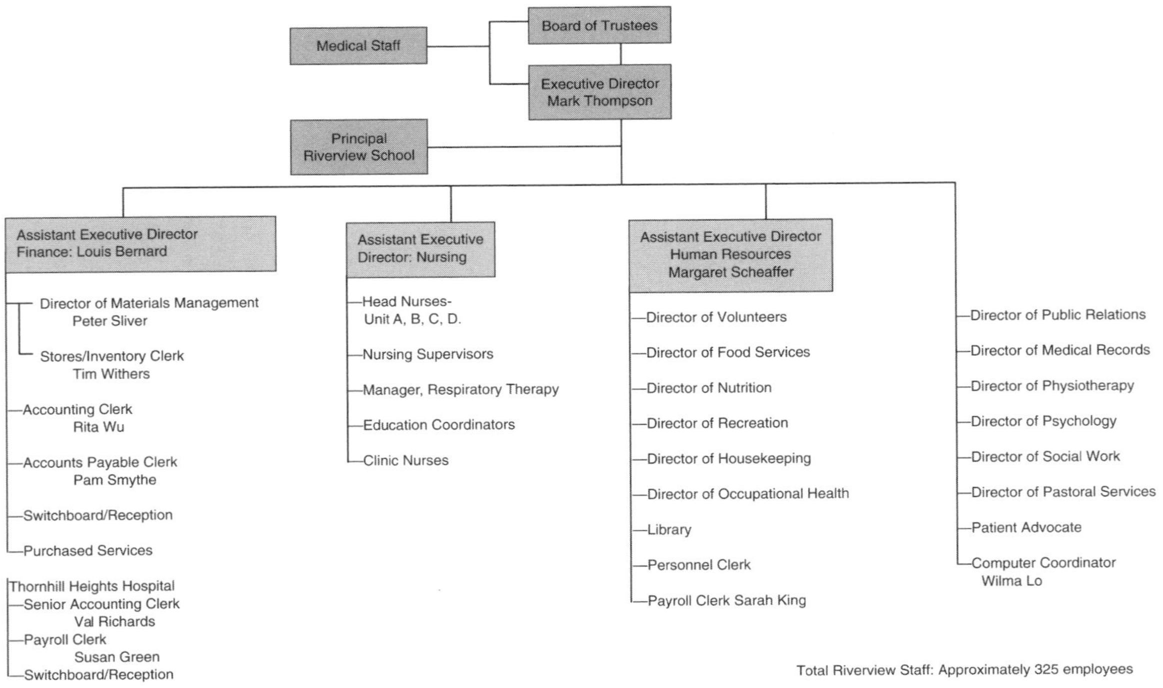

Board of Trustees

Medical Staff

Executive Director
Mark Thompson

Principal
Riverview School

Assistant Executive Director Finance: Louis Bernard
- Director of Materials Management
  Peter Silver
  - Stores/Inventory Clerk
    Tim Withers
- Accounting Clerk
  Rita Wu
- Accounts Payable Clerk
  Pam Smythe
- Switchboard/Reception
- Purchased Services
- Thornhill Heights Hospital
- Senior Accounting Clerk
  Val Richards
- Payroll Clerk
  Susan Green
- Switchboard/Reception

Assistant Executive Director: Nursing
- Head Nurses-
  Unit A, B, C, D.
- Nursing Supervisors
- Manager, Respiratory Therapy
- Education Coordinators
- Clinic Nurses

Assistant Executive Director Human Resources
Margaret Scheaffer
- Director of Volunteers
- Director of Food Services
- Director of Nutrition
- Director of Recreation
- Director of Housekeeping
- Director of Occupational Health
- Library
- Personnel Clerk
- Payroll Clerk Sarah King

- Director of Public Relations
- Director of Medical Records
- Director of Physiotherapy
- Director of Psychology
- Director of Social Work
- Director of Pastoral Services
- Patient Advocate
- Computer Coordinator
  Wilma Lo

Total Riverview Staff: Approximately 325 employees

Exhibit IV. *Biographical details of employees and Louis Bernard's comments on employees*

Louis Bernard's comments on the performance of selected employees are shown in *italics*.

Riverview Children's Hospital

Sarah King—Payroll clerk

- Five years at Riverview
- High school education
- Previous experience in payroll department in industry:

Poor management skills, non detail-oriented, weak comprehension skills, learned by copying procedures; good worker, but progressively poor attendance record; increasingly becoming flustered and missing details.
Part-time Assistant was necessary because of workload.

Rita Wu—Senior accounting clerk

- Six years at Riverview
- High school education
- Runs the current microcomputer accounting system:

Relatively independent: did not need much direction
Learned the new system well
Responsible for general ledger and management reports

Pam Smythe—Accounts payable clerk

- Six years at Riverview
- High school education:

Recently received Hospital Accounting Course Certificate
Not too confident—tends to hesitate
Procrastinates—somewhat insecure about the system

Peter Silver—Materials management supervisor

- Fifteen years at Riverview
- College in Portugal
- Community College—high marks—transcript posted on wall:

Not a delegator—runs the department very tightly
Workaholic—tries to do everything
Never used computer before but learns quickly

Tim Withers—Stores inventory clerk

- 10 years at Riverview
- Educated in Ireland:

Very laid back—likes to visit with the sales representatives
Prompt—arrives and leaves on time
Recent heart attack

(continues)

Exhibit IV.   *Biographical details of employees and Louis Bernard's comments on employees (Continued)*

Wilma Lo—Coordinator of computer and communication department

- Seven years at Riverview
- Diploma in computers from DeVry Institute
- Formerly supported the secretaries with their word processors
- Chosen (based on recommendations from ICS personnel) to head computer department:

Very busy worker, but poorly organized
Writes copious notes, but takes time for her to comprehend
Management skills are lacking—does not prioritize well

Thornhill Heights Hospital

Val Richards—Senior accounting clerk

- Nineteen years at Thornhill
- Bookkeeping training
- High school education:

Knows everyone, friendly
No computer experience

Susan Green—Payroll clerk

- Five years at Thornhill
- High school education
- Was junior clerk in payroll department at large employer:

Intelligent, learns quickly

---

technical management of the new computer system, in addition to her current word-processing support and telephone system management responsibilities. Wilma felt overextended by her new responsibilities:

> I feel I am so busy all the time, there is so much going on. I never have the time to do anything right. There is so much to learn about the new computer; I have never worked with such a large system before. Working with the new computer is enjoyable; there wasn't much challenge in my job of providing support for all the word-processing users. Now the word processor users always phone, often at awkward times, and expect me to come running to solve their problems. Mark tells me to not worry about them too much, but I used to be a word-processing clerk; they are my friends.

Wilma Lo also maintained the telephone system and reported to Louis Bernard, who had overall responsibility for the telephone system. The telephone system was not a big part of Wilma's job, but Louis felt that her priorities were not always logical.

> Whenever I asked Wilma to make a minor change to the telephone system, it would be done immediately, even if I specifically mentioned that this could wait. I thought her service was great, until my staff complained about Wilma's service. I realized she was doing everything I asked because I was the assistant executive director and her former supervisor.

The computer department at Riverview should logically have reported to the AED finance, but Louis had no time or desire to manage it at that stage. In Canada, hospital computer departments usually report to the AED finance, except the largest (500-plus beds) hospitals, where a chief information officer would manage the computer department. Louis felt that Wilma Lo was in over her head, and Mark Thompson was having to spend more time than he would like managing her and the computer problems. Louis remarked that he would need an accounting supervisor to look after the office and the day-to-day accounting issues before he could even consider managing the computer department.

Louis was concerned that the organizational structure of the administration departments did not make sense. Louis felt that the payroll department should not report to the same manager as the personnel department. More than 70 percent of a hospital's costs are salary expense. Proper audit and control practices dictate that the person who enters the hours worked into the system should work in the accounting department.

> However, I am not sure I would want Sarah King, the payroll clerk, reporting to me anyway. I have had numerous incidents with her and I do not have much confidence in her abilities. Once I even pushed to have her fired, but the AED of human resources, Margaret Scheaffer, protects her staff and supported Sarah.

PURCHASING THE NEW SYSTEM

Louis had been involved in the process of purchasing a new computer system soon after his arrival in 1987. Even though he did not have any experience installing computer systems, Louis was interested in introducing a hospital financial system into Riverview.

A computer evaluation committee had been formed to decide which computer system to purchase. The committee consisted of Mark Thompson, the three assistant executive directors, the director of Medical Records, and Wilma Lo.

There were six reasons for purchasing a system to replace the existing microcomputer–based accounting system and to automate other areas of the hospital. First, the existing system was inadequate and could not provide the department managers with anything beyond basic reports outlining the departmental expenses. Because the system had not been designed for hospitals, it could not produce the necessary statistical and budgeting reports that the department managers needed in an increasingly cost-conscious health care environment. Good management reports are important to enable managers to control costs.

Second, the Canadian Hospital Association had just finished the management information systems (MIS) guidelines. These guidelines covered how management information should be recorded, managed and disseminated in hospitals. While the guidelines were just recommendations, it would only be a matter of time until adherence to MIS guidelines would become a prerequisite of receiving hospital accreditation.

Third, the payroll deposits and earning statements had been processed by an off-site computer service bureau, which cost $1,300 a month. To update payroll and personnel information involved filling out forms that were couriered to the service bureau, where they were entered into the computer. The entire process was slow, error-prone, and cumbersome.

Fourth, the existing microcomputer system, purchased in 1982, was running out of capacity and was increasingly breaking down. The computer would have to be replaced or upgraded soon.

Fifth, the executive director realized that proper financial systems were an important factor that the Ontario Ministry of Health would consider before allowing Riverview to expand programs. The Ministry of Health encouraged all hospitals to install financial systems that would support the management of a hospital, in addition to simply maintaining the accounting ledgers. Louis felt that Riverview's installation of a new system had been a consideration in the recent approval of the new program for ventilator-dependent children.

Finally, the management of patient information was entirely manual. The patient record, a binder that contained a record of all treatments, diagnoses, and progress reports for a patient, was located in a central records room. Computerized patient care information would enhance productivity, reduce errors, and move Riverview into the 1990s by providing greater analysis of treatment outcomes, faster access to medical histories by medical staff, and the automatic output of the mandatory statistical reports for various governments.

In late autumn 1987, a consultancy firm specializing in Hospital Systems had been retained to prepare a needs analysis report of Riverview's systems requirements. After much discussion with the computer committee, the report was the basis of a Request For Proposal (RFP), which was sent to several computer systems vendors in May 1988. The 195-page RFP was analyzed by the vendors, who responded with elaborate proposals, addressing each question in the RFP. The proposals were reviewed by the committee with assistance from the consultants. A short list of three vendors was selected in June 1988. After numerous visits to other sites and further analysis of the proposals submitted by the vendors, a system was selected in August. A contract for purchasing the hardware, software and implementation services for $499,000 was finally signed in November 1988 with a major computer hardware vendor, Integrated Computer Systems (ICS).

ICS proudly advertised that it was the only single-source healthcare systems vendor in Canada. ICS was responsible in the contract for managing the training, hardware installation, software implementation, and hardware support for Riverview. Louis was impressed that the training and implementation costs were lower than other vendors' proposals. The hardware to be installed at Riverview was an ICS-A1 mainframe computer with twenty-five terminals connected to the computer.

The software comprised two parts: the first part was the patient-care system, which automated the patient information flow and computerized the Medical Records department. The patient-care software was developed by ICS in its Winnipeg office, and the software trainers were ICS employees. The second part was the financial system, which automated the materials management, payroll, and accounting departments. The financial system consisted of many interrelated subsystems, or modules (see Exhibit V), that had to be implemented in a coordinated fashion. There were few connections between the financial and patient-care systems; each system could be implemented independently of the other. The financial software was sold by ICS but written by Dovetail Software, a London, Ontario-based software firm, under contract to provide hospital financial software exclusively to ICS. The Dovetail financial software was one of the most advanced hospital financial systems in Canada. The software's many features, coupled with the fiscal control from the disciplined procedures the software required, made Dovetail software popular with larger, 400-plus-bed hospitals.

Implementing the System

A schedule for implementing the financial modules was agreed upon in the contract between the computer vendor and Riverview's computer committee (see Exhibit VI). The computer hardware was installed in the computer room during January and February 1989. A project manager was appointed in March 1989, and the ten-week implementation schedule was initiated. As executive director, Mark Thompson would oversee the entire project and liaise with ICS. Louis would direct the financial system portion, and the director of medical records would manage the patient-care system portion. The expected date for the financial system to be fully operational, or *live,* was early July. The patient accounting system would be implemented after the financial system. The entire implementation was expected to take six to seven months.

The implementation budget, to cover project management and miscellaneous technical support expenses to get the system live, was $64,000. This amount was included in the negotiated contract price and was estimated to cover a six- to eight-month implementation. Riverview relied on ICS to employ the appropriate project management candidate. ICS, in turn, subcontracted Sharon Picalle from a computer consultancy firm in March 1989. Sharon had worked extensively with

Exhibit V.   Financial system modules

General Ledger and Budgeting
The purpose of this application is to maintain and report the financial data arising from the operation of the hospital and its various departmental units. This system assists in cost control through more timely financial reports, prepared with less clerical effort.

The system provides an online auditing capability that allows the user and auditors to easily track the movements of data to and from the general ledger accounts.

The general ledger system accepts input transactions automatically from other applications, such as accounts payable, accounts receivable, payroll, purchasing and inventory control. Users can also input transactions manually, and inquire about the status of any general ledger account. The system allows use of the new chart of accounts described in the MIS Guidelines for Canadian Hospitals. A user-defined chart of accounts is also allowed.

The system generates financial reports as specified by the user. It can provide comparative reporting by period, by departmental unit or any other desired basis. The system also provides consolidated financial statements for the hospital.

The system monitors actual expenditures against the budget for the hospital and all departments. It generates monthly budget variance reports for each department and cost centre.

Accounts Payable
The purpose of this application is to assure the proper receipt of goods, to support the orderly payment of supplier accounts and to assure authorization of payments. This system allows the hospital staff to have current information on volume of purchases and minimizes the time required to find the status of a supplier order. It also helps the hospital staff avoid missing supplier discount dates.

The system regularly prints a list of invoices or accounts, which should be paid. The user can make modifications to this prepayment register. When a user is satisfied, the system prints the required checks.

The user can, at any time, request the printing of a single check, which is charged against a specified general ledger account (e.g., an expense account). The user could also write checks manually and enter the details, which the system uses to keep all account balances up to date. The system keeps track of outstanding checks and performs reconciliation with the monthly bank statements.

The system generates purchase analysis reports by department, product type, and supplier. It can also produce other useful reports such as product price histories and supplier delivery performance. On-line inquiry to all accounts payable information is available.

Payroll
The purpose of this application is to maintain time and attendance data for all hospital personnel and calculate the payroll. This system minimizes the clerical effort required to produce the payroll and other labor statistics. The system can generate a report of payroll costs and full-time equivalents used by department, cost center, and job description. It can produce consolidated reports for the hospital. It can also produce reports that monitor vacation days, sick leave, overtime hours, etc. by employee.

Time and attendance data for all full-time and permanent part-time staff are entered from time sheets. The system can handle multiple pay cycles. Some employees are paid on a weekly basis; others can be paid biweekly, semimonthly, etc. The hospital is a multi-union environment. The system should be flexible, allowing changes to union contracts and pay scales to be made with a minimum amount of effort.

In most cases, employees are paid through an automatic funds transfer to their bank accounts. The system can also issue cheques for those employees not on automatic deposit.

The system automatically prints records of employment forms for terminated employees. The system retains certain information on terminated employees for retroactive and reference purposes.

The system has the capability to calculate vacation pay, sick pay, bonuses, etc.

Exhibit V. Financial system modules (Continued)

Purchasing

The purpose of this application is to oversee the acquisition of commodities, parts, supplies, and any other material goods required by the hospital. The system captures requisition data from multiple departments, assists in the preparation of purchase orders and monitors the receipt of goods received.

The normal flow of operations within the purchasing application is:

1. A purchase requisition is created within a hospital department. This requisition is sent to the purchasing department for approval and creation of a purchase order.
2. The purchasing department will review the requisition. They will, when necessary, select the appropriate vendor. They may negotiate prices and discounts. They will ensure that delivery is for the required date. Once they approve the requisition, they will enter it into the computer.
3. The computer will generate the purchase order. A copy of this purchase order will be sent to the supplier. The system also automatically produces stock purchase orders based on inventory reorder points and economic order quantities.
4. The supplier will deliver the goods.
5. The receiving department will count the items received. Their receipt will update the inventory control and open purchase order records.

Inventory Control

The purpose of this application is to control the issue and stocking levels of most hospital stock items, including medical and surgical supplies, sterile supplies for nursing units, reusable linen items, dietary material and utensils, pharmaceutical supplies and paper supplies. The system attempts to prevent stockouts, while minimizing inventory carrying costs.

The inventory control system maintains a file of all stock items, including newly purchased, reusable and manufactured items. The system records all requisitions and issues. Details of supplier orders and receipt of goods are automatically received from the purchasing system. It assists in physical inventory taking and upon authorized clearance makes any necessary adjustments to the file.

The system calculates order points and quantities for all items. It regularly generates a report of all items near their order level.

The system prints product item catalogs for departments to use when requisitioning items from stores or nonstores.

The system allows for multiple stores locations. It maintains cart profiles for the multiple supply carts found throughout the hospital. It also provides information for charging supplies usage to the various cost centers.

The system allows on-line inquiry for inventory item status and purchase order status.

ICS computers and had project management experience in the banking industry. This was her first project working with patient-care and hospital financial software. Sharon reported to both Venkat Halambi, the support manager at ICS, and Mark Thompson at Riverview. Venkat actually employed Sharon, but any major decisions that Sharon referred to him were made after consultations with the ICS marketing account manager for Riverview.

After a few days on the job, Sharon realized that Wilma Lo, the computer coordinator, was not understanding the computer system and would require extra training and technical support. Sharon asked Venkat to provide an ICS technical

Exhibit VI. *Implementation schedule: Financial modules*

Task #		Weeks									
		1	2	3	4	5	6	7	8	9	10
1. a)	Initiate Team Formation	x									
1. b)	Identify Dovetail Modifications										
	Define Modifications for:										
	a) Accounts payable	x									
	b) General ledger	x									
	c) Financial reports	x									
2.	Analysis of Modifications Requested										
	a) Develop specifications		x								
	b) Provide cost estimates			x							
	c) Provide implementation estimates			x							
3.	Acceptance of Modifications										
	a) Priorities				x						
	b) Approval				x						
	c) Signoff/acceptance				x						
4.	Revise Implementation Plan										
	Revise Due to Modifications					x					
5.	Consultation for User Training										
	a) Core trainers assigned	x									
	b) Develop training plan	x									
6.	Customization & Programming										
	a) Programming commences					x	x	x			
	b) Testing						x	x	x		
	c) Incorporate in system							x	x	x	
	d) Documentation changes made						x	x			
7.	User Training										
	Implement Training Plan						x	x	x	x	
8.	Determine Conversion Methodology										
	Manual vs. tape-to-disk		x								
	Internal vs. contracted		x								
9.	Prepare Conversion Data										
	a) Define conversion specs.			x							
	b) Review & cost conversion specs.			x							
10.	Perform Conversion										
	a) Write conversion programs				x						
	b) Test validity of programs				x	x					
11.	Test Accounts Payable										
	a) Test for integrity of data						x				
	b) Test scripts						x				
12.	A/P Acceptance										
	Evaluate and accept A/P							x			
13.	Test General Ledger										
	Test for integrity of data						x				
	Test scripts						x				
14.	G/L Acceptance										
	Evaluate and accept G/L							x			
15.	Test Financial Reporting										
	Test For integrity of data						x				
	Test scripts						x				

Exhibit VI. Implementation schedule: Financial modules (Continued)

		Weeks									
Task #		1	2	3	4	5	6	7	8	9	10
16.	Financial Reporting Acceptance										
	Evaluate & Accept Financial Reporting								X		
17.	Test Payroll										
	Test integrity of data								X		
	Test scripts								X		
18.	Payroll Acceptance										
	Evaluate & accept payroll									X	
19.	Live Implementation										
	a) Provide conversion coverage										X
	b) Prepare for operations										X
	c) Implement plan (see 4)										X
20.	System Shakedown										
	Allow time post-implementation										
	to resolve any problems								X	X	
21.	Post-Implementation Review										
	a) A/P										
	b) G/L										
	c) F/R										
	d) Payroll/personnel										

support person to spend extra time with Wilma Lo to help her understand the system and enable her to solve the minor technical problems that invariably arose. The technical support person was billed to the project at $480 per day for time spent solving problems where ICS had determined they were not at fault or responsible. Fortunately for Riverview, there was not another system being installed at the same time. The technical support person was available to spend the extra time Wilma needed to learn the system.

Training for the first financial module (general ledger) started in April. The users complained to Sharon that the training was too rushed. After talking to the Dovetail trainers, Sharon realized that the days allocated for training in the contract had been cut roughly in half. Sharon learned from ICS's marketing department that the training days had been reduced because Riverview, at eighty-seven beds, was less than half the size of all the hospitals that had purchased the systems to date. Sharon reviewed the training days' shortfall with Mark Thompson and Louis. Louis felt that the training he attended for the general ledger module did not seem rushed. He wondered if days recommended by Dovetail were actually needed. Louis felt that Sharon should be able to help the staff with implementing the system. Sharon, Mark, and Louis agreed to keep to the original training plan rather than incur costs of $650 per day for a Dovetail trainer for the extra thirty days that were cut.

In May it became apparent to Sharon that the frequency of computer problems was not decreasing. Wilma Lo was still having trouble with the computer system, although the technical support person was spending two to three days per week at Riverview assisting Wilma. The computer was constantly going down, inconveniencing the users. Louis was not surprised when Sharon mentioned to him that she was postponing the second training sessions until the end of May to correct the hardware problems. The system live date was delayed by one month to July.

Many difficulties were created for Sharon by Riverview staff's lack of familiarity with computers. Although Sharon was spending extra time working with the staff, the users still complained that there was not enough time to learn the system. Sharon found that she had to be increasingly assertive and persistent to ensure that users completed any assigned project tasks. Payroll was the most complicated of all financial systems, and it become apparent to Margaret and Louis that the payroll clerk, Sarah King, would not be able to handle going live until some time in the fall. Louis was disappointed that Riverview would not be able to realize the monthly savings from implementing payroll earlier.

During June, several of Louis's staff mentioned that they felt increasingly uncomfortable working with Sharon. Louis saw Sharon only a couple of times per week and always asked how things were going. Sharon was positive about the system, the staff, and the prospects for going live with everything but payroll in August. Louis increasingly wondered why Sharon was spending so much time managing this project. Louis knew that the $64,000 budget was based on a part-time project manager.

Shortly after completion of the training session in June, Tim Withers, the stores inventory clerk, suffered a major heart attack. Tim would be off work, recovering until October. Peter Silver, the director of materials management, was taking holidays in September. Louis realized that the materials management module implementation would be delayed until November. Summer holidays interfered with the implementation plans of the other financial modules as well. Louis agreed with Sharon that the live dates for the other financial modules would have to be pushed back to October.

Louis's concerns about the amount of time Sharon was spending were realized in early July when he received an invoice from ICS for $59,000 for the implementation costs to date. The invoice did not provide a breakdown of the hours spent, and Louis requested, through Mark Thompson, that a breakdown be provided. Eventually, the hour totals revealed that over 400 hours, or 50 days, of billable technical support were charged between March and June. Louis and Mark met with Sharon and representatives from ICS and expressed their concern with the amount of the bill. Sharon felt confident that the implementation would not be delayed further.

Louis noticed that Sharon's style was becoming more controlling, demanding, and aggressive. Sharon annoyed Margaret, the AED of Human Resources, by her manner during meetings to discuss the payroll implementation. Louis learned that Wilma felt she was being treated "like a child" by Sharon. The situation reached a crisis in late August when ICS submitted an invoice for $20,000, which included an additional $15,000 charge over the remaining budget to cover extra implementation costs. Louis refused to pay the $15,000. Louis, Margaret, and Mark Thompson met to discuss the project and Sharon's role. They felt Sharon's handling of the project was inadequate, and Mark Thompson told ICS not to renew Sharon's contract, effective September 1.

The New Project Manager

At the end of August, Louis approached John Deans, the Dovetail trainer for the accounts payable and general ledger modules, about assuming the project management responsibilities. Louis and John had developed a good relationship from working together on the general ledger module over the past several months. Dovetail prepared a proposal that outlined the implementation dates, the project management days, and the extra training required (see Exhibit VII).

The Riverview board was very concerned about cost overruns. Louis was concerned how the board would react to a request for an additional $32,000 to cover project management costs. After discussions with Mark, Louis rearranged the computer budget by deferring a software purchase for a year. Louis was able to find enough funds to pay for the extra project management costs without requesting additional funds from the board. John Deans was appointed project manager in mid-September.

John realized how concerned Riverview was with its expenses. Riverview had a strict policy regarding overtime, and employees mentioned that several years back people had been asked to take unpaid leave in order to meet the budget. John submitted status reports every week detailing the days spent to date and was careful not to spend any unnecessary time at Riverview.

During October, John's visits focused on getting the general ledger and payroll system live. The payroll system live date of December 1 was delayed until January 1, 1990, because of problems in obtaining the specialized forms and making custom modifications to the software. Delaying the implementation of payroll past January 1 would cause more complications from converting tax and benefit deductions that were based on a calendar year.

The implementation of the general ledger went smoothly, and by the middle of November the closing balances from the old system were transferred to the new system and reconciled, and financial reports were prepared in time to be included in the 1990 budget packages for distribution to department managers by

Exhibit VII.   Proposal for project management and training

This proposal is for the balance of training required, the implementation assistance, and the project management required to implement the system successfully to meet the target dates.

Based on our daily rate of $525, the cost would be $32,025, excluding travel and lodging costs. We estimate our travel and lodging expenses to be approximately $150 per day. We look forward to discussing the details with you to ensure that we mutually understand the project requirements.

TRAINING REQUIREMENTS
Training required to implement Dovetail is outlined by module below.

Materials Management: **6 to 9 days**
Refresher training at both Thornhill and Riverview is highly recommended due to the delays between the original training and the implementation date. Recommended training is two to three days per site.

Implementation support is recommended when system goes live. This support is usually requested by hospitals and has proven extremely valuable to eliminate implementation problems and minimize future system issues. Suggested support is two to three days when system is going live.

Management Information: **2 to 3 days**
Outstanding training in management information will cover payroll and other complex reporting. This training would be scheduled over the next few months as data become available on the system.

General Ledger and Accounts Payable: **1 to 2 days**
Implementation support and review are recommended for these modules. This time would be scheduled concurrently with the implementation date.

Payroll and Personnel Training: **19 days**
There are five outstanding tasks before the payroll system can go live. The major items and their associated training days include:

Test with subset of 25 employees	5 days (Riverview only)
	4 days (Thornhill only)
Department head training	1 day
Parallel payroll	5 days
Miscellaneous payroll functions	3 days
Review outstanding issues	1 day

Operations Training: **5 days**
Consideration should be given to scheduling of operations training, two to three days when the system goes live and the remainder after several months of operation.

PROJECT MANAGEMENT

Schedule
In order to project the number of days required of project management, a summary schedule was developed based on conversations with hospital personnel coupled with past experience at other hospitals.

Key Milestone Dates
G/L, A/P implementation:	1-Oct-89
Material management live:	1-Nov-89
Conversion of payroll data:	?-Nov-89
Payroll/personnel:	1-Dec-89

Exhibit VII. *Proposal for project management and training (Continued)*

Projected Days Required

The days required for project management are listed below. These days are based on the above schedule and our previous experience with implementing Dovetail. These days are our best estimate and are, therefore, subject to mutually agreed revisions as the project progresses.

September	5 days
October	9 days
November	5 days
December	3 days
January	1 day
TOTAL:	23 days

Project Management Approach

Dovetail is committed to working with our customers toward the common goal of a smooth implementation of the Dovetail system. We believe that communication among all people involved in the implementation is vital. We will strive to keep communication open and as up-to-date as possible.

Successful projects are implemented in an environment of cooperation, communication, and teamwork. We believe that Dovetail's project management skill will be a positive addition to this implementation.

---

the end of November. The managers would review their results and prepare a budget that was to be submitted to finance by mid-January. Then the entire finance department faced four to five very busy weeks to consolidate the data, review, and prepare preliminary pro forma statements for budget meetings, then revise the statements and prepare final reports, meanwhile performing the required daily duties.

Each time John visited the hospital, he would check with the accounting and materials management staff to inquire about how the implementation was going and was always told that everything was going as planned. John's schedule of training and consulting at three other hospitals prevented him from traveling to Riverview during most of November and early December.

After an absence of five weeks, John arrived at Riverview on December 18. Louis told John that it appeared that everything was progressing as planned and the accounting and materials management staff were not having any major problems. Louis acknowledged, however, that he was not confident that everything was going as well as his staff let on. Louis remarked that he had been too busy to spend time down in Materials Management to learn what was really happening. John spent the morning investigating the status of the system and generating computer reports. After analyzing the computer output, John realized things were not going as planned and arranged a meeting with Louis.

> Louis, there are three concerns that I have with the system. First, the accounts payable invoices for November are not entered into the new system. Pam assured me the last time we talked that she was right on schedule. She has not spent any overtime doing this. Second, the materials management

department is not looking at their daily computer reports, and I believe the inventory balances are not accurate; they blame the delays in entering purchase orders into the system. This leads me to believe that they are not managing their inventory or their module well. Third, payroll will not be able to go live January 1, because the programmer who was writing the software to convert the files from the service centre to the new computer is very sick with pneumonia and nobody else will be able to finish the job in time.

Louis and John discussed the problems and arrived at an action plan to correct the problem. Louis agreed to encourage Pam Smythe, the accounts payable clerk, to spend time entering November's invoices into the system and to speak with Peter, the materials management manager, about his module. John mentioned that he would arrange for the Dovetail materials management trainer to come and review the system with the materials management staff as soon as possible.

In January, John worked with Pam to enter the November invoices into the system. Louis talked with Pam and realized that his earlier requests for working overtime went largely ignored because the staff thought they would not get paid for overtime. Louis assured them that they would get paid overtime.

Entering the invoices was complicated by the elapsed time since the goods had arrived and the many errors in purchase orders. Invoices could only be entered against a purchase order previously entered into the system, and then only when the order was marked by the receiver as having been received. When the system was notified that the goods had been received, a liability was created for the value of the goods, as per the purchase order. The invoice was entered into the system, and if it matched the liability exactly, a check was produced. In the conversion, all the purchase orders, receipts, and invoices were entered for one month as a parallel to ensure that the old and the new systems were matched, and this also provided an accounting trail for the auditors to follow.

THE DECISION

By Monday, February 12, all the November invoices were entered and matched to the purchase orders. The system was ready for the first month-end to be run overnight. The month-end failed because material management had not run its month-end first. After investigating further, John found that the materials management manager did not know how to run a month-end and thought that accounting was responsible for starting the run.

After both the inventory and accounts payable month-ends had been run on Tuesday evening, Louis requested the computer to generate the first set of financial statements Wednesday morning. The financial statements were worse than Louis imagined. First, the accounts payable liabilities for November were $1.4

million, not even close to the current system's liability of $50,000. Additionally, the inventory value was shown as over $1 million when it should have been about $70,000. Finally, three months of inventory issues had been entered into November's expenses.

After a long discussion with John, Louis realized he could either parallel the system for four months from November to February or start fresh after the audit and the year-end stock count sometime in April.

Louis spent most of Wednesday and Thursday working with the statements trying to understand the magnitude of the errors. His main concern was inventory, where there were some obvious mistakes, like photocopier paper inventoried at about $1 per sheet. With 200 items in inventory, correcting errors would require at least a half hour per item just to investigate the problems. Some of the discrepancies arose from mistakes on purchase orders or the corresponding goods receipts. With about 800 to 900 purchase orders generated over the past four months, it could take a couple of weeks of analysis to sort out the purchase order problems. If all accounts payable invoices were entered, the program to match invoices to purchase orders would help materials management to find a lot of problems. To date, only November invoices had been entered into the system; it could take Pam Smythe forty to fifty hours to enter a month's invoices.

Louis felt that his odds were 50–50 of being able to balance the statements to the old system in time. Louis did not feel confident that even with all invoices, purchase orders and receipts entered and checked that all errors could be found before the audit started in April. Louis wondered how the work would get done. Should he attempt to utilize temporary employees who would know nothing about the system, or spend the money and utilize the Dovetail trainers, at $650 per day, or should he rely on his staff to work enough overtime at an overtime rate of about $20 per hour?

Louis was concerned that delaying the implementation would prolong the frustration until next May or June. Louis knew the morale of the staff working with the new system was not good. The accounting staff had just finished working on the budget, a very busy time of the year. Wilma Lo was in especially bad shape; she was irritable, constantly blaming everyone else for each petty incident, and Louis noticed during one coffee break that she was shaking when she held her cup. The system was constantly going down; Louis believed that Wilma was too preoccupied to schedule preventive maintenance to catch problems in time. Pam Smythe was frustrated with all obstacles in both the new and the old systems. Louis was also worried about how much longer Tim Withers would be able to handle the stress, given his heart condition.

Louis wondered if the current microcomputer system would be able to last much longer. During the last month, the microcomputer had been broken for six days, and there were a number of corrupt files on the disk, which was overloaded. To work around the problems, Pam Smythe was having to trick the system into producing accounts payable checks.

Louis knew that the audit would keep his accounting staff very busy during the month of April, and it would be a messy and difficult audit if based on the old system. Louis could not imagine which was worse, trying to implement the system during April while doing an audit or implementing the system now. If he continued to implement the new system, and the system did not balance by March 30, his audit would become extremely complicated; the fees could double from the typical bill of $15,000. The auditors would not regard a botched implementation lightly. They could cast Louis in a most unfavorable light in the management report that was prepared by the auditors for the finance and audit committee.

Louis questioned why he had not known about the problems earlier, and wondered what changes were necessary to prevent this lack of communication from recurring.

The finance and audit committee would not be pleased with another delay in the system implementation. Implementing in April would push computer implementation expenses into another fiscal year, something Louis was sure the board would want to avoid.

Louis had to make a decision soon; the finance and audit committee of the board of directors was to meet on Tuesday, February 20. They expected an update on the status of the system implementation.

The Evolution of Project Management at Quixtar

COMPANY BACKGROUND

Quixtar is a business opportunity company, offering entrepreneurs the ability to have a business of their own through Quixtar's I-commerce business model. I-commerce empowers individuals to market products and manage their own business via the Internet, while being compensated by the low-cost, low-risk Independent Business Ownership Plan, and supported by the full-service infrastructure of Quixtar.

Since 1999, independent business owners (IBOs) powered by Quixtar have generated more than $4.2 billion in sales at www.quixtar.com plus nearly $320 million for Partner Stores, earning in excess of $1.37 billion in bonuses and other incentives. Their efforts have propelled Quixtar to be named the number-one online retailer in the drug/health & beauty category based on sales, and twelfth among all e-commerce sites, according to *Internet Retailer's* "Top 300 Guide."

Based near Grand Rapids, Michigan, Quixtar currently supports independent businesses in the United States, Canada, Puerto Rico, and various trust territories and independent island nations in the Pacific and Atlantic Oceans and Caribbean Sea. Quixtar Canada Corp. headquarters are located in London, Ontario, Canada.

THE NEED FOR PROJECT MANAGEMENT

A change in leadership can often bring about a change in the way that organizations get things done. Such was the case with the Quixtar communications department's adoption of project management (PM).

A new communications director, Beth Dornan, oversaw the department's shift from an internal service agency implementing mass communication vehicles to a *strategic partner* delivering more specific communications to targeted audiences. This change in philosophy meant that, rather than merely fulfilling the communication needs of other departments, Communications would now work closely with other areas to create strategies and implement communications tactics that would best help achieve Quixtar's business goals.

"Communications is the gateway by which Quixtar disseminates information and news to all its audiences, including independent business owners (IBOs), customers, and news media," says Beth. "We view our department's evolution as a shift from being all things to all people all the time to delivering the right message to the right audience at the right time."

This philosophical change led to a reevaluation of the role of the account executive (AE) in communications. The position of AE had existed to implement the direction set by other Quixtar departments. Now that communications played a part in setting that direction, would there be more value in having people responsible for planning, project managers, rather than people largely responsible for doing, account executives?

Beth asked a member of her management team, Gilann Vail-Boisvenue, to investigate the PM profession. Previously unfamiliar with the field, Gil discovered that it was exactly what Communications needed. The department had always excelled at executing but had fallen short at planning. Project management, Gil learned, could help manage budgets, schedules, project scope, workload, and resources.

In addition, PM seemed to support three of the company's core values of partnership, achievement, and personal responsibility. It also supported a Quixtar strategic goal to achieve initiatives on time and on budget to successfully impact business drivers like IBO productivity, recruiting and retention.

THE PLAN

Gil promptly earned her certification as a PM professional and applied the principles she'd learned to crafting a three- to five-year plan for transitioning AEs to project managers.

The plan proved to be successful for a number of reasons. First, management and senior management immediately understood the value of PM and supported it. Second, the plan was all about transitioning gradually through step-by-step phases into this new way of operating, rather than forcing the entire concept of PM on the communications group at once.

The plan included the following phases:

Year 1: Control budgets by aligning the department's operating budget with projects and individual teams, develop annual work plans for teams, offer broad-based PM training, and create formal PM job descriptions with salary standards.

Year 2: Control schedules by meeting deadlines, controlling scope, and managing resources; offer focused study groups for PMP certification; develop PM methodology.

Year 3: Communicate by building reports for upper management. Also, continue to manage resources and finalize a professional procurement group. *(This is the current year.)*

Year 4: Use risk planning and apply lessons learned to new projects.

Year 5: Use quality planning and refine components of the five-year plan as needed.

All aspects of the plan are based on the PMBOK® Guide, or *project management body of knowledge.* This plan has brought the PM discipline to a Level 3 on the PM Maturity Model within communications, while PM throughout the rest of Quixtar is at Level 1.

ADOPTING PROJECT MANAGEMENT ACROSS THE COMPANY

Communications is currently bringing PM to all of Quixtar through a training program designed to educate participants on standard PM processes and documentation. Approximately 140 people completed the program last year, including employees of Quixtar Canada and a sister company's Japanese affiliate.

"In just three short years we've gone from virtually no one at Quixtar knowing anything about PM to hearing many people say things like 'we need to get this project scoped' and 'do we have the money and time?'" says Gil. "Broad-based training helps everyone understand the concepts and the fact that a common language has really taken hold is proof of that understanding."

The project management professionals (PMPS) in communications also are hosting independent study groups to help other Quixtar employees who want to become certified. The study groups have a 100 percent success rate when it comes to participants passing the certification test, and are a major reason that Quixtar now boasts 25 PMPs across communications, IT, and finance. The 13 PMPs in communications have full job descriptions and a salary structure built around PM. The PMPs in other areas, such as IT, have other primary responsibilities like software development, and PM is secondary.

A SEVEN-STEP PROJECT MANAGEMENT METHODOLOGY

A result of the PM expertise within Quixtar communications is the creation of a seven-step project management methodology that can be used by any PMP across the company to effectively manage projects.

The methodology is laid out on the PM Web site on the corporate intranet and begins with a needs assessment—does the project in question relate to the company's strategic plan? The methodology takes the project through the charter and scope phases, execution and control, and closure.

"We can pair any new PMP with our seven-step methodology and they can manage projects effectively at Quixtar," says Gil. "There is about a two-month learning curve, but that's mostly to become familiar with our company culture. As long as they know PMBOK® Guide, they can do projects."

The methodology has been so effective in part because it's very streamlined.

"Many people dislike process, and with these seven steps, they don't even know they're in a process," says Gil. "Management often wants to see things get done, and how those things get done is up to us. The *how* is very important because that's where you spend time and money."

The seven-step methodology is also simple and flexible.

"You can use as many or as few of the steps as you like, depending on the complexity of the project," says Gil. "Some projects could be as simple as a scope statement defining the project and the assumptions and constraints. Others could be complicated enough to require a full project plan, including a risk mitigation plan, communication management plan, and tons of other components to ensure success."

LESSONS LEARNED

The success of introducing PM across Quixtar hasn't been without a few bumps along the road. Gil found that those implementing PM practices were at first too rigid, trying to wedge people into static processes and documentation methods.

"It's important to have standards but still be flexible. You don't want people to be so focused on the process that they think they can't do something that makes sense just because it's not the next step in their methodology," says Gil. "It's important to tailor the PM process to each individual project and the people working on it."

In addition to flexibility, reporting has been a success of PM at Quixtar. The PMs use a standard best practices reporting method to communicate project status to management:

- Red ligh—trouble
- Yellow light—jeopardy
- Green light— all clear

Such simple and effective reporting has helped Communications successfully manage all projects within budget for the past two years.

An aspect of PM at Quixtar that's currently under review is the control of *scope creep.* "Without a working scope management plan, many teams are incapable of meeting their schedules," says Gil. "It's part of the culture around here to always say yes to scope changes, so we've asked management to step in and assess changes to control scope creep."

Part 3

PROJECT MANAGEMENT CULTURES

Project management methodologies, regardless how good, are simply pieces of paper. What converts these pieces of paper into a world-class methodology is the culture of the organization and how quickly project management is accepted and used. Superior project management is attained when the organization has a culture based upon effective trust, communication, cooperation, and teamwork.

Creating a good culture cannot be done overnight. It may take years and strong executive leadership. Good project management cultures are leadership by example. Senior management must provide effective leadership in the same manner that they wish to see implemented by the corporate culture. If roadblocks exist, then senior management must take the initiative in overcoming these barriers.

Como Tool
and Die (A)[1]

Como Tool and Die was a second-tier component supplier to the auto industry. Their largest customer was Ford Motor Company. Como had a reputation for delivering a quality product. During the 1980s and the early 1990s, Como's business grew because of its commitment to quality. Emphasis was on manufacturing operations, and few attempts were made to use project management. All work was controlled by line managers who, more often than not, were overburdened with work.

The culture at Como underwent a rude awakening in 1996. In the summer of 1996, Ford Motor Company established four product development objectives for both tier one and tier two suppliers:

- Lead time: 25–35 percent reduction
- Internal resources: 30–40 percent reduction
- Prototypes: 30–35 percent reduction (time and cost)
- Continuous process improvement and cost reductions

The objectives were aimed at consolidation of the supply base with larger commitments to tier one suppliers, who would now have greater responsibility in vehicle development, launch, process improvement, and cost reduction. Ford had

---

[1]Fictitious case.

established a time frame of twenty-four months for achievement of the objectives. The ultimate goal for Ford would be the creation of one global, decentralized vehicle development system that would benefit from the efficiency and technical capabilities of the original equipment manufacturers (OEMs) and the subsupplier infrastructure.

STRATEGIC REDIRECTION: 1996

Como realized that it could no longer compete on quality alone. The marketplace had changed. The strategic plan for Como was now based upon maintaining an industry leadership position well into the twenty-first century. The four basic elements of the strategic plan included:

- First to market (faster development and tooling of the right products)
- Flexible processes (quickly adaptable to model changes)
- Flexible products (multiple niche products from shared platforms and a quick-to-change methodology)
- Lean manufacturing (low cost, high quality, speed, and global economies of scale)

The implementation of the strategy mandated superior project management performance, but changing a sixty-year culture to support project management would not be an easy task.

The president of the company established a task force to identify the cultural issues of converting over to an informal project management system. The president believed that project management would eventually become the culture and, therefore, that the cultural issues must be addressed first. The following list of cultural issues was identified by the task force:

- Existing technical, functional departments currently do not adequately support the systemic nature of projects as departmental and individual objectives are not consistent with those of the project and the customer.
- Senior management must acknowledge the movement away from traditional, "over the fence," management and openly endorse the significance of project management, teamwork, and delegation of authority as the future.
- The company must establish a system of project sponsorship to support project managers by trusting them with the responsibility and then empowering them to be successful.
- The company must educate managers in project and risk management and the cultural changes of cross-functional project support; it is in the manager's self interest to support the project manager by providing necessary resources and negotiating for adequate time to complete the work.

- The company must enhance information systems to provide cost and schedule performance information for decision-making and problem resolution.
- Existing informal culture can be maintained while utilizing project management to monitor progress and review costs. Bureaucracy, red tape, and lost time must be eliminated through project management's enhanced communications, standard practices, and goal congruence.

The task force, as a whole, supported the idea of informal project management and believed that all of the cultural issues could be overcome. The task force identified four critical risks and the method of resolution:

1. Trusting others and the system.
   - *Resolution:* Training in the process of project management and understanding of the benefits. Interpersonal training to learn to trust in each other and in keeping commitments will begin the cultural change.
2. Transforming sixty years of tradition in vertical reporting into horizontal project management.
   - *Resolution:* Senior management sponsor the implementation program, participate in training, and fully support efforts to implement project management across functional lines with encouragement and patience as new organizational relationships are forged.
3. Capacity constraints and competition for resources.
   - *Resolution:* Work with managers to understand constraints and to develop alternative plans for success. Develop alternative external capacity to support projects.
4. Inconsistency in application after introduction.
   - *Resolution:* Set the clear expectation that project management is the operational culture and the responsibility of each manager. Set the implementation of project management as a key measurable for management incentive plans. Establish a model project and recognize the efforts and successes as they occur.

The president realized that project management and strategic planning were related. The president wondered what would happen if the business base would grow as anticipated. Could project management excellence enhance the business base even further? To answer this question, the president prepared a list of competitive advantages that could be achieved through superior project management performance:

- Project management techniques and skills must be enhanced, especially for the larger, complex projects.
- Development of broader component and tooling supply bases would provide for additional capacity.

- Enhanced profitability would be possible through economies of scale to utilize project managers and skilled trades resources more efficiently through balanced workloads and level production.
- Greater purchasing leverage would be possible through larger purchasing volume and sourcing opportunities.
- Disciplined coordination, reporting of project status and proactive project management problem-solving must exist to meet timing schedules, budgets, and customer expectations.
- Effective project management of multitiered supply base will support sales growth beyond existing, capital intensive, internal tooling, and production capacities.

The wheels were set in motion. The president and his senior staff met with all of the employees of Como Tool and Die to discuss the implementation of project management. The president made it clear that he wanted a mature project management system in place within thirty-six months.

QUESTIONS

1. Does Como have a choice in whether to accept project management as a culture?
2. How much influence should a customer be able to exert on how the contractors manage projects?
3. Was Como correct in attacking the cultural issues first?
4. Does the time frame of thirty-six months seem practical?
5. What chance of success do you give Como?
6. What dangers exist when your customers are more knowledgeable than you are concerning project management?
7. Is it possible for your customers' knowledge of project management to influence the way that your organization performs strategic planning for project management?
8. Should your customer, especially if a powerful customer, have an input in the way that your organization performs strategic planning for project management? If so, what type of input should the customer have and on what subject matter?

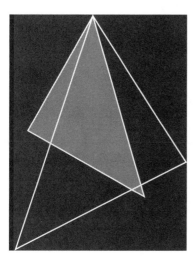

Como Tool and Die (B)[1]

By 1997, Como had achieved partial success in implementing project management. Lead times were reduced by 10 percent rather than the target of 25–35 percent. Internal resources were reduced by only 5 percent. The reduction in prototype time and cost was 15 percent rather than the expected 30–35 percent.

Como's automotive customers were not pleased with the slow progress and relatively immature performance of Como's project management system. Change was taking place, but not fast enough to placate the customers. Como was on target according to its thirty-six month schedule to achieve some degree of excellence in project management, but would its customers be willing to wait another two years for completion, or should Como try to accelerate the schedule?

FORD INTRODUCES "CHUNK" MANAGEMENT

In the summer of 1997, Ford announced to its suppliers that it was establishing a "chunk" management system. All new vehicle metal structures would be divided into three or four major portions with each chosen supplier (i.e., chunk manager)

---

[1]Fictitious case.

responsible for all components within that portion of the vehicle. To reduce lead time at Ford and to gain supplier commitment, Ford announced that advanced placement of new work (i.e., chunk managers) would take place without competitive bidding. Target agreements on piece price, tooling cost, and lead time would be established and equitably negotiated later with value engineering work acknowledged.

Chunk managers would be selected based on superior project management capability, including program management skills, coordination responsibility, design feasibility, prototypes, tooling, testing, process sampling, and start of production for components and subassemblies. Chunk managers would function as the second-tier component suppliers and coordinate vehicle build for multiple, different vehicle projects at varied stages in the development–tool–launch process.

STRATEGIC REDIRECTION: 1997

Ford Motor Company stated that the selection of the chunk managers would not take place for another year. Unfortunately, Como's plan to achieve excellence would not have been completed by then, and its chances to be awarded a chunk management slot were slim.

The automotive division of Como was now at a critical junction. Como's management believed that the company could survive as a low-level supplier of parts, but its growth potential would be questionable. Chunk managers might find it cost-effective to become vertically integrated and produce for themselves the same components that Como manufactured. This could have devastating results for Como. This alternative was unacceptable.

The second alternative required that Como make it clear to Ford Motor Company that Como wished to be considered for a chunk manager contract. If Como were to be selected, then Como's project management systems would have to:

- Provide greater coordination activities than previously anticipated
- Integrate concurrent engineering practices into the company's existing methodology for project management
- Decentralize the organization so as to enhance the working relationship with the customers
- Plan for better resource allocation so as to achieve a higher level of efficiency
- Force proactive planning and decision-making
- Drive out waste and lower cost while improving on-time delivery

There were also serious risks if Como were to become a chunk manager. The company would be under substantially more pressure to meet cost and delivery targets. Most of its resources would have to be committed to complex coordination activities rather than new product development. Therefore, value-added activities for its customers would be diminished. Finally, if Como failed to live up to its customers' expectations as a chunk manager, it might end up losing all automotive work.

The decision was made to inform Ford of Como's interest in chunk management. Now Como realized that its original three-year plan for excellence in project management would have to be completed in eighteen months. The question on everyone's mind was: "How?"

QUESTIONS

1. What was the driving force for excellence before the announcement of chunk management, and what is it now?
2. How can Como accelerate the learning process to achieve excellence in project management? What steps should management take based on its learning so far?
3. What are their chances for success? Justify your answer.
4. Should Como compete to become a chunk manager?
5. Can the decision to become a chunk supplier change the way Como performs strategic planning for project management?
6. Can the decision to become a chunk supplier cause an immediate change in Como's singular methodology for project management?
7. If a singular methodology for project management already exists, then how difficult will it be to make major changes to the methodology and what type of resistance, if any, should management expect?

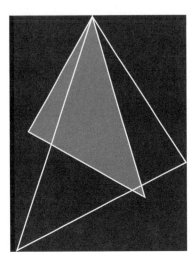

Apache Metals, Inc.

Apache Metals is an original equipment manufacturer of metal working equipment. The majority of Apache's business is as a supplier to the automotive, appliance, and building products industries. Each production line is custom-designed according to application, industry, and customer requirements.

Project managers are assigned to each purchase order only after the sales department has a signed contract. The project managers can come from anywhere within the company. Basically, anyone can be assigned as a project leader. The assigned project leaders can be responsible for as many as ten purchase orders at one time.

In the past, there has not been enough emphasis on project management. At one time, Apache even assigned trainees to perform project coordination. All failed miserably. At one point, sales dropped to an all-time low, and cost overruns averaged 20–25 percent per production line.

In January 2007, the board of directors appointed a new senior management team that would drive the organization to excellence in project management. Project managers were added through recruitment efforts and a close examination of existing personnel. Emphasis was on individuals with good people and communication skills.

The following steps were implemented to improve the quality and effectiveness of the project management system:

- Outside formal training for project managers
- Development of an apprenticeship program for future project managers

- Modification of the current methodology to put the project manager at the focal point
- Involvement of project managers to a greater extent with the customer

QUESTIONS

1. What problems can you see in the way project managers were assigned in the past?
2. Will the new approach taken in 2007 put the company on a path to excellence in project management?
3. What skill set would be ideal for the future project managers at Apache Metals?
4. What overall cultural issues must be considered in striving for excellence in project management?
5. What time frame would be appropriate to achieve excellence in project management? What assumptions must be made?

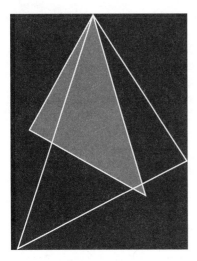

Haller Specialty
Manufacturing

For the past several years, Haller has been marginally successful as a specialty manufacturer of metal components. Sales would quote a price to the customer. Upon contract award, engineering would design the product. Manufacturing had the responsibility to produce the product as well as shipping the product to the customer. Manufacturing often changed the engineering design package to fit manufacturing capabilities.

The vice president of manufacturing was perhaps the most powerful position in the company next to the president. Manufacturing was considered to be the main contributor to corporate profits. Strategic planning was dominated by manufacturing.

To get closer to the customer, Haller implemented project management. Unfortunately, the vice president for manufacturing would not support project management for fear of a loss of power and authority.

QUESTIONS

1. If the vice president for manufacturing is a hindrance to excellence, how should this situation be handled?
2. Would your answer to the above question be different if the resistance came from middle or lower level management?

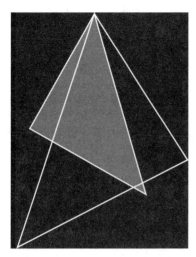

The NF3 Project: Managing Cultural Differences[1]

CREATING BEST PRACTICES OUT OF CULTURAL CLASHES

Mitsui Chemicals is one of the largest chemical companies in Japan and is among the largest twenty-five chemical companies in the world. Headquartered in Japan, Mitsui Chemicals has more than seventy-nine consolidated subsidiaries and ninety-seven companies in which it holds equity. Working with so many global companies making up the Mitsui Chemicals Group, the organization has been forced to address how it conducts project management and how it will overcome cultural differences.

A good example has been the global nitrogen trifluoride business. Nitrogen trifluoride (NF3) is a gas used for etching computer chips, cleaning CVD chambers, and making LCD panels. In 1990, the Shimonoseki Works began production at its facility in Shimonoseki, Japan. Due to tremendous growth and the need for production in the United States, a decision was made to conduct a technology transfer and build a new plant at its affiliate, Anderson Development Company, located in Adrian, Michigan. This project started in 1996 and was completed in

[1]© 2005 by Scott Tatro, PMP, NF3 Plant Manager & Responsible Care Coordinator, and Jessica Chen, PMP, NF3 Technical Manager & Special Projects Manager, Anderson Development Company; reproduced by permission

1997. Along the way, the two organizations had to learn how to adapt their different project management styles to complete the project.

There are many differences between American and Japanese culture as it relates to project management. The differences in project management practices and company values were vast and can be summarized in Exhibit I.

Anderson Development Company started out as an entrepreneurial company and still maintains some of that company culture today with a relatively flat hierarchy and medium level of formalization to promote quick decisions. The organization has around 150 employees, and 6 of them were assigned to the NF3 proj-ect. In contrast, the Shimonoseki Works project team was embedded with years of large Japanese business practices involving centralized authority with multiple levels of hierarchy. Every decision had to travel the proper hierarchical path and decisions were made by group consensus. Project management was conducted through a very rigid and formalized process, which often meant having many meetings to win support and approval of the initial ideas or project changes at each hierarchy level. Because it was a global project, there were shared accountabilities and multiple reporting requirements. Since it was also a technology transfer, the project was driven by the parent company.

During the initial construction of the Anderson Development Company NF3 plant, early clashes of these two project management styles hampered the overall schedule of the project. The Mitsui Chemical project manager required the ADC project manager to have frequent meetings, up to four per day, to update the project schedule. There was so much time spent in formal reporting that little progress was being made on the project. A review of the project GANTT chart revealed that the project was falling further behind schedule and budget. To get the project back on track, the companies eventually agreed to create a hybrid of the two project management styles.

The Mitsui and Anderson Development Company needed to find a way to use project management to overcome cultural differences while still satisfying each other's needs. Time was spent listening to and understanding the rigid project management reporting requirements of Mitsui's Project Management

Exhibit I.   American vs. Japanese project management

		ADC-American	MCI Shimonoseki-Japanese
Practices	Centralized	Decentralized authority	Centralized authority
	Formalized	Medium level of formalization	High level of formalization
	Hierarchy	Flatter hierarchy	Multiple level hierarchy
Values	Decision making	Individual decision making	Consensual decision making
	Communication	Individually-based	Group-based

system while also understanding the limited resources assigned to the project. With this understanding in place, a hybrid system was developed, as shown in Exhibit II.

To deal with these multinational cultural issues, guidelines had to be created and discussed. They addressed the following:

Integration Management
- Clearly defined roles of project sponsor(s), project manager, team members
- Agreement on project management methodology

Scope Management
- Well documented assumptions
- Well documented charter and scope statement

Time Management
- Agreement on working hours (8, 10, 12 hour days, weekends, holidays, etc.)
- Local understanding and determination of varying education, experience, and skill level as it relates to assignment of activities
- Understanding of missed milestones and consequences
- Agreement on format for reporting project progress (MS Project, MS Excel, etc.)

Cost Management
- Agreement on yen versus dollar currency exchange values and inflation rates

Procurement Management
- Negotiation of local and global procurements items
- Global Customs/shipping issues for declaration and transportation of goods/services
- Authority over specification interpretation

Exhibit II.   *Hybrid system for NF3 plant*

		MCI/ADC-Hybrid
Practices	Centralized	Decentralized authority
	Formalized	High level of formalization include project management templates and detailed roles and responsibility manuals
	Hierarchy	Flatter hierarchy
Values	Decision making	Consensual decision making in design phase.
		Individual decision making during implementation phase.
	Communication	Group/individually

Risk Management
- Understanding of global procurement issues
- Understanding of language barriers
- Understanding of engineering unit conversion and materials of construction

Quality Management
- Understanding of different codes and laws and impact on risks and design requirements
- Differing view of quality and development of agreed-upon quality metrics

Human Resource Management
- Differing value/policy systems and skill level sets. Union (Japan) versus Nonunion (America) issues
- Understanding different customs/holidays
- Understanding quantity and quality of resource capabilities internally and externally

Communications Management
- Agreed-upon project management communication templates and report timing
- Formal versus informal communication requirements
- Time zone differences
- Preferred method(s) of communicating (phone, fax, e-mail)
- Overcoming language differences
- Establishing trust

Based on the global project management framework started in 1997, the Anderson Development Company has gone through several additional projects involving expansion of the plant by more than 500 percent in the last five years. Mitsui Chemicals and Anderson Development Company continue to share best practices in project management and continuously improve the formalized project management template manuals.

As trust and communication improved between the management teams of Mitsui Chemicals and Anderson Development Company, the groundwork was set to bring the practices and culture of the operational workforce from Mitsui Chemicals to Anderson Development Company and see what practices, if any, would survive the cultural filter.

Mitsui Chemical's commitment to the development of its operational workforce is one of its best practices, and one that was brought to Anderson Development Company (ADC) as part of the technology transfer for its nitrogen trifluoride facility starting 1997. This was not, however, without significant effort on the part of all involved. There were several barriers to the integration of this best practice in conjunction with project management principles on an operations level. These barriers had to be gradually overcome in order to achieve the level of success currently seen, namely educational differences, union versus nonunion mentalities, and traditional manufacturing roles.

As is common in manufacturing facilities, many of the operators have a high school diploma at best, with typically no education or training beyond the mandatory requirements for OSHA and HAZMAT. Mitsui Chemicals Inc., however, has an exemplary training program that requires extensive training in math and chemistry. The requirements are no less for contractors, who must also be trained and pass certification exams to work in specific areas of the facility.

Japanese manufacturing facilities are historically union, as are Mitsui Chemical's facilities. Although the main ADC manufacturing facility in Adrian, Michigan, is located near the heart of automotive (and strongly union) manufacturing bases in Detroit, ADC is actually nonunion. The nearby and adjacent manufacturing facilities are also unionized, and ADC itself has shaken off a couple of union movements within its hourly personnel. A general characteristic of unions is that they can promote a separation between the "white hats" and the hourly work force, drawing a distinct line between what union employees should be empowered or allowed to do versus salaried personnel. Though this delineation can certainly serve a purpose for maintaining rules and regulations, it can be prohibitive to a team-based atmosphere, particularly when management is very hands-on. Japanese culture promotes a very strong respect for titles and the roles inherent to them, particularly uniformity—the phrase "the nail that sticks out will be hammered down" is quite applicable. ADC hourly employees were caught between the union mentality and the sudden requirement to empower themselves to make decisions and take on accountability that would have normally been given only to management.

Most significant among the barriers to a successful integration of Mitsui's operational style is the prevalent and accepted tradition of plant management, which ties in both of the issues just raised: that "blue collar" employees cannot be given the accountability, authority, and responsibility to make decisions that impact production and growth. Operators are usually limited to the basics in training, with the notion that a highly skilled operator is one who is experienced in the process and sticks tightly to the rules. The daily role of an operator can remain fairly unchanged except for the rare upgrade and update of a process, with scant training other than the requirements. Getting involvement from an operator usually comes in the form of process hazard analyses. Otherwise, from a project standpoint the only other objective is to complete the project and pass it off to the manufacturing group as soon as possible. Asking operators not only to be active participants, but to take on the work breakdown structures or even direct projects, is almost unheard of. Even more unusual is the thought that operators with a minimal level of education can take on what is thought to be fairly exclusive even in the engineering field—for example, software programming. Investment in training can be difficult to obtain among the salaried ranks, let alone the idea of sending an operator out of the plant for two weeks. The expense can be greater than that of a salaried person, as overtime coverage must also be arranged. Many will justifiably consider this to be a flight risk scenario, as the more highly trained personnel can either request higher pay or find a job elsewhere and take the training with them. All in all, a significant change in the mindset of both management and operations is necessary.

The approach to overcoming these barriers was certainly not an overnight process. The technology transfer from Mitsui to Anderson Development Co. at the NF3 plant was a wake-up call to the operators and engineers in terms of expectations and training requirements. Though the primary goal was to have a self-directed work force, this could not be done without proper long-term and continuous training, the provision of necessary tools, rewards, and, most critically, ownership of the process by the operators. The operators must understand the principles of scope planning, sorting out time, cost and quality constraints, resource planning, developing and carrying out work breakdown structures, and setting goals and milestones.

Establishing ownership is not just a matter of saying, "Run this plant or else." It means reinforcing the concept that there is a direct correlation on the plant's performance, daily work activities, and bonus structure tied in to how safely and efficiently the operators perform, and that they have direct impact on their own workload and pay. This shifts away from the attitude that only management or engineering can influence change or improvements, and instead focuses on putting the control and accountability in the hands of the operators.

The operators and contractors at the Mitsui facility must undergo rigorous training and develop a good background in engineering, chemistry, and math-related topics. At ADC, this was not typically a prerequisite. However, as part of the NF3 operator certification program, math, chemistry, and computer-based skills were integrated into the training and examinations. Additional formal training on math was also conducted at the plant. Oral and written exams are periodically administered during the four- to five-month training session which cumulates in a major written exam; the trainee has two chances to get a passing grade. Beyond standard HAZMAT and OSHA-required training, additional on-the-job training includes typical maintenance functions such as valve and instrumentation repair and replacement, quality control/SPC, root-cause analysis methodologies such as Kepner-Tregoe™, and troubleshooting, reading and understanding piping and instrument diagrams and project scheduling. With this foundation, areas in which the operators showed talent and the long-term need had value by the company proved to be ideal targets for further training and education. This included specialty welding courses, DCS programming, and obtaining and completing degrees. The operators themselves must initially express the interest and desire to receive the extra training, which places the onus upon them to declare their long-term goals and needs and take action to fulfill them. Note that the operators must be given opportunities to use their skills as frequently as possible; not only to maintain those skills but also to prevent frustration or boredom.

Management and engineering also had to undergo changes in appearance, attitude, and behavior. The NF3 plant adopted the same principle as that of the Mitsui sister plant in that all personnel wear uniforms, irregardless of position. This enables everyone to jump in and participate in all activities, whether it is taking out the trash, running the process, or packaging cylinders. This reinforces the attitude that everyone must be flexible and willing to take on whatever tasks in need of completion, and that rank or title should not be a barrier, nor should it be a buffer to accomplishments. The plant manager can just as easily be found in the control room temporarily substituting for an operator as he can be in a budget review meeting. The plant manager and technical manager also had to be willing to turn over activities in an increasing volume and scope, and to show trust in the capabilities of operations to handle issues. Among the most difficult activities was authorizing operators to proceed with additional training, which required funding and cooperation with other operators in order to provide coverage during their absence.

So, have the diligence, effort, and cost for additional training pay off? Absolutely! A company that shows interest and invests in its employees provides more incentive for people to stay with the company, thus retaining their skills and knowledge about the facility and eliminating costs for hiring and developing new people. On a major project level, operators have been able to present anywhere from 10 percent to 20 percent cost savings by taking over and managing specific work breakdown structures, including design, programming, installation, and fabrication. Overall manpower efficiency is increased by taking knowledgeable operators and putting them in charge, while reducing the risk of scope changes or errors. In terms of employee retention, none of the operators who received the additional training have left the company. The operators who received DCS programming are now charged with all programming activities, and one of the operators has moved into an instrumentation tech position. Some operators who have earned an associate or technical degree have moved into quality control roles. The savings on conducting these activities in-house versus the high cost of obtaining an outside programmer has already paid for all of the costs of training. The sense of pride in developing a program or graphic which will be used by the rest of the team, combined with the knowledge of how operators would like to have programs arranged rather than an engineer's or contractor's view of what is acceptable, is also invaluable. Operators with skills in certain types of welding have been able to take over segments or entire work breakdown structures in projects and pre-planned shutdowns. As the operators will ultimately be the ones forced to deal with workmanship, they are more apt to be vocal about poor quality, monitoring designs as they are installed, and offering suggestions for improvements in order to make designs more operator-friendly.

Furthermore, in situations of preplanned maintenance shutdowns, the need for engineering and plant management involvement has been nearly eliminated. The team coordinator sets up meetings with the necessary parties and establishes daily milestones and objectives for the operators. Better communication between maintenance and operations means there is less confusion on the prioritization and timing on work orders, better preparation on the details of the work to be done, and clarification on the roles each group will be performing. In preplanned maintenance shutdowns as well as in projects, there are often large numbers of resources trying to accomplish multiple tasks—usually in a limited space and time frame. Failure to appropriately plan and coordinate all of these resources and activities results in wasted time and therefore additional cost. Integrating project planning principles in top-to-bottom uniformity improves the consistency in planning and again transfers ownership to the operators while reducing manhours from engineering and management in such activities. In so doing, the operations team has been able to successfully reduce the duration of downtime required for preplanned maintenance shutdowns by more than 50 percent.

The benefits to the company in terms of operator development and cost reduction in projects and shutdowns extend beyond original targets. In smaller, resource-constricted facilities such as this nitrogen trifluoride plant, management and engineers often wear multiple hats, particularly when it comes to projects. By freeing these resources up to focus on long-term or other projects and goals, it similarly provides new opportunities and areas for growth that simply would not have been possible before due to time constrictions. The plant manager and the NF3 Technical Manager have been able to take on additional responsibilities and projects outside of the immediate NF3 facility and expand our roles in the company.

Still under development is a systematic way of rewarding people for ongoing improvements in daily activities and projects. Mitsui has a system that provides a monetary reward for suggestions that are related to improvements in safety, environmental, quality, and efficiency. If the suggestions are implemented and show actual improvement, additional rewards are provided. On a day-to-day level, operators at Anderson Development Company can freely make suggestions, but must also provide the scope of work, cost and time estimates along with the intended benefits as part of an informal project request. If approved, they often manage the project themselves including ordering materials and doing the actual work. Typically the reward is not a direct monetary bonus, but alternatives such as show or game tickets, gift certificates, or having special meals brought in to the facility for a team luncheon. Obtaining equipment or tools that can make a job more efficient or improve quality is also a good team-based reward.

The bottom line is that by dismissing the notion that only white-collar/ management employees can be entrusted with the skills and accountability required for leading projects and endowing blue-collar employees with the training and tools, companies can benefit considerably by involving them and literally handing over the reigns in projects and preplanned maintenance shutdowns. Providing the incentives in terms of the training, bonuses, and—most importantly—the opportunity for growth will only increase the likelihood of success in projects.

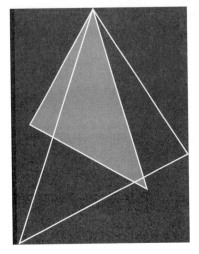

IVEY

Richard Ivey School of Business

The University of Western Ontario

An International Project Manager's Day (A)

SITUATION

The Maralinga–Ladawan Highway Project consists of fourteen expatriate families and the Sorongan counterpart personnel. Half expatriates are engineers from Hazelton. The other expatriates are mechanics, engineers, and other technical personnel from Beauval and MBI, the other two firms in the consortium. All expatriate personnel are under Hazelton's authority. This is the fifth largest project Hazelton has ever undertaken, with a fee of $1.63 million.

You arrived in Maralinga late on March 28 with your spouse. There was no chance for a briefing before you left. Head office had said John Anderson, the outgoing project manager, would fill you in on all you needed to know.[1] They had also arranged for you to meet people connected with the project in Kildona.

On March 29, you visited the project office briefly and met the accountant/administrative assistant, Tawi, the secretary, Julip, and the office messenger/driver, Satun. You then left immediately on a three-day site check of the 245-kilometer highway with John. Meanwhile, your spouse has started settling in and investigating job prospects in Maralinga.

On your trip you stopped at the field office in Corong. Chris Williams, second mechanic, and his wife, Beth, were living there. Chris was out at the timber company site to get help in recovering a grader that had toppled over the side of a ravine the night before, so you weren't able to see him. However, you met his Sorongan counterpart, and he advised you that everything was going well, although they could use more manpower.

You noted that Corong did not have any telephone facilities. The only communication link, a single side-band radio, had been unserviceable for the past few weeks. If you needed to contact Chris, it would involve a five-hour jeep ride to Corong to deliver the message.

You were able to see the haphazard way the work on the road was proceeding and witnessed the difficulty in finding appropriate gravel sites. Inspecting some of the bridges you had crossed made you shiver, too. Doing something about those would have to be a priority, before there was a fatality.

You returned to Maralinga on April 1 and met some of the staff and their families. Their comments made it clear that living conditions were less than ideal, the banking system make it difficult to get money transferred and converted into local currency (their salaries, paid in dollars, were deposited to their accounts at home), and the only school it was possible to send their children to was not appropriate for children who would have to return to the North American educational system.

That evening John left for another project on another continent. It is now Tuesday morning, April 2. This morning, while preparing breakfast with your spouse, the propane gas for your stove ran out. You have tried, unsuccessfully, on your way to work to get the gas cylinder filled, and have only now arrived at the office. It is 10 A.M. You have planned to have lunch with your spouse at noon and you are leaving for the airport at 2 P.M. for a week in Kildona to visit the Beauval office, the Sorongan Highway Department (SHD) people, and the International

---

[1]See Hazelton International Limited, 9A84C040.

Aid Agency (IAA) representative for discussions concerning the history and future of this project (it takes about one-half hour to drive to the airport). This trip has been planned as part of your orientation to the job (see Exhibit I). Since the IAA representative and the senior man in the Beauval office were both leaving for other postings at the end of the month, this may be the only opportunity you will have to spend time with them.

On your arrival at the office, Julip tells you that Jim, one of the surveyors, and his wife, Joyce, are arriving at 10:30 A.M. to discuss Joyce's medical problems with you. This is the first opportunity you have had to get into your office and do some work. You have about thirty minutes to go through the contents of your in-basket and take whatever action you feel is appropriate.

Exhibit I. Scheduling calendar

SUNDAY	MONDAY	TUESDAY	WEDNESDAY	THURSDAY	FRIDAY	SATURDAY
24	25	26	27	28 Arrival in Maralinga	29 ——— Site check with John ———	30
31	April 1 ———▶ Return	2 (TODAY) ———	3 Visit to Kildona ———	4	5	6
7	8	9 ———▶ Return to Maralinga	10	11	12	13
14	15	16	17	18	19	20
21	22	23	24	25	26	27
28	29	30	May 1	2	3	4

Note: You are in a Muslim area. People do not work Friday afternoons.
Saturday morning usually is a workday

INSTRUCTIONS

For the purpose of this exercise, you are to assume the position of Dan Simpson, the new project manager for the Maralinga-Ladawan Highway Project (see Exhibit II).

Please *write out* the action you choose on an action form. Your action may include writing letters, memos, telexes, or making phone calls. You may want to have meetings with certain individuals or receive reports from the office staff.

Exhibit II. *Organizational chart for Maralinga–Ladawan highway project*

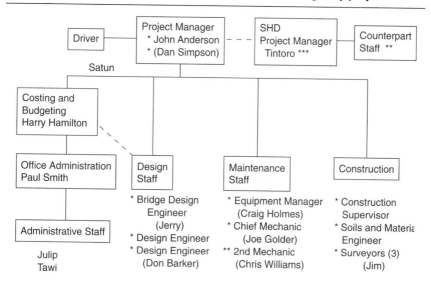

NOTES

*　　These people travel to Corong and other locations frequently.
**　　Stationed in Corong.
***　Located on the floor above Dan Simpson in the same building.
Note:　The 2 expatriates responsible for the training component had been sent home. The remaining 6 expatriates called for under the contract had not yet arrived in Soronga and the 2 construction supervisors recently requested by SHD would be in addition to these 6 people.

TRANSPORTATION AVAILABILITY:　(1) PROJECT OWNED— a) 1 Land Rover for administrative use by HQ staff, b) 1 car shared by all the families, c)most trucks are in Corong, however there usually are some around Maralinga(2) PUBLIC— a) peddle-cabs are available for short distances (like getting to work), b) local "taxis" are mini-van type vehicles which are usually very overcrowded and which expatriates usually avoid, c) there are a few flights to Kildona each week.

For example, if you decide to make a phone call, write out the purpose and content of the call on the action form. If you decide to have a meeting with one of the office staff or another individual, make a note of the basic agenda of things to be discussed and the date and time of the meeting. You also need to think about establishing priorities for the various issues.

To help you think of the time dimension, refer to Exhibit I. Also, Maralinga is twelve hours ahead of Eastern Standard Time.

IVEY

Richard Ivey School of Business

The University of Western Ontario

Ellen Moore (A): Living and Working in Korea

Chantell Nicholls and Gail Ellement prepared this case under the supervision of Professor Harry Lane solely to provide material for class discussion. The authors do not intend to illustrate either effective or ineffective handling of a managerial situation. The authors may have disguised certain names and other identifying information to protect confidentiality.

Ivey Management Services prohibits any form of reproduction, storage, or transmittal without its written permission. This material is not covered under authorization from CanCopy or any reproduction rights organization. To order copies or request permission to reproduce materials, contact Ivey Publishing, Ivey Management Services c/o Richard Ivey School of Business, The University of Western Ontario, London, Ontario, Canada, N6A 3K7; phone (519) 661-3208; fax (519) 661-3882; e-mail cases@ivey.uwo.ca.

Ellen Moore, a Systems Consulting Group (SCG) consultant, was increasingly concerned as she heard Andrew's voice grow louder through the paper-thin walls of the office next to her. Andrew Kilpatrick, the senior consultant on a joint North American and Korean consulting project for a government agency in Seoul, South Korea, was meeting with Mr. Song, the senior Korean project director, to discuss several issues, including the abilities of the Korean consultants. After four months on this Korean project, Ellen's evaluation of the assigned consultants suggested that they did not have the experience, background, or knowledge to complete the project within the allocated time. Additional resources would be required:

> I remember thinking, "I can't believe they are shouting at each other." I was trying to understand how their meeting had reached such a state. Andrew raised his voice and I could hear him saying, "I don't think you understand at all." Then, he shouted, "Ellen is not the problem!"

WSI IN KOREA

In 1990, Joint Venture Inc. (JVI) was formed as a joint venture between a Korean company, Korean Conglomerate Inc. (KCI), and a North American company, Western Systems Inc. (WSI) (Exhibit I). WSI, a significant information technology company with offices worldwide employing over 50,000 employees, included the Systems Consulting Group (SCG). KCI, one of the largest Korean *chaebols* (industrial groups), consisted of over forty companies, with sales in excess of U.S.$3.5 billion. The joint venture, in its eighth year, was managed by two regional directors—Mr. Cho, a Korean from KCI, and Robert Brown, an American from WSI.

The team working on Ellen's project was led by Mr. Park and consisted of approximately forty Korean consultants further divided into teams working on different areas of the project. The systems implementation (SI) team consisted of five Korean consultants, one translator, and three North American SCG consultants: Andrew Kilpatrick, Ellen Moore, and Scott Adams (see Exhibit II).

This consulting project was estimated to be one of the largest undertaken in South Korea to date. Implementation of the recommended systems into over one-hundred local offices was expected to take seven to ten years. The SCG consultants would be involved for the first seven months, to assist the Korean consultants with the system design and in creating recommendations for system implementation, an area in which the Korean consultants admitted they had limited expertise.

Andrew Kilpatrick became involved because of his experience with a similar systems implementation project in North America. Andrew had been a management consultant for nearly thirteen years. He had a broad and successful background in organizational development, information technology, and productivity improvement, and he was an early and successful practitioner of business process reengineering. Although Andrew had little international consulting experience, he was adept at change management and was viewed by both peers and clients as a flexible and effective consultant.

The degree of SCG's involvement had not been anticipated. Initially, Andrew had been asked by SCG's parent company, WSI, to assist JVI with the proposal

Exhibit I.   Organizational structure—Functional view

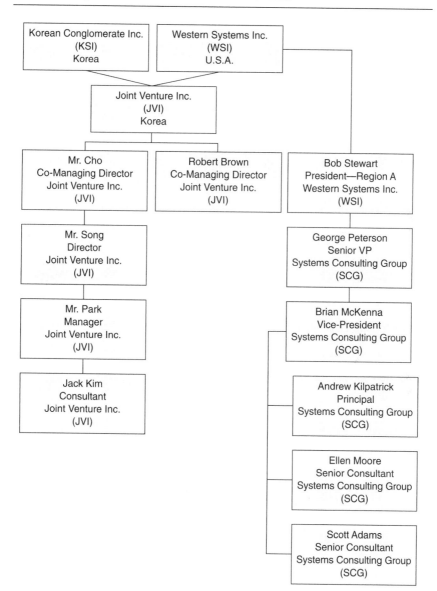

*Exhibit II. **Organizational structure—SI project team***

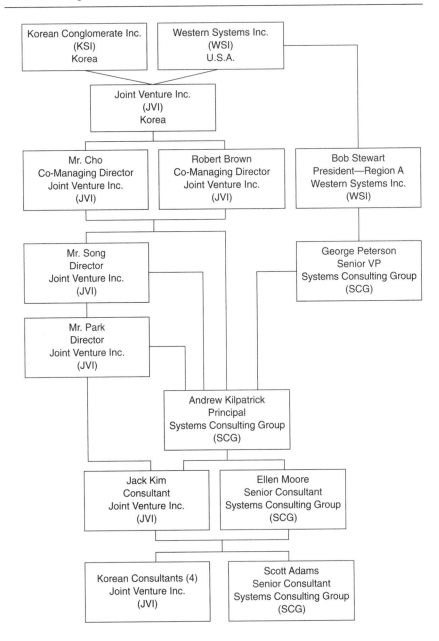

development. Andrew and his SCG managers viewed his assistance as a favor to WSI since SCG did not have plans to develop business in Korea. Andrew's work on the proposal in North America led to a request for his involvement in Korea to gather additional information for the proposal:

> When I arrived in Korea, I requested interviews with members of the prospective client's management team to obtain more information about their business environment. The Korean team at JVI was very reluctant to set up these meetings. However, I generally meet with client management prior to preparing a proposal. I also knew it would be difficult to obtain a good understanding of their business environment from a translated document. The material provided to me had been translated into English and was difficult to understand. The Korean and English languages are so different that conveying abstract concepts is very difficult.
>
> I convinced the Koreans at JVI that these meetings would help demonstrate our expertise. The meetings did not turn out exactly as planned. We met with the same management team at three different locations where we asked the same set of questions three times and got the same answers three times. We did not obtain the information normally provided at these fact-gathering meetings. However, they were tremendously impressed by our line of questioning because it reflected a deep interest and understanding of their business. They also were very impressed with my background. As a result, we were successful in convincing the government agency that we had a deep understanding of the nature and complexity of the agency's work and strong capabilities in systems development and implementation— key cornerstones of their project. The client wanted us to handle the project and wanted me to lead it.

JVI had not expected to get the contract, because its competitor for this work was a long-time supplier to the client. As a result, winning the government contract had important competitive and strategic implications for JVI. Essentially, JVI had dislodged an incumbent supplier to the client, one who had lobbied very heavily for this prominent contract. By winning the bid, JVI became the largest system implementer in Korea and received tremendous coverage in the public press.

The project was to begin in June 1995. However, the Korean project team convened in early May in order to prepare the team members. Although JVI requested Andrew to join the project on a full-time basis, he already had significant commitments to projects in North America. There was a great deal of discussion back and forth between WSI in North America, and JVI and the client in Korea. Eventually it was agreed that Andrew would manage the SI work on a part-time basis from North America, and he would send a qualified project management representative on a full-time basis. That person was Ellen Moore.

At that time, Andrew received immediate feedback from the American consultants with WSI in Korea that it would be impossible to send a woman to work in Korea. Andrew insisted that the Korean consultants be asked if they would accept a woman in the position. They responded that a woman would be acceptable if she were qualified. Andrew also requested that the client be consulted on this issue. He was again told that a woman would be acceptable if she were qualified. Andrew knew that Ellen had the skills required to manage the project:

> I chose Ellen because I was very impressed with her capability, creativity, and project management skills, and I knew she had worked successfully in Bahrain, a culture where one would have to be attuned to very different cultural rules from those prevalent in North America. Ellen lacked experience with government agencies, but I felt that I could provide the required expertise in this area.

ELLEN MOORE

After graduating as the top female student from her high school, Ellen worked in the banking industry, achieving the position of corporate accounts officer responsible for over twenty major accounts and earning a fellowship in the Institute of Bankers. Ellen went on to work for a former corporate client in banking and insurance, where she became the first female and youngest person to manage their financial reporting department. During this time, Ellen took university courses towards a Bachelor's degree at night. In 1983, she decided to stop working for two years, and completed her degree on a full-time basis. She graduated with a major in accounting and minors in marketing and management and decided to continue her studies for an MBA.

Two years later, armed with an MBA from a leading business school, Ellen Moore joined her husband in Manama, Bahrain, where she accepted a position as an expatriate manager for a large American financial institution.[1] Starting as a special projects coordinator, within one year Ellen was promoted to manager of business planning and development, a challenging position that she was able to design herself. In this role, she managed the quality assurance department, coordinated a product launch, developed a senior management information system, and participated actively in all senior management decisions. Ellen's position required her to interact daily with managers and staff from a wide range of cultures, including Arab nationals.

In March 1995, Ellen joined WSI working for SCG. After the highly successful completion of two projects with SCG in North America, Ellen was approached for the Korea project:

---

[1]For an account of Ellen's experience in Bahrain, see Ellen Moore (A): Living and Working in Bahrain, 9A90C019, and Ellen Moore (B), 9A90C020; Ivey Publishing, Ivey Management Services, c/o Richard Ivey School of Business, University of Western Ontario, London, Ontario, Canada, N6A 3K7.

I had never worked in Korea or East Asia before. My only experience in Asia had been a one-week trip to Hong Kong for job interviews. I had limited knowledge of Korea and received no formal training from my company. I was provided a twenty-page document on Korea. However, the information was quite basic and not entirely accurate.

After arriving in Korea, Ellen immediately began to familiarize herself with the language and proper business etiquette. She found that English was rarely spoken other than in some hotels and restaurants that catered to Western clientele. As a result, Ellen took advantage of every opportunity to teach herself the language basics:

> When Andrew and I were in the car on the way back to our hotel in the evening, we would be stuck in traffic for hours. I would use the time to learn how to read the Korean store signs. I had copied the Hangul symbols which form the Korean language onto a small piece of paper, and I kept this with me at all times. So, while sitting back in the car, exhausted at the end of each day, I would go over the symbols and read the signs.

SCOTT ADAMS

The third SCG consultant on the project, Scott Adams, arrived as planned three months after Ellen's start date. Upon graduation, Scott had begun his consulting career working on several international engagements (including Mexico, Puerto Rico, and Venezuela), and he enjoyed the challenges of working with different cultures. He felt that with international consulting projects the technical aspects of consulting came easy. What he really enjoyed was the challenge of communicating in a different language and determining how to modify Western management techniques to fit into the local business culture. Scott first met Ellen at a systems consulting seminar, unaware at the time that their paths would cross again. A few months later, he was asked to consider the Korea assignment. Scott had never travelled or worked in Asia, but he believed that the assignment would present a challenging opportunity that would advance his career.

Scott was scheduled to start work on the project in August 1995. Prior to arriving in Seoul, Scott prepared himself by frequently discussing the work being conducted with Ellen. Ellen also provided him with information on the culture and business etiquette aspects of the work:

> It was very fortunate for me that Ellen had arrived first in Korea. Ellen tried to learn as much as she could about the Korean language, the culture, mannerisms, and the business etiquette. She was able to interpret many of the subtleties and to prepare me for both business and social situations, right down to how to exchange a business card appropriately with a Korean, how to read behavior, and what to wear.

ABOUT KOREA[2]

Korea is a 600-mile-long peninsula stretching southward into the waters of the western Pacific, away from Manchuria and Siberia to the north on the Asian mainland. Facing eastward across the Sea of Japan, known to Koreans as the East Sea, Korea lies 120 miles from Japan. The Republic of Korea, or South Korea, consists of approximately 38,000 square miles, comparable in size to Virginia or Portugal. According to the 1990 census, the South Korean population is about 43 million, with almost 10 million residing in the capital city, Seoul.

Korea has an ancient heritage spanning 5,000 years. The most recent great historical era, the Yi Dynasty or Choson Dynasty, enlisted tremendous changes in which progress in science, technology, and the arts were achieved. Although Confucianism had been influential for centuries in Korea, it was during this time that Confucian principles permeated the culture as a code of morals and as a guide for ethical behavior. Confucian thought was designated as the state religion in 1392 and came to underpin education, civil administration, and daily conduct. During this time, Korean rulers began to avoid foreign contact and the monarchy was referred to as the *Hermit Kingdom* by outsiders. Lasting over 500 years and including twenty-seven rulers, the Yi Dynasty came to a close at the end of the nineteenth century. Today, in Korea's modern era, the nation is quickly modernizing and traditional Confucian values mix with Western lifestyle habits and business methods.

Although many Korean people, particularly in Seoul, have become quite Westernized, they often follow traditional customs. Confucianism dictates strict rules of social behavior and etiquette. The basic values of the Confucian culture are: (1) complete loyalty to a hierarchical structure of authority, whether based in the family, the company, or the nation; (2) duty to parents, expressed through loyalty, love, and gratitude; and (3) strict rules of conduct, involving complete obedience and respectful behavior within superiors–subordinate relationships, such as parents–children, old–young, male–female, and teacher–student. These values affect both social and work environments substantially.

MANAGING IN KOREA

Business etiquette in Korea was extremely important. Ellen found that everyday activities, such as exchanging business cards or replenishing a colleague's drink at dinner, involved formal rituals. For example, Ellen learned it was important to

---

[2]Some of the information in the "About Korea" and "Women in Korea" sections was obtained from *Fodor's Korea,* (New York: Fodor's Travel Publications, Inc., 1993) and Chris Taylor, *Seoul—City Guide,* (Hong Kong: Lonely Planet Publications: Colorcraft Ltd.,1993).

provide and to receive business cards in an appropriate manner, which included carefully examining a business card when received and commenting on it. If one just accepted the card without reading it, this behavior would be considered very rude. In addition, Ellen also found it important to know how to address a Korean by name. If a Korean's name was Y. H. Kim, non-Koreans would generally address him as either Y. H. or as Mr. Kim. Koreans would likely call him by his full name or by his title and name, such as Manager Kim. A limited number of Koreans, generally those who had lived overseas, took on Western names, such as Jack Kim.

WORK TEAMS

Teams were an integral part of the work environment in Korea. Ellen noted that the Korean consultants organized some special team building activities to bring together the Korean and North American team members:

> On one occasion, the Korean consulting team invited the Western consul-tants to a baseball game on a Saturday afternoon followed by a trip to the Olympic Park for a tour after the game, and dinner at a Korean restaurant that evening. An event of this nature is unusual and was very special. On another occasion, the Korean consultants gave up a day off with their families and spent it with the Western consultants. We toured a Korean palace and the palace grounds, and we were then invited to Park's home for dinner. It was very unusual that we, as Western folks, were invited to his home, and it was a very gracious event.

Ellen also found team-building activities took place on a regular basis, and that these events were normally conducted outside of the work environment. For example, lunch with the team was an important daily team event that everyone was expected to attend:

> You just couldn't work at your desk every day for lunch. It was important for everyone to attend lunch together in order to share in this social activity, as one of the means for team bonding.

Additionally, the male team members would go out together for food, drink, and song after work. Scott found these drinking activities to be an important part of his interaction with both the team and the client:

> Unless you had a medical reason, you would be expected to drink with the team members, sometimes to excess. A popular drink, soju, which is similar to vodka, would be poured into a small glass. Our glasses were never empty,

as someone would always ensure that an empty glass was quickly filled. For example, if my glass was empty, I learned that I should pass it to the person on my right and fill it for him as a gesture of friendship. He would quickly drink the contents of the glass, pass the glass back to me, and fill it for me to quickly drink. You simply had to do it. I recall one night when I really did not want to drink as I had a headache. We were sitting at dinner, and Mr. Song handed me his glass and filled it. I said to him "I really can't drink tonight. I have a terrible headache." He looked at me and said, "Mr. Scott, I have aspirin in my briefcase." I had about three or four small drinks that night.

Ellen found she was included in many of the team-building dinners, and soon after she arrived in Seoul, she was invited to a team dinner, which included client team members. Ellen was informed that although women were not normally invited to these social events, an exception was made because she was a senior team member.

During the dinner, there were many toasts and drinking challenges. During one such challenge, the senior client representative prepared a drink that consisted of one highball glass filled with beer and one shot glass filled to the top with whiskey. He dropped the whiskey glass into the beer glass and passed the drink to the man on his left. This team member quickly drank the cocktail in one swoop, and held the glass over his head, clicking the glasses to show both were empty. Everyone cheered and applauded. This man then mixed the same drink, and passed the glass to the man on his left, who also drank the cocktail in one swallow. It was clear this challenge was going around the table and would eventually get to me.

I don't generally drink beer and never drink whiskey. But it was clear, even without my translator present to assist my understanding, that this activity was an integral part of the team building for the project. As the man on my right mixed the drink for me, he whispered that he would help me. He poured the beer to the halfway point in the highball glass, filled the shot glass to the top with whiskey, and dropped the shotglass in the beer. Unfortunately, I could see that the beer didn't cover the top of the shot glass, which would likely move too quickly if not covered. I announced, "One moment, please, we are having technical difficulties." And to the amazement of all in attendance, I asked the man on my right to pour more beer in the glass. When I drank the concoction in one swallow, everyone cheered, and the senior client representative stood up and shouted, "You are now Korean. You are now Korean."

The norms for team management were also considerably different from the North American style of management. Ellen was quite surprised to find that the concept of saving face did not mean avoiding negative feedback or sharing failures:

It is important in Korea to ensure that team members do not lose face. However, when leading a team, it appeared just as important for a manager to demonstrate leadership. If a team member provided work that did not meet the stated requirements, a leader was expected to express disappointment in the individual's efforts in front of all team members. A strong leader was considered to be someone who engaged in this type of public demonstration when required.

In North America, a team leader often compliments and rewards team members for work done well. In Korea, leaders expressed disappointment in substandard work, or said nothing for work completed in a satisfactory manner. A leader was considered weak if he or she continuously provided compliments for work completed as required.

Hierarchy

The Koreans' respect for position and status was another element of the Korean culture that both Ellen and Scott found to have a significant influence over how the project was structured and how people behaved. The emphasis placed on hierarchy had an important impact upon the relationship between consultant and client that was quite different from their experience in North America. As a result, the North Americans' understanding of the role of a consultant differed vastly from their Korean counterparts.

Specifically, the North American consultants were familiar with managing client expectations. This activity involved informing the client of the best means to achieve their goals and included frequent communication with the client. Generally, a client's customer was also interviewed in order to understand how the client's system could better integrate with its customer's requirements. Ellen recalled, however, that the procedures were necessarily different in Korea:

> The client team members did not permit our team members to go to their offices unannounced. We had to book appointments ahead of time to obtain permission to see them. In part, this situation was a result of the formalities we needed to observe due to their rank in society, but I believe it was also because they wanted to be prepared for the topics we wanted to discuss.

The Korean consultants refused to interview the customers, because they did not want to disturb them. Furthermore, the client team members frequently came into the project office and asked the Korean consultants to work on activities not scheduled for that week or that were beyond the project scope. The Korean consultants accepted the work without question. Ellen and Scott found themselves powerless to stop this activity.

Shortly after arriving, Scott had a very confrontational meeting with one of the Korean consultants concerning this issue:

> I had been in Korea for about a week, and I was still suffering from jet lag. I was alone with one of the Korean consultants, and we were talking about

how organizational processes should be flowcharted. He was saying the client understands the process in a particular manner, so we should show it in that way. I responded that, from a technical standpoint, it was not correct. I explained that as a consultant, we couldn't simply do what the client requests if it is incorrect. We must provide value by showing why a different method may be taken by educating the client of the options and the reasons for selecting a specific method. There are times when you have to tell the client something different than he believes. That's what we're paid for. He said, "No, no, you don't understand. They're paying our fee." At that point I raised my voice: "You don't know what you are talking about. I have much more experience than you." Afterwards, I realized that it was wrong to shout at him. I pulled him aside and apologized. He said, "Well, I know you were tired." I replied that it was no excuse, and I should not have shouted. After that, we managed to get along just fine.

The behavior of subordinates and superiors also reflected the Korean's respect for status and position. Scott observed that it was very unusual for a subordinate to leave the office for the day unless his superior had already left:

I remember one day, a Saturday, when one of the young Korean consultants who had been ill for some time, was still at his desk. I made a comment: "Why don't you go home, Mr. Choi?" Although he was not working for me, I knew his work on the other team was done. He said, "I can't go home because several other team members have taken the day off. I have to stay." I repeated my observation that his work was done. He replied: "If I do not stay, I will be fired. My boss is still here, I have to stay." He would stay and work until his boss left, until late in the evening if necessary.

Furthermore, Scott found that the Korean consultants tended not to ask questions. Even when Scott asked the Korean consultants if they understood his instructions or explanation, they generally responded affirmatively, which made it difficult to confirm their understanding. He was advised that responding in a positive manner demonstrated respect for teachers or superiors. Asking a question would be viewed as inferring that the teacher or superior had not done a good job of explaining the material. As a result, achieving a coaching role was difficult for the North American consultants even though passing on their knowledge of SI to the Korean consultants was considered an important part of their function on this project.

WOMEN IN KOREA

Historically, Confucian values have dictated a strict code of behavior between men and women and husband and wife in Korea. Traditionally, there has been a clear delineation in the respective responsibilities of men and women. The male preserve can be defined as that which is public, whereas women are expected to cater to the private, personal world of the home. These values have lingered into the 1990s, with Korean public life very much dominated by men.

Nevertheless, compared to the Yi dynasty era, the position of women in society has changed considerably. There is now virtual equality in access to education for men and women, and a few women have embarked on political careers. As in many other areas of the world, the business world has until recently been accessible only to men. However, this is changing as Korean women are beginning to seek equality in the workplace. Young Korean men and women now often participate together in social activities such as evenings out and hikes, something that was extremely rare even ten years ago.

Dual-income families are becoming more common in South Korea, particularly in Seoul, although women generally hold lower-paid, more menial positions. Furthermore, working women often retain their traditional household responsibilities, while men are expected to join their male colleagues for late night drinking and eating events that exclude women. When guests visit a Korean home, the men traditionally sit and eat together separately from the women, who are expected to eat together while preparing the food.

Although the younger generation are breaking from such traditions, Scott felt that the gender differences were quite apparent in the work place. He commented:

> The business population was primarily male. Generally, the only women we saw were young women who were clerks, wearing uniforms. I suspected that these women were in the work force for only a few years, until they were married and left to have a family. We did have a few professional Korean women working with us. However, because we are a professional services firm, I believe it may have been more progressive than the typical Korean company.

THE SYSTEMS IMPLEMENTATION TEAM

Upon her arrival in Korea, Ellen dove into her work confident that the Korean consultants she would be working with had the skills necessary to complete the job in the time frame allocated. The project work was divided up among several work groups, each having distinct deliverables and due dates. The deliverables for the SI team were required as a major input to the other work groups on the project (see Exhibit III). As a result, delays with deliverables would impact the effectiveness of the entire project:

> JVI told us they had assigned experienced management consultants to work on the project. Given their stated skill level, Andrew's resource plan had him making periodic visits to Korea; I would be on the project on a full-time basis starting in May, and Scott would join the team about three to four months after the project start. We were informed that five Korean consultants were assigned. We believed that we had the resources needed to complete the project by December.

Exhibit III. *Project time frame*

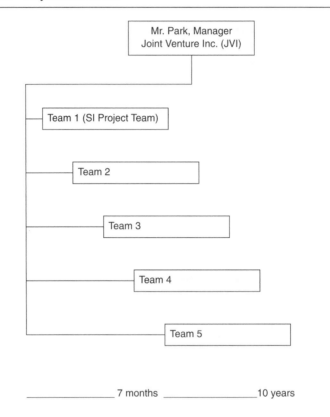

JACK KIM

J.T. Kim, whose Western name was Jack, was the lead Korean consultant report-ing to Mr. Park. Jack had recently achieved a Ph.D. in computer systems from a reputable American university and he spoke English fluently. When Andrew initially discussed the organizational structure of the SI team with Mr. Park and Jack, it was agreed that Jack and Ellen would be co-managers of the SI project.

Three weeks after her arrival, Jack informed Ellen, much to her surprise, that he had never worked on a systems implementation project. Additionally, Ellen soon learned that Jack had never worked on a consulting project:

> Apparently, Jack had been made the lead consultant of SI upon completing his Ph.D. in the United States. I believe Jack was told he was going to be the sole project manager for SI on a daily basis. However, I was informed I was going to be the co-project manager with Jack. It was confusing, particularly

for Jack, when I took on coaching and leading the team. We had a lot of controversy—not in the form of fights or heated discussions, but we had definite issues during the first few weeks because we were clearly stepping on each other's territory.

Given Jack's position as the lead Korean consultant, it was quite difficult for Ellen to redirect team members' activities. The Korean team members always followed Jack's instructions. Scott recalled:

> There were frequent meetings with the team to discuss the work to be completed. Often, following these meetings the Korean consultants would meet alone with Jack, and it appeared that he would instruct them to carry out different work. On one occasion, when both Andrew and Ellen were traveling away from the office, Andrew prepared specific instructions for the team to follow outlined in a memo.
>
> Andrew sent the memo to me so I could hand the memo to Jack directly, thereby ensuring he did receive these instructions. Upon his return, Andrew found the team had not followed his instructions. We were provided with the following line of reasoning: you told us to do A, B, and C, but you did not mention D. And, we did D. They had followed Jack's instructions. We had a very difficult time convincing them to carry out work as we requested, even though we had been brought onto the project to provide our expertise.

In July, a trip was planned for the Korean client team and some of the Korean consulting team to visit other project sites in North America. The trip would permit the Koreans to find out more about the capabilities of WSI and to discuss issues with other clients involved with similar projects. Jack was sent on the trip, leaving Ellen in charge of the SI project team in Korea. While Jack was away on the North American trip, Ellen had her first opportunity to work with and to lead the Korean consultants on a daily basis. She was very pleased that she was able to coach them directly, without interference, and advise them on how to best carry out the required work. Ellen felt that everyone worked together in a very positive manner, in complete alignment. When Jack returned, he saw that Ellen was leading the team and that they were accepting Ellen's directions. Ellen recalled the tensions that arose as a result:

> On the first day he returned, Jack instructed someone to do some work for him, and the person responded, "I cannot because I am doing something for Ellen." Jack did not say anything, but he looked very angry. He could not understand why anyone on the team would refuse his order's.

THE MARKETING RESEARCH PROJECT

A few days after Jack returned from the North American trip, the project team realized they did not have sufficient information about their client's customer.

Jack decided a market research study should be conducted to determine the market requirements. However, this type of study, which is generally a large undertaking on a project, was not within the scope of the contracted work. Ellen found out about the proposed market research project at a meeting held on a Saturday, which involved everyone from the entire project—about 40 people. The only person not at the meeting was Mr. Park. Jack was presenting the current work plans for SI, and he continued to describe a market research study:

> I thought to myself, "What market research study is he talking about?" I asked him to put aside his presentation of the proposed study until he and I had an opportunity to discuss the plans. I did not want to interrupt his presentation or disagree with him publicly, but I felt I had no choice.

DINNER WITH JACK

Two hours following the presentation, Ellen's translator, Susan Lim, informed her that there was a dinner planned for that evening and Jack wanted everyone on the SI team to attend. Ellen was surprised that Jack would want her present at the dinner. However, Susan insisted that Jack specifically said Ellen must be there. They went to a small Korean restaurant, where everyone talked about a variety of subjects in English and Korean, with Susan translating for Ellen as needed. After about one hour, Jack began a speech to the team, speaking solely in Korean. Ellen thought it was unusual for him to speak Korean when she was present, as everyone at the dinner also spoke English:

> Through the limited translations I received, I understood he was humbling himself to the team, saying, "I am very disappointed in my performance. I have clearly not been the project leader needed for this team." The team members were responding "No, no, don't say that." While Jack was talking to the team, he was consuming large quantities of beer. The pitchers were coming and coming. He was quite clearly becoming intoxicated. All at once, Susan stopped translating. I asked her what was wrong. She whispered that she would tell me later. Five minutes went by and I turned to her and spoke emphatically, "Susan, what is going on? I want to know now." She realized I was getting angry. She told me, "Jack asked me to stop translating. Please don't say anything, I will lose my job."
>
> I waited a couple of minutes before speaking, then I interrupted Jack's speech. I said, "Susan is having difficulty hearing you and isn't able to translate for me. I guess it is too noisy in this restaurant. Would it be possible for you to speak in English?" Jack did not say anything for about thirty seconds and then he started speaking in English. His first words were, "Ellen, I would like to apologize. I didn't realize you couldn't understand what I was saying.

Another thirty minutes of his speech and drinking continued. The Korean team members appeared to be consoling Jack, by saying: "Jack, we do respect you and the work you have done for our team. You have done your best." While they were talking, Jack leaned back, and appeared to pass out. Ellen turned to Susan and asked if they should help him to a taxi. Susan insisted it would not be appropriate. During the next hour, Jack appeared to be passed out or sleeping. Finally, one of the team members left to go home. Ellen asked Susan, "Is it important for me to stay, or is it important for me to go?" She said Ellen should go.

When Ellen returned to her hotel, it was approximately 11 p.m. on Saturday night. She felt the situation had reached a point where it was necessary to request assistance from senior management in North America. Andrew was on a wilderness camping vacation in the United States with his family, and could not be reached. Ellen decided to call the North American project sponsor, the senior vice president, George Peterson:

> I called George that Saturday night at his house and said: "We have a problem. They're trying to change the scope of the project. We don't have the available time, and we don't have the resources. It is impossible to do a market research study in conjunction with all the contracted work to be completed with the same limited resources. The proposed plan is to use our project team to handle this additional work. Our team is already falling behind the schedule, but due to their inexperience they don't realize it yet." George said he would find Andrew and send him to Korea to further assess the situation.

THE MEETING WITH THE DIRECTOR

When Andrew arrived in August, he conducted a very quick assessment of the situation. The project was a month behind schedule. It appeared to Andrew that the SI team had made limited progress since his previous visit:

> It was clear to me that the Korean team members weren't taking direction from Ellen. Ellen was a seasoned consultant and knew what to do. However, Jack was giving direction to the team which was leading them down different paths. Jack was requesting that the team work on tasks that were not required for the project deliverables, and he was not appropriately managing the client's expectations.

Andrew held several discussions with Mr. Park concerning these issues. Mr. Park insisted the problem was Ellen. He argued that Ellen was not effective, she did not assign work properly, and she did not give credible instructions to the team. However, Andrew believed the Korean consultants' lack of experience was the main problem.

Initially, we were told the Korean team consisted of experienced consultants, although they had not completed any SI projects. I felt we could work around it. I had previously taught consultants to do SI. We were also told that one of the Korean consultants had taught SI. This consultant was actually the most junior person on the team. She had researched SI by reading some texts and had given a presentation on her understanding of SI to a group of consultants.

Meanwhile, Andrew solicited advice from the WSI co-managing director, Robert Brown, who had over ten years' experience working in Korea. Robert suggested that Andrew approach Mr. Park's superior, Mr. Song, directly. He further directed Andrew to present his case to the Joint Venture committee if an agreement was not reached with Mr. Song. Andrew had discussed the issues with George Peterson and Robert Brown, and they agreed that there was no reason for Ellen to leave the project:

However, Robert's message to me was that I had been too compliant with the Koreans. It was very important for the project to be completed on time, and I would be the one held accountable for any delays. Addressing issues before the Joint Venture committee was the accepted dispute resolution process at JVI when an internal conflict could not be resolved. However, in most cases, the last thing a manager wants is to be defending his position before the Joint Venture committee. Mr. Song was in line to move into senior executive management. Taking the problem to the Joint Venture committee would be a way to force the issue with him.

Andrew attempted to come to a resolution with Mr. Park once again, but he refused to compromise. Andrew then tried to contact Mr. Song and was told he was out of the office. Coincidentally, Mr. Song visited the project site to see Mr. Park just as Ellen and Andrew were completing a meeting. Ellen recalls Mr. Song's arrival:

Mr. Song walked into the project office expecting to find Mr. Park. However, Mr. Park was out visiting another project that morning. Mr. Song looked around the project office for a senior manager, and he saw Andrew. Mr. Song approached Andrew and asked if Mr. Park was in the office. Andrew responded that he was not. Mr. Song proceeded to comment that he understood there were some concerns about the project work, and suggested that perhaps, sometime, they could talk about it. Andrew replied that they needed to talk about it immediately.

Andrew met with Mr. Song in Mr. Park's office, a makeshift set of thin walls that enclosed a small office area in one corner of the large open project office. Ellen was working in an area just outside the office when she heard Andrew's voice rise. She heard him shout, "Well, I don't think you're listening to what I am saying." Ellen was surprised to hear Andrew shouting. She knew Andrew was very sensitive to what should and should not be done in the Korean environment:

> Andrew's behavior seemed so confrontational. I believed this behavior was unacceptable in Korea. For a while, I heard a lot of murmuring, after which I heard Andrew speak adamantly, "No, I'm very serious. It doesn't matter what has been agreed and what has not been agreed because most of our agreements were based on inaccurate information. We can start from scratch." Mr. Song insisted that I was the problem.

The Richard Ivey School of Business gratefully acknowledges the generous support of The Richard and Jean Ivey Fund in the development of this case as part of the RICHARD AND JEAN IVEY FUND ASIAN CASE SERIES.

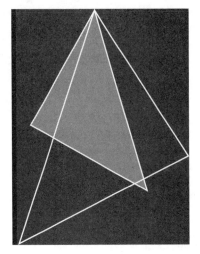

IVEY

Richard Ivey School of Business
The University of Western Ontario

Ji'nan Broadcasting Corporation

Ken Mark prepared this case under the supervision of Professor Michael Parent solely to provide material for class discussion. The authors do not intend to illustrate either effective or ineffective handling of a managerial situation. The authors may have disguised certain names and other identifying information to protect confidentiality.

INTRODUCTION

It was June 3, 2001. Zhou Jianglin, project manager for Ji'nan Broadcasting Corporation (JBC), was waiting for a meeting with Nortel Network's account sales manager, Frank Kang. There were many questions that Zhou intended to ask Kang regarding JBC's data and voice project (DVP): Could Nortel work in conjunction with Alcatel or Lucent to complete the project? Should the network be centrally controlled, or should adjustments be allowed remotely? How could JBC be assured that the project would be completed on time, as specified, and on budget?

Headquartered in Ji'nan, the capital of Shandong province in China, JBC planned to provide data and voice services to thousands of residential and business customers in sixty sites throughout the province. JBC had planned to go into competition with the national data and voice carriers in anticipation of liberalization of state-owned telecommunications markets. These services were scheduled to commence in January 2002, six months away. Based on the request of his managing director of Data and Voice Services, Han Xiaowei, Zhou had arranged an initial meeting with Eastern Postel (Postel) in late May 2001. Postel was a national manufacturer and installer of, among other products, data transmission equipment. To meet the requirement of the DVP, Postel had contacted Nortel for its Passport data and Meridian voice products. Nortel assured Postel that they could meet JBC's project goals, and a Nortel pre-sales team was dispatched from Hong Kong to perform a project scope.

In the meantime, however, Zhou had heard rumors of conflict between JBC's departments regarding the DVP. The managing director of radio and television programming, Liu Zhongshi was said to have voiced his objection to the project, insisting that providing telephone and Internet service was the domain of China Post & Telecom, the government-owned national provider. Shao Yangwei, managing director of corporate services had mentioned that after consulting colleagues in Beijing, he believed that Alcatel voice products "were more robust" than Nortel products. Perhaps, Shao insisted, Nortel and Alcatel products could be tested head-to-head to determine superiority. The president of JBC, Guang Chengmen, had listened to these comments and, expressing that he had little time to deal with them at the present, passed the suggestions on to Han in Internet Services. Guang requested that these suggestions be taken into consideration and expected answers within the week. Zhou, as Han's project manager, was subsequently given the task of finding answers.

JI'NAN BROADCASTING CORPORATION (JBC)

Created after WWII, JBC was Shandong province's public broadcaster. It operated one high-powered broadcast station and two low-powered broadcast stations, providing three television channels and seven radio channels for the province. Television and radio content were created in-house and purchased from national government-run networks. At the start of 2001, JBC had four business departments: television and radio programming, television and radio broadcasting, corporate services (human resources, publicity, business advertisement sales) and the newly created department of data and voice services. It employed 850 people in four locations, serving a population base of nine million people.

Currently, JBC received revenues of RMB108 million from the government for operational expenses.[1] Exhibit I is a breakdown of 2000 expenditures:

Exhibit I.   JBC expenditures, 2000 (in RMBs)

Internal program production costs	25 million
Purchased programming	25 million
Broadcasting costs	15 million
Salaries and wages	25 million
Capital expenditures and other expenses	18 million

In response to anticipated competition from selected foreign companies, the Chinese government announced publicly in February 2001 that JBC would be privatized before 2005, with shares listed on the Shenzhen Stock Exchange. This energized everyone at JBC—it was expected that employees would be granted stock options based on several factors, including department efficiency and seniority. Although the reward structure seemed straightforward, it was noted by JBC insiders that during similar privatization efforts in Beijing, responsibility for the division of employee stock options had been left to the company president.

The Growth Plan: Focus on Serving the Business Market

In early April 2001, Guang sat down with his three managing directors to lay out his new strategic growth plan. Central to JBC's growth was its focus on the under-served Shandong business market. There were estimates of more than 5,000 companies of all sizes operating within 250 kilometers of Ji'nan, and that number was expected to grow rapidly if China gained acceptance into the World Trade Organization. Many of these customers were exporters or export-ready firms awaiting the chance to conduct business with international customers. JBC, Guang proclaimed, would provide provincial and national advertising services, telephony, and Internet services. In addition to announcing the creation of a fourth department, Internet services, Guang laid out a reorganization plan and new department targets for the next four years (see Exhibit II).

---

[1]RMB8 = U.S.$1.

Exhibit II.   JBC targets (in RMBs)

Department	Metric	2001	2002	2003	2004
Corporate services	Broadcast advertising revenue	1 million	10 million	20 million	40 million
Data and voice services	Data and voice revenue	na*	5 million	15 million	30 million
Television & radio programming	Total programming costs	48 million	42 million	38 million	35 million
Television & radio broadcasting	General and administrative costs	48 million	42 million	38 million	35 million

*na = not available

Guang ended the meeting by announcing that JBC had been granted an additional lump sum of RMB150 million to prepare for its initial public offering (IPO), scheduled January 3, 2005.

ZHOU JIANGLIN

In 1997, Zhou joined JBC as a production assistant in the television and radio programming department. This was his first job after graduating from Tsinghua University, where he completed an undergraduate degree in electrical engineering. For his last year of university, because of his good grades and his fluency in English, Zhou was selected to be an exchange student in London, England. There, he was introduced to global television programming, and he marveled at the vast selection of programs available to Londoners. In addition, he encountered the Internet for the first time, using it to converse with contacts around the world. Convinced that his ideal career lay in the broadcast industry, Zhou relied on his network of contacts to secure an interview at JBC. A year later, he landed the job and moved to Ji'nan. During his spare time, he visited the local university to use the Internet for basic functions such as e-mail, news, and games. On May 15, 2001, Zhou was promoted to project manager. Because he was the only person in the company with both Internet and electrical engineering experience, he was assigned the task of managing the data and voice project implementation.

THE DATA AND VOICE PROJECT (DVP)

Intended to cover an area roughly 200 square kilometers, DVP could provide voice and data services to 90 percent of Shandong's businesses and inhabitants. No data services existed in the province, and telephone services—local and long-distance—were provided by China Post & Telecom.

Zhou had been informed that there existed a budget of RMB110 million to spend on this project. Han wanted this project to yield the following results by January 1, 2002:

- Capacity to provide up to 5,000 high-speed (10 megabits per second) data lines, and up to 3 million voice lines.
- Ensure that the equipment was *evergreen*—meaning that it could be used in conjunction with next generation equipment—and scalable.

JBC aimed to purchase blocks of long-distance and data capacity from the state-owned national carrier, China Post & Telecom, via China Unicom (a new national carrier), reselling it to Shangdong's businesses and general populace. With its new network, JBC would be able to service and bill customers for out-going data and voice, while incoming data and voice would continue to be handled by China Post & Telecom.

At this point, Zhou was still unsure if reselling long-distance capacity from China Post & Telecom was permitted under current government rules. Zhou was certain that Han had checked this detail—after all, no mention of this restriction had been made to Zhou. Another issue related to in-house telecommunications engineering expertise: JBC had none. Zhou wondered whether he should request a team be hired, retain an outside engineering firm, or rely on Postel. With many questions on his mind, he made his first telephone call to Postel.

EASTERN POSTEL (POSTEL)

Originally named Eastern Telecom Equipment Factory (ETEF) under the Ministry of Post & Telecommunications, Postel was one of the leading manufacturers of telecommunications equipment in China. In 1997, China Post & Telecom privatized Postel, issuing 100 million common shares on the Shenzhen Stock Exchange. In the three years following privatization, Postel had grown to 1,800 employees, including 700 senior- and middle-management staff. At the end of 2000, it had RMB1 billion in assets and income of more than RMB600 million.

Postel focused on the design, manufacturing, and marketing of telecommunications equipment, breaking its businesses into five categories:

1. Data communications
2. Wireless communications
3. Distribution equipment
4. Electrical equipment
5. Cabling systems

Postel stated that it would remain dedicated to serving China's telecommunications industry, supplying equipment to national and regional carriers. In addition, Postel exported its products to many countries, including Russia, Vietnam, Korea, Nepal, Cuba, Bangladesh, Pakistan, Colombia, Singapore and China's Hong Kong region.

A Postel Business Category: Data Communications and Transmission
To provide data services, networks required data communications and transmission equipment. Postel had a history of working with Nortel. In 1992, Postel had collaborated with Northern Telcom (Nortel's previous name) to produce DPN-100 packet switching equipment for data communications. This had resulted in Postel garnering more than 80 percent of China's packet switching network. In 1996, Postel continued to provide customers with multi-service integrated data solutions by constructing integrated information networks (internal, dedicated networks) for companies in both the broadcast and television industry and the broadband multimedia industry.

In addition, Postel had designed in-house products for data transmission, including plesiochronous digital hierarchy/synchronous digital hierarchy (PDH/SDH) equipment, PCM multiservice access equipment, E1 cross-connect equipment, converters and power distributors.

ZHOU'S FIRST MEETING WITH POSTEL

After exchanging pleasantries, Zhou requested a face-to-face meeting with Postel's network sales manager, Chin Anshang. To build the DVP, Chin explained to Zhou, Postel would have to use a combination of the latest generation of Nortel Passport data equipment, Nortel Meridian voice equipment and Postel data transmission equipment. Chin noted Zhou's other questions, promising to address them in a few days. The first move was to allow Postel's preferred supplier, Nortel, to perform a project scope. Zhou agreed, emphasizing that it was imperative they move quickly in order to meet the completion date.

THE DVP PROJECT SCOPE

Zhou was informed that a project scope was necessary before Nortel could submit a bid to provide equipment. Chin had added that he was considering allowing Alcatel and Lucent to bid, but was unsure if this would delay the project further. After all, Chin had estimated that preparing a network of the DVP's size required a lead time of eight to nine months. If Nortel was the sole supplier, Chin explained, the project would move much more smoothly than if two or three manufacturers were involved.

On June 3, 2001, Nortel's five-member team, headed by Enterprise Manager John Kang, arrived at JBC's headquarters and were allowed to inspect current equipment and gather information. They were informed that JBC had purchased Fujitsu data equipment for use between its offices. Nortel's team confirmed that the Fujitsu equipment could be integrated into the system with some engineering work. This integration required a specific connection and about two weeks of testing, with both Fujitsu and Nortel personnel present.

Nortel completed the scope in a day and presented general results to Postel and JBC. To meet JBC's requirements, Nortel would provide Passport and Meridian equipment for sixty nodes. Some of these nodes were as far as 90 kilometers from the central station. Kang added that he was confident Nortel and Postel would be able to provide the network equipment for the price at which JBC budgeted. Chin announced at the meeting that he would not ask Alcatel to perform a project scope. Instead, he would inquire about the price of voice equipment necessary to complement Nortel Passport data equipment.

Zhou's Questions

During a break in the meeting, Zhou presented his questions to Kang. Could Nortel work in conjunction with Alcatel or Lucent to complete the project? Should the network be centrally controlled or should adjustments be allowed remotely? How could JBC be assured that the project would be executed on time, as specified, and on budget?

First, Kang explained that Nortel would be willing to work with Alcatel or Lucent as long as demarcation points were identified. Demarcation points indicated where the work of one manufacturer stopped and another started. For example, a telephone service provider's demarcation point with a consumer ended at the phone jack installed into the consumer's house wall. From that point, the consumer was responsible for the function of his or her telephone equipment. Therefore, it would be possible for Nortel to provide data and Lucent to provide voice products as long as all four stakeholders (Nortel, Lucent, Postel, JBC) could agree on the demarcation point. In this scenario, Nortel would be responsible for testing its own equipment up to the agreed-upon demarcation point.

Kang continued by saying that the network could be centrally or remotely controlled. It was just a matter of whether JBC had the expertise to monitor the network remotely. Last, he assured Zhou that a project team consisting of JBC, Postel, and Nortel stakeholders would be struck, and timelines for the completion of the project would be set. In addition, Nortel could provide all the engineering support necessary to complete the project. Satisfied with these responses, Kang ended the meeting by stating that he looked forward to the start of the DVP.

REPORTING BACK TO THE MANAGING DIRECTOR

Han, Zhou's managing director, was pleased that the project was underway. He indicated to Zhou that he had consulted with his colleagues and that they continued to harbor some objections to the DVP. Han promised to keep his colleagues informed about the latest developments on the DVP, emphasizing to Zhou that this project carried significant political weight. Regarding the programming managing director's objection (that JBC had no right to provide data and voice services), Han informed Zhou that he would look into it personally. Since Shao Yangwei, managing director of corporate services, had requested that Alcatel products be used in place of Nortel products, Han asked if Zhou could develop a comparison between the rival products. Zhou diligently took notes.

A FOLLOW-UP MEETING WITH POSTEL

On June 5, 2001, Zhou had his second meeting with Postel. He received a proposal from Postel. Here was the summary page:

- *Recommended option:* Sixty nodes in broadcasting and TV centers located in various towns and villages. Trunk equipment would run standard Nortel Passport. Three first-degree nodes (Passport 15000 VSS), 528 sets, 14 second-degree nodes (Passport 7480 equipment, 148 sets), and 43 third-degree nodes (Passport 7440 equipment, 148 sets.) Each node would be equipped with a Postel 350 Ethernet exchange. 155M single module optical fiber would be applied between trunk nodes to connect relay chain. Nortel Meridian voice equipment matched with Passport throughout. Cost = RMB 109 million. Products had life spans of approximately eight years each. Engineering and installation extra. Delivery date, January 1, 2002.

- *Alternative:* Exchange Meridian voice equipment with Alcatel voice equipment. Reduces total equipment cost by RMB950,000. Note: Newest generation Alcatel voice product not yet 'Type Approved.[2]
- *Incorporating current equipment:* RMB850,000 for engineering work to link Fujitsu voice equipment to network. Reduces cost of Nortel-sourced equipment by RMB400,000.
- *Engineering and installation:* (managed by Postel) provided by Nortel, estimated at RMB22 million.
- *Routine maintenance support:* (managed by Postel) estimated at RMB15.4 million in the first year, full-cost (includes salaries, building, training, other expenses).

THE REACTION

Zhou was at a loss for words. Yes, his budget was sufficient—but only for the equipment. He knew that he would have to explain the current situation to Han. He began to prepare his response.

The Richard Ivey School of Business gratefully acknowledges the generous support of Canada Life in the development of these learning materials.

[2]Type approval allowed a company to import equipment into a country. The only way to get approval for new products was to pay the testing fees and submit the new equipment to a government testing facility for compatibility. Depending on the nature of the tests required, this process could take up to three months.

Part 4

PROJECT MANAGEMENT ORGANIZATIONAL STRUCTURES

In the early days of project management, there existed a common belief that project management had to be accompanied by organizational restructuring. Project management practitioners argued that some organizational structures, such as a matrix structure, were more conducive to good project management, while others were not quite effective. Every organizational structure comes with both advantages and disadvantages.

Today, we question whether organizational restructuring is necessary. Is it possible that project management can be implemented effectively in any organizational structure if we have a cooperative culture? Restructuring is often accompanied by a shift in authority and the balance of power. Can effective project management occur at the same time that the organization undergoes restructuring?

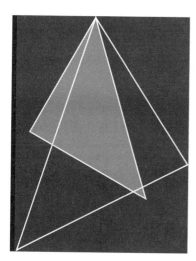

Quasar Communications, Inc.

Quasar Communications, Inc. (QCI), is a thirty-year-old, $350 million division of Communication Systems International, the world's largest communications company. QCI employs about 340 people of which more than 200 are engineers. Ever since the company was founded thirty years ago, engineers have held every major position within the company, including president and vice president. The vice president for accounting and finance, for example, has an electrical engineering degree from Purdue and a master's degree in business administration from Harvard.

QCI, up until 1996, was a traditional organization where everything flowed up and down. In 1996, QCI hired a major consulting company to come in and train *all* of their personnel in project management. Because of the reluctance of the line managers to accept formalized project management, QCI adopted an informal, fragmented project management structure where the project managers had lots of responsibility but very little authority. The line managers were still running the show.

In 1999, QCI had grown to a point where the majority of their business base revolved around twelve large customers and thirty to forty small customers. The time had come to create a separate line organization for project managers, where each individual could be shown a career path in the company and the company could benefit by creating a body of planners and managers dedicated to the completion of a project. The project management group was headed up by a vice president and included the following full-time personnel:

- Four individuals to handle the twelve large customers
- Five individuals for the thirty to forty small customers
- Three individuals for R&D projects
- One individual for capital equipment projects

The nine customer project managers were expected to handle two to three projects at one time if necessary. Because the customer requests usually did not come in at the same time, it was anticipated that each project manager would handle only one project at a time. The R&D and capital equipment project managers were expected to handle several projects at once.

In addition to the above personnel, the company also maintained a staff of four product managers who controlled the profitable off-the-shelf product lines. The product managers reported to the vice president of marketing and sales.

In October 1999, the vice president for project management decided to take a more active role in the problems that project managers were having and held counseling sessions for each project manager. The following major problem areas were discovered.

R&D PROJECT MANAGEMENT

Project manager: "My biggest problem is working with these diverse groups that aren't sure what they want. My job is to develop new products that can be introduced into the marketplace. I have to work with engineering, marketing, product management, manufacturing, quality assurance, finance, and accounting. Everyone wants a detailed schedule and product cost breakdown. How can I do that when we aren't even sure what the end-item will look like or what materials are needed? Last month I prepared a detailed schedule for the development of a new product, assuming that everything would go according to the plan. I worked with the R&D engineering group to establish what we considered to be a realistic milestone. Marketing pushed the milestone to the left because they wanted the product to be introduced into the marketplace earlier. Manufacturing then pushed the milestone to the right, claiming that they would need more time to verify the engineering specifications. Finance and accounting then pushed the milestone to the left asserting that management wanted a quicker return on investment. Now, how can I make all of the groups happy?"

Vice president: "Whom do you have the biggest problems with?"

Project manager: "That's easy—marketing! Every week marketing gets a copy of the project status report and decides whether to cancel the project. Several times marketing has canceled projects without even discussing it with me, and I'm supposed to be the project leader."

Vice president: "Marketing is in the best position to cancel the project because they have the inside information on the profitability, risk, return on investment, and competitive environment."

Project manager: "The situation that we're in now makes it impossible for the project manager to be dedicated to a project where he does not have all of the information at hand. Perhaps we should either have the R&D project managers report to someone in marketing or have the marketing group provide additional information to the project managers."

SMALL CUSTOMER PROJECT MANAGEMENT

Project manager: "I find it virtually impossible to be dedicated to and effectively manage three projects that have priorities that are not reasonably close. My low-priority customer always suffers. And even if I try to give all of my customers equal status, I do not know how to organize myself and have effective time management on several projects."

Project manager: "Why is it that the big projects carry all of the weight and the smaller ones suffer?"

Project manager: "Several of my projects are so small that they stay in one functional department. When that happens, the line manager feels that he is the true project manager operating in a vertical environment. On one of my projects I found that a line manager had promised the customer that additional tests would be run. This additional testing was not priced out as part of the original statement of work. On another project the line manager made certain remarks about the technical requirements of the project. The customer assumed that the line managers's remarks reflected company policy. Our line managers don't realize that only the project manager can make commitments (on resources) to the customer as well as on company policy. I know this can happen on large projects as well, but it is more pronounced on small projects."

LARGE CUSTOMER PROJECT MANAGEMENT

Project manager: "Those of us who manage the large projects are also marketing personnel, and occasionally, we are the ones who bring in the work. Yet, everyone appears to be our superior. Marketing always looks down on us, and when we bring in a large contract, marketing just looks down on us as if we're riding their coattails or as if we were just lucky. The engineering group outranks us because all managers and executives are promoted from there. Those guys never live up to commitments. Last month I sent an inflammatory memo to a line manager because of his poor response to my requests. Now, I get no support at all from him. This doesn't happen all of the time, but when it does, it's frustrating."

Project manager: "On large projects, how do we, the project managers, know when the project is in trouble? How do we decide when the project will fail? Some of our large projects are total disasters and should fail, but management comes to the rescue and pulls the best resources off of the good projects to cure the ailing projects. We then end up with six marginal projects and one partial catastrophe as opposed to six excellent projects and one failure. Why don't we just let the bad projects fail?"

Vice president: "We have to keep up our image for our customers. In most other companies, performance is sacrificed in order to meet time and cost. Here at QCI, with our professional integrity at stake, our engineers are willing to sacrifice time and cost in order to meet specifications. Several of our customers come to us because of this. Last year we had a project where, at the scheduled project termination date, engineering was able to satisfy only 75 percent of the customer's performance specifications. The project manager showed the results to the customer, and the customer decided to change his specification requirements to agree with the product that we designed. Our engineering people thought that this was a 'slap in the face' and refused to sign off the engineering drawings. The problem went all the way up to the president for resolution. The final result was that the customer would give us an additional few months if we would spend our own money to try to meet the original specification. It cost us a bundle, but we did it because our integrity and professional reputation were at stake."

CAPITAL EQUIPMENT PROJECT MANAGEMENT

Project manager: "My biggest complaint is with this new priority scheduling computer package we're supposedly considering to install. The way I understand it, the computer program will establish priorities for *all* of the projects in-house, based on the feasibility study, cost-benefit analysis, and return on investment. Somehow I feel as though my projects will always be the lowest priority, and I'll never be able to get sufficient functional resources."

Project manager: "Every time I lay out a reasonable schedule for one of our capital equipment projects, a problem occurs in the manufacturing area and the functional employees are always pulled off of my project to assist manufacturing.

And now I have to explain to everyone why I'm behind schedule. Why am I always the one to suffer?"

The vice president carefully weighed the remarks of his project managers. Now came the difficult part. What, if anything, could the vice president do to amend the situation given the current organizational environment?

Jones and Shephard Accountants, Inc.[1]

By 1990, Jones and Shephard Accountants, Inc. (J&S) was ranked a midsized company in size by the American Association of Accountants. In order to compete with the larger firms, J&S formed an Information Services Division designed primarily for studies and analyses. By 1995, the Information Services Division (ISD) had fifteen employees.

In 1997, the ISD purchased three minicomputers. With this increased capacity, J&S expanded its services to help satisfy the needs of outside customers. By September 1998, the internal and external work loads had increased to a point where the ISD now employed over fifty people.

The director of the division was very disappointed in the way that activities were being handled. There was no single person assigned to push through a project, and outside customers did not know who to call to get answers regarding project status. The director found that most of his time was being spent on day-to-day activities such as conflict resolution instead of strategic planning and policy formulation.

The biggest problems facing the director were the two continuous internal projects (called Project X and Project Y, for simplicity) that required month-end data collation and reporting. The director felt that these two projects were important enough to require a full-time project manager on each effort.

---

[1]Revised 2007.

In October 1998, corporate management announced that the ISD director would be reassigned on February 1, 1999, and that the announcement of his replacement would not be made until the middle of January. The same week that the announcement was made, two individuals were hired from outside the company to take charge of Project X and Project Y. Exhibit I shows the organizational structure of the ISD.

Within the next thirty days, rumors spread throughout the organization about who would become the new director. Most people felt that the position would be filled from within the division and that the most likely candidates would be the two new project managers. In addition, the associate director was due to retire in December, thus creating two openings.

On January 3, 1999, a confidential meeting was held between the ISD director and the systems manager.

ISD director: "Corporate has approved my request to promote you to division director. Unfortunately, your job will not be an easy one. You're going to have to restructure the organization somehow so that our employees will not have as many conflicts as they are now faced with. My secretary is typing up a confidential memo for you explaining my observations on the problems within our division.

"Remember, your promotion should be held in the strictest confidence until the final announcement later this month. I'm telling you this now so that you can

Exhibit I.   ISD organizational chart

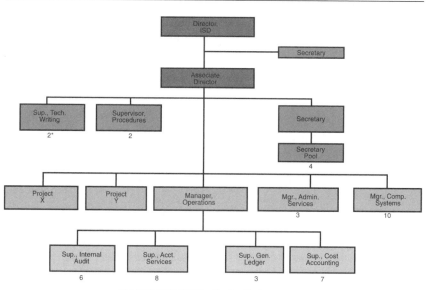

*Denotes The Number of Additional Functional Employees

begin planning the restructuring. My memo should help you." (See Exhibit II for the memo.)

The systems manager read the memo and, after due consideration, decided that some form of matrix would be best. To help him structure the organization properly, an outside consultant was hired to help identify the potential problems with changing over to a matrix. Six problem areas were identified by the consultant:

1. The operations manager controls more than 50 percent of the people resources. You might want to break up his empire. This will have to be done very carefully.
2. The secretary pool is placed too high in the organization.
3. The supervisors who now report to the associate director will have to be

---

Exhibit II. Confidential memo

From: ISD Director
To: Systems Manager
Date: January 3, 1999

Congratulations on your promotion to division director. I sincerely hope that your tenure will be productive both personally and for corporate. I have prepared a short list of the major obstacles that you will have to consider when you take over the controls.

1. Both Project X and Project Y managers are highly competent individuals. In the last four or five days, however, they have appeared to create more conflicts for us than we had previously. This could be my fault for not delegating them sufficient authority, or could be a result of the fact that several of our people consider these two individuals as prime candidates for my position. In addition, the operations manager does not like other managers coming into his "empire" and giving direction.
2. I'm not sure that we even need an associate director. That decision will be up to you.
3. Corporate has been very displeased with our inability to work with outside customers. You must consider this problem with any organizational structure you choose.
4. The corporate strategic plan for our division contains an increased emphasis on special, internal MIS projects. Corporate wants to limit our external activities for a while until we get our internal affairs in order.
5. I made the mistake of changing our organizational structure on a day-to-day basis. Perhaps it would have been better to design a structure that could satisfy advanced needs, especially one that we can grow into.

---

reassigned lower in the organization if the associate director's position is abolished.

4. One of the major problem areas will be trying to convince corporate management that their change will be beneficial. You'll have to convince them that this change can be accomplished without having to increase division manpower.

5. You might wish to set up a separate department or a separate project for customer relations.

6. Introducing your employees to the matrix will be a problem. Each employee will look at the change differently. Most people have the tendency of looking first at the shift in the balance of power—have I gained or have I lost power and status?

The systems manager evaluated the consultant's comments and then prepared a list of questions to ask the consultant at their next meeting.

QUESTIONS

1. What should the new organizational structure look like? Where should I put each person, specifically the managers?

2. When should I announce the new organizational change? Should it be at the same time as my appointment or at a later date?

3. Should I invite any of my people to provide input to the organizational restructuring? Can this be used as a technique to ease power plays?

4. Should I provide inside or outside seminars to train my people for the new organizational structure? How soon should they be held?

Fargo Foods[1]

Fargo Foods is a $2 billion a year international food manufacturer with canning facilities in twenty-two countries. Fargo products include meats, poultry, fish, vegetables, vitamins, and cat and dog foods. Fargo Foods has enjoyed a 12.5 percent growth rate each of the past eight years primarily due to the low overhead rates in the foreign companies.

During the past five years, Fargo had spent a large portion of retained earnings on capital equipment projects in order to increase productivity without increasing labor. An average of three new production plants have been constructed in each of the last five years. In addition, almost every plant has undergone major modifications each year in order to increase productivity.

In 2000, the president of Fargo Foods implemented formal project management for all construction projects using a matrix. By 2004, it became obvious that the matrix was not operating effectively or efficiently. In December 2004, the author consulted for Fargo Foods by interviewing several of the key managers and a multitude of functional personnel. What follows are the several key questions and responses addressed to Fargo Foods:

---

[1]Disguised case.

Q. Give me an example of one of your projects.

A. "The project begins with an idea. The idea can originate anywhere in the company. The planning group picks up the idea and determines the feasibility. The planning group then works 'informally' with the various line organizations to determine rough estimates for time and cost. The results are then fed back to the planning group and to the top management planning and steering committees. If top management decides to undertake the project, then top management selects the project manager and we're off and running."

Q. Do you have any problems with this arrangement?

A. "You bet! Our executives have the tendency of equating rough estimates as detailed budgets and rough schedules as detailed schedules. Then, they want to know why the line managers won't commit their best resources. We almost always end up with cost overruns and schedule slippages. To make matters even worse, the project managers do not appear to be dedicated to the projects. I really can't blame them. After all, they're not involved in planning the project, laying out the schedule, and establishing the budget. I don't see how any project manager can become dedicated to a plan in which the project manager has no input and may not even know the assumptions or considerations that were included. Recently, some of our more experienced project managers have taken a stand on this and are virtually refusing to accept a project assignment unless they can do their own detailed planning at the beginning of the project in order to verify the constraints established by the planning group. If the project managers come up with different costs and schedules (and you know that they will), the planning group feels that they have just gotten slapped in the face. If the costs and schedules are the same, then the planning group runs upstairs to top management asserting that the project managers are wasting money by continuously wanting to replan."

Q. Do you feel that replanning is necessary?

A. "Definitely! The planning group begins their planning with a very crude statement of work, expecting our line managers (the true experts) to read in between the lines and fill in the details. The project managers develop a detailed statement of work and a work breakdown structure, thus minimizing the chance that anything would fall through the crack. Another reason for replanning is that the ground rules have changed between the time that the project was originally adopted by the planning group and the time that the project begins implementation. Another possibility, of course, is that technology may have changed or people can be smarter now and can perform at a higher position on the learning curve."

Q. Do you have any problems with executive meddling?

A. "Not during the project, but initially. Sometimes executives want to keep the end date fixed but take their time in approving the project. As a result, the project manager may find himself a month or two behind scheduling before he even begins the project. The second problem is when the executive decides to arbitrarily change the end date milestone but keep the front end milestone fixed. On one of our projects it was necessary to complete the project in half the time. Our line managers worked like dogs to get the job done. On the next project, the same thing happened, and, once again, the line managers came to the rescue. Now, management feels that line managers cannot make good estimates and that they (the executives) can arbitrarily change the milestones on any project. I wish that they would realize what they're doing to us. When we put forth all of our efforts on one project, then all of the other projects suffer. I don't think our executives realize this."

Q. Do you have any problems selecting good project managers and project engineers?

A. "We made a terrible mistake for several years by selecting our best technical experts as the project managers. Today, our project managers are doers, not managers. The project managers do not appear to have any confidence in our line people and often try to do all of the work themselves. Functional employees are taking technical direction from the project managers and project engineers instead of the line managers. I've heard one functional employee say, 'Here come those project managers again to beat me up. Why can't they leave me alone and let me do my job?' Our line employees now feel that this is the way that project management is supposed to work. Somehow, I don't think so."

Q. Do you have any problems with the line manager–project manager interface?

A. "Our project managers are technical experts and therefore feel qualified to do all of the engineering estimates without consulting with the line managers. Sometimes this occurs because not enough time or money is allocated for proper estimating. This is understandable. But when the project managers have enough time and money and refuse to get off their ivory towers and talk to the line managers, then the line managers will always find fault with the project manager's estimate even if it is correct. Sometimes I just can't feel any sympathy for the project managers. There is one special case that I should mention. Many of our project managers do the estimating themselves but have courtesy enough to ask the line manager for his blessing. I've seen line managers who were so loaded with work that they look the estimate over for two seconds and say, 'It looks fine to me. Let's do it.' Then when the cost overrun appears, the project manager gets blamed."

Q. Where are your project engineers located in the organization?

A. "We're having trouble deciding that. Our project engineers are primarily responsible for coordinating the design efforts (i.e., electrical, civil, HVAC, etc).

The design manager wants these people reporting to him if they are responsible for coordinating efforts in his shop. The design manager wants control of these people even if they have their name changed to assistant project managers. The project managers, on the other hand, want the project engineers to report to them with the argument that they must be dedicated to the project and must be willing to complete the effort within time, cost, and performance. Furthermore, the project managers argue that project engineers will be more likely to get the job done within the constraints if they are not under the pressure of being evaluated by the design manager. If I were the design manager, I would be a little reluctant to let someone from outside of my shop integrate activities that utilize the resources under my control. But I guess this gets back to interpersonal skills and the attitudes of the people. I do not want to see a brick wall set up between project management and design."

Q. I understand that you've created a new estimating group. Why was that done?

A. "In the past we have had several different types of estimates such as first guess, detailed, 10 percent complete, etc. Our project managers are usually the first people at the job site and give a shoot-from-the-hip estimate. Our line managers do estimating as do some of our executives and functional employees. Because we're in a relatively slowly changing environment, we should have well-established standards, and the estimating department can maintain uniformity in our estimating policies. Since most of our work is approved based on first-guess estimates, the question is, 'Who should give the first-guess estimate?' Should it be the estimator, who does not understand the processes but knows the estimating criteria, or the project engineer, who understands the processes but does not know the estimates, or the project manager, who is an expert in project management? Right now, we are not sure where to place the estimating group. The vice president of engineering has three operating groups beneath him—project management, design, and procurement. We're contemplating putting estimating under procurement, but I'm not sure how this will work."

Q. How can we resolve these problems that you've mentioned?

A. "I wish I knew!"

Government Project Management

A major government agency is organized to monitor government subcontractors as shown in Exhibit I. Below are the vital characteristics of certain project office team members:

- *Project manager:* Directs all project activities and acts as the information focal point for the subcontractor.
- *Assistant project manager:* Acts as chairman of the steering committee and interfaces with both in-house functional groups and contractor.
- *Department managers:* Act as members of the steering committee for any projects that utilize their resources. These slots on the steering committee must be filled by the department managers themselves, not by functional employees.
- *Contracts officer:* Authorizes all work directed by the project office to in-house functional groups and to the customer, and ensures that all work requested is authorized by the contract. The contracts officer acts as the focal point for all contractor cost and contractual information.

1. Explain how this structure *should* work.
2. Explain how this structure *actually* works.
3. Can the project manager be a military type who is reassigned after a given tour of duty?
4. What are the advantages and disadvantages of this structure?
5. Could this be used in industry?

Exhibit I. ***Project team organizational structure***

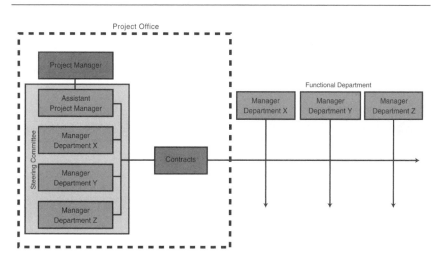

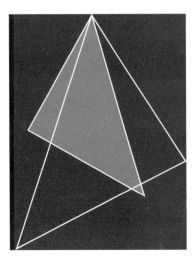

Falls Engineering

Located in New York, Falls Engineering is a $250-million chemical and materials operation employing 900 people. The plant has two distinct manufacturing product lines: industrial chemicals and computer materials. Both divisions are controlled by one plant manager, but direction, strategic planning, and priorities are established by corporate vice presidents in Chicago. Each division has its own corporate vice president, list of projects, list of priorities, and manpower control. The chemical division has been at this location for the past twenty years. The materials division is, you might say, the tenant in the landlord–tenant relationship, with the materials division manager reporting dotted to the plant manager and solid to the corporate vice president (see Exhibit I).

The chemical division employed 3,000 people in 1998. By 2003, there were only 600 employees. In 2004, the materials division was formed and located on the chemical division site with a landlord–tenant relationship. The materials division has grown from $50 million in 2000 to $120 million in 2004. Today, the materials division employs 350 people.

All projects originate in construction or engineering but usually are designed to support production. The engineering and construction departments have projects that span the entire organization directed by a project coordinator. The project coordinator is a line employee who is temporarily assigned to coordinate a project in his line organization in addition to performing his line responsibilities. Assignments are made by the division managers (who report to the plant manager)

Exhibit I. Falls Engineering organizational chart

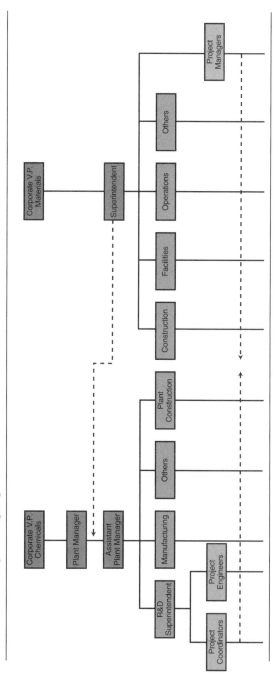

and are based on technical expertise. The coordinators have monitoring authority only and are not noted for being good planners or negotiators. The coordinators report to their respective line managers.

Basically, a project can start in either division with the project coordinators. The coordinators draw up a large scope of work and submit it to the project engineering group, who arrange for design contractors, depending on the size of the project. Project engineering places it on their design schedule according to priority and produces prints and specifications, and receives quotes. A construction cost estimate is then produced following 60–75 percent design completion. The estimate and project papers are prepared, and the project is circulated through the plant and in Chicago for approval and authorization. Following authorization, the design is completed, and materials are ordered. Following design, the project is transferred to either of two plant construction groups for construction. The project coordinators than arrange for the work to be accomplished in their areas with minimum interference from manufacturing forces. In all cases, the coordinators act as project managers and must take the usual constraints of time, money, and performance into account.

Falls Engineering has 300 projects listed for completion between 2006 and 2008. In the last two years, less than 10 percent of the projects were completed within time, cost, and performance constraints. Line managers find it increasingly difficult to make resource commitments because crises always seem to develop, including a number of fires.

Profits are made in manufacturing, and everyone knows it. Whenever a manufacturing crisis occurs, line managers pull resources off the projects, and, of course, the projects suffer. Project coordinators are trying, but with very little success, to put some slack onto the schedules to allow for contingencies.

The breakdown of the 300 plant projects is shown below:

Number of Projects	$ Range
120	less than $50,000
80	50,000–200,000
70	250,000–750,000
20	1–3 million
10	4–8 million

Corporate realized the necessity for changing the organizational structure. A meeting was set up between the plant manager, plant executives, and corporate executives to resolve these problems once and for all. The plant manager decided to survey his employees concerning their feelings about the present organizational structure. Below are their comments:

- "The projects we have the most trouble with are the small ones under $200,000. Can we use informal project management for the small ones and formal project management on the large ones?"

- Why do we persist in using computer programming to control our resources? These sophisticated packages are useless because they do not account for firefighting."
- "Project coordinators need access to various levels of management, in both divisions."
- "Our line managers do not realize the necessity for effective planning of resources. Resources are assigned based on emotions and not need."
- "Sometimes a line manager gives a commitment but the project coordinator cannot force him to keep it."
- "Line managers always find fault with project coordinators who try to develop detailed schedules themselves."
- "If we continuously have to 'crash' project time, doesn't that indicate poor planning?"
- "We need a career path in project coordination so that we can develop a body of good planners, communicators, and integrators."
- "I've seen project coordinators we have no interest in the job, cannot work with diverse functional disciplines, and cannot communicate. Yet, someone assigned them as a project coordinator."
- "Any organizational system we come up with has to be better than the one we have now."
- "Somebody has to have total accountability. Our people are working on projects and, at the same time, do not know the project status, the current cost, the risks, and the end date."
- "One of these days I'm going to kill an executive while he's meddling in my project."
- "Recently, management made changes requiring more paperwork for the project coordinators. How many hours a week do they expect me to work?"
- "I've yet to see any documentation describing the job description of the project coordinator."
- "I have absolutely no knowledge about who is assigned as the project coordinator until work has to be coordinated in my group. Somehow, I'm not sure that this is the way the system should work."
- "I know that we line managers are supposed to be flexible, but changing the priorities every week isn't exactly my idea of fun."
- "If the projects start out with poor planning, then management does not have the right to expect the line managers always to come to the rescue."
- "Why is it the line managers always get blamed for schedule delays, even if it's the result of poor planning up front?"
- "If management doesn't want to hire additional resources, then why should the line managers be made to suffer? Perhaps we should cut out some of these useless projects. Sometimes I think management dreams up some of these projects simply to spend the allocated funds."
- "I have yet to see a project I felt had a realistic deadline."

After preparing alternatives and recommendations as plant manager, try to do some role playing by putting yourself in the shoes of the corporate executives. Would you, as a corporate executive, approve the recommendation? Where does profitability, sales, return on investment, and so on enter in your decision?

White
Manufacturing

In 2004, White Manufacturing realized the necessity for project management in the manufacturing group. A three-man project management staff was formed. Although the staff was shown on the organizational chart as reporting to the manufacturing operations manager, they actually worked for the vice president and had sufficient authority to integrate work across all departments and divisions. As in the past, the vice president's position was filled by the manufacturing operations manager. Manufacturing operations was directed by the former manufacturing manager who came from manufacturing engineering (see Exhibit I).

In 2007, the manufacturing manager created a matrix in the manufacturing department with the manufacturing engineers acting as departmental project managers. This benefited both the manufacturing manager and the group project managers since all information could be obtained from one source. Work was flowing very smoothly.

In January 2008, the manufacturing manager resigned his position effective March, and the manufacturing engineering manager began packing his bags ready to move up to the vacated position. In February, the vice president announced that the position would be filled from outside. He said also that there would be an organizational restructuring and that the three project managers would now be staff to the manufacturing manager. When the three project managers confronted the manufacturing operations manager, he said, "We've hired the new man in at a very high salary. In order to justify this salary, we have to give him more responsibility."

Exhibit I.   White Manufacturing organizational structure

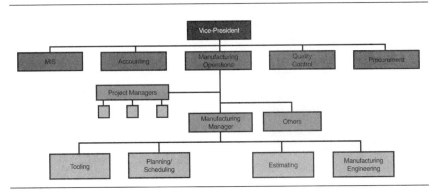

In March 2008, the new manager took over and immediately made two declarations:

1. The project managers will never go "upstairs" without first going through him.
2. The departmental matrix will be dissolved and the department manager will handle all of the integration.

QUESTIONS

1. How do you account for the actions of the new department manager?
2. What would you do if you were one of the project managers?

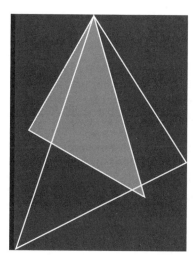

Martig Construction Company

Martig Construction was a family-owned mechanical subcontractor business that had grown from $5 million in 2006 to $25 million in 2008. Although the gross profit had increased sharply, the profit as a percentage of sales declined drastically. The question was, "Why the decline?" The following observations were made:

1. Since Martig senior died in July of 2008, Martig junior has tried unsuccessfully to convince the family to let him sell the business. Martig junior, as company president, has taken an average of eight days of vacation per month for the past year. Although the project managers are supposed to report to Martig, they appear to be calling their own shots and are in a continuous struggle for power.

2. The estimating department consists of one man, John, who estimates all jobs. Martig wins one job in seven. Once a job is won, a project manager is selected and is told that he must perform the job within the proposal estimates. Project managers are not involved in proposal estimates. They are required, however, to provide feedback to the estimator so that standards can be updated. This very seldom happens because of the struggle for power. The project managers are afraid that the estimator might be next in line for executive promotion since he is a good friend of Martig.

3. The procurement function reports to Martig. Once the items are ordered, the project manager assumes procurement responsibility. Several times in

the past, the project manager has been forced to spend hour after hour trying to overcome shortages or simply to track down raw materials. Most project managers estimate that approximately 35 percent of their time involves procurement.

4. Site superintendents believe they are the true project managers, or at least at the same level. The superintendents are very unhappy about not being involved in the procurement function and, therefore, look for ways to annoy the project managers. It appears that the more time the project manager spends at the site, the longer the work takes; the feedback of information to the home office is also distorted.

Mohawk
National Bank

"You're really going to have your work cut out for you, Randy," remarked Pat Coleman, vice president for operations. "It's not going to be easy establishing a project management organizational structure on top of our traditional structure. We're going to have to absorb the lumps and bruises and literally 'force' the system to work."

BACKGROUND

Between 1988 and 1998, Mohawk National matured into one of Maine's largest full-service banks, employing a full-time staff of some 1,200 employees. Of the 1,200 employees, approximately 700 were located in the main offices in downtown Augusta.

Mohawk matured along with other banks in the establishment of computerized information processing and decision-making. Mohawk leased the most up-to-date computer equipment in order to satisfy customer demands. By 1994, almost all departments were utilizing the computer.

By 1995, the bureaucracy of the traditional management structure was creating severe administrative problems. Mohawk's management had established many complex projects to be pursued, each one requiring the involvement of several departments. Each department manager was setting his or her own priorities for the work that had to be performed. The traditional organization was too weak structurally to handle problems that required integration across multiple departments. Work from department to department could not be tracked because there was no project manager who could act as focal point for the integration of work.

UNDERSTANDING THE CHANGEOVER PROBLEM

It was a difficult decision for Mohawk National to consider a new organizational structure, such as a matrix. Randy Gardner, director of personnel, commented on the decision:

> Banks, in general, thrive on traditionalism and regimentation. When a person accepts a position in our bank, he or she understands the strict rules, policies, and procedures that have been established during the last 30 years.
>
> We know that it's not going to be easy. We've tried to anticipate the problems that we're going to have. I've spent a great deal of time with our vice president of operations and two consultants trying to predict the actions of our employees.
>
> The first major problem we see is with our department managers. In most traditional organizations, the biggest functional department emerges as the strongest. In a matrix organization, or almost any other project form for that matter, there is a shift in the balance of power. Some managers become more important in their new roles and others not so important. We think our department managers are good workers and that they will be able to adapt.
>
> Our biggest concern is with the functional employees. Many of our functional people have been with us between twenty and thirty years. They're seasoned veterans. You must know that they're going to resist change. These people will fight us all the way. They won't accept the new system until they see it work. That'll be our biggest challenge: to convince the functional team members that the system will work.

Pat Coleman, the vice president for operations, commented on the problems that he would be facing with the new structure:

> Under the new structure, all project managers will be reporting to me. To be truthful, I'm a little scared. This changeover is like a project in itself. As with any project, the beginning is the most important phase. If the project starts out on the right track, people might give it a chance. But if we have trouble, people will be quick to revert back to the old system. Our people hate change. We cannot wait one and a half to two years for people to get

familiar with the new system. We have to hit them all at once and then go all out to convince them of the possibilities that can be achieved.

This presents a problem in that the first group of project managers must be highly capable individuals with the ability to motivate the functional team members. I'm still not sure whether we should promote from within or hire from the outside. Hiring from the outside may cause severe problems in that our employees like to work with people they know and trust. Outside people may not know our people. If they make a mistake and aggravate our people, the system will be doomed to failure.

Promoting from within is the only logical way to go, as long as we can find qualified personnel. I would prefer to take the qualified individuals and give them a lateral promotion to a project management position. These people would be on trial for about six months. If they perform well, they will be promoted and permanently assigned to project management. If they can't perform or have trouble enduring the pressure, they'll be returned to their former functional positions. I sure hope we don't have any inter- or intramatrix power struggles.

Implementation of the new organizational form will require good communications channels. We must provide all of our people with complete and timely information. I plan on holding weekly meetings with all of the project and functional managers. Good communications channels must be established between all resource managers. These team meetings will give people a chance to see each other's mistakes. They should be able to resolve their own problems and conflicts. I'll be there if they need me. I do anticipate several conflicts because our functional managers are not going to be happy in the role of a support group for a project manager. That's the balance of power problem I mentioned previously.

I have asked Randy Gardner to identify from within our ranks the four most likely individuals who would make good project managers and drive the projects to success. I expect Randy's report to be quite positive. His report will be available next week.

Two weeks later, Randy Gardner presented his report to Pat Coleman and made the following observations:

I have interviewed the four most competent employees who would be suitable for project management. The following results were found:

Andrew Medina, department manager for cost accounting, stated that he would refuse a promotion to project management. He has been in cost accounting for twenty years and does not want to make a change into a new career field.

Larry Foster, special assistant to the vice president of commercial loans, stated that he enjoyed the people he was working with and was afraid that a new job in project management would cause him to lose his contacts with upper level management. Larry considers his present position more powerful than any project management position.

Chuck Folson, personal loan officer stated that in the fifteen years he's been with Mohawk National, he has built up strong interpersonal ties with many members of the bank. He enjoys being an active member of the informal organization and does not believe in the applications of project management for our bank.

Jane Pauley, assistant credit manager, stated that she would like the position, but would need time to study up on project management. She feels a little unsure about herself. She's worried about the cost of failure.

Now Pat Coleman had a problem. Should he look for other bank employees who might be suitable to staff the project management functions or should he look externally to other industries for consultants and experienced project managers?

QUESTIONS

1. How do you implement change in a bank?
2. What are some of the major reasons why employees do not want to become project managers?
3. Should the first group of project managers be laterally assigned?
4. Should the need for project management first be identified from within the organization?
5. Can project management be forced upon an organization?
6. Does the bank appear to understand project management?
7. Should you start out with permanent or temporary project management positions?
8. Should the first group of project managers be found from within the organization?
9. Will people be inclined to support the matrix if they see that the project managers are promoted from within?
10. Suppose that the bank goes to a matrix, but without the support of top management. Will the system fail?
11. How do you feel about in-house workshops to soften the impact of project management?

Part 5

NEGOTIATING FOR RESOURCES

In most organizations, project management is viewed as multiple-boss reporting. It is possible for the employees to report to one line manager and several project managers at the same time. This multiple boss reporting problem can greatly influence the way that the project manager negotiates for resources. Project managers must understand the skill level needed to perform the work, whether the resource would be needed on a part-time or full-time basis, and the duration of the effort for this worker.

Some people argue that today's project managers no longer have a command of technology but possess more of an understanding of technology. If this is, in fact, the case, then the project managers might be better off negotiating for deliverables than for people. The argument is whether a project manager should manage people or manage deliverables.

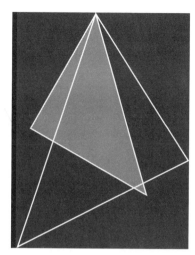

Ducor Chemical

In the fall of 2000, Ducor Chemical received a research and development contract from one of their most important clients. The client had awarded Ducor with a twelve-month, sole-source contract for the R&D effort to create a new chemical that the client required for one of its future products. If Ducor could develop the product, the long-term production contract that would follow could generate significant profits over the next several years.

In addition to various lab personnel who would be used as needed, the contract mandated that a senior chemist be assigned for the duration of the project. In the past, the senior chemists had been used mainly for internal rather than external customer projects. This would be the first time a senior chemist had been assigned to this client. With only four senior chemists on staff, the project manager expected the resource negotiation process with the lab manager to be an easy undertaking.

Project manager: "I understand you've already looked over the technical requirements, so you should understand the necessity for assigning your best senior scientist."

Lab manager: "All of my senior scientists are good. Any one of them can do the job. Based upon the timing of your project, I have decided to assign John Thornton."

Project manager: "Just my luck! You assigned the only one I cannot work with effectively. I have had the misfortune of working with him before. He's extremely arrogant and unpleasant to work with."

Lab manager: "Perhaps so, but he got the job done, didn't he?"

Project manager: "Yes, he did. Technically, he is capable. However, his arrogant attitude and sarcasm produced a demoralizing atmosphere for my team. That project was about three months in length. This project is at least a year. Also, if follow-on work is generated, as I expect it to be, I'll be stuck with him for a long time. That's unacceptable to me."

Lab manager: "I'll talk to John and see if I can put a gag in his mouth. Anyway, you're a good project manager and you should know how to work with these technical and scientific prima donnas."

Project manager: "I'll never be able to maintain my sanity having to work with him full-time for at least one year. Surely you can assign one of the other three senior chemists instead."

Lab manager: "Because of the nature of the other projects I have, John is the only senior chemist I can release for one full year. If your project were two or three months, then I might be able to give you one of the other senior chemists."

Project manager: "I feel like you are dumping Thornton on me without considering what is in the best interest of the project. Perhaps we should have the sponsor resolve this conflict."

Lab manager: "First of all, this is not a conflict. Second, threatening me with sponsor intervention will not help your case. Do you plan on asking for my resources or support ever again in the future? I'm like an elephant. I have a long memory. Third, my responsibility is to meet your deliverable in a manner that is in the best interest of the company.

"Try to look at resource assignments through my eyes. You're worried about the best interest of your project. I have to support some twenty projects and must make decisions in the best interest of the entire company. Benefiting one project at the expense of several other projects is not a good company decision. And I am paid to make sound *company* decisions, whereas you are paid to make a *project* decision."

Project manager: "My salary, promotion, and future opportunities rest solely on the success of this one project, not twenty."

Lab manager: "Our relationship must be a partnership based upon trust if project management is to succeed. You must trust me when I tell you that your deliverables will be accomplished within time, cost, and quality. It's my job to make that promise and to see that it is kept."

Project manager: "But what about morale? That should also be a factor. There is also another important consideration. The customer wants monthly team meetings, at our location, to assess progress."

Lab manager: "I know that. I read the requirements document. Why are the monthly meetings a problem?"

Project manager: "I have worked with this customer before. At the team meetings, they want to hear the technical status from the people doing the work rather than from the project manager. That means that John Thornton would be directly interfacing with the customer at least once a month. Thornton is a 'loose cannon,' and there's no telling what words will come out of his mouth. If it were not for the interface meetings, I might be agreeable to accept Thornton. But based upon previous experience, he simply does not know when to shut up! He could cause irrevocable damage to our project."

Lab manager: "I will take care of John Thornton. And to appease you, I will also attend each one of the customer interface meetings to keep Thornton in line. As far as I'm concerned, Thornton will be assigned and the subject is officially closed!"

THE PROJECT CONTINUED . . .

John Thornton was assigned to the project team. During the second interface meeting, Thornton stood up and complained to the customer that some of the tests that the customer had requested were worthless, serving no viable purpose. Furthermore, Thornton asserted that if he were left alone, he could develop a product far superior to what the customer had requested.

The customer was furious over Thornton's remarks and asserted that they would now evaluate the project performance to date, as well as Ducor's commitment to the project. After the evaluation they would consider whether the project should be terminated, or perhaps assigned to one of Ducor's competitors. The lab manager had not been present during either of the first two customer interface meetings. ·

QUESTIONS

1. How do we create a partnership between the project manager and line managers when project manager focuses only on the best interest of his/her project and the line manager is expected to make impartial company decisions?

2. Who should have more of a say during negotiations for resources: the project manager or the line manager?
3. How should irresolvable conflicts over staffing between the project and line managers be handled?
4. Should an external customer have a say in project staffing?
5. How do we remove an employee who is not performing as expected?
6. Should project managers negotiate for people or deliverables?

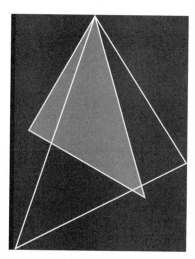

American Electronics International

On February 13, 2004, American Electronics International (AEI) was awarded a $30 million contract for R&D and production qualification for an advanced type of guidance system. During an experimental program that preceded this award and was funded by the same agency, AEI identified new materials with advanced capabilities, which could easily replace existing field units. The program, entitled The Mask Project, would be thirty months in length, requiring the testing of fifteen units. The Mask Project was longer than any other project that AEI had ever encountered. AEI personnel were now concerned about what kind of staffing problems there would be.

BACKGROUND

In June 2002, AEI won a one-year research project for new material development. Blen Carty was chosen as project manager. He had twenty-five years of experience with the company in both project management and project engineering positions. During the past five years Blen had successfully performed as the project manager on R&D projects.

AEI used the matrix approach to structuring project management. Blen was well aware of the problems that can be encountered with this organizational form.

When it became apparent that a follow-on contract would be available, Blen felt that functional managers would be reluctant to assign key personnel full-time to his project and lose their services for thirty months. Likewise, difficulties could be expected in staffing the project office.

During the proposal stage of the Mask Project, a meeting was held with Blen Carty, John Wallace, the director of project management, and Dr. Albert Runnels, the director of engineering. The purpose of the meeting was to satisfy a customer requirement that all key project members be identified in the management volume of the proposal.

John Wallace: "I'm a little reluctant to make any firm commitment. By the time your program gets off the ground, four of our other projects are terminating, as well as several new projects starting up. I think it's a little early to make firm selections."

Blen Carty: "But we have a proposal requirement. Thirty months is a long time to assign personnel for. We should consider this problem now."

Dr. Runnels: "Let's put the names of our top people into the proposal. We'll add several Ph.D.s from our engineering community. That should beef up our management volume. As soon as we're notified of contract go-ahead, we'll see who's available and make the necessary assignments. This is a common practice in the industry."

COMPLETION OF THE MATERIAL DEVELOPMENT PROJECT

The material development program was a total success. From its inception, everything went smoothly. Blen staffed the project office with Richard Flag, a Ph.D. in engineering, to serve as project engineer. This was a risky move at first, because Richard had been a research scientist during his previous four years with the company. During the development project, however, Richard demonstrated that he could divorce himself from R&D and perform the necessary functions of a project engineer assigned to the project office. Blen was pleased with the way that Richard controlled project costs and directed activities.

Richard had developed excellent working relations with development lab personnel and managers. Richard permitted lab personnel to work at their own rate of speed provided that schedule dates were kept. Richard spent ten minutes each week with each of the department managers informing them of the status of the project. The department managers liked this approach because they received firsthand (nonfiltered) information concerning the total picture, not necessarily on their own activities, and because they did not have to spend "wasted hours" in team meetings.

When it became evident that a follow-up contract might be available, Blen spent a large percentage of his time traveling to the customer, working out the details for future business. Richard then served as both project manager and project engineer.

The customer's project office was quite pleased with Richard's work. Information, both good and bad, was transmitted as soon as it became available. Nothing was hidden or disguised. Richard became familiar with all of the customer's project office personnel through the monthly technical interchange meetings.

At completion of the material development project, Blen and John decided to search for project office personnel and make recommendations to upper-level management. Blen wanted to keep Richard on board as chief project engineer. He would be assigned six engineers and would have to control all engineering activities within time, cost, and performance. Although this would be a new experience for him, Blen felt that he could easily handle it.

Unfortunately, the grapevine was saying that Larry Gilbert was going to be assigned as chief project engineer for the Mask Project.

SELECTION PROBLEMS

On November 15, Dr. Runnels and Blen Carty had a meeting to select the key members of the project team.

Dr. Runnels: "Well, Blen, the time has come to decide on your staff. I want to assign Larry Gilbert as chief engineer. He's a good man and has fifteen years' experience. What are your feelings on that?"

Blen Carty: "I was hoping to keep Richard Flag on. He has performed well, and the customer likes working with him."

Dr. Runnels: "Richard does not have the experience necessary for that position. We can still assign him to Larry Gilbert and keep him in the project office."

Blen Carty: "I'd like to have Larry Gilbert working for Richard Flag, but I don't suppose that we'd ever get approval to have a grade-9 engineer working for a grade-7 engineer. Personally, I'm worried about Gilbert's ability to work with people. He has been so regimented in his ways that our people in the functional units have refused to work with him. He treats them as kids, always walking around with a big stick. One department manager said that if Gilbert becomes the boss, then it will probably result in cutting the umbilical cord between the project office and his department. His people refuse to work for a dictator. I have heard the same from other managers."

Dr. Runnels: "Gilbert gets the job done. You'll have to teach him how to be a Theory Y manager. You know, Blen, we don't have very many grade-9 engineering positions in this company. I think we should have a responsibility to our employees. I can't demote Gilbert into a lower slot. If I were to promote Flag, and the project gets canceled, where would I reassign him? He can't go back to functional engineering. That would be a step down."

Blen Carty: "But Gilbert is so set in his ways. He's just totally inflexible. In addition, thirty months is a long time to maintain a project office. If he screws up we'll never be able to replace positions in time without totally upsetting the customer. There seem to be an awful lot of people volunteering to work on the Mask Project. Is there anyone else available?"

Dr. Runnels: "People always volunteer for long-duration projects because it gives them a feeling of security. This even occurs among our dedicated personnel. Unfortunately we have no other grade-9 engineers available. We could reassign one from another program, but I hate to do it. Our engineers like to carry a project through from start to finish. I think you had better spend some time with the functional managers making sure that you get good people."

Blen Carty: "I've tried that and am having trouble. The functional managers will not surrender their key people full-time for thirty months. One manager wants to assign two employees to our project so that they can get on-the-job training. I told him that this project is considered as strategic by our management and that we must have good people. The manager just laughed at me and walked away."

Dr. Runnels: "You know, Blen, you cannot have all top people. Our other projects must be manned. Also, if you were to use all seasoned veterans, the cost would exceed what we put into the proposal. You're just going to have to make do with what you can get. Prepare a list of the people you want and I'll see what I can do."

As Blen left the office, he wondered if Dr. Runnels would help him in obtaining key personnel.

QUESTIONS

1. Whose responsibility is it to staff the office?
2. What should be Blen Carty's role, as well as that of Dr. Runnels?
3. Should Larry Gilbert be assigned?
4. How would you negotiate with the functional managers?

The Carlson Project

"I sympathize with your problems, Frank," stated Joe McGee, manager of project managers. "You know as well as I do that I'm supposed to resolve conflicts and coordinate efforts among all projects. Staffing problems are your responsibility."

Frank: "Royce Williams has a resume that would choke a horse. I don't understand why he performs with a lazy, I-don't-care attitude. He has fifteen years of experience in a project organizational structure, with ten of those years being in project offices. He knows the work that has to be done."

McGee: "I don't think that it has anything to do with you personally. This happens to some of our best workers sooner or later. You can't expect guys to give 120 percent all of the time. Royce is at the top of his pay grade, and being an exempt employee, he doesn't get paid for overtime. He'll snap out of it sooner or later."

Frank: "I have deadlines to meet on the Carlson Project. Fortunately, the Carlson Project is big enough that I can maintain a full-time project office staff of eight employees, not counting myself.

"I like to have all project office employees assigned full-time and qualified in two or three project office areas. It's a good thing that I have someone else checked out in Royce's area. But I just can't keep asking this other guy to do his own work and that of Royce's. This poor guy has been working sixty to seventy hours a week and Royce has been doing only forty. That seems unfair to me."

McGee: "Look, Frank, I have the authority to fire him, but I'm not going to. It doesn't look good if we fire somebody because they won't work free overtime. Last year we had a case similar to this, where an employee refused to work on Monday and Wednesday evenings because it interfered with his MBA classes. Everyone knew he was going to resign the instant he finished his degree, and yet there was nothing that I could do."

Frank: "There must be other alternatives for Royce Williams. I've talked to him as well as to other project office members. Royce's attitude doesn't appear to be demoralizing the other members, but it easily could in a short period of time."

McGee: "We can reassign him to another project, as soon as one comes along. I'm not going to put him on my overhead budget. Your project can support him for the time being. You know, Frank, the grapevine will know the reason for his transfer. This might affect your ability to get qualified people to volunteer to work with you on future projects. Give Royce a little time and see if you can work it out with him. What about this guy, Harlan Green, from one of the functional groups?"

Frank: "Two months ago, we hired Gus Johnson, a man with ten years of experience. For the first two weeks that he was assigned to my project, he worked like hell and got the work done ahead of schedule. His work was flawless. That was the main reason why I wanted him. I know him personally, and he's one great worker.

"During weeks three and four, his work slowed down considerably. I chatted with him and he said that Harlan Green refused to work with him if he kept up that pace."

McGee: "Did you ask him why?"

Frank: "Yes. First of all, you should know that for safety reasons, all men in that department must work in two- or three-men crews. Therefore, Gus was not allowed to work alone. Harlan did not want to change the standards of performance for fear that some of the other employees would be laid off.

"By the end of the first week, nobody in the department would talk to Gus. As a matter of fact, they wouldn't even sit with him in the cafeteria. So, Gus had to either conform to the group or remain an outcast. I feel partially responsible for what has happened, since I'm the one who brought him here.

"I know that has happened before, in the same department. I haven't had a chance to talk to the department manager as yet. I have an appointment to see him next week."

McGee: "There are solutions to the problem, simple ones at that. But, again, it's not my responsibility. You can work it out with the department manager."

"Yeah," thought Frank. "But what if we can't agree?"

Part 6

PROJECT ESTIMATING

Some people believe the primary critical factor for project success is the quality of the estimate. Unfortunately, not all companies have estimating databases, nor do all companies have good estimates. Some companies are successful estimating at the top levels of the work breakdown structure, while others are willing to spend the time and money estimating at the lower levels of the work breakdown structure.

In organizations that are project-driven and survive on competitive bidding, good estimates are often "massaged" and then changed based on the belief by management that the job cannot be won without a lower bid. This built-in process can and does severely impact the project manager's ability to get people to be dedicated to the project's financial baseline.

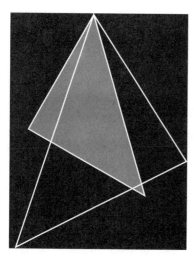

Capital
Industries

In the summer of 2006, Capital Industries undertook a material development program to see if a hard plastic bumper could be developed for medium-sized cars. By January 2007, Project Bumper (as it was called by management) had developed a material that endured all preliminary laboratory testing.

One more step was required before full-scale laboratory testing: a three-dimensional stress analysis on bumper impact collisions. The decision to perform the stress analysis was the result of a concern on the part of the technical community that the bumper might not perform correctly under certain conditions. The cost of the analysis would require corporate funding over and above the original estimates. Since the current costs were identical to what was budgeted, the additional funding was a necessity.

Frank Allen, the project engineer in the Bumper Project Office, was assigned control of the stress analysis. Frank met with the functional manager of the engineering analysis section to discuss the assignment of personnel to the task.

Functional manager: "I'm going to assign Paul Troy to this project. He's a new man with a Ph.D. in structural analysis. I'm sure he'll do well."

Frank Allen: "This is a priority project. We need seasoned veterans, not new people, regardless of whether or not they have Ph.D.s. Why not use some other project as a testing ground for your new employee?"

Functional manager: "You project people must accept part of the responsibility for on-the-job training. I might agree with you if we were talking about blue-collar workers on an assembly line. But this is a college graduate, coming to us with a good technical background."

Frank Allen: "He may have a good background, but he has no experience. He needs supervision. This is a one-man task. The responsibility will be yours if he fouls up."

Functional manager: "I've already given him our book for cost estimates. I'm sure he'll do fine. I'll keep in close communication with him during the project."

Frank Allen met with Paul Troy to get an estimate for the job.

Paul Troy: "I estimate that 800 hours will be required."

Frank Allen: "Your estimate seems low. Most three-dimensional analyses require at least 1,000 hours. Why is your number so low?"

Paul Troy: "Three-dimensional analysis? I thought that it would be a two-dimensional analysis. But no difference; the procedures are the same. I can handle it."

Frank Allen: "O.K. I'll give you 1,100 hours. But if you overrun it, we'll both be sorry."

Frank Allen followed the project closely. By the time the costs were 50 percent completed, performance was only 40 percent. A cost overrun seemed inevitable. The functional manager still asserted that he was tracking the job and that the difficulties were a result of the new material properties. His section had never worked with materials like these before.

Six months later Troy announced that the work would be completed in one week, two months later than planned. The two-month delay caused major problems in facility and equipment utilization. Project Bumper was still paying for employees who were "waiting" to begin full-scale testing.

On Monday mornings, the project office would receive the weekly labor monitor report for the previous week. This week the report indicated that the publications and graphics art department had spent over 200 man-hours (last week) in preparation of the final report. Frank Allen was furious. He called a meeting with Paul Troy and the functional manager.

Frank Allen: "Who told you to prepare a formal report? All we wanted was a go or no-go decision as to structural failure."

Paul Troy: "I don't turn in any work unless it's professional. This report will be documented as a masterpiece."

Frank Allen: "Your 50 percent cost overrun will also be a masterpiece. I guess your estimating was a little off!"

Paul Troy: "Well, this was the first time that I had performed a three-dimensional stress analysis. And what's the big deal? I got the job done, didn't I?"

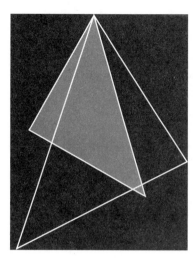

Polyproducts
Incorporated

Polyproducts Incorporated, a major producer of rubber components, employs 800 people and is organized with a matrix structure. Exhibit I shows the salary structure for the company, and Exhibit II identifies the overhead rate projections for the next two years.

Polyproducts has been very successful at maintaining its current business base with approximately 10 percent overtime. Both exempt and nonexempt employees are paid overtime at the rate of time and a half. All overtime hours are burdened at an overhead rate of 30 percent.

On April 16, Polyproducts received a request for proposal from Capital Corporation (see Exhibit III). Polyproducts had an established policy for competitive bidding. First, they would analyze the marketplace to see whether it would be advantageous for them to compete. This task was normally assigned to the marketing group (which operated on overhead). If the marketing group responded favorably, then Polyproducts would go through the necessary pricing procedures to determine a bid price.

On April 24, the marketing group displayed a prospectus on the four companies that would most likely be competing with Polyproducts for the Capital contract. This is shown in Exhibit IV.

Exhibit I. Salary structure

	Pay Scale	
	Grade	Hourly Rate
	1	8.00
	2	9.00
	3	11.00
	4	12.00
	5	14.00
	6	18.00
	7	21.00
	8	24.00
	9	28.00

Number of Employees per Grade

Department	1	2	3	4	5	6	7	8	9	Total
R&D			5	40	20	10	12	8	5	100
Design		3	5	40	30	10	10	2		100
Project engineering						30	15	10	5	60
Project management							10	10	10	30
Cost accounting				20	10	10	10	10		60
Contracts						3	4	2	1	10
Publications		3	5	3	3	3	3			20
Computers				2	3	3	1	1		10
Manufacturing engineering			2	7	7	3	1			20
Industrial engineering				4	3	2	1			10
Facilities				8	9	10	7	1		35
Quality control			3	4	5	5	2	1		20
Production line			55	50	50	30	10	5		200
Traffic			2	2	1					5
Procurement			2	2	2	2	1	1		10
Safety				2	2	1				5
Inventory control	2	2	2	2	1	1				10

At the same time, top management of Polyproducts made the following projections concerning the future business over the next eighteen months:

1. Salary increases would be given to all employees at the beginning of the thirteenth month.
2. If the Capital contract was won, then the overhead rates would go down 0.5 percent each quarter (assuming no strike by employees).
3. There was a possibility that the union would go out on strike if the salary increases were not satisfactory. Based on previous experience, the strike would last between one and two months. It was possible that, due to union demands, the overhead rates would increase by 1 percent per quarter for each quarter after the strike (due to increased fringe benefit packages).

Exhibit II. Overhead structure

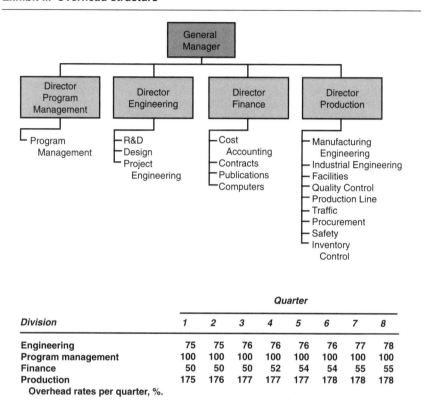

			Quarter					
Division	**1**	**2**	**3**	**4**	**5**	**6**	**7**	**8**
Engineering	75	75	76	76	76	76	77	78
Program management	100	100	100	100	100	100	100	100
Finance	50	50	50	52	54	54	55	55
Production	175	176	177	177	177	178	178	178

Overhead rates per quarter, %.

4. With the current work force, the new project would probably have to be done on overtime. (At least 75 percent of all man-hours were estimated to be performed on overtime). The alternative would be to hire additional employees.
5. All materials could be obtained from one vendor. It can be assumed that raw materials cost $200/unit (without scrap factors) and that these raw materials are new to Polyproducts.

On May 1, Roger Henning was selected by Jim Grimm, the director of project management, to head the project.

Grimm: "Roger, we've got a problem on this one. When you determine your final bid, see if you can account for the fact that we may lose our union. I'm not sure exactly how that will impact our bid. I'll leave that up to you. All I know is that a lot of our people are getting unhappy with the union. See what numbers you can generate."

Exhibit III. Request for proposal

Capital Corporation is seeking bids for 10,000 rubber components that must be manufactured according to specifications supplied by the customer. The contractor will be given sufficient flexibility for material selection and testing provided that all testing include latest developments in technology. All material selection and testing must be within specifications. All vendors selected by the contractor must be (1) certified as a vendor for continuous procurement (follow-on contracts will not be considered until program completion), and (2) operating with a quality control program that is acceptable to both the customer and contractor.

The following timetable must be adhered to:

Month after Go-ahead	Description
2	R&D completed and preliminary design meeting held
4	Qualification completed and final design review meeting held
5	Production setup completed
9	Delivery of 3,000 units
13	Delivery of 3,500 units
17	Delivery of 3,500 units
18	Final report and cost summary

The contract will be firm-fixed-price and the contractor can develop his own work breakdown structure on final approval by the customer.

Henning: "I've read the RFP and have a question about inventory control. Should I look at quantity discount buying for raw materials?"

Grimm: "Yes. But be careful about your assumptions. I want to know all of the assumptions you make."

Henning: "How stable is our business base over the next eighteen months?"

Grimm: "You had better consider both an increase and a decrease of 10 percent. Get me the costs for all cases. Incidentally, the grapevine says that there might be follow-on contracts if we perform well. You know what that means."

Henning: "Okay. I get the costs for each case and then we'll determine what our best bid will be."

On May 15, Roger Henning received a memo from the pricing department summing up the base case man-hour estimates. (This is shown in Exhibits V and VI.) Now Roger Henning wondered what people he could obtain from the functional departments and what would be a reasonable bid to make.

Exhibit IV. Prospectus

Company	Business Base $ Million	Growth Rate Last Year (%)	Profit %	R&D Personnel	Contracts In-House	Number of Employees	Overtime (%)	Personnel Turnover (%)
Alpha	10	10	5	Below avg.	6	30	5	1.0
Beta	20	10	7	Above avg.	15	250	30	0.25
Gamma	50	10	15	Avg.	4	550	20	0.50
Polyproducts	100	15	10	Avg.	30	800	10	1.0

Exhibit V

To: **Roger Henning**
From: **Pricing Department**
Subject: **Rubber Components Production**

1. All man-hours in the Exhibit (14–12) are based upon performance standards for a grade-7 employee. For each grade below 7, add 10 percent of the grade-7 standard and subtract 10 percent of the grade standard for each employee above grade 7. This applies to all departments as long as they are direct labor hours (i.e., not administrative support as in project 1).
2. Time duration is fixed at 18 months.
3. Each production run normally requires four months. The company has enough raw materials on hand for R&D, but must allow two months lead time for purchases that would be needed for a production run. Unfortunately, the vendors cannot commit large purchases, but will commit to monthly deliveries up to a maximum of 1,000 units of raw materials per month. Furthermore, the vendors will guarantee a fixed cost of $200 per raw material unit during the first 12 months of the project only. Material escalation factors are expected at month 13 due to renegotiation of the United Rubber Workers contracts.
4. Use the following work breakdown structure:

> Program: Rubber Components Production
>   Project 1: Support
>   TASK 1: Project office
>   TASK 2: Functional support
>   Project 2: Preproduction
>   TASK 1: R&D
>   TASK 2: Qualification
>   Project 3: Production
>   TASK 1: Setup
>   TASK 2: Production

Exhibit VI. Program: Rubber components production

Project	Task	Department	Month																	
			1	2	3	4	5	6	7	8	9	10	11	12	13	14	15	16	17	18
1	1	Proj. Mgt.	480	480	480	480	480	480	480	480	480	480	480	480	480	480	480	480	480	480
1	2	R&D	16	16	16	16	16	16	16	16	16	16	16	16	16	16	16	16	16	16
		Proj. Eng.	320	320	320	320	320	320	320	320	320	320	320	320	320	320	320	320	320	320
		Cost Acct.	80	80	80	320	320	320	320	320	320	320	320	320	320	320	320	320	320	320
		Contracts	320	320	320	320	320	320	320	320	320	320	320	320	320	320	320	320	320	320
		Manu. Eng.	320	320	320	320	320	320	320	320	320	320	320	320	320	320	320	320	320	320
		Quality Cont.	160	160	160	160	160	160	160	160	160	160	160	160	160	160	160	160	160	160
		Production	160	160	160	160	160	160	160	160	160	160	160	160	160	160	160	160	160	160
		Procurement	80	80	80	80	80	80	80	80	80	80	80	80	80	80	80	80	80	80
		Publications	80	80	80	80	80	80	80	80	80	80	80	80	80	80	80	80	80	80
		Invent. Cont.	80	80	80	80	80	80	80	80	80	80	80	80	80	80	80	80	80	80
2	1	R&D	480	480																
		Proj. Eng.	160	160																
		Manu. Eng.	160	160																
2	2	R&D			80	80														
		Proj. Eng.			160	160														
		Manu. Eng.			160	160														
		Ind. Eng.			40	40														
		Facilities			20	20														
		Quality Cont.			160	160														
		Production			600	600														
		Safety			20	20														
3	1	Proj. Eng.					160													
		Manu. Eng.					160													
		Facilities					80													
		Quality Cont.					160													
		Production					320													
3	2	Proj. Eng.						160	160	160	160	160	160	160	160	160	160	160	160	160
		Manu. Eng.						320	320	320	320	320	320	320	320	320	320	320	320	320
		Quality Cont.						320	320	320	320	320	320	320	320	320	320	320	320	320
		Production						1600	1600	1600	1600	1600	1600	1600	1600	1600	1600	1600	1600	1600
		Safety						20	20	20	20	20	20	20	20	20	20	20	20	20

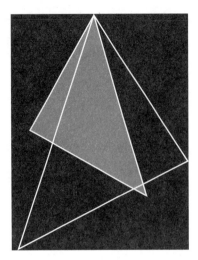

Small Project Cost Estimating at Percy Company

Paul graduated from college in June 2000 with a degree in industrial engineering. He accepted a job as a manufacturing engineer in the Manufacturing Division of Percy Company. His prime responsibility was performing estimates for the Manufacturing Division. Each estimate was then given to the appropriate project office for consideration. The estimation procedure history had shown the estimates to be valid.

In 2005, Paul was promoted to project engineer. His prime responsibility was the coordination of all estimates for work to be completed by all of the divisions. For one full year Paul went by the book and did not do any estimating except for project office personnel manager. After all, he was now in the project management division, which contained job descriptions including such words as "coordinating and integrating."

In 2006, Paul was transferred to small program project management. This was a new organization designed to perform low-cost projects. The problem was that these projects could not withstand the expenses needed for formal divisional cost estimates. For five projects, Paul's estimates were "right on the money." But the sixth project incurred a cost overrun of $20,000 in the Manufacturing Division.

In November 2007, a meeting was called to resolve the question of "Why did the overrun occur?" The attendees included the general manager, all division managers and directors, the project manager, and Paul. Paul now began to worry about what he should say in his defense.

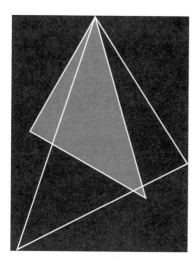

Cory Electric

"Frankly speaking, Jeff, I didn't think that we would stand a chance in winning this $20 million program. I was really surprised when they said that they'd like to accept our bid and begin contract negotiations. As chief contract administrator, you'll head up the negotiating team," remarked Gus Bell, vice president and general manager of Cory Electric. "You have two weeks to prepare your data and line up your team. I want to see you when you're ready to go."

Jeff Stokes was chief contract negotiator for Cory Electric, a $250-million-a-year electrical components manufacturer serving virtually every major U.S. industry. Cory Electric had a well-established matrix structure that had withstood fifteen years of testing. Job casting standards were well established, but did include some "fat" upon the discretion of the functional manager.

Two weeks later, Jeff met with Gus Bell to discuss the negotiation process:

Gus Bell: "Have you selected an appropriate team? You had better make sure that you're covered on all sides."

Jeff: "There will be four, plus myself, at the negotiating table; the program manager, the chief project engineer who developed the engineering labor package; the chief manufacturing engineer who developed the production labor package; and a pricing specialist who has been on the proposal since the kickoff meeting. We have a strong team and should be able to handle any questions."

Gus Bell: "Okay, I'll take your word for it. I have my own checklist for contract negotiations. I want you to come back with a guaranteed fee of $1.6 million for our stockholders. Have you worked out the possible situations based on the negotiated costs?"

Jeff: "Yes! Our minimum position is $20 million plus an 8 percent profit. Of course, this profit percentage will vary depending on the negotiated cost. We can bid the program at a $15 million cost; that's $5 million below our target, and still book a $1.6 million profit by overrunning the cost-plus-incentive-fee contract. Here is a list of the possible cases." (See Exhibit I.)

Gus Bell: "If we negotiate a cost overrun fee, make sure that cost accounting knows about it. I don't want the total fee to be booked as profit if we're going to need it later to cover the overrun. Can we justify our overhead rates, general and administrative costs, and our salary structure?"

Jeff: "That's a problem. You know that 20 percent of our business comes from Mitre Corporation. If they fail to renew our contract for another two-year follow-on effort, then our overhead rates will jump drastically. Which overhead rates should I use?"

Gus Bell: "Let's put in a renegotiation clause to protect us against a drastic change in our business base. Make sure that the customer understands that as part of the terms and conditions. Are there any unusual terms and conditions?"

Exhibit I. Cost positions

		Negotiated Fee			
Negotiated Cost	%	Target Fee	Overrun Fee	Total Fee	Total Package
15,000,000	14.00	1,600,000	500,000	2,100,000	17,100,000
16,000,000	12.50	1,600,000	400,000	2,000,000	18,000,000
17,000,000	11.18	1,600,000	300,000	1,900,000	18,900,000
18,000,000	10.00	1,600,000	200,000	1,800,000	19,800,000
19,000,000	8.95	1,600,000	100,000	1,700,000	20,700,000
20,000,000	8.00	1,600,000	0	1,600,000	21,600,000
21,000,000	7.14	1,600,000	−100,000	1,500,000	*22,500,000
22,000,000	6.36	1,600,000	−200,000	1,400,000	23,400,000
23,000,000	5.65	1,600,000	−300,000	1,300,000	24,300,000
24,000,000	5.00	1,600,000	−400,000	1,200,000	25,200,000

Assume total cost will be spent:

21,000,000	7.61		
22,000,000	7.27	Minimum position	= $20,000,000
23,000,000	6.96	Minimum fee	= 1,600,000 = 8% of minimum position
24,000,000	6.67	Sharing ratio	= 90%/10%

Jeff: "I've read over all terms and conditions, and so have all of the project office personnel as well as the key functional managers. The only major item is that the customer wants us to qualify some new vendors as sources for raw material procurement. We have included in the package the cost of qualifying two new raw material suppliers."

Gus Bell: "Where are the weak points in our proposal? I'm sure we have some."

Jeff: "Last month, the customer sent in a fact-finding team to go over all of our labor justifications. The impression that I get from our people is that we're covered all the way around. The only major problem might be where we'll be performing on our learning curve. We put into the proposal a 45 percent learning curve efficiency. The customer has indicated that we should be up around 50 to 55 percent efficiency, based on our previous contracts with him. Unfortunately, those contracts the customer referred to were four years old. Several of the employees who worked on those programs have left the company. Others are assigned to ongoing projects here at Cory. I estimate that we could put together about 10 percent of the people we used previously. That learning curve percentage will be a big point for disagreements. We finished off the previous programs with the customer at a 35 percent learning curve position. I don't see how they can expect us to be smarter, given these circumstances."

Gus Bell: "If that's the only weakness, then we're in good shape. It sounds like we have a foolproof audit trail. That's good! What's your negotiation sequence going to be?"

Jeff: "I'd like to negotiate the bottom line only, but that's a dream. We'll probably negotiate the raw materials, the man-hours and the learning curve, the overhead rate, and, finally, the profit percentage. Hopefully, we can do it in that order."

Gus Bell: "Do you think that we'll be able to negotiate a cost above our minimum position?"

Jeff: "Our proposal was for $22.2 million. I don't foresee any problem that will prevent us from coming out ahead of the minimum position. The 5 percent change in learning curve efficiency amounts to approximately $1 million. We should be well covered.

"The first move will be up to them. I expect that they'll come in with an offer of $18 to $19 million. Using the binary chop procedure, that'll give us our guaranteed minimum position."

Gus Bell: "Do you know the guys who you'll be negotiating with?"

Jeff: "Yes, I've dealt with them before. The last time, the negotiations took three days. I think we both got what we wanted. I expect this one to go just as smoothly."

Gus Bell: "Okay, Jeff. I'm convinced we're prepared for negotiations. Have a good trip."

The negotiations began at 9:00 A.M. on Monday morning. The customer countered the original proposal of $22.2 million with an offer of $15 million. After six solid hours of arguments, Jeff and his team adjourned. Jeff immediately called Gus Bell at Cory Electric:

Jeff: "Their counteroffer to our bid is absurd. They've asked us to make a counteroffer to their offer. We can't do that. The instant we give them a counteroffer, we are in fact giving credibility to their absurd bid. Now, they're claiming that, if we don't give them a counteroffer, then we're not bargaining in good faith. I think we're in trouble."

Gus Bell: "Has the customer done their homework to justify their bid?"

Jeff: "Yes. Very well. Tomorrow we're going to discuss every element of the proposal, task by task. Unless something drastically changes in their position within the next day or two, contract negotiations will probably take up to a month."

Gus Bell: "Perhaps this is one program that should be negotiated at the top levels of management. Find out if the person that you're negotiating with reports to a vice president and general manager, as you do. If not, break off contract negotiations until the customer gives us someone at your level. We'll negotiate this at my level, if necessary."

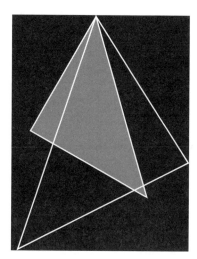

Camden Construction Corporation

"For five years I've heard nothing but flimsy excuses from you people as to why the competition was beating us out in the downtown industrial building construction business," remarked Joseph Camden, president. "Excuses, excuses, excuses; that's all I ever hear! Only 15 percent of our business over the past five years has been in this area, and virtually all of that was with our established customers. Our growth rate is terrible. Everyone seems to just barely outbid us. Maybe our bidding process leaves something to be desired. If you three vice presidents don't come up with the answers then we'll have three positions to fill by midyear.

"We have a proposal request coming in next week, and I want to win it. Do you guys understand that?"

BACKGROUND

Camden Construction Corporation matured from a $1 million to a $26 million construction company between 1989 and 1999. Camden's strength was in its ability to work well with the customer. Its reputation for quality work far exceeded the local competitor's reputation.

Most of Camden's contracts in the early 1990s were with long-time customers who were willing to go sole-source procurement and pay the extra price for quality and service. With the recession of 1995, Camden found that, unless it penetrated the competitive bidding market, its business base would decline.

In 1996, Camden was "forced" to go union in order to bid government projects. Unionization drastically reduced Camden's profit margin, but offered a greater promise for increased business. Camden had avoided the major downtown industrial construction market. But with the availability of multimillion-dollar skyscraper projects, Camden wanted its share of the pot of gold at the base of the rainbow.

MEETING OF THE MINDS

On January 17, 1999, the three vice presidents met to consider ways of improving Camden's bidding technique.

V.P. finance: "You know, fellas, I hate to say it, but we haven't done a good job in developing a bid. I don't think that we've been paying enough attention to the competition. Now's the time to begin."

V.P. operations: "What we really need is a list of who our competitors have been on each project over the last five years. Perhaps we can find some bidding trends."

V.P. engineering: "I think the big number we need is to find out the overhead rates of each of the companies. After all, union contracts specify the rate at which the employees will work. Therefore, except for the engineering design packages, all of the companies should be almost identical in direct labor man-hours and union labor wages for similar jobs."

V.P. finance: "I think I can hunt down past bids by our competitors. Many of them are in public records. That'll get us started."

V.P. operations: "What good will it do? The past is past. Why not just look toward the future?"

V.P. finance: "What we want to do is to maximize our chances for success and maximize profits at the same time. Unfortunately, these two cannot be met at the same time. We must find a compromise."

V.P. engineering: "Do you think that the competition looks at our past bids?"

V.P. finance: "They're stupid if they don't. What we have to do is to determine their target profit and target cost. I know many of the competitors personally and have a good feel for what their target profits are. We'll have to assume that their target direct costs equals ours; otherwise we will have a difficult time making a comparison."

V.P. engineering: "What can we do to help you?"

Exhibit I. **Proposal data summary (cost in tens of thousands)**

Year	Acme	Ajax	Pioneer	Camden Bid	Camden Cost
1990	270	244	260	283	260
1990	260	250	233	243	220
1990	355	340	280	355	300
1991	836	830	838	866	800
1991	300	288	286	281	240
1991	570	560	540	547	500
1992	240*	375	378	362	322
1992	100*	190	180	188	160
1992	880	874	883	866	800
1993	410	318	320	312	280
1993	220	170	182	175	151
1993	400	300	307	316	283
1994	408	300*	433	449	400
1995	338	330	342	333	300
1995	817	808	800	811	700
1995	886	884	880	904	800
1996	384	385	380	376	325
1996	140	148	158	153	130
1997	197	193	188	200	165
1997	750	763	760	744	640

*Buy-in contracts

V.P. Finance: "You'll have to tell me how long it takes to develop the engineering design packages, and how our personnel in engineering design stack up against the competition's salary structure. See if you can make some contacts and find out how much money the competition put into some of their proposals for engineering design activities. That'll be a big help.

"We'll also need good estimates from engineering and operations for this new project we're suppose to bid. Let me pull my data together, and we'll meet again in two days, if that's all right with you two."

REVIEWING THE DATA

The executives met two days later to review the data. The vice president for finance presented the data on the three most likely competitors (see Exhibit I). These companies were Ajax, Acme, and Pioneer. The vice president for finance made the following comments:

1. In 1993, Acme was contract-rich and had a difficult time staffing all of its projects.

2. In 1990, Pioneer was in danger of bankruptcy. It was estimated that it needed to win one or two in order to hold its organization together.
3. Two of the 1992 companies were probably buy-ins based on the potential for follow-on work.
4. The 1994 contract was for an advanced state-of-the-art project. It is estimated that Ajax bought in so that it could break into a new field.

The vice presidents for engineering and operations presented data indicating that the total project cost (fully burdened) was approximately $5 million. "Well," thought the vice president of finance, "I wonder what we should bid so it we will have at least a reasonable chance of winning the contract?"

Part 7

PROJECT PLANNING

Perhaps the most important phase of any project is planning. If the planning is performed effectively, and the workers participate in the development of the plan, the chances of success are greatly enhanced. Yet even with the best-prepared plan, changes will occur.

Good project planning begins with a definition of the requirements, such as the statement of work, work breakdown structure, specifications, timing, and spending curve. Effective planning also assumes that the project manager understands the business case and the accompanying assumptions and constraints.

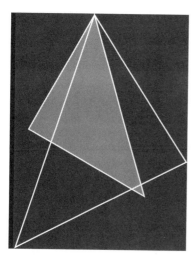

Greyson
Corporation

Greyson Corporation was formed in 1970 by three scientists from the University of California. The major purpose of the company was research and development for advanced military weaponry. Following World War II, Greyson became a leader in the field of research and development. By the mid-1980s, Greyson employed over 200 scientists and engineers.

The fact that Greyson handled only R&D contracts was advantageous. First of all, all of the scientists and engineers were dedicated to R&D activities, not having to share their loyalties with production programs. Second, a strong functional organization was established. The project management function was the responsibility of the functional manager whose department would perform the majority of the work. Working relationships between departments were excellent.

By the late 1980s Greyson was under new management. Almost all R&D programs called for establishment of qualification and production planning as well. As a result, Greyson decided to enter into the production of military weapons as well, and capture some of the windfall profits of the production market. This required a major reorganization from a functional to a matrix structure. Personnel problems occurred, but none that proved major catastrophes.

In 1994 Greyson entered into the aerospace market with the acquisition of a subcontract for the propulsion unit of the Hercules missile. The contract was projected at $200 million over a five-year period, with excellent possibilities for follow-on work. Between 1994 and 1998 Greyson developed a competent technical staff composed mainly of young, untested college graduates. The majority of the original employees who were still there were in managerial positions. Greyson never had any layoffs. In addition, Greyson had excellent career development programs for almost all employees.

Between 1997 and 2001 the Department of Defense procurement for new weapons systems was on the decline. Greyson relied heavily on their two major production programs, Hercules and Condor II, both of which gave great promise for continued procurement. Greyson also had some thirty smaller R&D contracts as well as two smaller production contracts for hand weapons.

Because R&D money was becoming scarce, Greyson's management decided to phase out many of the R&D activities and replace them with lucrative production contracts. Greyson believed that they could compete with anyone in regard to low-cost production. Under this philosophy, the R&D community was reduced to minimum levels necessary to support in-house activities. The director of engineering froze all hiring except for job-shoppers with special talents. All nonessential engineering personnel were transferred to production units.

In 2002, Greyson entered into competition with Cameron Aerospace Corporation for development, qualification, and testing of the Navy's new Neptune missile. The competition was an eight-motor shoot-off during the last ten months of 2003. Cameron Corporation won the contract owing to technical merit. Greyson Corporation, however, had gained valuable technical information in rocket motor development and testing. The loss of the Neptune Program made it clear to Greyson's management that aerospace technology was changing too fast for Greyson to maintain a passive position. Even though funding was limited, Greyson increased the technical staff and soon found great success in winning research and development contracts.

By 2005, Greyson had developed a solid aerospace business base. Profits had increased by 30 percent. Greyson Corporation expanded from a company with 200 employees in 1994 to 1,800 employees in 2005. The Hercules Program, which began in 1994, was providing yearly follow-on contracts. All indications projected a continuation of the Hercules Program through 2002.

Cameron Corporation, on the other hand, had found 1975 a difficult year. The Neptune Program was the only major contract that Cameron Corporation maintained. The current production buy for the Neptune missile was scheduled for completion in August 2005 with no follow-on work earlier than January 2006. Cameron Corporation anticipated that overhead rates would increase sharply prior to next buy. The cost per motor would increase from $55,000 to $75,000 for a January procurement, $85,000 for a March procurement, and $125,000 for an August procurement.

In February 2005, the Navy asked Greyson Corporation if they would be interested in submitting a sole-source bid for production and qualification of the Neptune missile. The Navy considered Cameron's position as uncertain, and wanted to maintain a qualified vendor should Cameron Corporation decide to get out of the aerospace business.

Greyson submitted a bid of $30 million for qualification and testing of thirty Neptune motors over a thirty-month period beginning in January 1976. Current testing of the Neptune missile indicated that the minimum motor age life would extend through January 2009. This meant that production funds over the next thirty months could be diverted toward requalification of a new vendor and still meet production requirements for 2009.

In August 2005, on delivery of the last Neptune rocket to the Navy, Cameron Corporation announced that without an immediate production contract for Neptune follow-on work it would close its doors and get out of the aerospace business. Cameron Corporation invited Greyson Corporation to interview all of their key employees for possible work on the Neptune Requalification Program.

Greyson hired thirty-five of Cameron's key people to begin work in October 2005. The key people would be assigned to ongoing Greyson programs to become familiar with Greyson methods. Greyson's lower-level management was very unhappy about bringing in these thirty-five employees for fear that they would be placed in slots that could have resulted in promotions for some of Greyson's people. Management then decreed that these thirty-five people would work solely on the Neptune Program, and other vacancies would be filled, as required, from the Hercules and Condor II programs. Greyson estimated that the cost of employing these thirty-five people was approximately $150,000 per month, almost all of which was being absorbed through overhead. Without these thirty-five people, Greyson did not believe that they would have won the contract as sole-source procurement. Other competitors could have "grabbed" these key people and forced an open-bidding situation.

Because of the increased overhead rate, Greyson maintained a minimum staff to prepare for contract negotiations and document preparation. To minimize costs, the directors of engineering and program management gave the Neptune program office the authority to make decisions for departments and divisions that were without representation in the program office. Top management had complete confidence in the program office personnel because of their past performances on other programs and years of experience.

In December 2005, the Department of Defense announced that spending was being curtailed sharply and that funding limitations made it impossible to begin the qualification program before July 2006. To make matters worse, consideration was being made for a compression of the requalification program to twenty-five motors in a twenty-month period. However, long-lead funding for raw materials would be available.

After lengthy consideration, Greyson decided to maintain its present position and retain the thirty-five Cameron employees by assigning them to in-house programs. The Neptune program office was still maintained for preparations to support contract negotiations, rescheduling of activities for a shorter program, and long-lead procurement.

In May 2006, contract negotiations began between the Navy and Greyson. At the beginning of contract negotiations, the Navy stated the three key elements for negotiations:

1. Maximum funding was limited to the 2005 quote for a thirty-motor/thirty-month program.
2. The amount of money available for the last six months of 2006 was limited to $3.7 million.
3. The contract would be cost plus incentive fee (CPIF).

After three weeks of negotiations there appeared a stalemate. The Navy contended that the production man-hours in the proposal were at the wrong level on the learning curves. It was further argued that Greyson should be a lot "smarter" now because of the thirty-five Cameron employees and because of experience learned during the 2001 shoot-off with Cameron Corporation during the initial stages of the Neptune Program.

Since the negotiation teams could not agree, top-level management of the Navy and Greyson Corporation met to iron out the differences. An agreement was finally reached on a figure of $28.5 million. This was $1.5 million below Greyson's original estimate to do the work. Management, however, felt that, by "tightening our belts," the work could be accomplished within budget.

The program began on July 1, 2006, with the distribution of the department budgets by the program office. Almost all of the department managers were furious. Not only were the budgets below their original estimates, but the thirty-five Cameron employees were earning salaries above the department mean salary, thus reducing total man-hours even further. Almost all department managers asserted that cost overruns would be the responsibility of the program office and not the individual departments.

By November 2006, Greyson was in trouble. The Neptune Program was on target for cost but 35 percent behind for work completion. Department managers refused to take responsibility for certain tasks that were usually considered to be joint department responsibilities. Poor communication between program office and department managers provided additional discouragement. Department managers refused to have their employees work on Sunday.

Even with all this, program management felt that catch-up was still possible. The thirty-five former Cameron employees were performing commendable work equal to their counterparts on other programs. Management considered that the potential cost overrun situation was not in the critical stage, and that more time should be permitted before considering corporate funding.

In December 2006, the Department of Defense announced that there would be no further buys of the Hercules missile. This announcement was a severe blow to Greyson's management. Not only were they in danger of having to lay off 500 employees, but overhead rates would rise considerably. There was an indication last year that there would be no further buys, but management did not consider the indications positive enough to require corporate strategy changes.

Although Greyson was not unionized, there was a possibility of a massive strike if Greyson career employees were not given seniority over the thirty-five former Cameron employees in the case of layoffs.

By February 2007, the cost situation was clear:

1. The higher overhead rates threatened to increase total program costs by $1 million on the Neptune Program.
2. Because the activities were behind schedule, the catch-up phases would have to be made in a higher salary and overhead rate quarter, thus increasing total costs further.
3. Inventory costs were increasing. Items purchased during long-lead funding were approaching shelf-life limits. Cost impact might be as high as $1 million.

The vice president and general manager considered the Neptune Program critical to the success and survival of Greyson Corporation. The directors and division heads were ordered to take charge of the program. The following options were considered:

1. Perform overtime work to get back on schedule.
2. Delay program activities in hopes that the Navy can come up with additional funding.
3. Review current material specifications in order to increase material shelf life, thus lowering inventory and procurement costs.
4. Begin laying off noncritical employees.
5. Purchase additional tooling and equipment (at corporate expense) so that schedule requirements can be met on target.

On March 1, 2007, Greyson gave merit salary increases to the key employees on all in-house programs. At the same time, Greyson laid off 700 employees, some of whom were seasoned veterans. By March 15, Greyson employees formed a union and went out on strike.

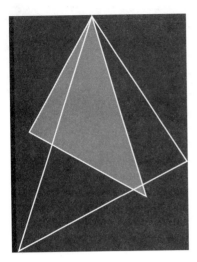

Teloxy
Engineering (A)

Teloxy Engineering has received a onetime contract to design and build 10,000 units of a new product. During the proposal process, management felt that the new product could be designed and manufactured at a low cost. One of the ingredients necessary to build the product was a small component that could be purchased for $60 in the marketplace, including quantity discounts. Accordingly, management budgeted $650,000 for the purchasing and handling of 10,000 components plus scrap.

During the design stage, your engineering team informs you that the final design will require a somewhat higher-grade component that sells for $72 with quantity discounts. The new price is substantially higher than you had budgeted for. This will create a cost overrun.

You meet with your manufacturing team to see if they can manufacture the component at a cheaper price than buying it from the outside. Your manufacturing team informs you that they can produce a maximum of 10,000 units, just enough to fulfill your contract. The setup cost will be $100,000 and the raw material cost is $40 per component. Since Teloxy has never manufactured this product before, manufacturing expects the following defects:

% defective	0	10	20	30	40
probability of occurrence (%)	10	20	30	25	15

All defective parts must be removed and repaired at a cost of $120 per part.

QUESTIONS

1. Using expected value, is it economically better to make or buy the component?
2. Strategically thinking, why might management opt for other than the most economical choice?

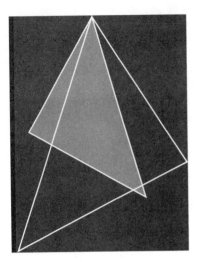

Teloxy
Engineering (B)

Your manufacturing team informs you that they have found a way to increase the size of the manufacturing run from 10,000 to 18,000 units, in increments of 2,000 units. However, the setup cost will be $150,000 and defects will cost the same $120 for removal and repair.

QUESTIONS

1. Calculate the economic feasibility of make or buy.
2. Should the probability of defects change if we produce 18,000 units as opposed to 10,000 units?
3. Would your answer to question 1 change if Teloxy management believes that follow-on contracts will be forthcoming? What would happen if the probability of defects changes to 15 percent, 25 percent, 40 percent, 15 percent, and 5 percent due to learning-curve efficiencies?

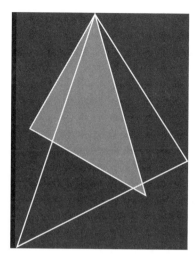

Payton
Corporation

Payton Corporation had decided to respond to a government RFP for the R&D phase on a new project. The statement of work specified that the project must be completed within ninety days after go-ahead, and that the contract would be at a fixed cost and fee.

The majority of the work would be accomplished by the development lab. According to government regulations, the estimated cost must be based on the *average* cost of the entire department, which was $19.00 per hour (unburdened).

Payton won the contract for a total package (cost plus fee) of $305,000. After the first weekly labor report was analyzed, it became evident that the development lab was spending $28.50 per hour. The project manager decided to discuss the problem with the manager of the development lab.

Project manager: "Obviously you know why I'm here. At the rate that you're spending money, we'll overrun our budget by 50 percent."

Lab manager: "That's your problem, not mine. When I estimate the cost to do a job, I submit only the hours necessary based on historical standards. The pricing department converts the hours to dollars based on department averages."

Project manager: "Well, why are we using the most expensive people? Obviously there must be lower-salaried people capable of performing the work."

Lab manager: "Yes, I do have lower-salaried people, but none who can complete the job within the two months required by the contract. I have to use people high on the learning curve, and they're not cheap. You should have told the pricing department to increase the average cost for the department."

Project manager: "I wish I could, but government regulations forbid this. If we were ever audited, or if this proposal were compared to other salary structures in other proposals, we would be in deep trouble. The only legal way to accomplish this would be to set up a new department for those higher-paid employees working on this project. Then the average department salary would be correct.

"Unfortunately the administrative costs of setting up a temporary unit for only two months is prohibitive. For long-duration projects, this technique is often employed.

"Why couldn't you have increased the hours to compensate for the increased dollars required?"

Lab manager: "I have to submit labor justifications for all hours I estimate. If I were to get audited, my job would be on the line. Remember, we had to submit labor justification for all work as part of the proposal.

"Perhaps next time management might think twice before bidding on a short-duration project. You might try talking to the customer to get his opinion."

Project manager: "His response would probably be the same regardless of whether I explained the situation to him before we submitted the proposal or now, after we have negotiated it. There's a good chance that I've just lost my Christmas bonus."

QUESTIONS

1. What is the basis for the problem?
2. Who is at fault?
3. How can the present situation be corrected?
4. Is there any way this situation can be prevented from recurring?
5. How would you handle this situation on a longer-duration project, say one year, assuming that multiple departments are involved and that no new departments were established other than possibly the project office?
6. Should a customer be willing to accept monetary responsibility for this type of situation, possibly by permitting established standards to be deviated from? If so, then how many months should be considered as a short-duration project?

IVEY

Richard Ivey School of Business
The University of Western Ontario

Spin Master Toys (A): Finding a Manufacturer for E-Chargers

In mid-July 1999, Alex Perez, operations manager of Spin Master Toys of Toronto, Ontario, was trying to decide from which supplier to purchase the design and production of the company's latest toy, an electrically powered airplane named E-Chargers. He had investigated a number of potential suppliers in southern China and had settled on two finalists, Wah Shing Electronics Co., Ltd. (Wah Shing) and Wai Lung Plastics Mfy., Ltd. (Wai Lung). With the anticipated date for the launch of this product just a few short months away, Perez had to make his choice quickly.

SPIN MASTER TOYS

In April 1994, Anton Rabie, Ronnen Harary, and Ben Varadi graduated from The University of Western Ontario, Rabie and Varadi from the Ivey Business School and Harary from political science. The three decided to forgo opportunities in the corporate world and strike out on their own. They were soon making Earth Buddy, a nylon stocking filled with sawdust and grass seed moulded into a head. After immersion in water, the grass seed would sprout to give the head a crop of grass—hair. Although Earth Buddy was clearly a fad item, the company managed to sell 1.5 million of them in just six months, making it one of the most popular gift items that year.

In February 1995, the company followed this success with the launch of Spin Master Devil Sticks, which consisted of two hand-held sticks used to manipulate a third. This product also became a resounding success. Eventually the company incorporated Spin Master into its name. The company's principals believed they had achieved their success through avant-garde, grassroots marketing savvy and a two-tier distribution network, which covered both the major and independent retail segments in North America.

In the following three years, Spin Master Toys produced an array of relatively low-technology, high-margin toys for the Canadian market. The product list included:

- Spin-A-Blo, spinning toys
- Radical Reptiles, foam reptiles attached to a metal leash
- Top-No-Sis, spinning board
- My First Kite, a starter kite for children
- Grow-Things, water-absorbent play animals

Although Spin Master Toys achieved notable success with these fad items, none reached the unit sales that Earth Buddy had produced. Following its success with Spin Master Devil Sticks, Spin Master Toys spent six months moving from being project-focused to building relationships with retailers and investors and creating a research and development department.

At a major 1996 toy show, two inventors approached Rabie and Harary, and many other toy companies, with the concept for a compressed-air-powered toy plane. Their initial design was a plastic soft-drink bottle with wings attached. Rabie and Harary and the major toy companies rejected the idea as being too ambitious. However, the inventors were persistent, and after the original prototype had been revised several times, Spin Master Toys decided to purchase the rights to the concept. After a frustrating two years and $500,000 in development, Spin Master Toys rolled out its Air Hogs line of compressed-air-powered planes, and, with outside engineering expertise, proceeded to manufacture them in China. The company used an innovative marketing campaign to generate a groundswell of excitement. Air Hogs became a top-selling toy for the 1998 North American Christmas season and was hailed by *Popular Science* as one of the 100 greatest inventions of the year, creating, as it did, a new category—compressed-air-powered planes. Spin Master Toys had to double production of Air Hogs just to keep up with demand, which was increased by the product shortage in the first few months after the initial shipments.

Following the success of Air Hogs, Spin Master Toys decided to develop a line of toys driven by compressed air. It subsequently launched a compressed-air-powered water rocket called the *Vector,* a car named the *Road Ripper,* and two new compressed-air-powered product-flanking planes, the *V-Wing Avenger* and the *Renegade.*

With over 50 people working in its Toronto head office, and a recently opened office in Hong Kong staffed by two project managers, Spin Master Toys was enjoying rapid expansion through its combination of speed to market and innovative marketing. Revenue had grown from nearly $525,000 in 1994 to a projected $45.8 million in 1999, earning it the tenth spot on the Profit 100 Canada's Fastest-Growing Companies list.

THE TOY INDUSTRY

The toy market included both hard and soft goods, as well as combinations. Hard goods included plastic and metal toys—water guns, construction toys, action figures, etc. Soft goods included plush toys, fabrics, and dolls. Either hard or soft toys increasingly used embedded electronic components as differentiators.

Southern China in and near Hong Kong accounted for a large percentage of the world's toy manufacturing industry; many manufacturers there had over 50 years of toy-making experience. Beginning with low-technology plastic and metal toys in the early years, toy makers in the area had developed sophisticated design, engineering, and manufacturing skills. Such factors could be important. Perez, who used to work for a large toy company, remembered a competitor that sourced from Thailand because production costs were slightly lower. Despite this advantage, the project was a dismal failure, in part because of the lack of toy-making expertise in that country.

Aside from experience, the Hong Kong market had English-speaking workers, a Western-style banking system, easy access to low-cost production facilities and workers in China, an entrepreneurial spirit, and major port facilities. Deciding to source toys from this region was relatively easy.

E-CHARGERS

E-Chargers were Spin Master Toy's next foray into the powered toy airplane market. Unlike the traditional toy airplane powered by a stretched rubber band, gasoline engine, or compressed air, E-Chargers were driven by electricity. The product came in two parts: a battery pack holding four AA dry-cell batteries, and a plastic foam airplane containing a small capacitor[1] connected to an electric motor. By inserting the battery pack into a special port on the airplane, the user both started the electric motor driving the plane's propeller and charged the capacitor. The user then disconnected the plane from the battery pack and launched it into the air. Spin Master Toys touted E-Chargers as being capable of flights of up to 90 meters and as "high performing, easy-to-use rechargeable planes that come with their own chargers—kids just have to let them charge for 10 seconds and then let them fly." In the company's view, the product line allowed it to extend the magic of real flight to children as young as five—younger than the user of Air Hogs. To encourage users to collect E-Chargers, the company planned to produce six different styles and promised high performance at a low price.

Spin Master Toys had sold the E-Chargers concept to retailers who subsequently placed endcap[2] orders for a December 7, 1999, delivery date to meet the spring planogram[3] shelving period. This was the first time that Spin Master Toys would ship products for a planogram. In the past, the company had been able to obtain special shelf space only because of its products' uniqueness. The main advantage in shipping to a set deadline was the guarantee of shelf space. Spin Master Toys now had to design and make the E-Chargers in time to meet the order date.

PRELIMINARY E-CHARGER PRODUCTION ESTIMATES

Working back from December 7, 1999, Perez developed a somewhat accelerated schedule that would allow delivery of the E-Chargers plane. Exhibit I shows the development schedule, delay in any step of which would make the project late.

---

[1] A capacitor is an electronic device used to store charge—in essence, it is like a rechargeable battery. It consists of an arrangement of conductors, separated by an insulator.

[2] Endcaps are the attractive, highly visible end spaces on shopping aisles. Executives of Spin Master Toys expected that an E-Chargers endcap order from a large retail customer would result in sales of about 150,000 units.

[3] Retailers took three weeks after Christmas to clear out old stock and put in new toys for the spring period. The layout of toys by aisle and shelf, known as a *planogram*, was determined in advance.

Exhibit I.   *Projected development schedule and current progress*

Item# 40004   Spin Master Toys Engineer: Alex
Item Name: E-Chargers Flying Machines (6 styles)
Pack: 12   Project Manager: Tammy
Target FOB HK (U.S.$) 1.75
FOB HK (U.S.$)   Date: June 30, 1999
Landed Cost (Estimated in U.S.$)

Description	Responsible	Planned	Current
Quote Package	Alex	July 1	
● General product profile	Tammy	June 23	June 30
● Product electronic schematics	Tammy	June 25	July 2
● Preliminary parts drawings	Tammy	June 25	July 2
● Assembly-exploded view drawings	Tammy	June 25	July 2
● Bill of materials/parts list	Tammy	June 25	July 2
● Rough engineering model	Tammy	June 15	June 22
Vendor Preliminary Quotes	Alex	July 10	July 17
● Final vendor decision	Ronnen	July 11	
● First engineering model	Tammy	July 1	
● Second engineering model	Tammy	July 3	
● Third engineering model	Tammy	July 5	
Final Design Release	Alex	July 1	
● Model ready (propeller)	Factory	July 10	
● Decision on gear	Factory	July 10	
● Recommend foam type	Factory	July 19	
● Approval on foam type	Alex	July 20	
● Samples of the motor and capacitor	Factory	July 22	
● Plastic housing evaluation	Alex	July 27	
● Verify motor specification is compatible with Mabuchi	Factory	July 31	
● Plastic housing resubmission	Factory	July 31	
Models Available	Factory	July 22	
Approved product quote (purchase order, material authorization release)	Tammy/Ronnen	July 26	
Tooling purchase order for airplane	James	July 22	
Tooling purchasing order release (all others)	James	August 4	
Tool start (35 days lead time)	Factory	August 4	
First test shot	Factory	September 8	
First engineering pilot	Factory	September 18	
Sales samples ready (from 1st shot)	Factory	September 23	
Second test shot	Factory	September 28	
Second engineering pilot	Factory	October 3	
Final shot	Factory	October 8	
Final engineering pilot	Factory	October 14	
Production pilot	Factory	October 21	

(continues)

Exhibit I. Projected development schedule and current progress (Continued)

Description	Responsible	Planned	Current
Production pilot tests completed	Factory	October 29	
Final production pilot approval	Ronnen	November 2	
Final quote approval	Ronnen	November 2	
Production start	**Factory**	**November 22**	
First on-board shipment	Factory	November 28	
Packaging Timeline			
English film and disk send to Hong Kong	Selene	July 20	
Packaging approval (7 days)	Tammy	July 27	
English package arrival (3 weeks)	Factory	August 17	
Bilingual package disk to Hong Kong	Selene	August 3	
Bilingual package approval in Hong Kong	Willy	August 10	
Bilingual package arrival	Factory	August 31	
TV commercial sample (quantity)			
TV commercial sample (date)			
Estimated sales forecast	Jennifer	July 17	
Consigned materials		N/A	
Motor and capacitor			
Material authorization or purchase order	Heather/James	August 3	

Ramp-up Schedule	Date	Produce	Cumulative	Changes
First week: Day 1	November 8	50	50	
Day 2	November 9	50	100	Ramp-up
Day 3	November 10	100	200	not yet
Day 4	November 11	150	350	confirmed
Day 5	November 12	250	600	
Day 6	November 13	400	1,000	
First on-board shipment	November 15	600	1,600	
Second week	November 22	9,000	10,600	
Third week	November 29	12,000	22,600	
Fourth week	December 6	18,000	40,600	
Fifth and subsequent weeks		18,000		

Source: Company files.

Rough Engineering Model

This stage involved the engineering work needed to craft a design to meet the desired specifications provided by the manufacturer. These specifications included, for example, that the toy would be capable of high-speed production while maintaining acceptable finished-product quality, that it was within the weight and size required, and that any electronic components involved would function within tolerances provided. Although design work normally took about eight weeks, Spin Master Toys allowed less than three weeks for E-Chargers; the design work would have to be completed no later than the middle of June.

On June 22, K-Development of Erie, Pennsylvania, the company to which Spin Master Toys had contracted the development engineering, transferred the completed engineering designs to Reh Kemper, a prototype designer based in Chicago, Illinois. Reh Kemper completed its work on July 2. According to Perez's timeline, the project was already a week behind schedule for the start of production.

Engineering Models

After one week of examination, study, and discussion of the prototype, Perez and his team approved it and issued a *final design release*. Spin Master Toys then returned it to K-Development, which had five days to improve the rough engineering model and produce three initial prototypes to ensure that the design was engineered correctly to the specified tolerances. This preliminary work showed that the weight of the plane would be of great concern. Initial tests showed that to achieve the expected flight times, E-Chargers had to weigh 17 grams. Once the third engineering model was ready, Perez released it to vendors, requesting preliminary quotes within five days.

Tooling

From this stage on, all work would be performed at the factory, with regular updates sent to Perez by phone or fax. The tool start involved creating the molds and other tooling required to produce the toy in mass quantity. Plastic parts such as those used in E-Chargers were normally made by injection molding in which a molten plastic was injected into the carefully machined cavity inside a two-piece block of metal (the mold). After applying pressure and cooling, the mold was opened to remove the part. In practice, molders might use large molds capable of making several parts simultaneously. This crucial step usually took four weeks; the time required was usually factored into the design component. Perez estimated that Spin Master Toys would need the first test samples by September 8.

Engineering Pilots

The next step was testing the molds and other tools, ideally with two engineering pilots. At least one engineering pilot had to be performed before the next stage, as it was almost inevitable that the molds would need some adjustments. A factory would count on three weeks to run both engineering pilots. The first and second engineering pilots and the shots from them had to be completed by October 8.

Final Engineering Pilot

In this two-week process, the final molds and other tools were finished. To have the product ready for the production pilot date of October 21, this step had to be completed in one week.

Production Pilot

This step tested whether the molds and other tools would withstand high-speed production while delivering product within the required tolerances. The production pilot tests and the final quote had to be approved by November 22.

Production Start

In the case of E-Chargers, Perez estimated that production would have to start at least two weeks before the shipping date to allow production of enough units to meet retailer demand. Thus, production would have to start on November 22 to just make the December 7 ship date.

SPIN MASTER TOYS' CONTRACT MANUFACTURERS

In the past, Spin Master Toys had obtained its products from various Chinese manufacturers. Because of the large differences between its previous toys, the company had treated each product separately. Consequently, Spin Master Toys had gained considerable experience with several suppliers, as each toy had been manufactured by a different factory. Spin Master Toys believed that its product closest in design to E-Chargers was Air Hogs. In May 1999, while working on Water Rocket, one of its second generation compressed-air-powered toys, Spin Master Toys had visited Kin Seng Ltd., the Air Hogs manufacturer. During a factory tour, Spin Master Toys discovered that the Kin-Seng factory was at capacity. Because of the tightness of its E-Chargers schedule, Spin Master Toys decided not to consider Kin Seng as a potential supplier.

Spin Master Toys thus searched for an alternative manufacturer, eventually creating a short list of two, Wai Lung and Wah Shing.

WAI LUNG

In early 1999, Harary had been introduced to the owner of privately owned Wai Lung Manufacturing Co. Harary believed that Spin Master Toys would receive more attention from an owner-operated factory than from a subsidiary of a public corporation. Reassuring Harary that he would provide personal attention to this project, Eric Lee, Wai Lung's owner seemed eager to strike a deal with Spin Master Toys. Harary subsequently initiated a toy project, Flick Trix Finger Bikes, with Wai Lung. Finger Bikes were miniature die-cast replicas of brand-name BMX bikes with fully functional parts. Already in a rushed situation, Harary had asked Wai Lung if it could engineer the Finger Bikes, produce and ship them in six weeks—it normally took other manufacturers six to ten weeks to perform these tasks. With Finger Bikes already engineered by Reh Kemper, Spin Master Toys would rely on Wai Lung's staff to beat a competitor to the market. Working at a break-neck pace, Wai Lung had been able not only to build the tools in the allotted time, but also to increase production very quickly with little lead time. Although Wai Lung had initially built tools to support a production rate of 10,000 bikes a day, once it was evident that demand was strong, the company was able to build additional tools in four weeks versus the previous six weeks, boosting Finger Bikes production to 40,000 bikes a day.

Not only had Wai Lung come through for Spin Master Toys, but it went on to produce a high-quality toy and increased production more steeply than Harary had thought possible. Perez expressed his thoughts:

> Wai Lung is highly committed and has put us at the top of its priority list. During our early experience with Finger Bikes, they returned calls promptly and answered all questions during the critical production period.
>
> Wai Lung's performance with Finger Bikes allowed us to beat a major competitor to the market. This prompted our competitor to drop the project in mid-design. We should look at Wai Lung as a supplier for E-Chargers because of our positive experience with them. However, their engineering workforce is fairly small and they haven't produced toys with electronic components. They have focused on die-casting and plastic action figures. E-Chargers have to be designed and produced to much more stringent tolerances than die-cast or plastic toys. To put it bluntly, flying toys would take a paradigm shift in Wai Lung's engineering expertise.
>
> We did plan to use a vendor survey report, but we don't have any engineering expertise at our Hong Kong office. And, in Canada, our manufacturing team includes me and Ronnen—with this in mind, I wonder if we can gather this information for Wai Lung and Wah Shing in time. We are already behind schedule as it stands.

Harary returned to visit Wai Lung in May 1999 and, while walking through the factory, estimated by observation that Wai Lung was at 40 percent of capacity. He also found out that Wai Lung had excess capacity to utilize because it had just lost a significant portion of its business during a disagreement with a large toy company. Harary was impressed by its size: It had 2,000 workers in its 100,000-square-foot factory in Shenzhen, about a one-hour journey by train and car from Hong Kong. Typical toy factories in this area averaged about 600 workers. He casually asked the owner of Wai Lung for a quick overview of the projects currently in progress. Wai Lung was working on plastic play sets and action figures for Hasbro. Another company with which Wai Lung had a contract had gone bankrupt. Pressing further on a different subject, Harary got the sense that Wai Lung would not begin many projects in the near future.

Lee, 48, had always been very accommodating to Harary and considered himself to be a self-made man, building up a successful factory. Still hungry to grow his business, he had recently hired three engineers. He was willing to extend favorable credit terms to Spin Master Toys, allowing for Finger Bikes production to commence with a simple wire transfer of funds versus a more formal letter of credit. Otherwise, a letter of credit from the bank, along with the requisite documentation, meant that up to 30 percent of the total invoice amount needed to be securely transferred before the start of production. Once production was started, payments would immediately be taken out of cash flow. With a wire transfer, however, funds would be wired to the supplier's bank account twenty-one days after the goods were shipped.

WAH SHING

Wah Shing was a subsidiary of a Hong Kong public toy manufacturer. It was a company with annual revenues of U.S.$40 million (the average Hong Kong toy company with product line similar to Wah Shing's earned about U.S.$30 million in revenues per year). While at his previous employer, Perez had worked with Wah Shing. Wah Shing had been one of the suppliers of choice for major toy companies such as Tiger and Hasbro, which needed electronic toys. These companies wanted to maintain their track record of successful electronic toy engineering development and manufacture in the electronic hand-held, feature electronic plush, radio control and IR interactive categories, including toys such as Shotgun and Skidzo, Furby, Laser Light Tennis, and Galactic Battle.

Wah Shing employed 3,500 people in its 100,000-square-foot factory, counting six engineers on its staff. Although Harary had toured the factory, during his visit, he had been unable to meet the owner, who was traveling. By observation, Harary estimated that Wah Shing was at 70 to 80 percent of capacity at its Chinese factory, which was located five hours away from Hong Kong. Perez expressed his thoughts:

> Before coming to Spin Master, I worked for a major toy company and got some experience with Wah Shing. Its upper and lower management are very committed. They are a nonhierarchial, action-oriented company. I have a personal friendship with the general manager.
>
> In my experience, Wah Shing provides products on time and within quality specifications. But it has been four years since my last contact.

During a visit a few weeks ago, I found out that the lower management had been changed. Also, there seemed to be less communication between upper and lower management than there used to be. However, they still have a good reputation in electronic toys, and their costs are comparable to similar companies.

Ronnen, who was with me on the tour, noted that they had put their North American account manager in charge of the tour. Ronnen is used to dealing directly with factory owners and wonders if we could expect the same commitment as we have had with our previous projects.

RONNEN HARARY'S CHOICE

Harary discussed the decision he faced:

We believe that retail sales for toy airplanes will peak from March to mid-May, after which water toys will dominate. For E-Chargers, we've been fortunate to have secured a sizable amount of shelf space in retail stores for this period and also have been awarded several large feature endcap orders! To meet this demand, we have to have 20,000 units ready to ship by December 1999 as shown in this schedule [see Exhibit I]. On top of the fact that the retailers need time to move our product through their distribution system, we've heard that a major competitor, a large toy company, is also working on the same E-Chargers concept. We have to beat them to market at all costs because, in this industry, it is hard to overcome the first mover advantage. While we would like to have a five- to six-month design-to-delivery window, we have four months, max.

But we also have to consider the tight tolerances we require. Our initial work revealed that we have to be very careful to balance weight shaving and structural integrity. Ideally, an E-Charger should weigh 17 grams. An increase of only one gram decreases the flying time by 15 seconds. Just painting the plane adds enough weight to affect the performance significantly. According to our preliminary tests, the plane will weigh 18 grams, and we have to work tremendously hard to reduce that figure. At 18.5 grams, this thing won't even fly.

We have to find a supplier who can deliver on engineering expertise. Not many manufacturers in Hong Kong had experience with flying toys and, to add to the complexity of this project, we are using materials that are not commonly available.

This is an unprecedented toy requiring design work for the engine and to accommodate the capacitor, not to mention the separate battery box. Our rough design calls for about fifty different parts! How should we compare the quality of work between the Wai Lung and Wah Shing factories? Although both have done projects for us in the past, this product is totally

Exhibit II.   Quote from Wai Lung Plastic Mfy., Ltd.

Quotation Submission Form (Summary)
Spin Master Toys

Attention:	Ronnen Harary
Item:	4004
Description:	E-Chargers
Reference:	Quotation Submission
From:	Wai Lung Plastics Mfy., Ltd

Description	Cost in HK$ Per 1,000 Toys[1]
1 Plastic	$540.50
2. Other parts	4,670.00
3. Packaging	3,620.00
4. Shipping carton	295.00
Total material cost	$9,125.50
Total labor cost	$2,380.00
Total materials plus labor	$11,505.50
Overhead and markup @ 16% (of materials and labor)	$1,840.88
Scrap allowance @ 1.5% (of materials)	136.88
Capacitor handling charge @ 3% (of capacitor cost)	150.74
Motor handling charge @ 3% (of motor cost)	197.12
Total	$13,831.12
Transportation FCL,[2] Hong Kong, FOB[3] Hong Kong, 40-foot FCL container	$487.00
Total	$14,318.12
Transportation LCL,[4] Hong Kong, FOB Hong Kong, 40-foot LCL container	$1,607.50
Total	$15,438.62[5]

Source: Company files.

[1] The Hong Kong dollar was pegged against the United States dollar at the rate of HK$7.75 = U.S.$1. In July 1999, a Canadian dollar was worth about HK$5.21.

[2] FCL: Full container load.

[3] FOB: Free on board. In essence, the location signifies the point at which the customer takes ownership, and thus financial responsibility.

[4] LCL: Less than container load.

[5] This price does not include the capacitor or the motor.

new. Price might play a factor in the decision, but it will not override our most pressing concern of getting to market quickly.

A concern is the quality of the suppliers' sources of raw materials and prefabricated components, most of which are based in mainland China. A large number of small- and mid-sized competitors vie for the world toy business—no one factory controls a significant portion of toy manufacturing.

Exhibit III. *Quote from Wah Shing Electronic Co., Ltd.*

Wah Shing Electronic Co., Ltd.

To: Alex Perez
From: John Yi
Subject: E-Flyer Quote: Ref "0" vs. Mattel

Cost Summary Sheet

Product Name: E-Flyer

Item	Cost description	FCL (HK$)	LCL (HK$)
1	Electronic parts (includes motor and capacitor)	15.7998	15.7998
2	Plastic material	0.2396	0.2396
3	Metal parts	0.8976	0.8976
4	Packaging material	2.5805	2.5805
5	Miscellaneous	4.2534	4.2534
6	Bonding	0.0000	0.0000
7	Labor cost	0.8000	0.8000
8	Decoration cost	0.0000	0.0000
9	Injection cost	0.5313	0.5313
10	Overhead and markup	3.3523	3.3523
11	Transportation	0.2914	1.0238
	Ex-factory price FOB Hong Kong	28.7459	29.4783

Source: Company files.

Clients like us have to be extra careful, because machinery and worker training in mainland China are generally inferior to those in Hong Kong.

We should consider many factors in making this decision: reputation, capacity, quality levels, capability in engineering, the capability of the factories' Chinese suppliers, speed to market, costs, tooling time needed (critical in this project), attention to your company. In the past, due to our small size and limited engineering expertise, we prioritized a close working relationship with the owner of the factory in question. Because the owner took a personal interest in our projects, it reassured us that our needs would be top priority, and he would do whatever it took to produce results. With E-Chargers, I still strongly believe that this is necessary to ensure we meet the December 7 deadline. A personal relationship is key. What could make that difficult is the fact that the owners of these private toy manufacturers, like many in Hong Kong, all seem to have several businesses going on at once.

We are very pressed. We might not have enough time to do proper due diligence on Wai Lung or Wah Shing. We just got these quotes from each of them [see Exhibits II and III]. Although we would like to have more time to qualify more suppliers in the Hong Kong area, we simply can't afford the time. We need engineering development work to start almost immediately! We need a factory to develop the wings and fuselage for E-Chargers, the rest of the fifty parts, prototype moulds, then sample shots for our inspection. We do not have the luxury of extra time. We're not even sure what our competitors are up to. Which factory should we choose?

The Richard Ivey School of Business gratefully acknowledges the generous support of the MBA 1989 class in the development of these learning materials.

Part 8

PROJECT SCHEDULING

Once project planning is completed, the next step is to schedule the project according to some timeline. This requires knowledge of the activities, the necessary depth of the activities, the dependencies between the activities, and the duration of the activities.

Effective scheduling allows us to perform what-if exercises, develop contingency plans, determine the risks in the schedule, perform trade-offs, and minimize paperwork during customer review meetings. Although there are four basic scheduling techniques, they all utilize the same basic principles and common terminology.

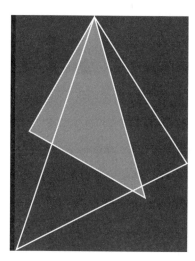

Crosby
Manufacturing
Corporation

"I've called this meeting to resolve a major problem with our management cost and control system (MCCS)," remarked Wilfred Livingston, president. "We're having one hell of a time trying to meet competition with our antiquated MCCS reporting procedures. Last year we were considered nonresponsive to three large government contracts because we could not adhere to the customer's financial reporting requirements. The government has recently shown a renewed interest in Crosby Manufacturing Corporation. If we can computerize our project financial reporting procedure, we'll be in great shape to meet the competition head-on. The customer might even waive the financial reporting requirements if we show our immediate intent to convert."

Crosby Manufacturing was a $250-million-a-year electronics component manufacturing firm in 2005, at which time Wilfred "Willy" Livingston became president. His first major act was to reorganize the 700 employees into a modified matrix structure. This reorganization was the first step in Livingston's long-range plan to obtain large government contracts. The matrix provided the customer focal point policy that government agencies prefer. After three years, the matrix seemed to be working. Now they could begin the second phase, an improved MCCS policy.

On October 20, 2007, Livingston called a meeting with department managers from project management, cost accounting, MIS, data processing, and planning.

Livingston: "We have to replace our present computer with a more advanced model so as to update our MCCS reporting procedures. In order for us to grow, we'll have to develop capabilities for keeping two or even three different sets of books for our customers. Our present computer does not have this capability. We're talking about a sizable cash outlay, not necessarily to impress our customers, but to increase our business base and grow. We need weekly, or even daily, cost data so as to better control our projects."

MIS manager: "I guess the first step in the design, development, and implementation process would be the feasibility study. I have prepared a list of the major topics which are normally included in a feasibility study of this sort" (see Exhibit I).

Livingston: "What kind of costs are you considering in the feasibility study?"

MIS manager: "The major cost items include input–output demands; processing; storage capacity; rental, purchase or lease of a system; nonrecurring expenditures; recurring expenditures; cost of supplies; facility requirements; and training requirements. We'll have to get a lot of this information from the EDP department."

EDP manager: "You must remember that, for a short period of time, we'll end up with two computer systems in operation at the same time. This cannot be helped. However, I have prepared a typical (abbreviated) schedule of my own (see Exhibit II). You'll notice from the right-hand column that I'm somewhat optimistic as to how long it should take us."

Livingston: "Have we prepared a checklist on how to evaluate a vendor?"

EDP manager: "Besides the benchmark test, I have prepared a list of topics that we must include in evaluation of any vendor (see Exhibit III). We should plan to call on or visit other installations that have purchased the same equipment and see the system in action. Unfortunately, we may have to commit real early and begin

---

Exhibit I. Feasibility study

- Objectives of the study
- Costs
- Benefits
- Manual or computer-based solution?
- Objectives of the system
- Input requirements
- Output requirements
- Processing requirements
- Preliminary system description
- Evaluation of bids from vendors
- Financial analysis
- Conclusions

Exhibit II. Typical schedule (in months)

Activity	Normal Time to Complete	Crash Time to Complete
Management go-ahead	0	0
Release of preliminary system specs.	6	2
Receipt of bids on specs.	2	1
Order hardware and systems software	2	1
Flow charts completed	2	2
Applications programs completed	3	6
Receipt of hardware and systems software	3	3
Testing and debugging done	2	2
Documentation, if required	2	2
Changeover completed	22	15*

*This assumes that some of the activities can be run in parallel, instead of series.

developing software packages. As a matter of fact, using the principle of concurrency, we should begin developing our software packages right now."

Livingston: "Because of the importance of this project, I'm going to violate our normal structure and appoint Tim Emary from our planning group as project leader. He's not as knowledgeable as you people are in regard to computers, but he does know how to lay out a schedule and get the job done. I'm sure your people will give him all the necessary support he needs. Remember, I'll be behind this project all the way. We're going to convene again one week from today, at which time I expect to see a detailed schedule with all major milestones, team meetings, design review meetings, etc., shown and identified. I'd like the project to be complete in eighteen months, if possible. If there are risks in the schedule, identify them. Any questions?"

Exhibit III. Vendor support evaluation factors

- Availability of hardware and software packages
- Hardware performance, delivery, and past track record
- Vendor proximity and service-and-support record
- Emergency backup procedure
- Availability of applications programs and their compatibility with our other systems
- Capacity for expansion
- Documentation
- Availability of consultants for systems programming and general training
- Who burdens training cost?
- Risk of obsolescence
- Ease of use

Part 9

PROJECT EXECUTION

The best prepared plans can result in a project failure because of poor execution. Project execution involves the working relationships among the participants and whether or not they support project management. There are two critical working relationships: the project–line manager interface and the project–executive management interface.

There are other factors that can affect the execution of a project. These include open communications, honesty, and integrity in dealing with customers, truth in negotiations, and factual status reporting. Execution can also be influenced by the quality of the original project plan. A project plan based on faulty or erroneous assumptions can destroy morale and impact execution.

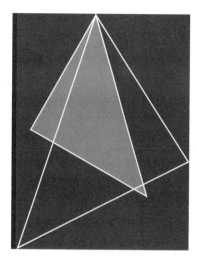

The Blue Spider Project

"This is impossible! Just totally impossible! Ten months ago I was sitting on top of the world. Upper-level management considered me one of the best, if not the best, engineer in the plant. Now look at me! I have bags under my eyes, I haven't slept soundly in the last six months, and here I am, cleaning out my desk. I'm sure glad they gave me back my old job in engineering. I guess I could have saved myself a lot of grief and aggravation had I not accepted the promotion to project manager."

HISTORY

Gary Anderson had accepted a position with Parks Corporation right out of college. With a Ph.D. in mechanical engineering, Gary was ready to solve the world's most traumatic problems. At first, Parks Corporation offered Gary little opportunity to do the pure research that he eagerly wanted to undertake. However, things soon changed. Parks grew into a major electronics and structural design corporation during the big boom of the late 1950s and early 1960s when Department of Defense (DoD) contracts were plentiful.

Parks Corporation grew from a handful of engineers to a major DoD contractor, employing some 6,500 people. During the recession of the late 1960s, money became scarce and major layoffs resulted in lowering the employment level to 2,200 employees. At that time, Parks decided to get out of the R&D business and compete as a low-cost production facility while maintaining an engineering organization solely to support production requirements.

After attempts at virtually every project management organizational structure, Parks Corporation selected the matrix form. Each project had a program manager who reported to the director of program management. Each project also maintained an assistant project manager—normally a project engineer—who reported directly to the project manager and indirectly to the director of engineering. The program manager spent most of his time worrying about cost and time, whereas the assistant program manager worried more about technical performance.

With the poor job market for engineers, Gary and his colleagues began taking coursework toward MBA degrees in case the job market deteriorated further.

In 1995, with the upturn in DoD spending, Parks had to change its corporate strategy. Parks had spent the last seven years bidding on the production phase of large programs. Now, however, with the new evaluation criteria set forth for contract awards, those companies winning the R&D and qualification phases had a definite edge on being awarded the production contract. The production contract was where the big profits could be found. In keeping with this new strategy, Parks began to beef up its R&D engineering staff. By 1998, Parks had increased in size to 2,700 employees. The increase was mostly in engineering. Experienced R&D personnel were difficult to find for the salaries that Parks was offering. Parks was, however, able to lure some employees away from the competitors, but relied mostly upon the younger, inexperienced engineers fresh out of college.

With the adoption of this corporate strategy, Parks Corporation administered a new wage and salary program that included job upgrading. Gary was promoted to senior scientist, responsible for all R&D activities performed in the mechanical engineering department. Gary had distinguished himself as an outstanding production engineer during the past several years, and management felt that his contribution could be extended to R&D as well.

In January 1998, Parks Corporation decided to compete for Phase I of the Blue Spider Project, an R&D effort that, if successful, could lead into a $500 million program spread out over 20 years. The Blue Spider Project was an attempt to improve the structural capabilities of the Spartan missile, a short-range tactical missile used by the Army. The Spartan missile was exhibiting fatigue failure after six years in the field. This was three years less than what the original design specifications called for. The Army wanted new materials that could result in a longer life for the Spartan missile.

Lord Industries was the prime contractor for the Army's Spartan Program. Parks Corporation would be a subcontractor to Lord if they could successfully bid and win the project. The criteria for subcontractor selection were based not only on low bid, but also on technical expertise as well as management performance on other projects. Park's management felt that it had a distinct advantage over most of the other competitors because they had successfully worked on other projects for Lord Industries.

THE BLUE SPIDER PROJECT KICKOFF

On November 3, 1997, Henry Gable, the director of engineering, called Gary Anderson into his office.

Henry Gable: "Gary, I've just been notified through the grapevine that Lord will be issuing the RFP for the Blue Spider Project by the end of this month, with a 30-day response period. I've been waiting a long time for a project like this to come along so that I can experiment with some new ideas that I have. This project is going to be my baby all the way! I want you to head up the proposal team. I think it must be an engineer. I'll make sure that you get a good proposal manager to help you. If we start working now, we can get close to two months of research in before proposal submittal. That will give us a one-month's edge on our competitors."

Gary was pleased to be involved in such an effort. He had absolutely no trouble in getting functional support for the R&D effort necessary to put together a technical proposal. All of the functional managers continually remarked to Gary, "This must be a biggy. The director of engineering has thrown all of his support behind you."

On December 2, the RFP was received. The only trouble area that Gary could see was that the technical specifications stated that all components must be able to operate normally and successfully through a temperature range of −65 °F to 145 °F. Current testing indicated the Parks Corporation's design would not function above 130 °F. An intensive R&D effort was conducted over the next three weeks. Everywhere Gary looked, it appeared that the entire organization was working on his technical proposal.

A week before the final proposal was to be submitted, Gary and Henry Gable met to develop a company position concerning the inability of the preliminary design material to be operated above 130 °F.

Gary Anderson: "Henry, I don't think it is going to be possible to meet specification requirements unless we change our design material or incorporate new materials. Everything I've tried indicates we're in trouble."

Gable: "We're in trouble only if the customer knows about it. Let the proposal state that we expect our design to be operative up to 155 °F. That'll please the customer."

Anderson: "That seems unethical to me. Why don't we just tell them the truth?"

Gable: "The truth doesn't always win proposals. I picked you to head up this effort because I thought that you'd understand. I could have just as easily selected one of our many moral project managers. I'm considering you for program manager after we win the program. If you're going to pull this conscientious crap on me like the other project managers do, I'll find someone else. Look at it this way; later we can convince the customer to change the specifications. After all, we'll be so far downstream that he'll have no choice."

After two solid months of sixteen-hour days for Gary, the proposal was submitted. On February 10, 1998, Lord Industries announced that Parks Corporation would be awarded the Blue Spider Project. The contract called for a ten-month effort, negotiated at $2.2 million at a firm-fixed price.

SELECTING THE PROJECT MANAGER

Following contract award, Henry Gable called Gary in for a conference.

Gable: "Congratulations, Gary! You did a fine job. The Blue Spider Project has great potential for ongoing business over the next ten years, provided that we perform well during the R&D phase. Obviously you're the most qualified person in the plant to head up the project. How would you feel about a transfer to program management?"

Anderson: "I think it would be a real challenge. I could make maximum use of the MBA degree I earned last year. I've always wanted to be in program management."

Gable: "Having several masters' degrees, or even doctorates for that matter, does not guarantee that you'll be a successful project manager. There are three requirements for effective program management: You must be able to communicate both in writing and orally; you must know how to motivate people; and you must be willing to give up your car pool. The last one is extremely important in that program managers must be totally committed and dedicated to the program, regardless of how much time is involved.

"But this is not the reason why I asked you to come here. Going from project engineer to program management is a big step. There are only two places you can go from program management—up the organization or out the door. I know of very, very few engineers who failed in program management and were permitted to return."

Anderson: "Why is that? If I'm considered to be the best engineer in the plant, why can't I return to engineering?"

Gable: "Program management is a world of its own. It has its own formal and informal organizational ties. Program managers are outsiders. You'll find out. You might not be able to keep the strong personal ties you now have with your fellow employees. You'll have to force even your best friends to comply with your standards. Program managers can go from program to program, but functional departments remain intact.

"I'm telling you all this for a reason. We've worked well together the past several years. But if I sign the release so that you can work for Grey in program management, you'll be on your own, like hiring into a new company. I've already signed the release. You still have some time to think about it."

Anderson: "One thing I don't understand. With all of the good program managers we have here, why am I given this opportunity?"

Gable: "Almost all of our program managers are over forty-five years old. This resulted from our massive layoffs several years ago when we were forced to lay off the younger, inexperienced program managers. You were selected because of your age and because all of our other program managers have worked only on production-type programs. We need someone at the reins who knows R&D. Your counterpart at Lord Industries will be an R&D type. You have to fight fire with fire.

"I have an ulterior reason for wanting you to accept this position. Because of the division of authority between program management and project engineering, I need someone in program management whom I can communicate with concerning R&D work. The program managers we have now are interested only in time and cost. We need a manager who will bend over backwards to get performance also. I think you're that man. You know the commitment we made to Lord when we submitted that proposal. You have to try to achieve that. Remember, this program is my baby. You'll get all the support you need. I'm tied up on another project now. But when it's over, I'll be following your work like a hawk. We'll have to get together occasionally and discuss new techniques.

"Take a day or two to think it over. If you want the position, make an appointment to see Elliot Grey, the director of program management. He'll give you the same speech I did. I'll assign Paul Evans to you as chief project engineer. He's a seasoned veteran and you should have no trouble working with him. He'll give you good advice. He's a good man."

THE WORK BEGINS

Gary accepted the new challenge. His first major hurdle occurred in staffing the project. The top priority given to him to bid the program did not follow through for staffing. The survival of Parks Corporation depended on the profits received from the production programs. In keeping with this philosophy Gary found that engineering managers (even his former boss) were reluctant to give up their key people to the Blue Spider Program. However, with a little support from Henry Gable, Gary formed an adequate staff for the program.

Right from the start Gary was worried that the test matrix called out in the technical volume of the proposal would not produce results that could satisfy specifications. Gary had ninety days after go-ahead during which to identify the raw materials that could satisfy specification requirements. Gary and Paul Evans held a meeting to map out their strategy for the first few months.

Anderson: "Well, Paul, we're starting out with our backs against the wall on this one. Any recommendations?"

Paul Evans: "I also have my doubts about the validity of this test matrix. Fortunately, I've been through this before. Gable thinks this is his project and he'll sure as hell try to manipulate us. I have to report to him every morning at 7:30 A.M. with the raw data results of the previous day's testing. He wants to see it before you do. He also stated that he wants to meet with me alone.

"Lord will be the big problem. If the test matrix proves to be a failure, we're going to have to change the scope of effort. Remember, this is an FFP contract. If we change the scope of work and do additional work in the earlier phases of the program, then we should prepare a trade-off analysis to see what we can delete downstream so as to not overrun the budget."

Anderson: "I'm going to let the other project office personnel handle the administrating work. You and I are going to live in the research labs until we get some results. We'll let the other project office personnel run the weekly team meetings."

For the next three weeks Gary and Paul spent virtually twelve hours per day, seven days a week, in the research and development lab. None of the results showed any promise. Gary kept trying to set up a meeting with Henry Gable but always found him unavailable.

During the fourth week, Gary, Paul, and the key functional department managers met to develop an alternate test matrix. The new test matrix looked good. Gary and his team worked frantically to develop a new workable schedule that would not have impact on the second milestone, which was to occur at the end of 180 days. The second milestone was the final acceptance of the raw materials and preparation of production runs of the raw materials to verify that there would be no scale-up differences between lab development and full-scale production.

Gary personally prepared all of the technical handouts for the interchange meeting. After all, he would be the one presenting all of the data. The technical interchange meeting was scheduled for two days. On the first day, Gary presented all of the data, including test results, and the new test matrix. The customer appeared displeased with the progress to date and decided to have its own in-house caucus that evening to go over the material that was presented.

The following morning the customer stated its position: "First of all, Gary, we're quite pleased to have a project manager who has such a command of technology. That's good. But every time we've tried to contact you last month, you were unavailable or had to be paged in the research laboratories. You did an acceptable job presenting the technical data, but the administrative data was presented by your project office personnel. We, at Lord, do not think that you're maintaining the proper balance between your technical and administrative responsibilities. We prefer that you personally give the administrative data and your chief project engineer present the technical data.

"We did not receive any agenda. Our people like to know what will be discussed, and when. We also want a copy of all handouts to be presented at least three days in advance. We need time to scrutinize the data. You can't expect us to walk in here blind and make decisions after seeing the data for ten minutes.

"To be frank, we feel that the data to date is totally unacceptable. If the data does not improve, we will have no choice but to issue a work stoppage order and look for a new contractor. The new test matrix looks good, especially since this is a firm-fixed-price contract. Your company will bear the burden of all costs for the additional work. A trade-off with later work may be possible, but this will depend on the results presented at the second design review meeting, 90 days from now.

"We have decided to establish a customer office at Parks to follow your work more closely. Our people feel that monthly meetings are insufficient during R&D activities. We would like our customer representative to have daily verbal meetings with you or your staff. He will then keep us posted. Obviously, we had expected to review much more experimental data than you have given us.

"Many of our top-quality engineers would like to talk directly to your engineering community, without having to continually waste time by having to go through the project office. We must insist on this last point. Remember, your effort may be only $2.2 million, but our total package is $100 million. We have a lot more at stake than you people do. Our engineers do not like to get information that has been filtered by the project office. They want to help you.

"And last, don't forget that you people have a contractual requirement to prepare complete minutes for all interchange meetings. Send us the original for signature before going to publication."

Although Gary was unhappy with the first team meeting, especially with the requests made by Lord Industries, he felt that they had sufficient justification for their comments. Following the team meeting, Gary personally prepared the complete minutes. "This is absurd," thought Gary. "I've wasted almost one entire week doing nothing more than administrative paperwork. Why do we need such detailed minutes? Can't a rough summary suffice? Why is it that customers want everything documented? That's like an indication of fear. We've been completely cooperative with them. There has been no hostility between us. If we've gotten this much paperwork to do now, I hate to imagine what it will be like if we get into trouble."

A NEW ROLE

Gary completed and distributed the minutes to the customer as well as to all key team members.

For the next five weeks testing went according to plan, or at least Gary thought that it had. The results were still poor. Gary was so caught up in administrative paperwork that he hadn't found time to visit the research labs in over a month. On a Wednesday morning, Gary entered the lab to observe the morning testing. Upon arriving in the lab, Gary found Paul Evans, Henry Gable, and two technicians testing a new material, JXB-3.

Gable: "Gary, your problems will soon be over. This new material, JXB-3, will permit you to satisfy specification requirements. Paul and I have been testing it for two weeks. We wanted to let you know, but were afraid that if the word leaked out to the customer that we were spending their money for testing materials that were not called out in the program plan, they would probably go crazy and might cancel the contract. Look at these results. They're super!"

Anderson: "Am I supposed to be the one to tell the customer now? This could cause a big wave."

Gable: "There won't be any wave. Just tell them that we did it with our own IR&D funds. That'll please them because they'll think we're spending our own money to support their program."

Before presenting the information to Lord, Gary called a team meeting to present the new data to the project personnel. At the team meeting, one functional manager spoke out: "This is a hell of a way to run a program. I like to be kept informed about everything that's happening here at Parks. How can the project office expect to get support out of the functional departments if we're kept in the dark until the very last minute? My people have been working with the existing materials for the last two months and you're telling us that it was all for nothing. Now you're giving us a material that's so new that we have no information on it whatsoever. We're now going to have to play catch-up, and that's going to cost you plenty."

One week before the 180-day milestone meeting, Gary submitted the handout package to Lord Industries for preliminary review. An hour later the phone rang.

Customer: "We've just read your handout. Where did this new material come from? How come we were not informed that this work was going on? You know, of course, that our customer, the Army, will be at this meeting. How can we explain this to them? We're postponing the review meeting until all of our people have analyzed the data and are prepared to make a decision.

"The purpose of a review or interchange meeting is to exchange information when *both* parties have familiarity with the topic. Normally, we (Lord Industries) require almost weekly interchange meetings with our other customers because we don't trust them. We disregard this policy with Parks Corporation based on past working relationships. But with the new state of developments, you have forced us to revert to our previous position, since we now question Parks Corporation's integrity in communicating with us. At first we believed this was due to an inexperienced program manager. Now, we're not sure."

Anderson: "I wonder if the real reason we have these interchange meetings isn't to show our people that Lord Industries doesn't trust us. You're creating a hell of a lot of work for us, you know."

Customer: "You people put yourself in this position. Now you have to live with it."

Two weeks later Lord reluctantly agreed that the new material offered the greatest promise. Three weeks later the design review meeting was held. The Army was definitely not pleased with the prime contractor's recommendation to put a new, untested material into a multimillion-dollar effort.

THE COMMUNICATIONS BREAKDOWN

During the week following the design review meeting Gary planned to make the first verification mix in order to establish final specifications for selection of the raw materials. Unfortunately, the manufacturing plans were a week behind schedule, primarily because of Gary, since he had decided to reduce costs by accepting the responsibility for developing the bill of materials himself.

A meeting was called by Gary to consider rescheduling of the mix.

Anderson: "As you know we're about a week to ten days behind schedule. We'll have to reschedule the verification mix for late next week."

Production manager: "Our resources are committed until a month from now. You can't expect to simply call a meeting and have everything reshuffled for the Blue Spider Program. We should have been notified earlier. Engineering has the responsibility for preparing the bill of materials. Why aren't they ready?"

Engineering integration: "We were never asked to prepare the bill of materials. But I'm sure that we could get it out if we work our people overtime for the next two days."

Anderson: "When can we remake the mix?"

Production manager: "We have to redo at least 500 sheets of paper every time we reschedule mixes. Not only that, we have to reschedule people on all three shifts. If we are to reschedule your mix, it will have to be performed on overtime. That's going to increase your costs. If that's agreeable with you, we'll try it. But this will be the first and last time that production will bail you out. There are procedures that have to be followed."

Testing engineer: "I've been coming to these meetings since we kicked off this program. I think I speak for the entire engineering division when I say that the role that the director of engineering is playing in this program is suppressing individuality among our highly competent personnel. In new projects, especially those involving R&D, our people are not apt to stick their necks out. Now our people are becoming ostriches. If they're impeded from contributing, even in their own slight way, then you'll probably lose them before the project gets completed. Right now I feel that I'm wasting my time here. All I need are minutes of the team meetings and I'll be happy. Then I won't have to come to these pretend meetings anymore."

The purpose of the verification mix was to make a full-scale production run of the material to verify that there would be no material property changes in scale-up from the small mixes made in the R&D laboratories. After testing, it became obvious that the wrong lots of raw materials were used in the production verification mix.

A meeting was called by Lord Industries for an explanation of why the mistake had occurred and what the alternatives were.

Lord: "Why did the problem occur?"

Anderson: "Well, we had a problem with the bill of materials. The result was that the mix had to be made on overtime. And when you work people on overtime, you have to be willing to accept mistakes as being a way of life. The energy cycles of our people are slow during the overtime hours."

Lord: "The ultimate responsibility has to be with you, the program manager. We, at Lord, think that you're spending too much time doing and not enough time managing. As the prime contractor, we have a hell of a lot more at stake than you do. From now on we want documented weekly technical interchange meetings and closer interaction by our quality control section with yours."

Anderson: "These additional team meetings are going to tie up our key people. I can't spare people to prepare handouts for weekly meetings with your people."

Lord: "Team meetings are a management responsibility. If Parks does not want the Blue Spider Program, I'm sure we can find another subcontractor. All you (Gary) have to do is give up taking the material vendors to lunch and you'll have plenty of time for handout preparation."

Gary left the meeting feeling as though he had just gotten raked over the coals. For the next two months, Gary worked sixteen hours a day, almost every day. Gary did not want to burden his staff with the responsibility of the handouts, so he began preparing them himself. He could have hired additional staff, but with such a tight budget, and having to remake verification mix, cost overruns appeared inevitable.

As the end of the seventh month approached, Gary was feeling pressure from within Parks Corporation. The decision-making process appeared to be slowing down, and Gary found it more and more difficult to motivate his people. In fact, the grapevine was referring to the Blue Spider Project as a loser, and some of his key people acted as though they were on a sinking ship.

By the time the eighth month rolled around, the budget had nearly been expended. Gary was tired of doing everything himself. "Perhaps I should have stayed an engineer," thought Gary. Elliot Grey and Gary Anderson had a meeting to see what could be salvaged. Grey agreed to get Gary additional corporate funding to complete the project. "But performance must be met, since there is a lot riding on the Blue Spider Project," asserted Grey. He called a team meeting to identify the program status.

Anderson: "It's time to map out our strategy for the remainder of the program. Can engineering and production adhere to the schedule that I have laid out before you?"

Team member, engineering: "This is the first time that I've seen this schedule. You can't expect me to make a decision in the next ten minutes and commit the resources of my department. We're getting a little unhappy being kept in the dark until the last minute. What happened to effective planning?"

Anderson: "We still have effective planning. We must adhere to the original schedule, or at least try to adhere to it. This revised schedule will do that."

Team member, engineering: "Look, Gary! When a project gets in trouble it is usually the functional departments that come to the rescue. But if we're kept in the dark, then how can you expect us to come to your rescue? My boss wants to know, well in advance, every decision that you're contemplating with regard to our departmental resources. Right now, we . . ."

Anderson: "Granted, we may have had a communications problem. But now we're in trouble and have to unite forces. What is your impression as to whether your department can meet the new schedule?"

Team member, engineering: "When the Blue Spider Program first got in trouble, my boss exercised his authority to make all departmental decisions regarding the program himself. I'm just a puppet. I have to check with him on everything."

Team member, production: "I'm in the same boat, Gary. You know we're not happy having to reschedule our facilities and people. We went through this once before. I also have to check with my boss before giving you an answer about the new schedule."

The following week the verification mix was made. Testing proceeded according to the revised schedule, and it looked as though the total schedule milestones could be met, provided that specifications could be adhered to.

Because of the revised schedule, some of the testing had to be performed on holidays. Gary wasn't pleased with asking people to work on Sundays and holidays, but he had no choice, since the test matrix called for testing to be accomplished at specific times after end-of-mix.

A team meeting was called on Wednesday to resolve the problem of who would work on the holiday, which would occur on Friday, as well as staffing Saturday and Sunday. During the team meeting Gary became quite disappointed. Phil Rodgers, who had been Gary's test engineer since the project started, was assigned to a new project that the grapevine called Gable's new adventure. His replacement was a relatively new man, only eight months with the company. For an hour and a half, the team members argued about the little problems and continually avoided the major question, stating that they would first have to coordinate commitments with their bosses. It was obvious to Gary that his team members were afraid to make major decisions and therefore "ate up" a lot of time on trivial problems.

On the following day, Thursday, Gary went to see the department manager responsible for testing, in hopes that he could use Phil Rodgers this weekend.

Department manager: "I have specific instructions from the boss (director of engineering) to use Phil Rodgers on the new project. You'll have to see the boss if you want him back."

Anderson: "But we have testing that must be accomplished this weekend. Where's the new man you assigned yesterday?"

Department manager: "Nobody told me you had testing scheduled for this weekend. Half of my department is already on an extended weekend vacation, including Phil Rodgers and the new man. How come I'm always the last to know when we have a problem?"

Anderson: "The customer is flying down his best people to observe this weekend's tests. It's too late to change anything. You and I can do the testing."

Department manager: "Not on your life. I'm staying as far away as possible from the Blue Spider Project. I'll get you someone, but it won't be me. That's for sure!"

The weekend's testing went according to schedule. The raw data was made available to the customer under the stipulation that the final company position would be announced at the end of the next month, after the functional departments had a chance to analyze it.

Final testing was completed during the second week of the ninth month. The initial results looked excellent. The materials were within contract specifications, and although they were new, both Gary and Lord's management felt that there would be little difficulty in convincing the Army that this was the way to go. Henry Gable visited Gary and congratulated him on a job well done.

All that now remained was the making of four additional full-scale verification mixes in order to determine how much deviation there would be in material properties between full-sized production-run mixes. Gary tried to get the customer to concur (as part of the original trade-off analysis) that two of the four production runs could be deleted. Lord's management refused, insisting that contractual requirements must be met at the expense of the contractor.

The following week, Elliot Grey called Gary in for an emergency meeting concerning expenditures to date.

Elliot Grey: "Gary, I just received a copy of the financial planning report for last quarter in which you stated that both the cost and performance of the Blue Spider Project were 75 percent complete. I don't think you realize what you've done. The target profit on the program was $200,000. Your memo authorized the vice president and general manager to book 75 percent of that, or $150,000, for corporate profit spending for stockholders. I was planning on using all $200,000 together with the additional $300,000 I personally requested from corporate headquarters to bail you out. Now I have to go back to the vice president and general manager and tell them that we've made a mistake and that we'll need an additional $150,000."

Anderson: "Perhaps I should go with you and explain my error. Obviously, I take all responsibility."

Grey: "No, Gary. It's our error, not yours. I really don't think you want to be around the general manager when he sees red at the bottom of the page. It takes an act of God to get money back once corporate books it as profit. Perhaps you should reconsider project engineering as a career instead of program management. Your performance hasn't exactly been sparkling, you know."

Gary returned to his office quite disappointed. No matter how hard he worked, the bureaucratic red tape of project management seemed always to do him in. But late that afternoon, Gary's disposition improved. Lord Industries called to say that, after consultation with the Army, Parks Corporation would be awarded a sole-source contract for qualification and production of Spartan missile components using the new longer-life raw materials. Both Lord and the Army felt that the sole-source contract was justified, provided that continued testing showed the same results, since Parks Corporation had all of the technical experience with the new materials.

Gary received a letter of congratulations from corporate headquarters, but no additional pay increase. The grapevine said that a substantial bonus was given to the director of engineering.

During the tenth month, results were coming back from the accelerated aging tests performed on the new materials. The results indicated that although the new materials would meet specifications, the age life would probably be less than five years. These numbers came as a shock to Gary. Gary and Paul Evans had a conference to determine the best strategy to follow.

Anderson: "Well, I guess we're now in the fire instead of the frying pan. Obviously, we can't tell Lord Industries about these tests. We ran them on our own. Could the results be wrong?"

Evans: "Sure, but I doubt it. There's always margin for error when you perform accelerated aging tests on new materials. There can be reactions taking place that we know nothing about. Furthermore, the accelerated aging tests may not even correlate well with actual aging. We must form a company position on this as soon as possible."

Anderson: "I'm not going to tell anyone about this, especially Henry Gable. You and I will handle this. It will be my throat if word of this leaks out. Let's wait until we have the production contract in hand."

Evans: "That's dangerous. This has to be a company position, not a project office position. We had better let them know upstairs."

Anderson: "I can't do that. I'll take all responsibility. Are you with me on this?"

Evans: "I'll go along. I'm sure I can find employment elsewhere when we open Pandora's box. You had better tell the department managers to be quiet also."

Two weeks later, as the program was winding down into the testing for the final verification mix and final report development, Gary received an urgent phone call asking him to report immediately to Henry Gable's office.

Gable: "When this project is over, you're through. You'll never hack it as a program manager, or possibly a good project engineer. We can't run projects around here without honesty and open communications. How the hell do you expect top management to support you when you start censoring bad news to the top? I don't like surprises. I like to get the bad news from the program manager and project engineers, not secondhand from the customer. And of course, we cannot forget the cost overrun. Why didn't you take some precautionary measures?"

Anderson: "How could I when you were asking our people to do work such as accelerated aging tests that would be charged to my project and was not part of program plan? I don't think I'm totally to blame for what's happened."

Gable: "Gary, I don't think it's necessary to argue the point any further. I'm willing to give you back your old job, in engineering. I hope you didn't lose too many friends while working in program management. Finish up final testing and the program report. Then I'll reassign you."

Gary returned to his office and put his feet up on the desk. "Well," thought Gary, "perhaps I'm better off in engineering. At least I can see my wife and kids once in a while." As Gary began writing the final report, the phone rang:

Functional manager: "Hello, Gary. I just thought I'd call to find out what charge number you want us to use for experimenting with this new procedure to determine accelerated age life."

Anderson: "Don't call me! Call Gable. After all, the Blue Spider Project is his baby."

QUESTIONS

1. If you were Gary Anderson, would you have accepted this position after the director stated that this project would be his baby all the way?
2. Do engineers with MBA degrees aspire to high positions in management?
3. Was Gary qualified to be a project manager?
4. What are the moral and ethical issues facing Gary?
5. What authority does Gary Anderson have and to whom does he report?
6. Is it true when you enter project management, you either go up the organization or out the door?
7. Is it possible for an executive to take too much of an interest in an R&D project?
8. Should Paul Evans have been permitted to report information to Gable before reporting it to the project manager?
9. Is it customary for the project manager to prepare all of the handouts for a customer interchange meeting?
10. What happens when a situation of mistrust occurs between the customer and contractor?
11. Should functional employees of the customer and contractor be permitted to communicate with one another without going through the project office?
12. Did Gary demonstrate effective time management?
13. Did Gary understand production operations?

14. Are functional employees authorized to make project decisions?
15. On R&D projects, should profits be booked periodically or at project termination?
16. Should a project manager ever censor bad news?
17. Could the above-mentioned problems have been resolved if there had been a singular methodology for project management in place?
18. Can a singular methodology for project management specify morality and ethics in dealing with customers? If so, how do we then handle situations where the project manager violates protocol?
19. Could the lessons learned on success and failure during project debriefings cause a major change in the project management methodology?

Corwin Corporation[1]

By June 2003, Corwin Corporation had grown into a $950 million per year corporation with an international reputation for manufacturing low-cost, high-quality rubber components. Corwin maintained more than a dozen different product lines, all of which were sold as off-the-shelf items in department stores, hardware stores, and automotive parts distributors. The name *Corwin* was now synonymous with "quality." This provided management with the luxury of having products that maintained extremely long life cycles.

Organizationally, Corwin had maintained the same structure for more than fifteen years (see Exhibit I). The top management of Corwin Corporation was highly conservative and believed in using a marketing approach to find new markets for existing product lines rather than exploring for new products. Under this philosophy, Corwin maintained a small R&D group whose mission was simply to evaluate state-of-the-art technology and its application to existing product lines.

Corwin's reputation was so good that it continually received inquiries about the manufacturing of specialty products. Unfortunately, the conservative nature of Corwin's management created a "do not rock the boat" atmosphere opposed to taking any type of risks. A management policy was established to evaluate all specialty-product requests. The policy required answering yes to the following questions:

- Will the specialty product provide the same profit margin (20 percent) as existing product lines?

[1]Revised 2007.

Exhibit I. Organizational chart for Corwin Corporation

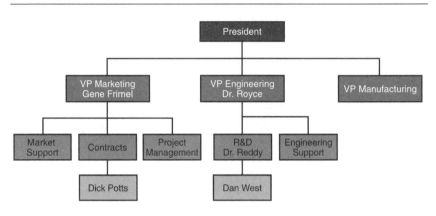

- What is the total projected profitability to the company in terms of follow-on contracts?
- Can the specialty product be developed into a product line?
- Can the specialty product be produced with minimum disruption to existing product lines and manufacturing operations?

These stringent requirements forced Corwin to no-bid more than 90 percent of all specialty-product inquiries.

Corwin Corporation was a marketing-driven organization, although manufacturing often had different ideas. Almost all decisions were made by marketing with the exception of product pricing and estimating, which was a joint undertaking between manufacturing and marketing. Engineering was considered as merely a support group to marketing and manufacturing.

For specialty products, the project managers would always come out of marketing even during the R&D phase of development. The company's approach was that if the specialty product should mature into a full product line, then there should be a product line manager assigned right at the onset.

THE PETERS COMPANY PROJECT

In 2000, Corwin accepted a specialty-product assignment from Peters Company because of the potential for follow-on work. In 2001 and 2002, and again in 2003, profitable follow-on contracts were received, and a good working relationship developed, despite Peters' reputation for being a difficult customer to work with.

On December 7, 2002, Gene Frimel, the vice president of marketing at Corwin, received a rather unusual phone call from Dr. Frank Delia, the marketing vice president at Peters Company.

Frank Delia: "Gene, I have a rather strange problem on my hands. Our R&D group has $250,000 committed for research toward development of a new rubber product material, and we simply do not have the available personnel or talent to undertake the project. We have to go outside. We'd like your company to do the work. Our testing and R&D facilities are already overburdened."

Gene Frimel: "Well, as you know, Frank, we are not a research group even though we've done this once before for you. And furthermore, I would never be able to sell our management on such an undertaking. Let some other company do the R&D work and then we'll take over on the production end."

Delia: "Let me explain our position on this. We've been burned several times in the past. Projects like this generate several patents, and the R&D company almost always requires that our contracts give them royalties or first refusal for manufacturing rights."

Frimel: "I understand your problem, but it's not within our capabilities. This project, if undertaken, could disrupt parts of our organization. We're already operating lean in engineering."

Delia: "Look, Gene! The bottom line is this: We have complete confidence in your manufacturing ability to such a point that we're willing to commit to a five-year production contract if the product can be developed. That makes it extremely profitable for you."

Frimel: "You've just gotten me interested. What additional details can you give me?"

Delia: "All I can give you is a rough set of performance specifications that we'd like to meet. Obviously, some trade-offs are possible."

Frimel: "When can you get the specification sheet to me?"

Delia: "You'll have it tomorrow morning. I'll ship it overnight express."

Frimel: "Good! I'll have my people look at it, but we won't be able to get you an answer until after the first of the year. As you know, our plant is closed down for the last two weeks in December, and most of our people have already left for extended vacations."

Delia: "That's not acceptable! My management wants a signed, sealed, and delivered contract by the end of this month. If this is not done, corporate will reduce our budget for 2003 by $250,000, thinking that we've bitten off more than we can chew. Actually, I need your answer within 48 hours so that I'll have some time to find another source."

Frimel: "You know, Frank, today is December 7, Pearl Harbor Day. Why do I feel as though the sky is about to fall in?"

Delia: "Don't worry, Gene! I'm not going to drop any bombs on you. Just remember, all that we have available is $250,000, and the contract must be a firm-fixed-price effort. We anticipate a six-month project with $125,000 paid on contract signing and the balance at project termination."

Frimel: "I still have that ominous feeling, but I'll talk to my people. You'll hear from us with a go or no-go decision within 48 hours. I'm scheduled to go on a cruise in the Caribbean, and my wife and I are leaving this evening. One of my people will get back to you on this matter."

Gene Frimel had a problem. All bid and no-bid decisions were made by a four-man committee composed of the president and the three vice presidents. The president and the vice president for manufacturing were on vacation. Frimel met with Dr. Royce, the vice president of engineering, and explained the situation.

Royce: "You know, Gene, I totally support projects like this because it would help our technical people grow intellectually. Unfortunately, my vote never appears to carry any weight."

Frimel: "The profitability potential as well as the development of good customer relations makes this attractive, but I'm not sure we want to accept such a risk. A failure could easily destroy our good working relationship with Peters Company."

Royce: "I'd have to look at the specification sheets before assessing the risks, but I would like to give it a shot."

Frimel: "I'll try to reach our president by phone."

By late afternoon, Frimel was fortunate enough to be able to contact the president and received a reluctant authorization to proceed. The problem now was how to prepare a proposal within the next two or three days and be prepared to make an oral presentation to Peters Company.

Frimel: "The Boss gave his blessing, Royce, and the ball is in your hands. I'm leaving for vacation, and you'll have total responsibility for the proposal and presentation. Delia wants the presentation this weekend. You should have his specification sheets tomorrow morning."

Royce: "Our R&D director, Dr. Reddy, left for vacation this morning. I wish he were here to help me price out the work and select the project manager. I assume that, in this case, the project manager will come out of engineering rather than marketing."

Frimel: "Yes, I agree. Marketing should not have any role in this effort. It's your baby all the way. And as for the pricing effort, you know our bid will be for $250,000. Just work backwards to justify the numbers. I'll assign one of our contracting people to assist you in the pricing. I hope I can find someone who has experience in this type of effort. I'll call Delia and tell him we'll bid it with an unsolicited proposal."

Royce selected Dan West, one of the R&D scientists, to act as the project leader. Royce had severe reservations about doing this without the R&D director, Dr. Reddy, being actively involved. With Reddy on vacation, Royce had to make an immediate decision.

On the following morning, the specification sheets arrived and Royce, West, and Dick Potts, a contracts man, began preparing the proposal. West prepared the direct labor man-hours, and Royce provided the costing data and pricing rates. Potts, being completely unfamiliar with this type of effort, simply acted as an observer and provided legal advice when necessary. Potts allowed Royce to make all decisions even though the contracts man was considered the official representative of the president.

Finally completed two days later, the proposal was actually a ten-page letter that simply contained the cost summaries (see Exhibit II) and the engineering intent. West estimated that 30 tests would be required. The test matrix described the test conditions only for the first five tests. The remaining 25 test conditions would be determined at a later date, jointly by Peters and Corwin personnel.

On Sunday morning, a meeting was held at Peters Company, and the proposal was accepted. Delia gave Royce a letter of intent authorizing Corwin

Exhibit II. Proposal cost summaries

Direct labor and support	$ 30,000
Testing (30 tests at $2,000 each)	60,000
Overhead at 100%	90,000
Materials	30,000
G&A (general and administrative, 10%)	21,000
Total	$231,000
Profit	19,000
Total	$250,000

Corporation to begin working on the project immediately. The final contract would not be available for signing until late January, and the letter of intent simply stated that Peters Company would assume all costs until such time that the contract was signed or the effort terminated.

West was truly excited about being selected as the project manager and being able to interface with the customer, a luxury that was usually given only to the marketing personnel. Although Corwin Corporation was closed for two weeks over Christmas, West still went into the office to prepare the project schedules and to identify the support he would need in the other areas, thinking that if he presented this information to management on the first day back to work, they would be convinced that he had everything under control.

THE WORK BEGINS

On the first working day in January 2003, a meeting was held with the three vice presidents and Dr. Reddy to discuss the support needed for the project. (West was not in attendance at this meeting, although all participants had a copy of his memo.)

Reddy: "I think we're heading for trouble in accepting this project. I've worked with Peters Company previously on R&D efforts, and they're tough to get along with. West is a good man, but I would never have assigned him as the project leader. His expertise is in managing internal rather than external projects. But, no matter what happens, I'll support West the best I can."

Royce: "You're too pessimistic. You have good people in your group and I'm sure you'll be able to give him the support he needs. I'll try to look in on the project every so often. West will still be reporting to you for this project. Try not to burden him too much with other work. This project is important to the company."

West spent the first few days after vacation soliciting the support that he needed from the other line groups. Many of the other groups were upset that they had not been informed earlier and were unsure as to what support they could provide. West met with Reddy to discuss the final schedules.

Reddy: "Your schedules look pretty good, Dan. I think you have a good grasp on the problem. You won't need very much help from me. I have a lot of work to do on other activities, so I'm just going to be in the background on this project. Just drop me a note every once in a while telling me what's going on. I don't need anything formal. Just a paragraph or two will suffice."

By the end of the third week, all of the raw materials had been purchased, and initial formulations and testing were ready to begin. In addition, the contract was ready for signature. The contract contained a clause specifying that Peters Company had the right to send an in-house representative into Corwin Corporation for the duration of the project. Peters Company informed Corwin that Patrick Ray would be the in-house representative, reporting to Delia, and would assume his responsibilities on or about February 15.

By the time Pat Ray appeared at Corwin Corporation, West had completed the first three tests. The results were not what was expected, but gave promise that Corwin was heading in the right direction. Pat Ray's interpretation of the tests was completely opposite to that of West. Ray thought that Corwin was "way off base," and that redirection was needed.

Pat Ray: "Look, Dan! We have only six months to do this effort and we shouldn't waste our time on marginally acceptable data. These are the next five tests I'd like to see performed."

Dan West: "Let me look over your request and review it with my people. That will take a couple of days, and, in the meanwhile, I'm going to run the other two tests as planned."

Ray's arrogant attitude bothered West. However, West decided that the project was too important to "knock heads" with Ray and simply decided to cater to Ray the best he could. This was not exactly the working relationship that West expected to have with the in-house representative.

West reviewed the test data and the new test matrix with engineering personnel, who felt that the test data was inconclusive as yet and preferred to withhold their opinion until the results of the fourth and fifth tests were made available. Although this displeased Ray, he agreed to wait a few more days if it meant getting Corwin Corporation on the right track.

The fourth and fifth tests appeared to be marginally acceptable just as the first three had been. Corwin's engineering people analyzed the data and made their recommendations.

West: "Pat, my people feel that we're going in the right direction and that our path has greater promise than your test matrix."

Ray: "As long as we're paying the bills, we're going to have a say in what tests are conducted. Your proposal stated that we would work together in developing the other test conditions. Let's go with my test matrix. I've already reported back to my boss that the first five tests were failures and that we're changing the direction of the project."

West: "I've already purchased $30,000 worth of raw materials. Your matrix uses other materials and will require additional expenditures of $12,000."

Ray: "That's your problem. Perhaps you shouldn't have purchased all of the raw materials until we agreed on the complete test matrix."

During the month of February, West conducted 15 tests, all under Ray's direction. The tests were scattered over such a wide range that no valid conclusions could be drawn. Ray continued sending reports back to Delia confirming that Corwin was not producing beneficial results and there was no indication that the situation would reverse itself. Delia ordered Ray to take any steps necessary to ensure a successful completion of the project.

Ray and West met again as they had done for each of the past 45 days to discuss the status and direction of the project.

Ray: "Dan, my boss is putting tremendous pressure on me for results, and thus far I've given him nothing. I'm up for promotion in a couple of months and I can't let this project stand in my way. It's time to completely redirect the project."

West: "Your redirection of the activities is playing havoc with my scheduling. I have people in other departments who just cannot commit to this continual rescheduling. They blame me for not communicating with them when, in fact, I'm embarrassed to."

Ray: "Everybody has their problems. We'll get this problem solved. I spent this morning working with some of your lab people in designing the next 15 tests. Here are the test conditions."

West: "I certainly would have liked to be involved with this. After all, I thought I was the project manager. Shouldn't I have been at the meeting?"

Ray: "Look, Dan! I really like you, but I'm not sure that you can handle this project. We need some good results immediately, or my neck will be stuck out for the next four months. I don't want that. Just have your lab personnel start on these tests, and we'll get along fine. Also, I'm planning on spending a great deal of time in your lab area. I want to observe the testing personally and talk to your lab personnel."

West: "We've already conducted 20 tests, and you're scheduling another 15 tests. I priced out only 30 tests in the proposal. We're heading for a cost overrun condition."

Ray: "Our contract is a firm-fixed-price effort. Therefore, the cost overrun is your problem."

West met with Dr. Reddy to discuss the new direction of the project and potential cost overruns. West brought along a memo projecting the costs through the end of the third month of the project (see Exhibit III).

Exhibit III. *Projected cost summary at the end of the third month*

	Original Proposal Cost Summary for Six-Month Project	Total Project Costs Projected at End of Third Month
Direct labor/support	$ 30,000	$ 15,000
Testing	60,000 (30 tests)	70,000 (35 tests)
Overhead	90,000 (100%)	92,000 (120%)*
Materials	30,000	50,000
G&A	21,000 (10%)	22,700 (10%)
Totals	$231,000	$249,700

*Total engineering overhead was estimated at 100 percent, whereas the R&D overhead was 120 percent.

Reddy: "I'm already overburdened on other projects and won't be able to help you out. Royce picked you to be the project manager because he felt that you could do the job. Now, don't let him down. Send me a brief memo next month explaining the situation, and I'll see what I can do. Perhaps the situation will correct itself."

During the month of March, the third month of the project, West received almost daily phone calls from the people in the lab stating that Pat Ray was interfering with their job. In fact, one phone call stated that Ray had changed the test conditions from what was agreed on in the latest test matrix. When West confronted Ray on his meddling, Ray asserted that Corwin personnel were very unprofessional in their attitude and that he thought this was being carried down to the testing as well. Furthermore, Ray demanded that one of the functional employees be removed immediately from the project because of incompetence. West stated that he would talk to the employee's department manager. Ray, however, felt that this would be useless and said, "Remove him or else!" The functional employee was removed from the project.

By the end of the third month, most Corwin employees were becoming disenchanted with the project and were looking for other assignments. West attributed this to Ray's harassment of the employees. To aggravate the situation even further, Ray met with Royce and Reddy, and demanded that West be removed and a new project manager be assigned.

Royce refused to remove West as project manager, and ordered Reddy to take charge and help West get the project back on track.

Reddy: "You've kept me in the dark concerning this project, West. If you want me to help you, as Royce requested, I'll need all the information tomorrow, especially the cost data. I'll expect you in my office tomorrow morning at 8:00 A.M. I'll bail you out of this mess."

West prepared the projected cost data for the remainder of the work and presented the results to Dr. Reddy (see Exhibit IV). Both West and Reddy agreed that the project was now out of control, and severe measures would be required to correct the situation, in addition to more than $250,000 in corporate funding.

Reddy: "Dan, I've called a meeting for 10:00 A.M. with several of our R&D people to completely construct a new test matrix. This is what we should have done right from the start."

West: "Shouldn't we invite Ray to attend this meeting? I'm sure he'd want to be involved in designing the new test matrix."

Reddy: "I'm running this show now, not Ray!! Tell Ray that I'm instituting new policies and procedures for in-house representatives. He's no longer authorized to visit the labs at his own discretion. He must be accompanied by either you or me. If he doesn't like these rules, he can get out. I'm not going to allow that guy to disrupt our organization. We're spending our money now, not his."

West met with Ray and informed him of the new test matrix as well as the new policies and procedures for in-house representatives. Ray was furious over the new turn of events and stated that he was returning to Peters Company for a meeting with Delia.

On the following Monday, Frimel received a letter from Delia stating that Peters Company was officially canceling the contract. The reasons given by Delia were as follows:

1. Corwin had produced absolutely no data that looked promising.
2. Corwin continually changed the direction of the project and did not appear to have a systematic plan of attack.
3. Corwin did not provide a project manager capable of handling such a project.

Exhibit IV.   *Estimate of total project completion costs*

Direct labor/support	$ 47,000*
Testing (60 tests)	120,000
Overhead (120%)	200,000
Materials	103,000
G&A	47,000
	$517,000
Peters contract	250,000
Overrun	$267,000

*Includes Dr. Reddy.

4. Corwin did not provide sufficient support for the in-house representative.
5. Corwin's top management did not appear to be sincerely interested in the project and did not provide sufficient executive-level support.

Royce and Frimel met to decide on a course of action in order to sustain good working relations with Peters Company. Frimel wrote a strong letter refuting all of the accusations in the Peters letter, but to no avail. Even the fact that Corwin was willing to spend $250,000 of their own funds had no bearing on Delia's decision. The damage was done. Frimel was now thoroughly convinced that a contract should not be accepted on "Pearl Harbor Day."

QUESTIONS

1. What were the major mistakes made by Corwin?
2. Should Corwin have accepted the assignment?
3. Should companies risk bidding on projects based upon rough draft specifications?
4. Should the shortness of the proposal preparation time have required more active top management involvement before the proposal went out-of-house?
5. Are there any risks in not having the vice president for manufacturing available during the go or no-go bidding decision?
6. Explain the attitude of Dick Potts during the proposal activities.
7. None of the executives expressed concern when Dr. Reddy said, "I would never have assigned him (West) as project leader." How do you account for the executives' lack of concern?
8. How important is it to inform line managers of proposal activities even if the line managers are not required to provide proposal support?
9. Explain Dr. Reddy's attitude after go-ahead.
10. How should West have handled the situation where Pat Ray's opinion of the test data was contrary to that of Corwin's engineering personnel?
11. How should West have reacted to the remarks made by Ray that he informed Delia that the first five tests were failures?
12. Is immediate procurement of all materials a mistake?
13. Should Pat Ray have been given the freedom to visit laboratory personnel at any time?
14. Should an in-house representative have the right to remove a functional employee from the project?
15. Financially, how should the extra tests have been handled?
16. Explain Dr. Reddy's attitude when told to assume control of the project.
17. Delia's letter, stating the five reasons for canceling the project, was refuted by Frimel, but with no success. Could Frimel's early involvement as a project sponsor have prevented this?

18. In retrospect, would it have been better to assign a marketing person as project manager?
19. Your company has a singular methodology for project management. You are offered a special project from a powerful customer that does not fit into your methodology. Should a project be refused simply because it is not a good fit with your methodology?
20. Should a customer be informed that only projects that fit your methodology would be accepted?

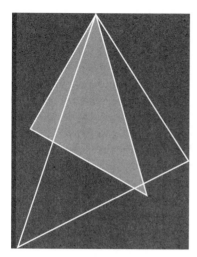

Quantum Telecom

In June of 1998, the executive committee of Quantum Telecom reluctantly approved two R&D projects that required technical breakthroughs. To make matters worse, the two products had to be developed by the summer of 1999 and introduced into the marketplace quickly. The life expectancy of both products was estimated to be less than one year because of the rate of change in technology. Yet, despite these risks, the two projects were fully funded. Two senior executives were assigned as the project sponsors, one for each project.

Quantum Telecom had a world-class project management methodology with five life cycle phases and five gate review meetings. The gate review meetings were go/no-go decision points based upon present performance and future risks. Each sponsor was authorized and empowered to make any and all decisions relative to projects, including termination.

Company politics always played an active role in decisions to terminate a project. Termination of a project often impacted the executive sponsor's advancement opportunities because the projects were promoted by the sponsors and funded through the sponsor's organization.

During the first two gate review meetings, virtually everyone recommended the termination of both projects. Technical breakthroughs seemed unlikely, and the schedule appeared unduely optimistic. But terminating the projects this early would certainly not reflect favorably upon the sponsors. Reluctantly, both sponsors agreed to continue the projects to the third gate in hopes of a "miracle."

During the third gate review, the projects were still in peril. Although the technical breakthrough opportunity now seemed plausible, the launch date would have to be slipped, thus giving Quantum Telecom a window of only six months to sell the products before obsolescence would occur.

By the fourth gate review, the technical breakthrough had not yet occurred but did still seem plausible. Both project managers were still advocating the cancellation of the projects, and the situation was getting worse. Yet, in order to "save face" within the corporation, both sponsors allowed the projects to continue to completion. They asserted that, "If the new products could not be sold in sufficient quantity to recover the R&D costs, then the fault lies with marketing and sales, not with us." The sponsors were now off the hook, so to speak.

Both projects were completed six months late. The salesforce could not sell as much as one unit, and obsolescence occurred quickly. Marketing and sales were blamed for the failures, not the project sponsors.

QUESTIONS

1. How do we eliminate politics from gate review meetings?
2. How can we develop a methodology where termination of a project is not viewed as a failure?
3. Were the wrong people assigned as sponsors?
4. What options are available to a project manager when there exists a disagreement between the sponsor and the project manager?
5. Can your answer to the above question be outlined as part of the project management methodology?

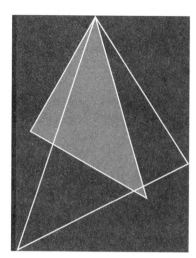

The Trophy Project

The ill-fated Trophy Project was in trouble right from the start. Reichart, who had been an assistant project manager, was involved with the project from its conception. When the Trophy Project was accepted by the company, Reichart was assigned as the project manager. The program schedules started to slip from day one, and expenditures were excessive. Reichart found that the functional managers were charging direct labor time to his project but working on their own pet projects. When Reichart complained of this, he was told not to meddle in the functional manager's allocation of resources and budgeted expenditures. After approximately six months, Reichart was requested to make a progress report directly to corporate and division staffs.

Reichart took this opportunity to bare his soul. The report substantiated that the project was forecasted to be one complete year behind schedule. Reichart's staff, as supplied by the line managers, was inadequate to stay at the required pace, let alone make up any time that had already been lost. The estimated cost at completion at this interval showed a cost overrun of at least 20 percent. This was Reichart's first opportunity to tell his story to people who were in a position to correct the situation. The result of Reichart's frank, candid evaluation of the Trophy Project was very predictable. Nonbelievers finally saw the light, and the line managers realized that they had a role to play in the completion of the project. Most of the problems were now out in the open and could be corrected by providing adequate staffing and resources. Corporate staff ordered immediate remedial action and staff support to provide Reichart a chance to bail out his program.

The results were not at all what Reichart had expected. He no longer reported to the project office; he now reported directly to the operations manager. Corporate staff's interest in the project became very intense, requiring a 7:00 A.M. meeting every Monday morning for complete review of the project status and plans for recovery. Reichart found himself spending more time preparing paperwork, reports, and projections for his Monday morning meetings than he did administering the Trophy Project. The main concern of corporate was to get the project back on schedule. Reichart spent many hours preparing the recovery plan and establishing manpower requirements to bring the program back onto the original schedule.

Group staff, in order to closely track the progress of the Trophy Project, assigned an assistant program manager. The assistant program manager determined that a sure cure for the Trophy Project would be to computerize the various problems and track the progress through a very complex computer program. Corporate provided Reichart with twelve additional staff members to work on the computer program. In the meantime, nothing changed. The functional managers still did not provide adequate staff for recovery, assuming that the additional manpower Reichart had received from corporate would accomplish that task.

After approximately $50,000 was spent on the computer program to track the problems, it was found that the program objectives could not be handled by the computer. Reichart discussed this problem with a computer supplier and found that $15,000 more was required for programming and additional storage capacity. It would take two months for installation of the additional storage capacity and the completion of the programming. At this point, the decision was made to abandon the computer program.

Reichart was now a year and a half into the program with no prototype units completed. The program was still nine months behind schedule with the overrun projected at 40 percent of budget. The customer had been receiving his reports on a timely basis and was well aware of the fact that the Trophy Project was behind schedule. Reichart had spent a great deal of time with the customer explaining the problems and the plan for recovery. Another problem that Reichart had to contend with was that the vendors who were supplying components for the project were also running behind schedule.

One Sunday morning, while Reichart was in his office putting together a report for the client, a corporate vice president came into his office. "Reichart," he said, "in any project I look at the top sheet of paper and the man whose name appears at the top of the sheet is the one I hold responsible. For this project your name appears at the top of the sheet. If you cannot bail this thing out, you are in serious trouble in this corporation." Reichart did not know which way to turn or what to say. He had no control over the functional managers who were creating the problems, but he was the person who was being held responsible.

After another three months the customer, becoming impatient, realized that the Trophy Project was in serious trouble and requested that the division general manager and his entire staff visit the customer's plant to give a progress and "get well" report within a week. The division general manager called Reichart into his office and said, "Reichart, go visit our customer. Take three or four functional line people with you and try to placate him with whatever you feel is necessary." Reichart and four functional line people visited the customer and gave a four-and-a-half-hour presentation defining the problems and the progress to that point. The customer was very polite and even commented that it was an excellent presentation, but the content was totally unacceptable. The program was still six to eight months late, and the customer demanded progress reports on a weekly basis. The customer made arrangements to assign a representative in Reichart's department to be "on-site" at the project on a daily basis and to interface with Reichart and his staff as required. After this turn of events, the program became very hectic.

The customer representative demanded constant updates and problem identification and then became involved in attempting to solve these problems. This involvement created many changes in the program and the product in order to eliminate some of the problems. Reichart had trouble with the customer and did not agree with the changes in the program. He expressed his disagreement vocally when, in many cases, the customer felt the changes were at no cost. This caused a deterioration of the relationship between client and producer.

One morning Reichart was called into the division general manager's office and introduced to Mr. "Red" Baron. Reichart was told to turn over the reins of the Trophy Project to Red immediately. "Reichart, you will be temporarily reassigned to some other division within the corporation. I suggest you start looking outside the company for another job." Reichart looked at Red and asked, "Who did this? Who shot me down?"

Red was program manager on the Trophy Project for approximately six months, after which, by mutual agreement, he was replaced by a third project manager. The customer reassigned his local program manager to another project. With the new team the Trophy Project was finally completed one year behind schedule and at a 40 percent cost overrun.

QUESTIONS

1. Did the project appear to be planned correctly?
2. Did functional management seem to be committed to the project?
3. Did senior management appear supportive and committed?
4. Can a singular methodology for project management be designed to "force" cooperation to occur between groups?
5. Is it possible or even desirable for strategic planning for project management to include ways to improve cooperation and working relationships, or is this beyond the scope of strategic planning for project management?

Concrete Masonry Corporation

INTRODUCTION

The Concrete Masonry Corporation (CMC), after being a leader in the industry for over twenty-five years, decided to get out of the prestressed concrete business. Although there had been a boom in residential construction in recent years, commercial work was on the decline. As a result, all the prestressed concrete manufacturers were going farther afield to big jobs. In order to survive, CMC was forced to bid on jobs previously thought to be out of their geographical area. Survival depended upon staying competitive.

In 1975, the average selling price of a cubic foot of concrete was $8.35, and in 1977, the average selling price had declined to $6.85. As CMC was producing at a rate of a million cubic feet a year, not much mathematics was needed to calculate they were receiving one-and-a-half million dollars per year less than they had received a short two years before for the same product.

Product management was used by CMC in a matrix organizational form. CMC's project manager had total responsibility from the design to the completion of the construction project. However, with the declining conditions of the market and the evolution that had drastically changed the character of the marketplace, CMC's previously successful approach was in question.

HISTORY—THE CONCRETE BLOCK BUSINESS

CMC started in the concrete block business in 1946. At the beginning, CMC became a leader in the marketplace for two reasons: (1) advanced technology of manufacturing and (2) an innovative delivery system. With modern equipment, specifically the flat pallet block machine, CMC was able to make different shapes of block without having to make major changes in the machinery. This change, along with the pioneering of the self-unloading boom truck, which permitted efficient, cost-saving delivery, contributed to the success of CMC's block business. Consequently, the block business success provided the capital needed for CMC to enter the prestressed concrete business.

HISTORY—THE PRESTRESSED CONCRETE BUSINESS

Prestressed concrete is made by casting concrete around steel cables that are stretched by hydraulic jacks. After the concrete hardens, the cables are releasd, thus compressing the concrete. Concrete is strongest when it is compressed. Steel is strongest when it is stretched, or in tension. In this way, CMC combined the two strongest qualities of the two materials. The effectiveness of the technique can be readily demonstrated by lifting a horizontal row of books by applying pressure at each end of the row at a point below the center of gravity.

Originally, the concrete block manufacturing business was a natural base from which to enter the prestressed concrete business because the very first prestressed concrete beams were made of a row of concrete block, prestressed by using high tension strength wires through the cores of the block. The wire was pulled at a high tension, and the ends of the beams were grouted. After the grout held the wires or cables in place, the tension was released on the cables, with resultant compression on the bottom portion of the beams. Thus the force on the bottom of the beam would tend to counteract the downward weight put on the top of the beam. By this process, these prestressed concrete beams could cover three to four times the spans possible with conventional reinforced concrete.

In 1951, after many trips to Washington, DC, and an excellent selling job by CMC's founder, T. L. Goudvis, CMC was able to land their first large-volume prestressed concrete project with the Corps of Engineers. The contract authorized the use of prestressed concrete beams, as described, with concrete block for the roofs of warehouses in the large Air Force depot complex being built in Shelby, Ohio. The buildings were a success, and CMC immediately received prestige and notoriety as a leader in the prestressed concrete business.

Wet-cast beams were developed next. For wet-cast beams, instead of concrete block, the cables were placed in long forms and pulled to the desired tension, after which concrete was poured in the forms to make beams. As a result of wet-cast beams, prestressed concrete was no longer dependent on concrete block.

At first, prestressed concrete was primarily for floors and roofs, but, in the early 1960s, precasters became involved in more complicated structures. CMC started designing and making not only beams, but columns and whatever other components it took to put together a whole structure. Parking garages became a natural application for prestressed concrete structures. Eventually an entire building could be precast out of prestressed concrete.

PROJECT MANAGEMENT

Constructing the entire building, as in the case of a parking garage, meant that jobs were becoming more complex with respect to interdependence of detailed task accomplishment. Accordingly, in 1967, project management was established at CMC. The functional departments did the work, but the project managers saw to it that the assigned projects were completed on schedule and within budget and specifications. A matrix organization, as illustrated in Exhibit I, was adopted and used effectively by CMC. The concept of a matrix organization, as applied at CMC, entailed an organizational system designed as "web of relationships" rather than a line and staff relationship for work performance.

Each project manager was assigned a number of personnel with the required qualification from the functional departments for the duration of the project. Thus the project organization was composed of the project manager and functional personnel groups. The project manager had not only the responsibility and accountability for the successful completion of the contract, but also the delegated authority for work design, assignments of functional group personnel, and the determination of procedural relationships.

The most important functional area for the project manager was the engineering department, since prestressed concrete is a highly engineered product. A great deal of coordination and interaction was required between the project manager and the engineering department just to make certain that everything fit together and was structurally sound. A registered engineer did the design. The project manager's job was to see that the designing was done correctly and efficiently. Production schedules were made up by the project manager subject, of course, to minor modifications by the plant. The project manager was also required to do all the coordination with the customer, architect, general contractor, and the erection force. The project manager was also required to have interaction with the distribution manager to be certain that the product designed could be shipped by trucks. Finally, there had to be interaction between the project manager and the sales department to determine that the product the project manager was making was what the sales department had sold.

Exhibit I. *Matrix organization of Concrete Masonry Corporation*

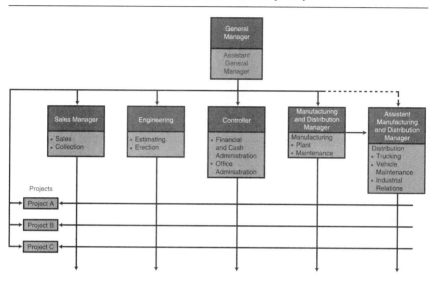

ESTIMATING—WHICH DEPARTMENT?

At one time or another during CMC's history, the estimating function had been assigned to nearly every functional area of the organization, including sales, engineering, manufacturing, and administration. Determining which functional area estimating was to be under was a real problem for CMC. There was a short time when estimating was on its own, reporting directly to the general manager.

Assignment of this function to any one department carried with it some inherent problems, not peculiar to CMC, but simply related to human nature. For example, when the estimating was supervised in the sales department, estimated costs would tend to be low. In sales, the estimator knows the boss wants to be the low bidder on the job and therefore believes he or she is right to say, "It is not going to take us ten days to cast this thing; we could run three at a time."

When estimating was performed by production, the estimate would tend to be high. This was so because the estimator did not want the boss, the production manager, coming back and saying, "How come you estimated this thing at $5 a cubic foot and it's costing us $6? It's not the cost of production that's wrong, it's the estimate."

W. S. Lasch, general manager of CMC, had this comment about estimating in a project management situation:

> It is very difficult to get accountability for estimating a project. When many
> of your projects are new ballgames, a lot of your information has to come

from . . . well, let's just say there is a lot of art to it as well as science. You never can say with 100 percent certainty that costs were high because you could have just as easily said the estimate was too low.

So, as a compromise, most of the time we had our estimating done by engineering. While it solved some problems, it also created others. Engineers would tend to be more fair; they would call the shots as they saw them. However, one problem was that they still had to answer to sales as far as their workload was concerned. For example, an engineer is in the middle of estimating a parking garage, a task that might take several days. All of a sudden, the sales department wants him to stop and estimate another job. The sales department had to be the one to really make that decision because they are the ones that know what the priorities are on the bidding. So even though the estimator was working in engineering, he was really answering to the sales manager as far as his workload was concerned.

ESTIMATING—COSTING

Estimating was accomplished through continual monitoring and comparison of actual versus planned performance, as shown in Exhibit II.

The actual costing process was not a problem for CMC. In recent years, CMC had eliminated as much as possible the actual dollars and cents from the estimator's control. A great deal of the "drudge work" was done on the computer. The estimator, for example, would predict how much the prestressed concrete

Exhibit II. Actual versus planned performance

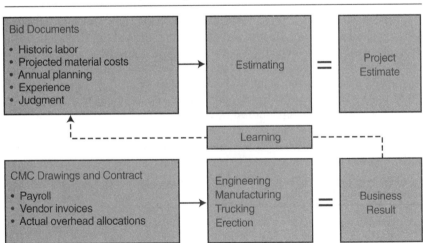

must span, and how many cubic feet of concrete was needed. Once that information was in hand, the estimator entered it in the computer. The computer would then come up with the cost. This became an effective method because the estimator would not be influenced by either sales or production personnel.

THE EVOLUTION OF THE PRESTRESSED CONCRETE MARKETPLACE

During the twenty or more years since prestressing achieved wide acceptance in the construction industry, an evolution has been taking place that has drastically changed the character of the marketplace and thus greatly modified the role of the prestresser.

Lasch had the following comments about these changes that occurred in the marketplace:

In the early days, designers of buildings looked to prestressers for the expertise required to successfully incorporate the techniques and available prestressed products into their structures. A major thrust of our business in those days was to introduce design professionals, architects, and engineers to our fledgling industry and to assist them in making use of the many advantages that we could offer over other construction methods. These advantages included fire resistance, long spans, permanence, factory-controlled quality, speed of erection, aesthetic desirability, virtual elimination of maintenance costs, and, last but of prime importance, the fact that we were equipped to provide the expertise and coordination necessary to successfully integrate our product into the building. Many of our early jobs were bid from sketches. It was then up to our in-house experts, working closely with the owner's engineer and architect, to develop an appropriate, efficient structure that satisfied the aesthetic and functional requirements and hopefully maximized production and erection efficiency, thereby providing maximum financial return to CMC. It should be noted that, although our contract was normally with the project's general contractor, most of our design coordination was through the owner's architect or engineer and, more often than not, it was our relationship with the owner and his design professional that determined our involvement in the project in the first place. It should be readily seen that, in such an environment, only organizations with a high degree of engineering background and a well-organized efficient team of professionals, could compete successfully. CMC was such an organization.

There are, however, few, if any, proprietary secrets in the prestressing industry, and it was inevitable that this would in later years be largely responsible for a dramatic change in the marketplace. The widespread acceptance of the product, which had been achieved through the success of companies like CMC, carried with it a proliferation of the technical knowledge and production techniques which design professionals had previously relied upon the producer to provide. In the later 1960s, some colleges and universities began to include prestressed concrete design as a part of their structural engineering programs. Organizations, such as the Portland Cement Association, offered seminars for architects and engineers to promote the prestressing concept. As a result, it is now common for architects and engineers to incorporate prestressed concrete products in bid drawings for their projects, detailing all connections, reinforcement, mix designs, and so on. This, obviously, makes it possible for any organization capable of reading drawings and filling forms to bid on the project. We have found ourselves bidding against companies with a few molds in an open field and, in several cases, a broker with no equipment or organization at all! The result of all this, of course, is a market price so low as to prohibit the involvement of professional prestressing firms with the depth of organization described earlier.

OBTAINING A PRESTRESSED CONCRETE JOB

The author believes the following example demonstrates the change in market conditions and best illustrates one of the reasons CMC decided not to remain in the prestressed concrete business. A large insurance company in Columbus, Ohio, was planning a parking garage for 2,500 cars. CMC talked to the owner and owner's representative (a construction management firm) about using prestressed concrete in the design of their project rather than the poured-in-place concrete, steel, or whatever options they had. Just by doing this, CMC had to give away some knowledge. You just cannot walk in and say, "Hey, how about using prestressed concrete?" You have to tell them what is going to be saved and how, because the architect has to make the drawings. Once CMC felt there was an open door, and that the architect and owner would possibly incorporate their product, then sales would consult engineering to come up with a proposal. A proposal in the early stages was simply to identify what the costs were going to be, and to show the owner and architect photographs or sketches of previous jobs. As time went by, CMC had to go into more detail and provide more and more information, including detailed drawings of several proposed layouts. CMC illustrated connection details, reinforcing details, and even computer design of some of the pieces for the parking garage. Receiving all this engineering information, the owner and the construction management firm became convinced that using this product was the most inexpensive way for them to go. In fact, CMC demonstrated to the insurance company that they could save over $1 million over any other product. At this point, CMC had spent thousands of dollars to come up with the solution for the problem of designing the parking garage.

Months and years passed until the contract manager chose to seek bids from other precasters, who, up to this time, had little or no investment in the project. CMC had made available an abundance of free information that could be used by the competition. The competition only had to put the information together, make a material takeoff, calculate the cost, and put a price on it. Without the costly depth of organization required to support the extensive promotional program conducted by CMC, the competition could naturally bid the job lower.

Lasch felt that, as a result of present-day market conditions, there were only two ways that one survives in the prestressed concrete business:

> Face the fact that you are going to be subservient to a general contractor and that you are going to sell not your expertise but your function as a 'job shop' manufacturer producing concrete products according to someone else's drawings and specifications. If you do that, then you no longer need, for example, an engineering department or a technically qualified sales organization. All you are going to do is look at drawings, have an estimator who can read the drawings, put a price tag on them, and give a bid. It is going to be a low bid because you have eliminated much of your overhead. We simply do not choose to be in business in this manner.
>
> The other way to be in the business is that you are not going to be subservient to a general contractor, or owner's architect, or engineer. What you are going to do is to deal with owners or users. That way a general contractor may end up as a subcontractor to the prestresser. We might go out and build a parking garage or other structure and assume the role of developer or builder or even owner/leaser. In that way, we would control the whole job. After all, in most cases the precast contract on a garage represents more than half the total cost. It could be argued with great justification that the conventional approach (i.e., precaster working for general contractor) could be compared to the tail wagging the dog.
>
> With complete control of design, aesthetics, and construction schedule, it would be possible to achieve maximum efficiency of design, plant usage, and field coordination which, when combined, would allow us to achieve that most important requirement—that of providing the eventual user with maximum value for a minimum investment. Unless this can be achieved, the venture would not be making a meaningful contribution to society, and there would be no justification for being in business.

SYNOPSIS

Concrete Masonry Corporation's (CMC) difficulties do not arise from the fact that the organization employs a matrix structure, but rather from the failure of the corporation's top management to recognize, in due time, the changing nature (with respect to the learning curve of the competition and user of the product and services of CMC) of the prestressed concrete business.

At the point in time when prestressed concrete gained wide industry acceptance, and technical schools and societies began offering courses in the techniques for utilizing this process, CMC should have begun reorganizing its prestressed concrete business activities in two separate functional costing groups. Marketing and selling CMC's prestressed concrete business services and utilizing the company's experience, technical expertise, judgment, and job estimating abilities should satisfy the responsibilities of one of these groups, to perform the actual prestressed concrete engineering and implementation of the other.

With the responsibilities and functions separated as noted above, the company is able to determine more precisely how competitive they really are and which (if either) phase of the concrete business to divest themselves of.

Project management activities are best performed when complex tasks are of a limited life. Such is not the case in securing new or continuing business in the prestressed concrete business but rather is an effort or activity that should continue as long as CMC is in the business. This phase of the business should therefore be assigned to a functional group. However, it may be advantageous at times to form or utilize a project management structure in order to assist the functional group in satisfying a task's requirements when the size of the task is large and complex.

The engineering and implementation phase of the business should continue to be performed through the project management–matrix structure because of the limited life of such tasks and the need for concentrated attention to time, cost, and performance constraints inherent in these activities.

QUESTIONS

1. Did CMC have long-range planning?
2. What are the problems facing CMC?
3. Did CMC utilize the matrix effectively?
4. Where should project estimating be located?
5. Does the shifting of the estimating function violate the ground rules of the matrix?
6. What are the alternatives for CMC?
7. Will they be successful as a job shop?
8. Should companies like CMC utilize a matrix?
9. How does the company plan to recover R&D and bid and proposal costs?
10. Has CMC correctly evaluated the marketplace?
11. Do they respond to changes in the marketplace?
12. With what speed is monitoring done? (Exhibit II). How many projects must be estimated, bid, and sold before actuals catch up to and become historical data?

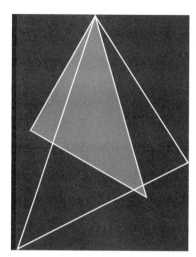

Margo Company

"I've called this meeting, gentlemen, because that paper factory we call a computer organization is driving up our overhead rates," snorted Richard Margo, president, as he looked around the table at the vice presidents of project management, engineering, manufacturing, marketing, administration, and information systems. "We seem to be developing reports faster than we can update our computer facility. Just one year ago, we updated our computer and now we're operating three shifts a day, seven days a week. Where do we go from here?"

V.P. information: "As you all know, Richard asked me, about two months ago, to investigate this gigantic increase in the flow of paperwork. There's no question that we're getting too many reports. The question is, are we paying too much money for the information that we get? I've surveyed all of our departments and their key personnel. Most of the survey questionnaires indicate that we're getting too much information. Only a small percentage of each report appears to be necessary. In addition, many of the reports arrive too late. I'm talking about scheduled reports, not planning, demand, or exception reports."

V.P. project management: "Every report people may receive is necessary for us to make decisions effectively with regard to planning, organizing, and controlling each project. My people are the biggest users and we can't live with fewer reports."

V.P. information: "Can your people live with less information in each report? Can some of the reports be received less frequently?"

V.P. project management: "Some of our reports have too much information in them. But we need them at the frequency we have now."

V.P. engineering: "My people utilize about 20 percent of the information in most of our reports. Once our people find the information they want, the report is discarded. That's because we know that each project manager will retain a copy. Also, only the department managers and section supervisors read the reports."

V.P. information: "Can engineering and manufacturing get the information they need from other sources, such as the project office?"

V.P. project management: "Wait a minute! My people don't have time to act as paper pushers for each department manager. We all know that the departments can't function without these reports. Why should we assume the burden?"

V.P. information: "All I'm trying to say is that many of our reports can be combined into smaller ones and possibly made more concise. Most of our reports are flexible enough to meet changes in our operating business. We have two sets of reports: one for the customer and one for us. If the customer wants the report in a specific fashion, he pays for it. Why can't we act as our own customer and try to make a reporting system that we can all use?"

V.P. engineering: "Many of the reports obviously don't justify the cost. Can we generate the minimum number of reports and pass it on to someone higher or lower in the organization?"

V.P. project management: "We need weekly reports, and we need them on Monday mornings. I know our computer people don't like to work on Sunday evenings, but we have no choice. If we don't have those reports on Monday mornings, we can't control time, cost, and performance."

V.P. information: "There are no reports generated from the pertinent data in our original computer runs. This looks to me like every report is a one-shot deal. There has to be room for improvement.

"I have prepared a checklist for each of you with four major questions. Do you want summary or detailed information? How do you want the output to look? How many copies do you need? How often do you need these reports?"

Richard Margo: "In project organizational forms, the project exists as a separate entity except for administrative purposes. These reports are part of that administrative purpose. Combining this with the high cost of administration in our project structure, we'll never remain competitive unless we lower our overhead. I'm going to leave it up to you guys. Try to reduce the number of reports, but don't sacrifice the necessary information you need to control the projects and your resources."

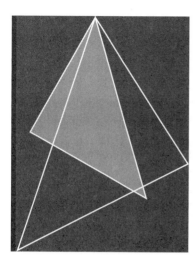

Project
Overrun

The Green Company production project was completed three months behind schedule and at a cost overrun of approximately 60 percent. Following submittal of the final report, Phil Graham, the director of project management, called a meeting to discuss the problems encountered on the Green Project.

Phil Graham: "We're not here to point the finger at anyone. We're here to analyze what went wrong and to see if we can develop any policies and/or procedures that will prevent this from happening in the future. What went wrong?"

Project manager: "When we accepted the contract, Green did not have a fixed delivery schedule for us to go by because they weren't sure when their new production plant would be ready to begin production activities. So, we estimated 3,000 units per month for months five through twelve of the project. When they found that the production plant would be available two months ahead of schedule, they asked us to accelerate our production activities. So, we put all of our production people on overtime in order to satisfy their schedule. This was our mistake, because we accepted a fixed delivery date and budget before we understood everything."

Functional manager: "Our problem was that the customer could not provide us with a fixed set of specifications, because the final set of specifications depended on OSHA and EPA requirements, which could not be confirmed until initial testing of the new plant. Our people, therefore, were asked to commit to man-hours before specifications could be reviewed.

"Six months after project go-ahead, Green Company issued the final specifications. We had to remake 6,000 production units because they did not live up to the new specifications."

Project manager: "The customer was willing to pay for the remake units. This was established in the contract. Unfortunately, our contract people didn't tell me that we were still liable for the penalty payments if we didn't adhere to the original schedule."

Phil Graham: "Don't you feel that misinterpretation of the terms and conditions is your responsibility?"

Project manager: "I guess I'll have to take some of the blame."

Functional manager: "We need specific documentation on what to do in case of specification changes. I don't think that our people realize that user approval of specification is not a contract agreed to in blood. Specifications can change, even in the middle of a project. Our people must understand that, as well as the necessary procedures for implementing change."

Phil Graham: "I've heard that the functional employees on the assembly line are grumbling about the Green Project. What's their gripe?"

Functional manager: "We were directed to cut out all overtime on all projects. But when the Green Project got into trouble, overtime became a way of life. For nine months, the functional employees on the Green Project had as much overtime as they wanted. This made the functional employees on other projects very unhappy.

"To make matters worse, the functional employees got used to a big take-home paycheck and started living beyond their means. When the project ended, so did their overtime. Now, they claim that we should give them the opportunity for more overtime. Everybody hates us."

Phil Graham: "Well, now we know the causes of the problem. Any recommendations for cures and future prevention activities?"

The Automated Evaluation Project[1]

"No deal!" said the union. "The current method of evaluating government employees at this agency is terrible, and if a change doesn't occur, we'll be in court seeking damages."

In 1984, a government agency approved and initiated an ambitious project, part of which was to develop an updated, automated evaluation system for the 50,000 employees located throughout the United States. The existing evaluation system was antiquated. Although there were forms used for employee evaluation, standardization was still lacking. Not all promotions were based on performance. Often, it was based on time in grade, the personal whims of management, or friendships. Some divisions seemed to promote employees faster than others. The success or failure of a project could also seriously impact performance opportunities. Some type of standardization was essential.

In June 1985, a project manager was finally assigned and brought on board. The assignment of the project manager was based upon rank and availability at that time rather than the requirements of the project. Team members often possessed a much better understanding of the project than did the project manager.

[1]Copyright © 2005 by Harold Kerzner. This case study is fictitious and was prepared as the basis for classroom discussion rather than to illustrate an effective or ineffective handling of an administrative situation.

The project manager, together with his team, quickly developed an *action plan*. The action plan did *not* contain a work breakdown structure, but did contain a statement of work which called out high-level deliverables that would be essential for structured analyses, design and programming. The statement of work and deliverables were more so in compliance with agency requirements for structured analyses, design, and programming than for the project's requirement. The entire action plan was prepared by the project office, which was composed of eight employees.

Bids from outside vendors were solicited for the software packages, with the constraint that all deliverables must be operational on existing agency hardware. In October 1985, the award was made by the project office to Primco Corporation with work scheduled to begin in December 1985.

In the spring of 1986, it became apparent that the project was running into trouble and disaster was imminent. There were three major problems facing the project manager. As stated by the project manager:

1. The requirements for the project had to be changed because of new regulations for government worker employee evaluation.
2. Primco did not have highly skilled personnel assigned to the project.
3. The agency did not have highly skilled personnel from the functional areas assigned to the project.

The last item was argumentative. The line managers at the agency contended that they had assigned some of their best people and that the real problem was that the project manager was trying to make all of the decisions himself without any input from the assigned personnel. The employees contended that proper project management practices were not being used. The project was being run like a dictatorship rather than a democracy. Several employees felt as though they were not treated as part of the project team.

According to one of the team members,

> The project manager keeps making technical decisions without any solid foundation to support his views. Several of us in the line organization have significantly more knowledge than does the project manager, yet he keeps overriding our recommendations and decisions. Perhaps he has that right, but I dislike being treated as a second-class citizen. If the project manager has all of this technical knowledge, then why does he need us?

In June 1986, the decision was made by the project manager to ask one of the assistant agency directors to tell the union that the original commitment date of January 1987 would not be met. A stop workage order was issued to Primco, thus canceling the contract.

The original action plan called for the use of existing agency hardware. However, because of unfavorable publicity about hardware and software problems at the agency during the spring of 1986, the agency felt that the UNIVAC System would not support the additional requirements, and system overload might occur. Now hardware, as well as software, would be needed.

To help maintain morale, the project manager decided to perform as much of the work as possible in-house, even though the project lacked critical resources and was already more than one year late. The project office took what was developed thus far and tried to redefine the requirements.

With the support of senior management at the agency, the original statement of work was thrown away and a new statement of work was prepared. "It was like starting over right from the beginning," remarked one of the employees. "We never looked back at what was accomplished thus far. It was a whole new project!" With the support of the agency's personnel office, the new requirements were finally completed in February of 1987.

The union, furious over the schedule slippage, refused to communicate with the project office and senior management. The union's contention was that an "illegal" evaluation system was in place, and the current system could not properly validate performance review requirements. The union initiated a lawsuit against the agency seeking damages in excess of $21 million.

In November 1986, procurement went out for bids for both hardware and a database management system. The procurement process continued until June 1987, when it was canceled by another government agency responsible for procurement. No reason was ever provided for the cancellation.

Seeking alternatives, the following decisions were made:

1. Use rented equipment to perform the programming.
2. Purchase a database management system from ITEKO Corporation, provided that some customization could be accomplished. The new database management system was scheduled to be released to the general public in about two months.

The database management system was actually in the final stages of development and ITEKO Corporation promised the agency that a fully operational version, with the necessary customization, could be provided quickly. Difficulties arose with the use of the ITEKO package. After hiring a consultant from ITEKO, it was found that the ITEKO package was a beta rather than a production version. Despite these setbacks, personnel kept programming on the leased equipment with the hope of eventually purchasing a Micronet Hardware System. ITEKO convinced the agency that the Micronet hardware system was the best system available to support the database management system. The Micronet hardware was then added to the agency's equipment contract but later disallowed on September 29, 1987, because it was not standard agency equipment.

On October 10, 1987, the project office decided to outsource some of the work using a small/minority business procurement strategy for hardware to support the ITEKO package. The final award was made in November 1987, subject to software certification by the one of the agency's logistics centers. Installation in all of the centers was completed between November and December 1987.

QUESTIONS

1. Is there anything in the case that indicates the maturity level of project management at the agency around 1985–1986?
2. What are the major problems in the case?
3. Who was at fault?
4. How do you prevent this from occurring on other projects?

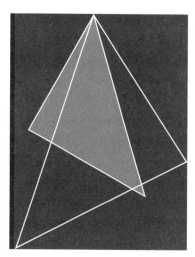

The Rise, Fall, and Resurrection of Iridium: A Project Management Perspective

The Iridium Project was designed to create a worldwide wireless handheld mobile phone system with the ability to communicate anywhere in the world at any time. Executives at Motorola regarded the project as the eighth wonder of the world. But more than a decade later and after investing billions of dollars, Iridium had solved a problem that very few customers needed solved. What went wrong? How did the Iridium Project transform from a leading-edge technical marvel to a multi-billion-dollar blunder? Could the potential catastrophe have been prevented?[1]

> What it looks like now is a multibillion-dollar science project. There are fundamental problems: The handset is big, the service is expensive, and the customers haven't really been identified.
>
> **—Chris Chaney, Analyst, A.G. Edwards, 1999**

> There was never a business case for Iridium. There was never market demand. The decision to build Iridium wasn't a rational business decision. It was more of a religious decision. The remarkable thing is that this happened at a big corporation, and

---

[1] © 2007 by Harold Kerzner. Some of the material has been adapted from Sydney Finkelstein and Shade H. Sanford, "Learning from Corporate Mistakes: The Rise and Fall of Iridium," *Organizational Dynamics*, vol. 29, no. 2, pp.138–148, 2000. © 2000 by Elsevier Sciences, Inc. Reproduced by permission.

that there was not a rational decision-making process in place to pull the plug. Technology for technology's sake may not be a good business case."[2]

—Herschel Shosteck, Telecommunication Consultant

Iridium is likely to be some of the most expensive space debris ever.

—William Kidd, Analyst, C.E. Unterberg, Towbin

In 1985, Bary Bertiger, chief engineer in Motorola's strategic electronics division, and his wife Karen were on a vacation in the Bahamas. Karen tried unsuccessfully to make a cellular telephone call back to her home near the Motorola facility in Chandler, Arizona, to close a real-estate transaction. Unsuccessful, she asked her husband why it would not be possible to create a telephone system that would work anywhere in the world, even in remote locations.

At this time, cell technology was in its infancy but was expected to grow at an astounding rate. AT&T projected as many as 40 million subscribers by 2000.[3] Cell technology was based upon tower-to-tower transmission as shown in Exhibit 1. Each tower or "gateway" ground station reached a limited geographic area or cell and had to be within the satellite's field of view. Cell phone users likewise had to be near a gateway that would uplink the transmission to a satellite. The satellite would then downlink the signal to another gateway that would connect the transmission to a ground telephone system. This type of communication is often referred to as bent pipe architecture. Physical barriers between the senders/receivers and the gateways, such as mountains, tunnels, and oceans created interference problems and therefore limited service to high-density communities. Simply stated, cell phones couldn't leave home. And, if they did, there would be additional "roaming" charges. To make matters worse, every country had its own standards, and some cell phones were inoperable when traveling in other countries.

Communications satellites, in use since the 1960s, were typically geostationary satellites that orbited at altitudes of more than 22,300 miles. At this altitude, three geosynchronous satellites and just a few gateways could cover most of the Earth. But satellites at this altitude meant large phones and annoying quarter-second voice delays. Comsat's Planet 1 phone, for example, weighed in at a computer-case-sized 4.5 pounds. Geosynchronous satellites require signals with a great deal of power. Small mobile phones, with a one-watt signal, could not work with satellites positioned at this altitude. Increasing the power output of the mobile phones would damage human tissue. The alternative was therefore to move the satellites closer to Earth such that less power would be needed. This would require significantly more satellites the closer we get to Earth as well as additional gateways. Geosynchronous satellites, which are 100 times further away

---

[2]Stephanie Paterik, "Iridium Alive and Well," *The Arizona Republic*, April 27, 2005, p. D5.
[3]Judith Bird, "Cellular Technology in Telephones," *data processing*, vol. 27, no. 8, October 1985, p. 37.

Exhibit 1. *Typical satellite communication architecture*

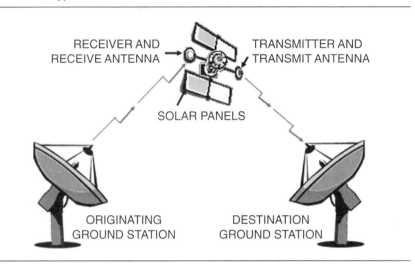

RECEIVER AND
RECEIVE ANTENNA

TRANSMITTER AND
TRANSMIT ANTENNA

SOLAR PANELS

ORIGINATING
GROUND STATION

DESTINATION
GROUND STATION

from Earth than low-Earth-orbiting (LEO) satellites, could require almost 10,000 times as much power as LEOs, if everything else were the same.[4]

When Bary Bertiger returned to Motorola, he teamed up with Dr. Raymond Leopold and Kenneth Peterson to see if such a worldwide system could be developed while overcoming all of the limitations of existing cell technology. There was also the problem that LEO satellites would be orbiting the Earth rapidly and going through damaging temperature variations—from the heat of the sun to the cold shadow of Earth.[5] The LEO satellites would most likely need to be replaced every 5 years. Numerous alternative terrestrial designs were discussed and abandoned. In 1987 research began on a constellation of LEO satellites moving in polar orbits that could communicate directly with telephone systems on the ground and with one another.

Iridium's innovation was to use a large constellation of low-orbiting satellites approximately 400–450 miles in altitude. Because Iridium's satellites were closer to Earth, the phones could be much smaller and the voice delay imperceptible. But there were still major technical design problems. With the existing design, a large number of gateways would be required, thus substantially increasing the cost of the system. As they left work one day in 1988, Dr. Leopold proposed a critical design element. The entire system would be inverted whereby the transmission would go from satellite to satellite until the transmission reached the satellite directly above the person who would be receiving the message. With this approach, only one gateway Earth station would be required to connect mobile-to-landline calls to

[4]Bird, p. 37.
[5]Bruce Gerding, "Personal Communications via Satellite: An Overview," *Telecommunications*, vol. 30, no. 2, February 1996, pp. 35, 77.

existing land-based telephone systems. This was considered to be the sought-after solution and was immediately written in outline format on a whiteboard in a security guard's office. Thus came forth the idea behind a worldwide wireless handheld mobile phone with the ability to communicate anywhere and anytime.

NAMING THE PROJECT "IRIDIUM"

Motorola cellular telephone system engineer, Jim Williams, from the Motorola facility near Chicago, suggested the name, Iridium. The proposed 77-satellite constellation reminded him of the electrons that encircle the nucleus in the classical Bohr model of the atom. When he consulted the periodic table of the elements to discover which atom had 77 electrons, he found iridium—a creative name that had a nice ring. Fortunately, the system had not yet been scaled back to 66 satellites, or else he might have suggested the name Dysprosium.

OBTAINING EXECUTIVE SUPPORT

Initially, Bertiger's colleagues and superiors at Motorola had rejected the Iridium concept because of its cost. Originally, the Iridium concept was considered perfect for the U.S. government. Unfortunately, the era of lucrative government-funded projects was coming to an end, and it was unlikely that the government would fund a project of this magnitude. However, the idea behind the Iridium concept intrigued Durrell Hillis, the general manager of Motorola's Space and Technology Group. Hillis believed that Iridium was workable if it could be developed as a commercial system. Hillis instructed Bertiger and his team to continue working on the Iridium concept but to keep it quiet.

> "I created a bootleg project with secrecy so no one in the company would know about it," Hillis recalls. He was worried that if word leaked out, the ferociously competitive business units at Motorola, all of which had to fight for R&D funds, would smother the project with nay-saying.[6]

After 14 months of rewrites on the commercialized business plan, Hillis and the Iridium team leaders presented the idea to Robert Galvin, Motorola's chairman at the time, who gave approval to go ahead with the project. Robert Galvin, and later his successor and son Christopher Galvin, viewed Iridium as a potential symbol of Motorola's technological prowess and believed that this would become the eighth wonder in the world. In one of the initial meetings, Robert Galvin turned to John Mitchell, Motorola's president and chief operating officer, and said, "If you don't write out a check for this John, I will, out of my own pocket."[7]

---

[6]David S. Bennahum, "The United Nations of Iridium," *Wired*, issue 6.10, October 1998, p. 194.
[7]Quentin Hardy, "How a Wife's Question Led Motorola to Chase a Global Cell-Phone Plan," *Wall Street Journal* (Eastern edition), New York, December 16, 1996, p. A1.

To the engineers at Motorola, the challenge of launching Iridium's constellation provided considerable motivation. They continued developing the project that resulted in initial service in November 1998 at a total cost of over $5 billion.

LAUNCHING THE VENTURE

On June 26, 1990, Hillis and his team formally announced the launch of the Iridium Project to the general public. The response was not very pleasing to Motorola with skepticism over the fact that this would be a new technology, the target markets were too small, the revenue model was questionable, obtaining licenses to operate in 170 countries could be a problem, and the cost of a phone call might be overpriced. Local phone companies that Motorola assumed would buy into the project viewed Iridium as a potential competitor since the Iridium system bypassed traditional landlines. In many countries, postal, telephone, and telegraph (PTT) operators are state owned and a major source of revenue because of the high profit margins. Another issue was that the Iridium Project was announced before permission was granted by the Federal Communications Commission (FCC) to operate at the desired frequencies.

Both Mitchell and Galvin made it clear that Motorola would not go it alone and absorb the initial financial risk for a hefty price tag of about $3.5 billion. Funds would need to be obtained from public markets and private investors. In order to minimize Motorola's exposure to financial risk, Iridium would need to be set up as a project-financed company. Project financing involves the establishment of a legally independent project company where the providers of funds are repaid out of cash flow and earnings, and where the assets of the unit (and only the unit) are used as collateral for the loans. Debt repayment would come from the project company only rather than from any other entity. A risk with project financing is that the capital assets may have a limited life. The potential limited life constraint often makes it difficult to get lenders to agree to long-term financial arrangements.

Another critical issue with project financing especially for high-tech projects is that the projects are generally long-term. It would be nearly 8 years before service would begin, and in terms of technology, 8 years is an eternity. The Iridium Project was certainly a "bet on the future." And if the project were to fail, the company could be worth nothing after liquidation.

In 1991, Motorola established Iridium Limited Liability Corporation (Iridium LLC) as a separate company. In December of 1991, Iridium promoted Leo Mondale to vice president of Iridium International. Financing the project was still a critical issue. Mondale decided that, instead of having just one gateway, there should be as many as 12 regional gateways that plugged into local, ground-based telephone lines. This would make Iridium a truly global project rather than appear as an American-based project designed to seize market share from state-run telephone companies. This would also make it easier to get regulatory approval to operate in 170 countries. Investors would pay $40 million for the right to own their own regional gateway. As stated by Flower:

> The motive of the investors is clear: They are taking a chance on owning a slice of a de-facto world monopoly. Each of them will not only have a piece of the company, they will own the Iridium gateways and act as the local distributors in their respective home markets. For them it's a game worth playing.[8]

There were political ramifications with selling regional gateways. What if in the future the U.S. government forbids shipment of replacement parts to certain gateways? What if sanctions are imposed? What if Iridium were to become a political tool during international diplomacy because of the number of jobs it creates?

In addition to financial incentives, gateway owners were granted seats on the board of directors. As described by David Bennahum, reporter for *Wired*:

> Four times a year, 28 Iridium board members from 17 countries gather to coordinate overall business decisions. They met around the world, shuttling between Moscow, London, Kyoto, Rio de Janeiro, and Rome, surrounded by an entourage of assistants and translators. Resembling a United Nations in miniature, board meetings were conducted with simultaneous translation in Russian, Japanese, Chinese, and English.[9]

The partner with the largest equity share was Motorola. For its contribution of $400 million, Motorola originally received an equity stake of 25 percent, and 6 of the 28 seats on Iridium's board. Additionally, Motorola made loan guarantees to Iridium of $750 million, with Iridium holding an option for an additional $350 million loan.

For its part, Iridium agreed to $6.6 billion in long-term contracts with Motorola that included $3.4 billion for satellite design and launch and $2.9 billion for operations and maintenance. Iridium also exposed Motorola to developing satellite technology that would provide the latter with significant expertise in building satellite communications systems, as well as vast intellectual property.

THE IRIDIUM SYSTEM[10]

The Iridium system is a satellite-based, wireless personal communications network providing a robust suite of voice features to virtually any destination anywhere on Earth.

---

[8]Joe Flower, "Iridium," *Wired*, issue 1.05, November, 1993.
[9]Bennahum,1998, p. 136.
[10]This is the operational version of the Iridium system today taken from the Iridium website, www.Iridium.com.

The Iridium system comprises three principal components: the satellite network, the ground network, and the Iridium subscriber products including phones and pagers. The design of the Iridium network allows voice and data to be routed virtually anywhere in the world. Voice and data calls are relayed from one satellite to another until they reach the satellite above the Iridium subscriber unit (handset) and the signal is relayed back to Earth.

THE TERRESTIAL AND SPACE-BASED NETWORK[11]

The Iridium constellation consists of 66 operational satellites and 11 spares orbiting in a constellation of 6 polar planes. Each plane has 11 mission satellites performing as nodes in the telephony network. The remaining 11 satellites orbit as spares ready to replace any unserviceable satellite. This constellation ensures that every region on the globe is covered by at least one satellite at all times.

The satellites are in a near-polar orbit at an altitude of 485 miles (780 km). They circle the Earth once every 100 minutes traveling at a rate of 16,832 miles per hour. The satellite weight is 1500 pounds. Each satellite is approximately 40 feet in length and 12 feet in width. In addition, each satellite has 48 spot beams, 30 miles in diameter per beam.

Each satellite is cross-linked to four other satellites; two satellites in the same orbital plane and two in an adjacent plane. The ground network is comprised of the System Control Segment and telephony gateways used to connect into the terrestrial telephone system. The System Control Segment is the central management component for the Iridium system. It provides global operational support and control services for the satellite constellation, delivers satellite-tracking data to the gateways, and performs the termination control function of messaging services. The System Control Segment consists of three main components: four Telemetry Tracking and Control sites, the Operational Support Network, and the Satellite Network Operation Center. The primary linkage between the System Control Segment, the satellites, and the gateways is via K-band feeder links and cross-links throughout the satellite constellation.

Gateways are the terrestrial infrastructure that provides telephony services, messaging, and support to the network operations. The key features of gateways are their support and management of mobile subscribers and the interconnection of the Iridium network to the terrestrial phone system. Gateways also provide network management functions for their own network elements and links.

---

[11]See note 10.

PROJECT INITIATION: DEVELOPING THE BUSINESS CASE

For the Iridium Project to be a business success rather than just a technical success there had to exist an established customer base. Independent studies conducted by A.T. Kearney, Booz, Allen & Hamilton, and Gallup indicated that 34 million people had a demonstrated need for mobile satellite services, with that number expected to grow to 42 million by 2002. Of these 42 million, Iridium anticipated 4.2 million to be satellite-only subscribers, 15.5 million satellite and world terrestrial roaming subscribers, and 22.3 million terrestrial roaming-only subscribers.

A universal necessity in conducting business is ensuring that you are never out of touch. Iridium would provide this unique solution to business with the essential communications tool. This proposition of one phone, one number with the capability to be accessed anywhere, anytime was a message that target markets—the global traveler, the mining, rural, maritime industries, government, disaster relief, and community aid groups—would readily embrace.

Also at the same time of Iridium's conception, there appeared to be another potentially lucrative opportunity in the telecommunications marketplace. When users of mobile or cellular phones crossed international borders, they soon discovered that there existed a lack of common standards, thus making some phones inoperable. Motorola viewed this as an opportunity to create a worldwide standard allowing phones to be used anywhere in the world.

The expected breakeven market for Iridium was estimated between 400,000 and 600,000 customers globally, assuming a reasonable usage rate per customer per month. With a launch date for Iridium service established for 1998, Iridium hoped to recover all of its investment within one year. By 2002, Iridium anticipated a customer base of 5 million users. The initial Iridium target market had been the vertical market, those of the industry, government, and world agencies that have defended needs and far-reaching communication requirements. Also important would be both industrial and public sector customers. Often isolated in remote locations outside of cellular coverage, industrial users were expected to use handheld Iridium satellite services to complement or replace their existing radio or satellite communications terminals. The vertical markets for Iridium would include:

- Aviation
- Construction
- Disaster relief/emergency
- Forestry
- Government

- Leisure travel
- Maritime
- Media and entertainment
- Military
- Mining
- Oil and gas
- Utilities

Using its own marketing resources, Iridium appeared to have identified an attractive market segment after having screened over 200,000 people, interviewed 23,000 people from 42 countries, and surveyed over 3000 corporations.

Iridium would also need regional strategic partners, not only for investment purposes and to share the risks, but to provide services throughout their territories. The strategic regional partners or gateway operating companies would have exclusive rights to their territories and were obligated to market and sell Iridium services. The gateways would also be responsible for end-user sales, activation and deactivation of Iridium services, account maintenance, and billing.

Iridium would need each country to grant full licenses for access to the Iridium system. Iridium would need to identify the "priority" countries that account for the majority of the business plan.

Because of the number of countries involved in the Iridium network, Iridium would need to establish global Customer Care Centers for support services in all languages. No matter where an Iridium user was located, he or she would have access to a customer service representative in their native language. The Customer Care Centers would be strategically located to offer 24-hours-a-day, 7-days-a-week, and 365-days-a-year support.

THE "HIDDEN" BUSINESS CASE

The decision by Motorola to invest heavily into the Iridium Project may have been driven by a secondary or hidden business case. Over the years, Motorola achieved a reputation of being a first mover (i.e., first to market). With the Iridium Project, Motorola was poised to capture first-mover advantage in providing global telephone service via LEO satellites. In addition, even if the Iridium Project never resulted in providing service, Motorola would still have amassed valuable intellectual property that would make Motorola possibly the major player for years to come in satellite communications. There may have also been the desire of Robert and Christopher Galvin to have their names etched in history as the pioneers in satellite communication.

RISK MANAGEMENT

Good business cases identify the risks that the project must consider. For simplicity sake, the initial risks associated with the Iridium Project could be classified as follows.

Technology Risks: Although Motorola had some technology available for the Iridium Project, there was still the need to develop additional technology, specifically satellite communications technology. The development process was expected to take years and would eventually result in numerous patents.

Mark Gercenstein, Iridium's vice president of operations, explains the system's technological complexity:

> More than 26 completely impossible things had to happen first, and in the right sequence (before we could begin operations)—like getting capital, access to the marketplace, global spectrum, the same frequency band in every country of operations.[12]

While there was still some risk in the development of new technology, Motorola had the reputation of being a high-tech, can-do company. The engineers at Motorola believed that they could bring forth miracles in technology. Motorola also had a reputation for being a first mover with new ideas and products, and there was no reason to believe that this would not happen on the Iridium Project. There was no competition for Iridium at its inception.

Because the project schedule was more than a decade in duration, there was the risk of technology obsolescence. This required that certain assumptions be made concerning technology a decade downstream. Developing a new product is relatively easy if the environment is stable. But in a high-tech environment that is both turbulent and dynamic, it is extremely difficult to determine how customers will perceive and evaluate the product 10 years later.

Development Risks: The satellite communication technology, once developed, had to be manufactured, tested, and installed in the satellites and ground equipment. Even though the technology existed or would exist, there was still the transitional or development risks from engineering to manufacturing to implementation that would bring with it additional problems that were not contemplated or foreseen.

Financial Risks: The cost of the Iridium Project would most certainly be measured in the billions of dollars. This would include the costs for technology development and implementation, the manufacture and launch of satellites, the

---

[12]Peter Grams and Patrick Zerbib, "Caring for Customers in a Global Marketplace," *Satellite Communications*, October 1998, p. 25.

construction of ground support facilities, marketing, and supervision. Raising money from Wall Street's credit and equity markets was years away. Investors were unlikely to put up the necessary hundreds of millions of dollars on merely an idea or a vision. The technology needed to be developed and possibly accompanied by the launch of a few satellites before the credit and equity markets would come on board.

Private investors were a possibility, but the greatest source of initial funding would have to come from the members of the Iridium consortium. While sharing the financial risks among the membership seemed appropriate, there was no question that bank loans and lines of credit would be necessary. Since the Iridium Project was basically an idea, the banks would require some form of collateral or guarantee for the loans. Motorola, being the largest stakeholder (and also with the "deepest pockets"), would need to guarantee the initial loans.

Marketing Risks: The marketing risks were certainly the greatest risks facing the Iridium membership. Once again, the risks were shared among its membership where each member was expected to sign up customers in its geographic area.

Each consortium member had to aggressively sign up customers for a product that didn't exist yet, no prototypes existed to be shown to the customers, limitations on the equipment were unknown as yet, and significant changes in technology could occur between the time the customer signed up and the time the system was ready for use. Companies that see the need for Iridium today may not see the same need 10 years later.

Motivating the consortium partners to begin marketing immediately would be extremely difficult since marketing material was nonexistent. There was also the very real fear that the consortium membership would be motivated more so by the technology rather than the necessary size of the customer base required.

The risks were interrelated. The financial risks were highly dependent upon the marketing risks. If a sufficient customer base could not be signed up, there could be significant difficulty in raising capital.

THE COLLECTIVE BELIEF

Although the literature doesn't clearly identify it, there was most likely a collective belief among the workers assigned to the Iridium Project. The collective belief is a fervent, and perhaps blind, desire to achieve that can permeate the entire team, the project sponsor, and even the most senior levels of management. The collective belief can make a rational organization act in an irrational manner.

When a collective belief exists, people are selected based upon their support for the collective belief. Nonbelievers are pressured into supporting the collective belief and team members are not allowed to challenge the results. As the collective belief grows, both advocates and nonbelievers are trampled. The pressure of the collective belief can outweigh the reality of the results.

There are several characteristics of the collective belief, which is why some large, high-tech projects are often difficult to kill:

- Inability or refusal to recognize failure
- Refusing to see the warning signs
- Seeing only what you want to see
- Fearful of exposing mistakes
- Viewing bad news as a personal failure
- Viewing failure as a sign of weakness
- Viewing failure as damage to one's career
- Viewing failure as damage to one's reputation

THE EXIT CHAMPION

Project champions do everything possible to make their project successful. But what if the project champions, as well as the project team, have blind faith in the success of the project? What happens if the strongly held convictions and the collective belief disregard the early warning signs of imminent danger? What happens if the collective belief drowns out dissent?

In such cases, an exit champion must be assigned. The exit champion sometimes needs to have some direct involvement in the project in order to have credibility. Exit champions must be willing to put their reputation on the line and possibly face the likelihood of being cast out from the project team. According to Isabelle Royer:[13]

> Sometimes it takes an individual, rather than growing evidence, to shake the collective belief of a project team. If the problem with unbridled enthusiasm starts as an unintended consequence of the legitimate work of a project champion, then what may be needed is a countervailing force—an exit champion. These people are more than devil's advocates. Instead of simply raising questions about a project, they seek objective evidence showing that problems in fact exist. This allows them to challenge—or, given the ambiguity of existing data, conceivably even to confirm—the viability of a project. They then take action based on the data.

The larger the project and the greater the financial risk to the firm, the higher up the exit champion should reside. On the Iridium Project, the collective belief

---

[13] Isabelle Royer, "Why Bad Projects Are So Hard to Kill," *Harvard Business Review*, February 2003, p.11:

originated with Galvin, Motorola's CEO. Therefore, who could possibly function as the exit champion on the Iridium Project? Since it most likely should be someone higher up than Galvin, the exit champion should have been someone on the board of directors or even the entire Iridium board of directors.

Unfortunately, the entire Iridium board of directors was also part of the collective belief and shirked its responsibility for oversight on the Iridium Project. In the end, Iridium had no exit champion. Large projects incur large cost overruns and schedule slippages. Making the decision to cancel such a project, once it has started, is very difficult, according to David Davis.[14]

> The difficulty of abandoning a project after several million dollars have been committed to it tends to prevent objective review and recosting. For this reason, ideally an independent management team—one not involved in the projects development—should do the recosting and, if possible, the entire review. . . . If the numbers do not hold up in the review and recosting, the company should abandon the project. The number of bad projects that make it to the operational stage serves as proof that their supporters often balk at this decision.
>
> . . . Senior managers need to create an environment that rewards honesty and courage and provides for more decision making on the part of project managers. Companies must have an atmosphere that encourages projects to succeed, but executives must allow them to fail.

The longer the project, the greater the necessity for the exit champions and project sponsors to make sure that the business plan has "exit ramps" such that the project can be terminated before massive resources are committed and consumed. Unfortunately, when a collective belief exists, exit ramps are purposefully omitted from the project and business plans.

IRIDIUM'S INFANCY YEARS

By 1992, the Iridium Project attracted such stalwart companies as General Electric, Lockheed, and Raytheon. Some companies wanted to be involved to be part of the satellite technology revolution while others were afraid of falling behind the technology curve. In any event, Iridium was lining up strategic partners, but slowly.

The Iridium Plan, submitted to the FCC in August, 1992, called for a constellation of 66 satellites, expected to be in operation by 1998, more powerful than originally proposed, thus keeping the project's cost at the previously estimated

---

[14]David Davis, "New Projects: Beware of False Economics," *Harvard Business Review*, March–April 1985, pp.100–101. Copyright © 1985 by the President and Fellows of Harvard College. All rights reserved.

$3.37 billion. But the Iridium Project, while based on lofty forecasts of available customers, was now attracting other companies competing for FCC approval on similar satellite systems including Loral Corp., TRW Inc., and Hughes Aircraft Co., a unit of General Motors Corp. There were at least nine companies competing for the potential billions of dollars in untapped revenue possible from satellite communications.

Even with the increased competition, Motorola was signing up partners. Motorola had set an internal deadline of December 15, 1992, to find the necessary funding for Iridium. Signed letters of intent were received from the Brazilian government and United Communications Co., of Bangkok, Thailand, to buy 5 percent stakes in the project, each now valued at about $80 million. The terms of the agreement implied that the Iridium consortium would finance the project with roughly 50 percent equity and 50 percent debt.

When the December 15 deadline arrived, Motorola was relatively silent on the signing of funding partners, fueling speculation that it was having trouble. Motorola did admit that the process was time-consuming because some investors required government approval before proceeding. Motorola was expected to announce at some point, perhaps in the first half of 1993, whether it was ready to proceed with the next step, namely receiving enough cash from its investors, securing loans, and ordering satellite and group equipment.

As the competition increased, so did the optimism about the potential size of the customer base.

> "We're talking about a business generating billions of dollars in revenue," says John F. Mitchell, Vice Chairman at Motorola. "Do a simple income extrapolation," adds Edward J. Nowacki, a general manager at TRW's Space & Electronics Group, Redondo Beach, Calif., which plans a $1.3 billion, 12-satellite system called Odyssey. "You conclude that even a tiny fraction of the people around the world who can afford our services will make them successful." Mr. Mitchell says that if just 1% to 1.5% of the expected 100 million cellular users in the year 2000 become regular users at $3 a minute, Iridium will breakeven. How does he know this? "Marketing studies," which he won't share. TRW's Mr. Nowacki says Odyssey will blanket the Earth with two-way voice communication service priced at "only a slight premium" to cellular. "With two million subscribers we can get a substantial return on our investment," he says. "Loral Qualcomm Satellite Services, Inc. aims to be the 'friendly' satellite by letting phone-company partners use and run its system's ground stations," says Executive Vice President Anthony Navarra. "By the year 2000 there will be 15 million unserved cellular customers in the world," he says.[15]

But while Motorola and other competitors were trying to justify their investment with "inflated market projections" and a desire from the public for faster and clearer reception, financial market analysts were not so benevolent. First,

---

[15]John J. Keller, "Telecommunications: Phone Space Race Has Fortune at Stake," *Wall Street Journal* (Eastern edition), New York, January 18, 1993, p. B1.

market analysts questioned the size of the customer base that would be willing to pay $3000 or more for a satellite phone in addition to $3–$7 per minute for a call. Second, the system required a line-of-sight transmission, which meant that the system would not work in buildings or in cars. If a businessman were attending a meeting in Bangkok and needed to call his company, he must exit the building, raise the antenna on his $3000 handset, point the antenna toward the heavens, and then make the call. Third, the low-flying satellites would eventually crash into the Earth's atmosphere every 5–7 years because of atmospheric drag and would need to be replaced. That would most likely result in high capital costs. And fourth, some industry analysts believed that the startup costs would be closer to $6–$10 billion rather than the $3.37 billion estimated by Iridium. In addition, the land-based cellular phone business was expanding in more countries, thus creating another competitive threat for Iridium.

The original business case needed to be reevaluated periodically. But with strong collective beliefs and no exit champions, the fear of a missed opportunity, irrespective of the cost, took center stage.

Reasonably sure that 18 out of 21 investors were on board, Motorola hoped to start launching test satellites in 1996 and begin commercial service by 1998. But critics argued that Iridium might be obsolete by the time it actually started working.

Eventually, Iridium was able to attract financial support from 19 strategic partners:

- AIG Affiliated Companies
- China Great Wall Industry Corporation (CGWIC)
- Iridium Africa Corporation (based in Cape Town)
- Iridium Canada, Inc.
- Iridium India Telecom Private Ltd, (ITIL)
- Iridium Italia S.p.A.
- Iridium Middle East Corporation
- Iridium SudAmerica Corporation
- Khrunichev State Research and Production Space Center
- Korea Mobile TELECOM
- Lockheed Martin
- Motorola
- Nippon Iridium Corporation
- Pacific Electric Wire & Cable Co. Ltd (PEWC)
- Raytheon
- STET
- Sprint
- Thai Satellite Telecommunications Co., Ltd.
- Verbacom.

Seventeen of the strategic partners also participated in gateway operations with the creation of operating companies.

The Iridium board of directors consisted of 28 telecommunications executives. All but one board member was a member of the consortium as well. This made it very difficult for the board to fulfill its oversight obligation, effectively giving the members' vested/financial interest in the Iridium Project.

In August 1993, Lockheed announced that it would receive $700 million in revenue for satellite construction. Lockheed would build the satellite structure, solar panels, attitude and propulsion systems, along with other parts, and engineering support. Motorola and Raytheon Corp. would build the satellite's communications gear and antenna.

In April 1994, McDonnell Douglas Corp. received from Iridium a $400 million contract to launch 40 satellites for Iridium. Other contracts for launch services would be awarded to Russia's Khrunichev Space Center and China's Great Wall Industry Corporation, both members of the consortium. The lower-cost contracts with Russia and China were putting extraordinary pressure on U.S. providers to lower their costs.

Also at the same time, one of Iridium's competitors, the Globalstar system, which was a 48-satellite mobile telephone system led by Loral Corporation, announced that it intended to charge 65 cents per minute in the areas it served. Iridium's critics were arguing that Iridium would be too pricey to attract a high volume of callers.[16]

DEBT FINANCING

In September 1994, Iridium said that it had completed its equity financing by raising an additional $733.5 million. This brought the total capital committed to Iridium through equity financing to $1.57 billion. The completion of equity financing permitted Iridium to enter into debt financing to build the global wireless satellite network.

In September 1995, Iridium announced that it would be issuing $300 million 10-year senior-subordinated discounted notes rated Caa by Moody's and CCC+ by Standard & Poor's, via the investment banker Goldman Sachs Inc. The bonds were considered to be high-risk, high-yield "junk" bonds after investors concluded that the rewards weren't worth the risk.

---

[16]Jeff Cole, "McDonnell Douglas Said to Get Contract to Launch 40 Satellites for Iridium Plan," *Wall Street Journal* (Eastern edition), New York, April 12, 1994, p. A4.

The rating agencies cited the reasons for the low rating to be yet unproven sophisticated technology, and the fact that a significant portion of the system's hardware would be located in space. But there were other serious concerns:

- The ultimate cost of the Iridium Project would be more like $6 billion or higher rather than $3.5 billion, and it was unlikely that Iridium would recover that cost.
- Iridium would be hemorrhaging cash for several more years before service would begin.
- The optimistic number of potential customers for satellite phones may not choose the Iridium system.
- The number of competitors had increased since the Iridium concept was first developed.
- If Iridium defaulted on its debt, the investors could lay claim to Iridium's assets. But what would investors do with more than 66 satellites in space, waiting to disintegrate upon reentering the atmosphere?

Iridium was set up as "project financing" in which case, if a default occurred, only the assets of Iridium could be attached. With project financing, the consortium's investors would be held harmless for any debt incurred from the stock and bond markets and could simply walk away from Iridium. These risks associated with project financing were well understood by those that invested in the equity and credit markets.

Goldman Sachs & Co., the lead underwriter for the securities offering, determined that for the bond issue to be completed successfully, there would need to exist a completion guarantee from investors with deep pockets, such as Motorola. Goldman Sachs cited a recent $400 million offering by one of Iridium's competitors, Globalstar, which had a guarantee from the managing general partner, Loral Corp.[17]

Because of the concern by investors, Iridium withdrew its planned $300 million debt offering. Also, Globalstar, even with its loan guarantee, eventually withdrew its $400 million offering. Investors wanted both an equity position in Iridium and a 20 percent return. Additionally, Iridium would need to go back to its original 17-member consortium and arrange for internal financing.

In February 1996, Iridium had raised an additional $315 million from the 17-member consortium and private investors. In August 1996, Iridium had secured a $750 million credit line with 62 banks co-arranged by Chase Securities Inc., a unit of Chase Manhattan Corp. and the investment banking division of Barclays Bank PLC. The credit line was oversubscribed by more than double its

---

[17]Quentin Hardy, "Iridium Pulls $300 Million Bond Offer; Analysts Cite Concerns about Projects," *Wall Street Journal* (Eastern edition), New York, September 22, 1995, p. A5.

original goal because the line of credit was backed by a financial guarantee by Motorola and its AAA credit rating. Because of the guarantee by Motorola, the lending rate was slightly more than the 5.5 percent baseline international commercial lending rate and significant lower than the rate in the $300 million bond offering that was eventually recalled.

Despite this initial success, Iridium still faced financial hurdles. By the end of 1996, Iridium planned on raising more than $2.65 billion from investors. It was estimated that more than 300 banks around the globe would be involved, and that this would be the largest private debt placement ever. Iridium believed that this debt placement campaign might not be that difficult since the launch date for Iridium services was getting closer.

THE M-STAR PROJECT

In October 1996, Motorola announced that it was working on a new project dubbed M-Star, which would be a $6.1 billion network of 72 low-orbit satellites capable of worldwide voice, video, and high-speed data links targeted at the international community. The project was separate from the Iridium venture and was expected to take 4 years to complete after FCC approval. According to Bary Bertiger, now corporate vice president and general manager of Motorola's satellite communications group, "Unlike Iridium, Motorola has no plans to detach M-Star as a separate entity. We won't fund it ourselves, but we will have fewer partners than in Iridium."[18]

The M-Star Project raised some eyebrows in the investment community. Iridium employed 2000 people but M-Star had only 80. The Iridium Project generated almost 1100 patents for Motorola, and that intellectual property would most likely be transferred to M-Star. Also, Motorola had three contracts with Iridium for construction and operation of the global communication system providing for approximately $6.5 billion in payments to Motorola over a 10-year period that began in 1993. Was M-Star being developed at the expense of Iridium? Could M-Star replace Iridium? What would happen to the existing 17-member consortium at Iridium if Motorola were to withdraw its support in lieu of its own internal competitive system?

A NEW CEO

In 1996, Iridium began forming a very strong top management team with the hiring of Dr. Edward Staiano as CEO and vice chairman. Prior to joining Iridium in

---

[18]Quentin Hardy, "Motorola is Plotting New Satellite Project—M-Star Would Be Faster Than the Iridium System, Pitched to Global Firms," *Wall Street Journal* (Eastern edition), New York, October 14, 1996, p. B4.

1996, Staiano had worked for Motorola for 23 years, during which time he developed a reputation for being hard-nosed and unforgiving. During his final 11 years with Motorola, Staiano led the company's General Systems Sector to record growth levels. In 1995, the division accounted for approximately 40 percent of Motorola's total sales of $27 billion. In leaving Motorola's payroll for Iridium's, Staiano gave up a $1.3 million per year contract with Motorola for a $500,000 base salary plus 750,000 Iridium stock options that vested over a 5-year period. Staiano commented,

> I was spending 40 percent to 50 percent of my time [at Motorola] on Iridium anyway ... If I can make Iridium's dream come true, I'll make a significant amount of money.[19]

SATELLITE LAUNCHES

At 11:28 AM on a Friday morning the second week of January 1997, a Delta 2 rocket carrying a Global Positioning System (GPS) exploded upon launch, scattering debris above its Cape Canaveral launch pad. The launch, which was originally scheduled for the third quarter of 1996, would certainly have an impact on Iridium's schedule, while an industry board composed of representatives from McDonnell-Douglas and the Air Force determined the cause of the explosion. Other launches had already been delayed for a variety of technical reasons.

In May of 1997, after six failed tries, the first five Iridium satellites were launched. Iridium still believed that the target date for launch of service, September 1998, was still achievable but that all slack in the schedule had been eliminated due to the earlier failures.

By this time, Motorola had amassed tremendous knowledge on how to mass-produce satellites. As described by Bennahum:

> The Iridium constellation was built on an assembly line, with all the attendant reduction in risk and cost that comes from doing something over and over until it is no longer an art but a process. At the peak of this undertaking, instead of taking 18 to 36 months to build one satellite, the production lines disgorged a finished bird every four and a half days, sealed it in a container, and placed it on the flatbed of an idling truck that drove it to California or Arizona, where a waiting Boeing 747 carried it to a launchpad in the mountains of Taiyuan, China, or on the steppes of Baikonur in Kazakhstan.[20]

---

[19]Quentin Hardy, "Staiano Is Leaving Motorola to Lead Firm's Iridium Global Satellite Project," *Wall Street Journal* (Eastern edition), New York, December 10, 1996, p. B8.
[20]Bennahum, 1998.

AN INITIAL PUBLIC OFFERING (IPO)

Iridium was burning cash at the rate of $100 million per month. Iridium filed a preliminary document with the Security and Exchange Commission (SEC) for an initial public offering of 10 million shares to be offered at $19–$21 a share. Because of the launch delays, the IPO was delayed.

In June of 1997, after the first five satellites were placed in orbit, Iridium filed for an IPO of 12 million shares priced at $20 per share. This would cover about 3 months of operating expenses including satellite purchases and launch costs. The majority of the money would go to Motorola.

SIGNING UP CUSTOMERS

The reality of the Iridium concept was now at hand. All that was left to do was to sign up 500,000–600,000 customers, as predicted, to use the service. Iridium set aside $180 million for a marketing campaign including advertising, public relations, and worldwide, direct mail effort. Part of the advertising campaign included direct mail translated into 13 languages, ads on television and on airlines, airport booths, and Internet web pages.

How to market Iridium was a challenge. People would certainly hate the phone. According to John Windolph, executive director of marketing communications at Iridium, "It's huge! It will scare people. It is like a brick-size device with an antenna like a stout bread stick. If we had a campaign that featured our product, we'd lose." The decision was to focus on the fears of being out of touch. Thus, the marketing campaign began. But Iridium still did not have a clear picture of who would subscribe to the system. An executive earning $700,000 would probably purchase the bulky phone, have his or her assistant carry the phone in his or her briefcase, be reimbursed by the company for the use of the phone, and pay $3–$7 per minute for calls, also a business expense. But are there 600,000 executives worldwide that need the service?

There were several other critical questions that needed to be addressed. How do we hide or downplay the $3400 purchase price of the handset and the usage cost of $7 per minute? How do we avoid discussions about competitors that are offering similar services at a lower cost? With operating licenses in about 180 countries, do we advertise in all of them? Do we take out ads in *Oil and Gas Daily*? Do we advertise in girlie magazines? Do we use full-page or double-page spreads?

Iridium had to rely heavily upon its "gateway" partners for marketing and sales support. Iridium itself would not be able to reach the entire potential audience. Would the gateway partners provide the required marketing and sales support? Do the gateway partners know how to sell the Iridium system and the associated products?

The answer to these questions appeared quickly.

Over a matter of weeks, more than one million sales inquiries poured into Iridium's sales offices. They were forwarded to Iridium's partners—and many of them promptly disappeared, say several Iridium insiders. With no marketing channels and precious few sales people in place, most global partners were unable to follow up on the inquiries. A mountain of hot sales tips soon went cold.[21]

IRIDIUM'S RAPID ASCENT

On November 1, 1998, the Iridium system was officially launched. It was truly a remarkable feat that the 11-year project was finally launched, just a little more than a month late.

After 11 years of hard work, we are proud to announce that we are open for business. Iridium will open up the world of business, commerce, disaster relief, and humanitarian assistance with our first-of-its-kind global communications service . . . The potential use of Iridium products is boundless. Business people who travel the globe and want to stay in touch with home and office, industries that operate in remote areas—all will find Iridium to be the answer to their communications needs.[22]

On November 2, 1998, Iridium began providing service. With the Iridium system finally up and running, most financial analysts issued "buy" recommendations for Iridium stock with expected yearly revenues of $6–$7 billion within 5 years. On January 25, 1999, Iridium held a news conference to discuss its earnings for the fourth quarter of 1998. Ed Staiano, CEO of Iridium stated:

In the fourth quarter of 1998, Iridium made history as we became the first truly global mobile telephone company. Today, a single wireless network, the Iridium Network, covers the planet. And we have moved into 1999 with an aggressive strategy to put a large number of customers on our system, and quickly transform Iridium from a technological event to a revenue generator. We think the prospects for doing this are excellent. Our system is performing at a level beyond expectations.

Financing is now in place through projected cash flow positives. Customer interest remains very high and a number of potentially large customers have now evaluated our service and have given it very high ratings. With all of this going for us, we are in position to sell the service and that is precisely where we are focusing the bulk of our efforts.[23]

---

[21]Leslie Cauley, "Losses in Space—Iridium's Downfall: The Marketing Took a Back Seat to Science—Motorola and Partners Spent Billions on Satellite Links for a Phone Few Wanted," Wall Street Journal (Eastern edition), New York, August 18, 1999, p. A1.
[22]Excerpts from the Iridium press release, November 1, 1998.
[23]Excerpts from the Iridium conference call, January 25, 1999.

Roy Grant, CEO of Iridium, stated:

> Last week Iridium raised approximately $250 million through a very successful 7.5 million-share public offering. This offering had three major benefits. It provided $250 million of cash to our balance sheet. It increased our public float to approximately 20 million shares. And it freed up restrictions placed on $300 million of the $350 million of Motorola guarantees. These restrictions were placed on that particular level of guarantees by our bankers in our $800 million secured credit facility.
>
> With this $250 million, combined with the $350 million of additional guarantees from Motorola, this means we have approximately $600 million of funds in excess of what we need to break cash flow breakeven. This provides a significant contingency for the company.[24]

DECEMBER, 1998

In order to make its products and services known to travelers, Iridium agreed to acquire Claircom Corporation from AT&T and Rogers Cantel Mobile Communications for about $65 million. Claircom provided in-flight telephone systems for U.S. planes as well as equipment for international carriers. The purchase of Claircom would be a marketing boost for Iridium.

The problems with large, long-term technology projects were now appearing in the literature. As described by Bennahum:

> "This system does not let you do what a lot of wired people want to do," cautions Professor Heather Hudson, who runs the telecommunications program at the University of San Francisco and studies the business of wireless communications. "Nineteen-nineties technologies are changing so fast that it is hard to keep up. Iridium is designed from a 1980s perspective of a global cellular system. Since then, the Internet has grown and cellular telephony is much more pervasive. There are many more opportunities for roaming than were assumed in 1989. So there are fewer businesspeople who need to look for an alternative to a cell phone while they are on the road."[25]

Additionally, toward the late 1990s, some industry observers felt that Motorola had additional incentive to ensure that Iridium succeeded, irrespective of the costs—namely, protecting its reputation. Between 1994 and 1997, Motorola had suffered slowing sales growth, a decline in net income, and declining margins. Moreover, the company had experienced several previous business mishaps, including a failure to anticipate the cellular industry's switch to digital cell phones, which played a major role in Motorola's more than 50 percent share-price decline in 1998.

---

[24]See note 23.
[25]Bennehum,1998.

IRIDIUM'S RAPID DESCENT

It took more than a decade for the Iridium Project to ascend and only a few months for descent. In the first week of March, almost 5 weeks after the January teleconference, Iridium's financial woes began to surface. Iridium had expected 200,000 subscribers by the end of 1998 and additional subscribers at a rate of 40,000 per month. Iridium's bond covenants stated a target of 27,000 subscribers by the end of March. Failure to meet such a small target could send investor confidence spiraling downward. Iridium had only 10,000 subscribers. The market that was out there 10 years ago was not the market that was there today. Also, 10 years ago there was little competition for Iridium.

Iridium cited the main cause of the shortfall in subscriptions as being shortages of phones, glitches in some of the technology, software problems, and, most important, a lack of trained sales channels. Iridium found out that it had to train a sales staff and that Iridium itself would have to sell the product, not its distributors. The investor community did not appear pleased with the sales problem that should have been addressed years ago, not 4 months into commercial service.

Iridium's advertising campaign was dubbed "Calling Planet Earth" and promised that you had the freedom to communicate anytime and anywhere. This was not exactly true because the system could not work within buildings or even cars. Furthermore, Iridium underestimated the amount of time subscribers would require to examine and test the system before signing on. In some cases, this would be six months.

Many people blamed marketing and sales for Iridium's rapid descent:

> True, Iridium committed so many marketing and sales mistakes that its experiences could form the basis of a textbook on how not to sell a product. Its phones started out costing $3,000, were the size of a brick, and didn't work as promised. They weren't available in stores when Iridium ran a $180 million advertising campaign. And Iridium's prices, which ranged from $3.00 to $7.50 a call, were out of this world.[26]

Iridium's business plan was flawed. With service beginning on November 2, 1998, it was unlikely that 27,000 subscribers would be on board by March of 1999, given the time required to test the product. The original business plan required that the consortium market and sell the product prior to the onset of service. But selling the service from just a brochure was almost impossible. Subscribers want to touch the phone, use it, and test it prior to committing to a subscription.

---

[26]James Surowieckipp, "The Latest Satellite Startup Lifts Off. Will It Too Explode?" *Fortune Magazine*, October 25, 1999, pp. 237–254.

Iridium announced that it was entering into negotiations with its lenders to alter the terms of an $800 million secured credit agreement due to the weaker-than-expected subscriber and revenue numbers. Covenants on the credit agreement included the following[27]:

Date	Cumulative Cash Revenue ($ Millions)	Cumulative Accrued Revenue ($ Millions)	Number of Satellite Phone Subscribers	Number of System Subscribers[28]
March 31, 1999	$ 4	$ 30	27,000	52,000
June 30, 1999	50	150	88,000	213,000
Sept. 30, 1999	220	470	173,000	454,000

The stock, which had traded as high as almost $73 per share, was now at approximately $20 per share. And, in yet another setback, the chief financial officer, Roy T. Grant, resigned.

April, 1999

Iridium's CEO, Ed Staiano, resigned at the April 22 board meeting. Sources believed that Staiano resigned when the board nixed his plan requesting additional funds to develop Iridium's own marketing and distribution team rather than relying on its strategic partners. Sources also stated another issue in that Staiano had cut costs to the barebones at Iridium but could not get Motorola to reduce its lucrative $500 million service contract with Iridium. Some people believed that Staiano wanted to reduce the Motorola service contract by up to 50 percent. John Richardson, the CEO of Iridium Africa Corp., was assigned as interim CEO. Richardson's expertise was in corporate restructuring. For the quarter ending March, Iridium said it had a net loss of $505.4 million, or $3.45 a share. The stock fell to $15.62 per share. Iridium managed to attract just 10,294 subscribers 5 months after commercial rollout.

One of Richardson's first tasks was to revamp Iridium's marketing strategy. Iridium was unsure as to what business it was in. According to Richardson,

The message about what this product was and where it was supposed to go changed from meeting to meeting. . . . One day, we'd talk about cellular applications, the next

---

[27]Iridium World Communications Ltd., 1998 Annual Report.
[28]Total system subscribers include users of Iridium's phone, fax, and paging services.

day it was a satellite product. When we launch in November, I'm not sure we had a clear idea of what we wanted to be.[29]

May, 1999

Iridium officially announced that it did not expect to meet its targets specified under the $800 million loan agreement. Lenders granted Iridium a 2-month extension. The stock dropped to $10.44 per share, party due to a comment by Motorola that it might withdraw from the ailing venture.

Wall Street began talking about the possibility of bankruptcy. But Iridium stated that it was revamping its business plan and by month's end hoped to have chartered a new course for its financing. Iridium also stated in a regulatory filing that it was uncertain whether it would have enough cash to complete the agreement to purchase Claircom Communications Group Inc., an in-flight telephone-service provider, for the promised $65 million in cash and debt.

Iridium had received extensions on debt payments because the lending community knew that it was no small feat transforming from a project plan to an operating business. Another reason why the banks and creditors were willing to grant extensions was because bankruptcy was not a viable alternative. The equity partners owned all of the Earth stations, all distribution, and all regulatory licenses. If the banks and creditors forced Iridium into bankruptcy, they could end up owning a satellite constellation that could not talk to the ground or gateways.

June, 1999

Iridium received an additional 30-day extension beyond the 2-month extension it had already received. Iridium was given until June 30 to make a $90 million bond payment. Iridium began laying off 15 percent of its 550 employee workforce including two senior officers. The stock had now sunk to $6 per share and the bonds were selling at 19 cents on the dollar. John Richardson, CEO of Iridium, said: "We did all of the difficult stuff well, like building the network, and did all of the no-brainer stuff at the end poorly."[30]

In a later interview John Richardson stated[31]:

> Iridium's major mistake was a premature launch for a product that wasn't ready. People became so obsessed with the technical grandeur of the project that they missed fatal marketing traps . . . Iridium's international structure has proven almost

---

[29]Carleen Hawn, "High Wireless Act," *Forbes*, June 14, 1999, pp. 60–62.
[30]Hawn, 1999.
[31]Leslie Cauley, "Losses in Space—Iridium's Downfall: The Marketing Took a Back Seat to Science," *Wall Street Journal* (Eastern edition), New York, August 18, 1999, p. A1.

impossible to manage: the 28 members of the board speak multiple languages, turning meetings into mini-U.N. conferences complete with headsets translating the proceedings … into five languages.

We're a classic MBA case study in how not to introduce a product. First we created a marvelous technological achievement. Then we asked how to make money on it.

Iridium was doing everything possible to avoid bankruptcy. Time was what Iridium needed. Some industrial customers would take 6–9 months to try out a new product, but would be reluctant to subscribe if it appeared that Iridium would be out of business in 6 months. In addition, Iridium's competitors were lowering their prices significantly-putting further pressure on Iridium. Richardson then began providing price reductions of up to 65 percent off of the original price for some of Iridium's products and services.

July, 1999

The banks and investors agreed to give Iridium yet a third extension to August 11 to meet its financial covenants. Everyone seemed to understand that the restructuring effort was much broader than originally contemplated.

Motorola, Iridium's largest investor and general contractor, admitted that the project may have to be shut down and liquidated as part of bankruptcy proceedings unless a restructuring agreement could be reached. Motorola also stated that if bankruptcy occurred, Motorola would continue to maintain the satellite network, but for a designated time period only.

Iridium had asked its consortium investors and contractors to come up with more money. But to many consortium members, it looked like they would be throwing good money after bad. Several partners made it clear that they would simply walk away from Iridium rather than providing additional funding. That could have a far-reaching effect on the service at some locations. Therefore, all partners had to be involved in the restructuring. Wall Street analysts expected Iridium to be allowed to repay its cash payments on its debt over several years or offer debt holders an equity position in Iridium. It was highly unlikely that Iridium's satellites orbiting the Earth would be auctioned off in bankruptcy court.

August, 1999

On August 12, Iridium filed for bankruptcy protection. This was like having "a dagger stuck in their heart" for a company that a few years earlier had predicted financial breakeven in just the first year of operations. This was one of the 20 largest bankruptcy filings up to this time. The stock, which had been trading as little as $3 per share, was suspended from the NASDAQ on August 13, 1999.

Iridium's phone calls had been reduced to around $1.40–$3 per minute and the handsets were reduced to $1500 per unit.

There was little hope for Iridium. Both the business plan and the technical plan were flawed. The business plan for Iridium seemed like it came out of the film "Field of Dreams" where an Iowa corn farmer was compelled to build a baseball field in the middle of a corn crop. A mysterious voice in his head said, "Build it and they will come." In the film, he did, and they came. While this made for a good plot for a Hollywood movie, it made a horrible business plan. In 1992, Herschel Shosteck, a telecommunications consultant said: "If you build Iridium, people may come. But what is more likely is, if you build something cheaper, people will come to that first."

The technical plan was designed to build the holy grail of telecommunications. Unfortunately, after spending billions, the need for the technology changed over time. The engineers that designed the system, many of whom had worked previously on military projects, lacked an understanding of the word "affordability" and the need for marketing a system to more than just one customer, namely the Department of Defense. "Satellite systems are always far behind the technology curve. Iridium was completely lacking the ability to keep up with Internet time,"[32] Stated Bruce Egan, senior fellow at Columbia University's Institute for Tele-Information.

September, 1999

Leo Mondale resigned as Iridium's chief financial officer. Analysts believed that Mondale's resignation was the result of a successful restructuring no longer being possible. According to one analyst, "If they [Iridium] were close [to a restructuring plan], they wouldn't be bringing in a whole new team."

THE IRIDIUM "FLU"

The bankruptcy of Iridium was having a flulike effect on the entire industry. ICO Global Communications, one of Iridium's major competitors, also filed for bankruptcy protection just 2 weeks after the Iridium filing. ICO failed to raise $500 million it sought from public-rights offerings that had already been extended twice. Another competitor, the Globalstar Satellite Communications System, was still financially sound. Anthony Navarro, Globalstar's chief operating officer, stated "They [Iridium] set everybody's expectations way too high."[33]

---

[32]Stephanie Paterik, "Iridium Alive and Well," *The Arizona Republic*, April 27, 2005, p. D5.
[33]Quentin Hardy, "Surviving Iridium," *Forbes*, September 6, 1999, pp. 216–217.

SEARCHING FOR A WHITE KNIGHT

Iridium desperately needed a qualified bidder who would function as a white knight. It was up to the federal bankruptcy court to determine whether someone was a qualified bidder. A qualified bidder was required to submit a refundable cash deposit or letter of credit issued by a respected bank that would equal the greater of $10 million or 10 percent of the value of the amount bid to take control of Iridium.

According to bankruptcy court filing, Iridium was generating revenue of $1.5 million per month. On December 9, 1999, Motorola agreed to a $20 million cash infusion for Iridium. Iridium desperately needed a white knight quickly or it could run out of cash by February 15, 2000. With a monthly operating cost of $10 million, and a staggering cost of $300 million every few years for satellite replenishment, it was questionable if anyone could make a successful business from Iridium's assets because of asset specificity.

The cellular-phone entrepreneur Craig McCaw planned on a short-term cash infusion while he considered a much larger investment to rescue Iridium. He was also leading a group of investors who pledged $1.2 billion to rescue the ICO satellite system that filed for bankruptcy protection shortly after the Iridium filing.[34]

Several supposedly white knights came forth, but Craig McCaw's group was regarded as the only credible candidate. Although McCaw's proposed restructuring plan was not fully disclosed, it was expected that Motorola's involvement would be that of a minority stakeholder. Also, under the restructuring plan, Motorola would reduce its monthly fee for operating and maintaining the Iridium system from $45 million to $8.8 million.[35]

DEFINITION OF FAILURE (OCTOBER, 1999)

The Iridium network was an engineering marvel. Motorola's never-say-die attitude created technical miracles and overcame NASA-level technical problems. Iridium overcame global political issues, international regulatory snafus, and a range of other geopolitical issues on seven continents. The Iridium system was, in fact, what Motorola's Galvin called the eighth wonder of the world.

But did the bankruptcy indicate a failure for Motorola? Absolutely not! Motorola collected $3.65 billion in Iridium contracts. Assuming $750 million in

---

[34]"Craig McCaw Plans Cash Infusion to Support Cash-Hungry Iridium," *Wall Street Journal* (Eastern edition), New York, February 7, 2000, p.1.
[35]"Iridium Set to Get $75 Million from Investors Led by McCaw," *Wall Street Journal* (Eastern edition), New York, February 10, 2000, p.1.

profit from these contracts, Motorola's net loss on Iridium was about $1.25 billion. Simply stated, Motorola spent $1.25 billion for a project that would have cost it perhaps as much as $5 billion out of its own pocket had it wished to develop the technology itself. Iridium provided Motorola with more than 1000 patents in building satellite communication systems. Iridium allowed Motorola to amass a leadership position in the global satellite industry. Motorola was also signed up as the prime contractor to build the 288-satellite "Internet in the Sky," dubbed the Teledesic Project. Backers of the Teledesic Project, which had a price tag of $15 billion to transmit data, video, and voice, included Boeing, Microsoft's Chairman Bill Gates, and cellular magnate Craig McCaw. Iridium had enhanced Motorola's reputation for decades to come.

Motorola stated that it had no intention of providing additional funding to ailing Iridium, unless of course other consortium members followed suit. Several members of the consortium stated that they would not provide any additional investment and were considering liquidating their involvement in Iridium.[36]

In March 2000 McCaw withdrew its offer to bail out Iridium even at a deep discount asserting that his efforts would be spent on salvaging the ICO satellite system instead. This, in effect, signed Iridium's death warrant. One of the reasons for McCaw's reluctance to rescue Iridium may have been the discontent by some of the investors who would have been completely left out as part of the restructuring effort, thus losing perhaps their entire investment.

THE SATELLITE DEORBITING PLAN

With the withdrawal of McCaw's financing, Iridium notified the U.S. Bankruptcy Court that Iridium had not been able to attract a qualified buyer by the deadline assigned by the court. Iridium would terminate its commercial service after 11:59 PM on March 17, 2000, and that it would begin the process of liquidating its assets.

Immediately following the Iridium announcement, Motorola issued the following press release:

> Motorola will maintain the Iridium satellite system for a limited period of time while the deorbiting plan is being finalized. During this period, we also will continue to work with the subscribers in remote locations to obtain alternative communications. However, the continuation of limited Iridium service during this time will depend on whether the individual gateway companies, which are separate operating companies, remain open.

---

[36]Scott Thurm, "Motorola Inc., McCaw Shift Iridium Tactics," *Wall Street Journal* (Eastern edition), New York, February 18, 2000, p.1.

> In order to support those customers who purchased Iridium service directly from Motorola, Customer Support Call Centers and a website that are available 24 hours a day, seven days a week have been established by Motorola. Included in the information for customers is a list of alternative satellite communications services.

The deorbiting plan would likely take 2 years to complete at a cost of $50–$70 million. This would include all 66 satellites and the other 22 satellites in space serving as spare or decommissioned failures. Iridium would most likely deorbit the satellites four at a time by firing their thrusters to drop them into the atmosphere where they would burn up.

IRIDIUM IS RESCUED FOR $25 MILLION

In November 2000, a group of investors led by an airline executive won bankruptcy court approval to form Iridium Satellite Corporation and purchase all remaining assets of failed Iridium Corporation. The purchase was at a fire-sale price of $25 million, which was less than a penny on the dollar. As part of the proposed sale, Motorola would turn over responsibility for operating the system to Boeing. Although Motorola would retain a 2 percent stake in the new system, Motorola would have no further obligations to operate, maintain, or decommission the constellation.

Almost immediately after the announcement, Iridium Satellite was awarded a $72 million contract from the Defense Information Systems Agency, which is part of the Department of Defense (DoD). Dave Oliver, principal deputy undersecretary of Defense for Acquisition stated:

> "Iridium will not only add to our existing capability, it will provide a commercial alternative to our purely military systems. This may enable real civil/military dual use, keep us closer to leading edge technologically, and provide a real alternative for the future."[37]

Iridium had been rescued from the brink of extinction. As part of the agreement, the newly formed company acquired all of the assets of the original Iridium and its subsidiaries. This included the satellite constellation, the terrestrial network, Iridium real estate, and the intellectual property originally developed by Iridium. Because of the new company's significantly reduced cost structure, it was able to develop a workable business model based upon a targeted market for Iridium's products and services. Weldon Knape, WCC chief executive officer stated: "Everyone thinks the Iridium satellites crashed and burned, but they're all still up there."[38]

---

[37]"DoD Awards $72 Million to Revamp Iridium," *Satellite Today*. Potomac: December 7, 2000, vol.3, iss. 227, p. 1.
[38]Stephanie Paterik, "Iridium Alive and Well," *The Arizona Republic*, April 27, 2005.

A new Iridium phone costs $1495 and is the size of a cordless home phone. Older, larger models start at $699 or one can be rented for about $75 per week. Service costs $1–$1.60 a minute.[39]

EPILOGUE

February 6, 2006, Iridium Satellite declared that 2005 was the best year ever. The company had 142,000 subscribers, which was a 24 percent increase from 2004, and the 2005 revenue was 55 percent greater than in 2004. According to Carmen Lloyd, Iridium's CEO, "Iridium is on an exceptionally strong financial foundation with a business model that is self-funding."[40]

For the year ending 2006, Iridium had $212 million in sales and $54 million in profit. Iridium had 180,000 subscribers and a forecasted growth rate of 14–20 percent per year. Iridium had changed its business model, focusing on sales and marketing first and hype second. This allowed it to reach out to new customers and new markets.[41]

SHAREHOLDER LAWSUITS

The benefit to Motorola, potentially at the expense of Iridium and its investors, did not go unnoticed. At least 20 investor groups filed suit against Motorola and Iridium, citing:

- Motorola milked Iridium and used the partners' money to finance its own foray into satellite communication technology.
- By using Iridium, Motorola ensured that its reputation would not be tarnished if the project failed.
- Most of the money raised through the IPOs went to Motorola for designing most of the satellite and ground-station hardware and software.
- Iridium used the proceeds of its $1.45 billion in bonds, with interest rates from 10.875 to 14 percent, mainly to pay Motorola for satellites.
- Defendants falsely reported achievable subscriber numbers and revenue figures.
- Defendants failed to disclose the seriousness of technical issues.
- Defendants failed to disclose delays in handset deliveries.

---

[39]Paterik, 2005.
[40]Iridium Press Release, February 6, 2006.
[41]Adapted from Reena Jana, "Companies Known for Inventive Tech Were Dubbed the Next Big Thing and Then Disappeared. Now They're Back and Growing," **Business Week**, Innovation, April 10, 2007.

- Defendants violated covenants between itself and its lenders.
- Defendants delayed disclosure of information, provided misleading information, and artificially inflated Iridium's stock price.
- Defendants took advantage of the artificially inflated price to sell significant amounts of their own holdings for millions of dollars in personal profit.

THE BANKRUPTCY COURT RULING

On September 4, 2007, after almost 10 months, the Bankruptcy Court in Manhattan ruled in favor of Motorola and irritated the burned creditors that had hoped to get a $3.7 billion judgment against Motorola. The judge ruled that even though the capital markets were "terribly wrong" about Iridium's hopes for huge profits, Iridium was "solvent" during the critical period when it successfully raised rather impressive amounts of debt and equity in the capital markets.

The court said that even though financial experts now know that Iridium was a hopeless one-way cash flow, flawed technology project, and doomed business model, Iridium was solvent at the critical period of fundraising. Even when the bad news began to appear, Iridium's investors and underwriters still believed that Iridium had the potential to become a viable enterprise.

The day after the court ruling, newspapers reported that Iridium LLC, the now privately held company, was preparing to raise about $500 million in a private equity offering to be followed by an IPO within the next year or two.

AUTOPSY[42]

There were several reasons for Iridium's collapse:

Cellular Build-out Dramatically Reduced the Target Market's Need for Iridium's Service.   Iridium knew its phones would be too large and too expensive to compete with cellular service, forcing the company to play in areas where cellular was unavailable. With this constraint in mind, Iridium sought a target market by focusing on international business executives who frequently traveled to remote areas where cellular phone service wasn't available. Although this market plan predated the rise of cell phones, Iridium remained focused on the business traveler group through the launch of its service. As late as 1998, CEO Staiano predicted Iridium would have 500,000 subscribers by the end of 1999.

---

[42]Sydney Finkelstein and Shade H. Sanford, "Learning from Corporate Mistakes: The Rise and Fall of Iridium," *Organizational Dynamics*, vol. 29, no. 2, 2000, pp. 138–148. © 2000 by Elsevier Sciences, Inc. Reproduced by permission.

One of the main problems with Iridium's offering was that terrestrial cellular had spread faster than the company had originally expected. In the end, cellular was available. Due to Iridium's elaborate technology, the concept-to-development time was 11 years—during this period, cellular networks spread to cover the overwhelming majority of Europe and even migrated to developing countries such as China and Brazil. In short, Iridium's marketing plan targeted a segment—business travelers—whose needs were increasingly being met by cell phones that offered significantly better value than Iridium.

Iridium's Technological Limitations and Design Stifled Adoption. Because Iridium's technology depended on line-of-sight between the phone antenna and the orbiting satellite, subscribers were unable to use the phone inside moving cars, inside buildings, and in many urban areas. Moreover, even in open fields users had to align the phone just right in order to get a good connection. As a top industry consultant said to us in an interview, "you can't expect a CEO traveling on business in Bangkok to leave a building, walk outside on a street corner, and pull out a $3,000 phone." Additionally, Iridium lacked adequate data capabilities, an increasingly important feature for business users. Making matters worse were annoyances such as the fact that battery recharging in remote areas required special solar-powered accessories. These limitations made the phone a tough sell to Iridium's target market of high-level traveling businessmen.

The design of Iridium's phone also hampered adoption. In November 1997, John Windolph, Iridium director of marketing communications, described the handset in the following manner: "It's huge! It will scare people. If we had a campaign that featured our product, we'd lose." Yet a year later Iridium went forward with essentially the same product. The handset, although smaller than competitor Comsat's Planet 1, was still literally the size of a brick.

Poor Operational Execution Plagued Iridium. Manufacturing problems also caused Iridium's launch to stumble out of the gate. Management launched the service before enough phones were available from one of its two main suppliers, Kyocera, which was experiencing software problems at the time. Ironically, this manufacturing bottleneck meant that Iridium couldn't even get phones to the few subscribers that actually wanted one. The decision to launch service in November 1998, in spite of the manufacturing problems, was made by CEO Staiano, although not without opposition. As one report put it, "[John Richardson] claimed to be vociferous in board meetings, arguing against the November launch. Neither the service, nor the service providers, were ready. Supply difficulties meant that there were few phones available in the market."

Iridium's Partners Did Not Provide Adequate Sales and Marketing Support. Although at first Motorola had difficulty attracting investors for Iridium, by 1994

Iridium LLC had partnerships with 18 companies including Sprint, Raytheon, Lockheed Martin, and a variety of companies from China, the Middle East, Africa, India, and Russia. In exchange for investments of $3.7 billion, the partners received equity and seats on Iridium LLC's board of directors. In 1998, 27 of the 28 directors on Indium's board were either Iridium employees or directly appointed by Iridium's partners.

Iridium's partners would ultimately control marketing, pricing, and distribution when the service came on line. Iridium's revenues came from wholesale rates for its phone service. Unfortunately for Iridium, its partners, outside the United States in particular, delayed setting up marketing teams and distribution channels. "The gateways were very often huge telecoms," said Stephane Chard, chief analyst at Euroconsult, a Paris-based research firm. "To them, Iridium was a tiny thing." So tiny, in fact, that Iridium's partners failed to build sales teams, create marketing plans, or set up distribution channels for their individual countries. As the *Wall Street Journal* reported, "with less than six months to go before the launch of the service, time became critical . . . Most partners didn't reveal they were behind schedule."

FINANCIAL IMPACT OF THE BANKRUPTCY

At the time of the bankruptcy, equity investments in Iridium totaled approximately $2 billion. Most analysts, however, considered the stock worthless. Iridium's stock price, which had IPOed at $20 per share in June 1997, and reached an all time high of $72.19 in May 1998, had plummeted to $3.06 per share by the time Iridium declared bankruptcy in August 1999. Moreover, the NASDAQ exchange reacted to the bankruptcy news by immediately halting trading of the stock, and actually delisted Iridium in November 1999. Iridium's partners—who had also made investments by building ground stations, assembling management teams, and marketing Iridium services—were left with little to show for their equity. Iridium's bondholders didn't fare much better than its equity holders. After Iridium declared bankruptcy, its $1.5 billion in bonds were trading for around 15 cents to the dollar as the company entered restructuring talks with its creditors.

WHAT REALLY WENT WRONG?

Iridium will go down in history as one of the most significant business failures of the 1990s. That its technology was breathtakingly elegant and innovative is without question. Indeed, Motorola and Iridium leaders showed great vision in directing the development and launch of an incredibly complex constellation of satellites. Equally as amazing, however, was the manner in which these same leaders led Iridium into bankruptcy by supporting an untenable business plan.

Over the past several years, there have been perhaps thousands of articles written about Iridium's failure to attract customers and its resulting bankruptcy. Conventional wisdom often argues that Iridium was simply caught off guard by the spread of terrestrial cellular. By focusing almost strictly on what happened, such an analysis provides little in the way of valuable learning. A more interesting question is why Iridium's failure happened—namely, why the company continued to press forward with an increasingly flawed business plan.

Three Forces Combined to Create Iridium's Failure

Three forces combined to create Iridium's business failure. First, an "escalating commitment," particularly among Motorola executives who pushed the project forward in spite of known and potentially fatal technology and market problems. Second, for personal and professional reasons Iridium's CEO was unwilling to cut losses and abandon the project. And third, Iridium's board was structured in a way that prevented it from performing its role of corporate governance.

Problem 1: Escalating Commitment. During the 11 years that passed between Indium's initial concept to its actual development, its business plan eroded. First, the gradual build-out of cellular dramatically shrank Iridium's target market—international executives who regularly traveled to areas not covered by terrestrial cellular. Second, it became apparent over time that Iridium's phones would have significant design, operational, and cost problems that would further limit usage.

Motorola's decision to push Iridium forward in spite of a deeply flawed business plan is a classic example of the pitfalls of "escalating commitment." The theory behind escalating commitment is based in part on the "sunk cost fallacy"—making decisions based on the size of previous investments rather than on the size of the expected return. People tend to escalate their commitment to a project when they (a) believe that future gains are available, (b) believe they can turn a project around, (c) are publicly committed or identified with the project, and (d) can recover a large part of their investment if the project fails.

Motorola's involvement in the Iridium Project met all four of these conditions. In spite of known problems, top executives maintained blind faith in Iridium. To say that Iridium's top management was unaware of Iridium's potential problems would be wholly inaccurate. In fact, Iridium's prospectus written in 1998 listed 25 full pages of risks including:

- A highly leveraged capital structure
- Design limitations—including phone size
- Service limitations—including severe degradation in cars, buildings, and urban areas

- High handset and service pricing
- The build-out of cellular networks
- A lack of control over partners' marketing efforts

During Iridium's long concept-to-development time, there is little evidence to suggest that Motorola or Iridium made any appreciable progress in addressing any of these risks. Yet Iridium went forward, single-mindedly concentrating on satellite design and launch while discounting the challenges in sales and marketing the phones. The belief that innovative technology would eventually attract customers, in fact, was deeply ingrained in Motorola's culture.

Indeed, Motorola's history was replete with examples of spectacular innovations that had brought the company success and notoriety. In the 1930s, Paul Gavin developed the first affordable car radio. In the 1940s, Motorola rose to preeminence when it developed the first handheld two-way radio, which was used by the Army Signal Corps during World War II. In the 1950s, Motorola manufactured the first portable television sets. In the 1969, Neil Armstrong's first words from the Moon were sent by a transponder designed and manufactured by the company. In the 1970s and 1980s, Motorola enjoyed success by developing and manufacturing microprocessors and cellular phones.

By the time it developed the concept for Iridium in the early 1990s, Motorola had experienced over 60 years of success in bringing often startling new technology to consumers around the world. Out of this success, however, came a certain arrogance and biased faith in the company's own technology. Just as Motorola believed in the mid-1990s that cellular customers would be slow to switch from Motorola's analog phones to digital phones produced by Ericsson and Nokia, their faith in Iridium and its technology was unshakable.

Problem 2: Staiano's Leadership Was a Double-Edged Sword.   Dr. Edward Staiano became CEO of Iridium in late 1996—before the company had launched most of its satellites. During his previous tenure with Motorola, Staiano had developed a reputation as intimidating and demanding—imposing in both stature, at 6 feet 4 inches, and in temperament. Staiano combined his leadership style with an old Motorola ethic that argued leaders had a responsibility to support their projects. Staiano also had significant financial incentives to push the project forward, rather than cutting losses and moving on. In both 1997 and 1998, he received 750,000 Iridium stock options that vested over a 5-year period. Indeed, this fact didn't escape Staiano's attention when he took the CEO position in late 1996, stating: "If I can make Indium's dream come true, I'll make a significant amount of money."

Ironically, the demanding leadership style, commitment to the project at hand, and financial incentives that made Ed Staiano such an attractive leader for a startup company such as Iridium turned out to be a double-edged sword. Indeed, these same characteristics also made him unwilling to abandon a project with a failed business plan and obsolete technology.

Problem 3: Indium's Board Did Not Provide Adequate Corporate Governance. In 1997, Iridium's board had 28 directors—27 of whom were either Iridium employees or directors designated by Iridium's partners. The composition, not to mention size, of Iridium's board created two major problems. First, the board lacked the insight of outside directors who could have provided a diversity of expertise and objective viewpoints. Second, the fact that most of the board was comprised of partner appointees made it difficult for Iridium to apply pressure to its partners in key situations—such as when many partners were slow to set up the necessary sales and marketing infrastructure prior to service launch. In the end, Iridium's board failed to provide proper corporate oversight and limited Iridium's ability to work with its partners effectively.

LESSONS LEARNED

Executives Should Evaluate Projects such as Iridium as Real Options

Projects with long concept-to-development times pose unique problems for executives. These projects may seem like good investments during initial concept development; but by the time the actual product or service comes on line, both the competitive landscape and the company's ability to provide the service or product have often changed significantly.

To deal with long concept-to-development times, executives should evaluate these projects as real options. A simple model would be a two-stage project. The first stage is strategic in nature and provides the opportunity for a further investment and increased return in the second stage. When the initial stage is complete, however, the company must reevaluate the expected return of future investments based on a better understanding of the product/service and the competitive landscape.

Iridium is a textbook example of a project that would have benefited from this type of analysis. The Iridium Project itself essentially consisted of two stages. During stage one (1987–1996), Motorola developed the technology behind Iridium. During stage two (1996–1999), Motorola built and launched the satellites—and the majority of Iridium's costs occurred during this part of the project.

Investment in R&D for Iridium Was Appropriate—Follow-on Investment Was Not

Looking back, it would be unfair to assert that the initial decision to invest in R&D for Iridium was a mistake. In the late 1980s, Iridium appeared to have a sound business plan. Travel among business executives was increasing, and terrestrial cellular networks didn't cover many of their destinations. It was certainly not unreasonable to foresee a large demand for a wireless phone that had no geographic boundaries. In turn, the investment in R&D was reasonable as it provided the option to deploy (or not deploy) the complex Iridium satellite system 9 years later.

By 1996, however, when Iridium had to make the decision of whether to invest in building and launching satellites, much had changed. Not only had the growth in cellular networks drastically eroded Iridium's target market, but Iridium's own technology was never able to overcome key design, cost, and operational problems. Put simply, Iridium didn't have a viable business plan. Armed with this additional insight, a reasonable evaluation of the project would have precluded further investment.

Executives Must Build Option Value Assessments into Their Business Plans

The key to using the option value approach is to include it in the business plan. Specifically, executives must specify a priori when they will reevaluate the project and its merits. During this evaluation, the company should objectively evaluate updated market data and its own ability to satisfy changing customer demands. The board of directors plays a key role in this process by making sure that inertia doesn't carry a failed project beyond its useful life. This is particularly important when company executives have ancillary reasons, such as concerns about personal reputation or compensation, to press forward in spite of a flawed business plan.

Top executives were publicly committed to, and identified with, Iridium. Just as important as its financial investment in Iridium was Motorola's psychological investment in the project. Motorola's chairmen, Robert Galvin and later his son Christopher Galvin, publicly expressed support for Iridium and looked to it as an example of Motorola's technological might. Indeed, it was Robert Galvin, Motorola's chairman at the time, who first gave Bary Bertiger approval to go ahead with Iridium, after Bertiger's superiors had rejected the project as being too costly. In the end, both Galvins staked much of Motorola's reputation on Iridium's success, and the project provided Motorola and the rest of its partners with a great deal of cachet.

Costs of Risky Projects Can Be Reduced via Opportunities for Contracting and Learning

Motorola did gain important benefits from its relationship with Iridium. In fact, Motorola signed $6.6 billion in contracts to design, launch, and operate Iridium's

66 satellites and manufacture a portion of its handsets. David Copperstein of Forrester Research described Motorola's deal with Iridium as "a pretty crafty way of creating a no-lose situation." Other analysts were less complimentary: "That contract [Motorola's $50 million a month agreement with Iridium to provide operational satellite support] is absurdly lucrative for Motorola," said Armand Mussey, an analyst who followed the industry for Bank of America Securities, "Iridium needs to cut that by half."

These contracts—while lucrative—also gave Motorola an incentive to push Iridium forward regardless of its business plan. Even if Iridium failed, Motorola would still generate significant new revenues along the way. In quantifying the importance of Motorola's contracts with Iridium, in May 1999 Wojtek Uzdelewicz of SG Cowen estimated that Motorola had already earned and collected $750 million in profits from its dealings with the company. Based on these offsetting profits, he placed Motorola's total exposure in Iridium to be between $1.0 and $1.15 billion—much less than many observers realized.

Further, Iridium would ultimately expose Motorola to developing satellite technology and the patent protection that came with it. This exposure came at a time when Motorola was interested in entering the satellite communications industry beyond Iridium, in projects such as Craig McCaw's Teledesic—a $9 billion project consisting of a complex constellation of LEO satellites designed to provide global high-speed Internet access.

Strategic Leadership of CEOs and Boards Can Make, or Break, Strategic Initiatives

In an era where executive compensation is dominated by stock options, the Iridium story should give pause to those who see only the benefits of options-based pay. Financial incentives are extremely powerful, and companies that rely on them for motivation must be particularly careful to consider both intended and unintended consequences. Would CEO Staiano have been more attentive to the numerous warning signs with Iridium if stock options didn't play such a large role in his compensation package? The heavy emphasis on options gave Staiano an incentive to persist with the Iridium strategy; it was the only opportunity he had to make the options pay.

The lessons of the board of directors at Iridium are just as stark. Surely few boards can operate with 28 members, most representing different constituencies surely holding different goals. That all but one board member was a member of the Iridium consortium similarly speaks volumes about the vigilance of the board in fulfilling its oversight function. Actually, this type of board, consisting as it does of representatives of investors, is becoming more common in high-tech startups. Companies such as General Magic, Excite At Home, and Net2Phone have all had multiple investors, typically represented on the board and not always agreeing on strategic direction. In fact, General Magic's development of a personal digital assistant was severely hampered by its dependence on investors such as Apple, Sony, IBM, and AT&T. With Iridium, the magnitude of the ancillary contractual benefits Motorola derived from Iridium appear rather out-sized given Iridium's financial condition. An effective board should be simultaneously vigilant and supportive, a tall order for an insider-dominated, multiple-investor board.

CONCLUSION

What is fascinating about studying cases such as Iridium is that what look like seemingly incomprehensible blunders are really windows into the world of managerial decision-making, warts and all. In-depth examinations of strategy in action can highlight how such processes as escalating commitment are real drivers of managerial action. When organizations stumble, observers often wonder why the company, or the top management, did something so "dumb." Much more challenging is to start the analysis by assuming that management is both competent and intelligent and then ask, why did it stumble? The answers one gets with this approach tend to be at once both more interesting and revealing. Students of strategy and organization can surely benefit from such a probing analysis.

Richard Ivey School of Business
The University of Western Ontario

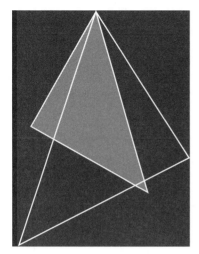

Missing Person— Peter Leung

On April 8, 2001, Mark Armstrong, angry and in a state of disbelief, sat at his desk and stared at the closed office door in front him. Moments before, he had finished a meeting with Melissa Cartwright, the testing team lead, who informed him of some disturbing news regarding his lead developer, Peter Leung. Melissa and other team members had been trying to contact Peter for the past few days to help them with some critical system testing and nobody had been able to find him. They would see Peter for an hour or two in the late morning and then wouldn't be able to find him until late afternoon, or sometimes not at all. Melissa was afraid that Peter's absenteeism was putting the system testing, and ultimately the whole project, at risk. She was very surprised when she learned that Peter's absence had not been approved and that Mark wasn't even aware of the problem.

Mark couldn't believe what he had just heard from Melissa. Peter was one of the top performers on his team and he had always been very dependable, always keeping Mark informed of any reasons he needed to be away from the office. Mark knew he had to act quickly to rectify the situation with Peter if he wanted to save the project. The testing phase was one of the most critical of the project, and Peter's knowledge was essential to successfully complete the system testing, within the already constrained project timelines.

COMPANY INFORMATION

Mark, Peter, and Melissa all worked for the Toronto office of Excel Consulting, a leading, global provider of management and technology services and solutions. With more than 35,000 employees, a global reach including over fifty-five offices in twenty countries and serving mainly Fortune 100 and Fortune 500 companies, Excel was one of the world's largest and most reputable consulting firms.

Excel's primary focuses were delivering innovative solutions to clients and providing exceptional client service. Excel delivered its services and solutions by organizing its professionals into focused industry groups. This industry focus allowed the firm's professionals to develop a thorough understanding of the client's industry, business issues, and applicable technologies to deliver tailored solutions to each client. Each professional was further aligned to a specific service function, such as strategy, human performance, or technology. This alignment allowed an individual to develop specialized skills and knowledge in an area of expertise. The organizational structure encouraged a collaborative and team-based atmosphere. Most client teams included professionals from a similar industry focus, but from several different service function specialties.

Excel's professionals had a variety of educational, cultural, and geographical backgrounds, but most shared similar skills and attributes, including leadership, intelligence, innovation, integrity, and dedication. Many professionals began their careers with Excel directly after completing an undergraduate degree; however, Excel hired experienced professionals, as well. The career path at Excel generally made the following progression: analyst, consultant, manager, associate partner, and partner. A new hire, directly after graduating with a university undergraduate degree, would begin as an analyst, and then, typically after two years, would be promoted to consultant. After consultant, it would typically take another three years before a promotion to a manager position within the firm.

Along with its focus on client service, Excel emphasized employee satisfaction. It conducted regular satisfaction surveys and had many corporate policies in place to help ensure work–life balance. These policies included flexible working hours and the assignment of professionals in the city of their home office to limit their travel. Partners and managers had the flexibility and discretion to implement these policies on their individual projects. In practice, most projects were unable to effectively implement these policies due to constrained project timelines and budgets. Balancing client service and employee satisfaction was a challenge at the firm, and one that it shared with most consulting companies.

CLIENT AND PROJECT INFORMATION

Nayacom, a telecommunications provider for the corporate customer, offered cost-effective, next-generation services across Canada, throughout the United States, and around the world. Nayacom owned and operated a coast-to-coast Internet protocol (IP) broadband network that had more than 100 points of presence in key locations throughout Canada and the United States. Nayacom was a subsidiary of a leading Canadian telecommunications company, Vextel, headquartered in Montreal. Nayacom was headquartered in Toronto and employed more than 1,600 people.

Excel Consulting and Vextel had been working together on several projects since the early 1980s. This specific project relationship with Nayacom began in August 1999. Together, Excel and Nayacom were developing a new process and technology solution to increase operational efficiencies in Nayacom's order-management process. This solution would eventually enable several Nayacom systems, each holding various pieces of data for a customer order, to share information through Enterprise Application Integration (EAI) technology. Since EAI would enable the order to automatically distribute to work positions, several steps in the order process would be carried out simultaneously instead of moving one position at a time. Any keyed information would automatically update (or *auto-populate*) all related systems, eliminating rekeying and improving data accuracy. The solution would also offer many valuable tracking and measurement capabilities to enable more thorough order management and analysis.

The project consisted of several smaller-phased business requirement–driven projects, each delivering information technology (IT) and associated business capabilities. Program development began in June 2000, and the first phase was targeted for implementation on May 1, 2001.

The Excel and Nayacom team worked together in Nayacom's head office in downtown Toronto. The team included approximately forty-seven Excel professionals and twenty-two Nayacom employees. All members were distributed among several smaller teams, including process design, technical design, development, testing, and training/deployment. Each of these teams included an Excel manager, several Excel consultants and analysts, and Nayacom employees. The entire project was overseen by Dave Fisher, an Excel associate partner, who worked directly with, and reported to, the Nayacom project manager and executives.

Due to the long-term potential of this project (the entire project could take from four to six years) and the number of Excel professionals this project employed locally in Toronto, Excel Consulting had negotiated a fixed-fee contract with Nayacom. Instead of the usual contracts, where clients are billed per consulting hour, Nayacom was paying Excel a fixed amount regardless of the number of consulting personnel and hours. The May 1, 2001, target date for implementation was very important to Excel for two reasons: the project was already over-budget, and as the first phase of the project, Excel wanted it delivered on time to ensure the project's continuation. Additionally, Excel highly valued its long-term relationship with Vextel, Nayacom's parent company. Excel had done a lot of work with Vextel over the years and it was one of Excel's largest clients in Canada. As a consequence, this project was very closely managed; there were regular Excel partner visits to the site and all project details were closely scrutinized by all the team managers and Dave Fisher.

Peter Leung

Peter was a consultant with Excel in the telecommunications industry group, within the technology service function area. Peter was about to celebrate his third-year anniversary with the firm. He had begun working with Excel in May 1997, after graduating from the University of Waterloo with a degree in computer science. Peter married his classmate and university sweetheart, Cynthia, in July 1999. They had recently purchased a condominium in downtown Toronto and were anxiously waiting for its expected completion date of March 2002.

Peter was one of the first employees staffed on the Nayacom project; he began working for the technical design team in June 2000. Peter became involved in all aspects of the technical designs and developed knowledge about all of the Nayacom systems and data. In order to leverage his design knowledge, in October 2000, he was transferred to the development team managed by Mark Armstrong and was promoted to lead developer.

The development team consisted of three analysts and one consultant, in addition to Peter. All of the team members reported to Mark Armstrong. As lead developer, Peter was responsible for developing parts of the system, integrating all development efforts, and assisting his team members with any technical design and development questions. Although, there were no formal supervisory responsibilities for Peter or his fellow consultant, Trent Gartner, Peter was often regarded as an informal mentor by the team. He had been on the project the longest and could contribute from his extensive design knowledge. Peter worked closely with the client and with other teams to ensure that any pertinent information was shared across the teams. He was also responsible for creating weekly status reports for Mark and updating him on any problems or issues that the team was facing.

Peter enjoyed his work and he was very good at it. He was considered by others to be very knowledgeable about technology and conscientious about his work. He also had more specific knowledge about the Nayacom system design and details than anyone else. Therefore, he was a valued and important team member. Socially, Peter was considered by his co-workers to be a very quiet person who kept to himself. The Excel team was very close and often went out socially after work and on weekends. Peter was always invited; however, he had never gone out with the team. As a result, although he was liked by his co-workers, he had not established any friendships or close relationships with his colleagues.

Mark Armstrong

Mark was a manager with Excel in the telecommunications industry group, in the technology service function area. He was married and had two young children, Dave and Theresa. Mark had begun working with Excel in April 1996, after he had been laid off from Excel's main competitor, Alta Consulting.

Mark began working on the Nayacom project in August 2000, as manager of the development team. His responsibilities included managing all development efforts for the project, including managing a team of five professionals and ensuring project deadlines and deliverables were met. In addition, Mark was responsible for communicating with all team managers and ensuring the dissemination of appropriate project information across all of the teams. He spent most of his time in client and cross-team strategy meetings; therefore, he had little time to help the team with detailed development of the system. Mark relied on the strength of his team members and expected them to bring any concerns promptly to him. He usually called development team meetings once a week, every Monday morning at 8 A.M., to disseminate important project information and to answer any questions. These meetings normally lasted fifteen to twenty minutes. Aside from this formal meeting, Mark was involved very little in the team's day-to-day affairs. He asked for weekly status reports to be completed by all members. Mark used these status reports to create a report for Dave Fisher and the other Excel team managers, in order to inform them of his team's progress.

Peter joined Mark's team in October 2000, a few months after Mark had started. Mark saw Peter as dependable and very intelligent, and he felt comfortable with Peter on his team. He knew that Peter was also available to help the team with their questions. Mark trusted that Peter and the other members would bring any questions or concerns promptly to him. He liked all the people on his team, and knew he had some of the best developers within Excel. So far, Mark had not had any problems with any of his team members since they had been meeting the appropriate deadlines and deliverables.

RECENT EVENTS

Design and development of the system had been completed and the first phase of the project was currently in the testing portion of the project development cycle. Testing is the process by which the technical system/solution is checked, in detail, against expected scenarios and outcomes. During this process any bugs or errors within the system are recorded and rectified before the system can "go live" and be launched for the actual client users. Testing is usually a very busy and stressful time since managers often budget too little time for this process. Testing is also one of the most important periods on a project and it requires all members of the project team to be available to help the testing team quickly fix any errors or bugs that are detected. With only three weeks left until the expected launch date of May 1, 2001, the testing team was detecting many bugs that needed to be fixed. Specifically, there were many problems in the areas of data integration where the Nayacom systems were pulling or pushing out inaccurate or different data compared to the data that was expected in the technical designs.

The team really needed Peter Leung's help to resolve most of the issues since he knew the designs and development better than most members on the team. Unfortunately, Melissa Cartwright, the testing team lead, and many of her other team members had been unable to locate Peter recently. His laptop was on his desk and his papers were all out, but Peter was never there. Occasionally, someone would spot him for an hour in the morning or in the evening, but there was never enough time to really work through the system problems with him.

Melissa and her testing team were getting very frustrated and worried about being able to make the deadline that was only three weeks away. Melissa decided to approach Mark and talk to him about Peter's absenteeism. Mark was shocked to find out about this problem. He had not approved Peter's absences and he had not heard anything from his other development team members either. Melissa was angry that Mark had no idea about the situation on his team. She told him that she would be unable to meet the testing deadline unless she had Peter's help. Mark told her that he would find out what was going on and try to rectify the situation within the next twenty-four hours.

THE INVESTIGATION

Minutes after Melissa left his office, Mark stormed out towards Peter's desk. When he got there he saw Peter's laptop on and a pile of papers spread out across the desk. It definitely looked like he was there; however, when he asked the other development team members if they had seen Peter, they mentioned that he had not been in yet that day. Supposedly, he left his desk this way all the time to look like he was in the office.

Mark immediately called an emergency development team meeting. All of the members were present except for Peter. He asked if anyone knew where Peter was. Trent responded that he did not know, and that Peter had not been coming to the office regularly for the past three weeks. Usually, he came in for an hour in the morning and then left, returning later for another hour in the afternoon. Rumors were that Peter was having personal problems and that he and his wife were on the verge of a divorce; however, since Peter kept to himself, nobody on the team had asked him about his personal issues.

When Mark asked why the team had not mentioned anything to him before, they said they assumed both that he was aware of the absences and that Peter had probably asked Mark for time off. They also felt uncomfortable approaching Mark about a colleague who was going through personal difficulties. In addition, since development was complete, the team had little work besides helping the testing team, and they had not noticed the team's performance being affected by Peter's absence.

NEXT STEPS

Mark knew that he had to rectify this situation promptly, in order to avoid it developing into an even bigger issue. Melissa and the testing team needed Peter, and it was Peter's responsibility to be there to help with the project. Mark wondered how he would deal with Peter's personal issues. He sympathized that Peter was having problems with his marriage, but at the same time, the project's success was in jeopardy. How would he approach the first meeting with Peter? He did not want to drive Peter away since he was too important to the project and was irreplaceable, especially due to the short timeframes, but he also felt angry that Peter had betrayed his trust. Why did Peter not come and talk to him about his personal problems and indicate that he needed time off?

Mark also wondered how to handle the situation with his development team. He understood the reasons that Trent and the others had given for not coming to speak to him, but he had hoped that they would have approached him about this situation. He wondered what he could do to make sure that the team felt comfortable talking to him about other issues in the future.

Finally, Mark was concerned about his own performance as a manager. It had taken another manager's inquiries for him to find out about the situation. He questioned what he could have done to avoid this. How was he supposed to balance his client obligations and manager meetings with the day-to-day detailed management of his team? Also, he wondered how many other people on the project team knew about the situation. How would he communicate and rectify the situation with the other project team members and managers? How would he approach this situation with Dave Fischer? Should Mark talk to Dave, or should he just resolve it quickly and quietly, hoping that he wouldn't hear about it from someone else?

Mark had given Melissa his word that he would rectify this situation within twenty-four hours. He quickly reached toward the phone directory to look for Peter's home number. The first thing that Mark needed to do was to locate Peter. He hoped he would be able to find him soon.

IVEY

Richard Ivey School of Business
The University of Western Ontario

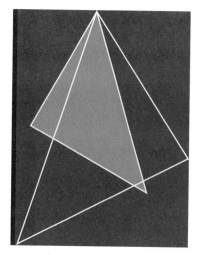

Zhou Jianglin,
Project Manager

INTRODUCTION

"The Data and Voice Project (DVP) is behind schedule, over budget, and under specification," explained Zhou Jianglin, project manager for Ji'nan Broadcasting Corporation (JBC). It was November 3, 2001, in Ji'nan, China, and Zhou's determination was fading. He still had two months to go before the DVP was set to go live. Looking back, he tried to determine what went wrong and what could be done to save the project.

THE DATA AND VOICE PROJECT (DVP)

Intended to cover a 200-square-kilometers area, DVP could provide voice and data services to 90 percent of Shandong's businesses and inhabitants. Currently, no data services existed in the province, and voice telephone services (local and long-distance), were provided by China Post & Telecom.

Zhou had been given a budget of RMB110 million[1] to spend on this project. Its goals were:

- To provide up to 5,000 high-speed (100 megabits per second) data lines, and up to three million voice lines.
- Ensure that the equipment is *evergreen* and *scalable*.
- To complete the project by January 1, 2002.

There was a menu of choices that was presented to JBC by Eastern Postel (Postel), the telecommunications manufacturer and lead contractor chosen for DVP. Han Xiaowei, managing director of data and voice services for JBC (and Zhou's superior) intervened and insisted on this set:

- Use Nortel Passport data equipment and Alcatel's new VIT voice equipment to go with Postel's locally manufactured data communications equipment.
- Incorporate current Fujitsu voice equipment. Make the change cost-neutral to JBC.
- Implement a centrally controlled network
- Engineer and install services with a capacity to serve two million customers.
- Provide routine maintenance and monitoring for twelve weeks. Include in the price the training of new hires to maintain the equipment.

Postel and its supplier Nortel were surprised at the choices made. They insisted that it would not be possible to include Alcatel and Fujitsu equipment and still be cost-neutral. In addition, they objected to the training of non-technical people to maintain the network. Han countered by reiterating that other contractors could be chosen. Besides, he offered, the new hires would be computer science or engineering graduates.

---

[1]RMB8 = US$1.

After a week of protracted negotiating, both Postel and JBC agreed to a scaled-down version of the DVP. With the agreement signed, Zhou returned to leading the project. The project would be able to serve only 2,500 business clients and one million residential customers, a compromise from the original goals. It was the June 20, 2001.

EARLY SETBACKS FOR THE DVP

Five weeks later, Zhou encountered the first of his problems: Type Approval was taking longer than the four weeks needed by the government-run testing agency. Located in Shanghai, China Standards Approval Agency (CSAA) had received Alcatel's new voice equipment on the June 25, 2001. Along with the equipment was a note that indicated Alcatel was willing to pay the posted rate of RMB30,000 for CSAA to test the entire batch of VIT equipment. On August 1, 2001, Zhou was informed by Postel that CSAA had not even begun to test the equipment. CSAA had insisted that Alcatel pay the fee for each piece of equipment to be tested, RMB150,000 in total. Alcatel refused.

Fujitsu equipment was taking longer to engineer into the system than previously estimated. It was taking up too much of Nortel's engineering time, John Lian, enterprise sales manager for Nortel, explained. Nortel would honor its end of the deal—to provide cost-neutral inclusion of Fujitsu equipment—but Nortel engineers would work on it "as time constraints allowed." Further, they estimated that inclusion of Fujitsu equipment might cause network outages up to three percent of the time. Lian also mentioned that the go live date would be pushed back, to approximately February 5, 2002.

At JBC, Zhou's request for additional funds to hire a trained project manager had been rejected. In addition, no extra staff members would be assigned to DVP. Han insisted that Zhou complete this "simple" project on his own.

ANOTHER POINT OF VIEW

Through his personal network of friends, Zhou requested the help of a project manager, Paul Scott, who worked for an English telecommunications equipment manufacturer. Scott offered this view:

> Nortel has been going through a transition to formalize project management as a proper discipline in order to achieve consistent application of project management skills worldwide. If project management approaches were consistent across companies worldwide, less misunderstanding and conflict would occur.

The official text explaining the need for type approval testing for telecommunications equipment involves ensuring that the equipment protocols work as stated; ensuring the ISO layers are adhered to; for safety purposes to reduce possibility of unexpected electrical damage or danger to personnel working on the equipment; and for legitimate customs income.

There are three unofficial explanations: First, this is a way for unscrupulous officials to receive bribes. This is more overt in some countries (a *facilitating* charge), can be more creative (a manufacturer would have to take equipment to a testing company owned by a relative of the government customs official overseeing Type Approval), and nonexistent in others. Second, forcing a company to subject its equipment for testing delays the entry of this equipment into a market, possibly forcing time-strapped companies to use locally produced products. Last, testing is a way to conduct research. One government testing agency had been known to the telecommunications industry to have taken apart and cloned new data or voice equipment in as little as three months. Shortly thereafter, one could find a strikingly similar national version (of a company's latest generation equipment) on the international market.

With the issues facing the inclusion of Fujitsu equipment, I would term that *scope creep,* and lay the blame on JBC. Making equipment from two manufacturers function together is akin to taking parts from two different cars in an attempt to create a third—something will go wrong. Frankly, integrating existing equipment with new equipment causes more problems than it solves.

ZHOU LOOKS INTERNALLY

Thanking Scott, Zhou took a few minutes to think about his own organization. Was political infighting at fault? In the drive toward an initial public offering, managing directors were certainly positioning themselves in the eyes of the president. Could Zhou avoid being the scapegoat if this project failed? Was JBC committed to create a data and voice network from scratch? After all, though they were skilled broadcasters and programmers, experienced with creating and airing content, JBC had no previous experience with data or voice products. The DVP was important to many stakeholders. Zhou believed that he could navigate his way through the obstacles that had arisen. He needed the DVP to be a success.

Part 10

CONTROLLING PROJECTS

Controlling projects is a necessity such that meaningful and timely information can be obtained to satisfy the needs of the project's stakeholders. This includes measuring resources consumed, measuring status and accomplishments, comparing measurements to projections and standards, and providing effective diagnosis and replanning.

For cost control to be effective, both the scheduling and estimating systems must be somewhat disciplined in order to prevent arbitrary and inadvertent budget or schedule changes. Changes must be disciplined and result only from a deliberate management action. This includes distribution of allocated funds and redistribution of funds held in reserve.

The Two-Boss Problem

On May 15, 2001, Brian Richards was assigned full-time to Project Turnbolt by Fred Taylor, manager of the thermodynamics department. All work went smoothly for four and one-half of the five months necessary to complete this effort. During this period of successful performance Brian Richards had good working relations with Edward Compton (the Project Turnbolt engineer) and Fred Taylor.

Fred treated Brian as a Theory Y employee. Once a week Fred and Brian would chat about the status of Brian's work. Fred would always conclude their brief meeting with, "You're doing a fine job, Brian. Keep it up. Do anything you have to do to finish the project."

During the last month of the project Brian began receiving conflicting requests from the project office and the department manager as to the preparation of the final report. Compton told Brian Richards that the final report was to be assembled in viewgraph format (i.e., "bullet" charts) for presentation to the customer at the next technical interchange meeting. The project did not have the funding necessary for a comprehensive engineering report.

The thermodynamics department, on the other hand, had a policy that all engineering work done on new projects would be documented in a full and comprehensive report. This new policy had been implemented about one year ago when Fred Taylor became department manager. Rumor had it that Fred wanted formal reports so that he could put his name on them and either publish or present them at technical meetings. All work performed in the thermodynamics department required Taylor's signature before it could be released to the project office as an official company position. Upper-level management did not want its people to publish and therefore did not maintain a large editorial or graphic arts department. Personnel desiring to publish had to get the department manager's approval and, on approval, had to prepare the entire report themselves, without any "overhead" help. Since Taylor had taken over the reins as department head, he had presented three papers at technical meetings.

A meeting was held between Brian Richards, Fred Taylor, and Edward Compton.

Edward: "I don't understand why we have a problem. All the project office wants is a simple summary of the results. Why should we have to pay for a report that we don't want or need?"

Fred: "We have professional standards in this department. All work that goes out must be fully documented for future use. I purposely require that my signature be attached to all communications leaving this department. This way we obtain uniformity and standarization. You project people must understand that, although you can institute or own project policies and procedures (within the constraints and limitations of company policies and procedures), we department personnel also have standards. Your work must be prepared within our standards and specifications."

Edward: "The project office controls the purse strings. We (the project office) specified that only a survey report was necessary. Furthermore, if you want a more comprehensive report, then you had best do it on your own overhead account. The project office isn't going to foot the bill for your publications."

Fred: "The customary procedure is to specify in the program plan the type of report requested from the departments. Inasmuch as your program plan does not specify this, I used my own discretion as to what I thought you meant."

Edward: "But I told Brian Richards what type of report I wanted. Didn't he tell you?"

Fred: "I guess I interpreted the request a little differently from what you had intended. Perhaps we should establish a new policy that all program plans must specify reporting requirements. This would alleviate some of the misunderstandings, especially since my department has several projects going on at one time. In addition, I am going to establish a policy for my department that all requests for interim, status, or final reports be given to me directly. I'll take personal charge of all reports."

Edward: "That's fine with me! And for your first request I'm giving you an order that I want a survey report, not a detailed effort."

Brian: "Well, since the meeting is over, I guess I'll return to my office (and begin updating my résumé just in case)."

The Bathtub Period

The award of the Scott contract on January 3, 1987, left Park Industries elated. The Scott Project, if managed correctly, offered tremendous opportunities for follow-on work over the next several years. Park's management considered the Scott Project as strategic in nature.

The Scott Project was a ten-month endeavor to develop a new product for Scott Corporation. Scott informed Park Industries that sole-source production contracts would follow, for at least five years, assuming that the initial R&D effort proved satisfactory. All follow-on contracts were to be negotiated on a year-to-year basis.

Jerry Dunlap was selected as project manager. Although he was young and eager, he understood the importance of the effort for future growth of the company. Dunlap was given some of the best employees to fill out his project office as part of Park's matrix organization. The Scott Project maintained a project office of seven full-time people, including Dunlap, throughout the duration of the project. In addition, eight people from the functional department were selected for representation as functional project team members, four full-time and four half-time.

Although the workload fluctuated, the manpower level for the project office and team members was constant for the duration of the project at 2,080 hours per month. The company assumed that each hour worked incurred a cost of $60.00 per person, fully burdened.

At the end of June, with four months remaining on the project, Scott Corporation informed Park Industries that, owing to a projected cash flow problem, follow-on work would not be awarded until the first week in March (1988). This posed a tremendous problem for Jerry Dunlap because he did not wish to break up the project office. If he permitted his key people to be assigned to other projects, there would be no guarantee that he could get them back at the beginning of the follow-on work. Good project office personnel are always in demand.

Jerry estimated that he needed $40,000 per month during the "bathtub" period to support and maintain his key people. Fortunately, the bathtub period fell over Christmas and New Year's, a time when the plant would be shut down for seventeen days. Between the vacation days that his key employees would be taking, and the small special projects that this people could be temporarily assigned to on other programs, Jerry revised his estimate to $125,000 for the entire bathtub period.

At the weekly team meeting, Jerry told the program team members that they would have to "tighten their belts" in order to establish a management reserve of $125,000. The project team understood the necessity for this action and began rescheduling and replanning until a management reserve of this size could be realized. Because the contract was firm-fixed-price, all schedules for administrative support (i.e., project office and project team members) were extended through February 28 on the supposition that this additional time was needed for final cost data accountability and program report documentation.

Jerry informed his boss, Frank Howard, the division head for project management, as to the problems with the bathtub period. Frank was the intermediary between Jerry and the general manager. Frank agreed with Jerry's approach to the problem and requested to be kept informed.

On September 15, Frank told Jerry that he wanted to "book" the management reserve of $125,000 as excess profit since it would influence his (Frank's) Christmas bonus. Frank and Jerry argued for a while, with Frank constantly saying, "Don't worry! You'll get your key people back. I'll see to that. But I want those uncommitted funds recorded as profit and the program closed out by November 1."

Jerry was furious with Frank's lack of interest in maintaining the current organizational membership.

QUESTIONS

1. Should Jerry go to the general manager?
2. Should the key people be supported on overhead?
3. If this were a cost-plus program, would you consider approaching the customer with your problem in hopes of relief?

4. If you were the customer of this cost-plus program, what would your response be for additional funds for the bathtub period, assuming cost overrun?
5. Would your previous answer change if the program had the money available as a result of an underrun?
6. How do you prevent this situation from recurring on all yearly follow-on contracts?

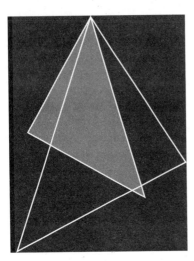

Ford Motor Co.: Electrical/Electronic Systems Engineering

Ford Motor Co. has revenues of $164.196 billion and 327,531 employees world-wide. The Electrical/Electronic Systems Engineering department develops electrical systems valued at $800 to $1,000 at cost to more than eighty vehicle programs. The department consists of approximately 740 staff resources, with electrical program management teams comprising about twenty-five engineering resources each.

The Electrical/Electronic Systems Engineering department has four functional engineering areas, each with its own chief engineer:

1. North America Truck
2. North America Car
3. Commodity and Application Engineering
4. E/E Software and Modeling

This department is aligned with the product creation mission of Ford—"Great Products ... More Products ... Faster"—that outlines the priorities for the department:

● Improve quality.
● Improve quality (intentionally repeated).
● Develop exciting products.

- Achieve competitive cost and revenue.
- Build relationships.

Additionally, Ford's Electrical/Electronic Systems Engineering department has aligned with the company's key focus areas for department communication and processes. To "intensify communications," the department stresses the following:

- Communicate consistently.
- Focus on vital few priorities.
- Keep the message simple.
- Help people prioritize.
- Remove barriers.

The department also emphasizes that each team member should improve working processes by simplifying, stabilizing, standardizing, setting cadence, and sustaining.

By using these principles across all product development commodities, the department has achieved a reduction in engineering errors, as well as higher engineer engagement.

OVERALL BEST PRACTICES

Examining project management at Ford revealed three best practices. First is Ford's executive sponsorship of an Electrical/Electronic Systems Engineering project management office. This office standardizes project management and engineering processes across its internal functional areas and the electrical program management team. It also acts as a single governance board for the project management office framework. The department's directors, chiefs, and the electrical business planning and technology office participate in the governance board through weekly project management meetings to provide support and shift priorities as required.

Second, professional project managers consult on the implementation, execution, and maintenance of the project management office, as well as assisting with the transfer of project management knowledge for the organization.

Additionally, the Electrical/Electronic Systems Engineering department has internalized project management as a discipline in engineering and provided training to the entire organization, with follow-up auditing processes in place for implemented projects. It has always been Ford's intent for engineers to develop competencies in the area and build an in-house project management discipline.

Managing Resistance

Transferring the leadership and ownership of project management from professional project managers to the engineering division has allowed further entrenching of the organization's goal of increasing project management maturity and has produced positive results.

Senior-level managers in the organization expect 100 percent compliance with the project management tools and methodologies developed by the project management office and approved by the governance board. They approached the changes as sustained continuous improvement and took the time to listen to comments and criticism from the people in the framework, which resulted in less overall resistance than was expected.

Another method used by the Electrical/Electronic Systems Engineering department to counter resistance was to design the project management office framework around stakeholder participation. All organization personnel can participate in the project management office tools and methodology discussions at the management level, as well as the project management office working level meetings. This level of participation in the organization helps build the best practice process.

DRIVING CONSISTENCY IN PROJECT MANAGEMENT

Project Definition

The Electrical/Electronic Systems Engineering project management office acts as the central project manager to standardize projects. The office engages defined projects that usually have a short time frame with a clearly defined scope and a clear allocation of resources. Long-term technical or business planning projects are handled outside of the project management office. Although these projects may interact with the office, it does not directly manage them.

Project Management Organization/Methodology

The Electrical/Electronic Systems Engineering department's project management office comprises three levels.

1. *The governance board of executive directors and engineering chiefs.* This small body prioritizes projects according to the corporate scorecard. The group includes two executive directors and four engineering chiefs and sets the tone for the department's overall level of project management excellence.

2. *Stakeholders.* This group includes members of the department that participate either regularly or sporadically in approved projects, usually as subject matter experts. These resources provide technical knowledge regarding the various engineering disciplines and tools.
3. *Professional project managers.* These staff members are from the project management-consulting firm retained by the department. Their duties include participating in cascaded/prioritized projects, developing project execution plans and work plans, performing audit processes, and facilitating team formation and execution of deliverables in a specified timeline and scope as approved by the governance board. The professional project managers also developed a change management process for updating existing project management tools on an as-needed basis.

The professional firm of Pcubed Inc. is considered the owner of the project management methodology employed at Ford Motor Company. This methodology is aligned with the Project Management Body of Knowledge (PMBOK® Guide), PMI, and PM Berkeley Maturity models, which are the recognized industry standards. The approach comprises three phases.

1. *Discover and define.* The objective during this phase is to assess the overall health and baseline project management process.
2. *Develop and deliver.* The Phase 2 objective is to develop and pilot the recommended solutions to address the needs identified in Phase 1.
3. *Deploy and drive.* The last objective is to ensure solutions are fully implemented across the department.

Project Managers/Teams

Five to eight full-time professional project managers staff the Electrical/Electronic Systems Engineering project management office per quarter, depending on the project needs. The project management office reports its general project scope recommendations or issues to the Electrical/Electronic Systems Engineering department business office manager prior to those recommendations/issues being elevated to the governance board review process, where they are then reviewed by directors and engineering chiefs.

The relationship between the project management office and the functional areas is clearly structured, with the project management office as the focal point for all project management processes. The functional teams do not have the authority to influence or overrule the directives managed by the project management office. In 2004, the project management office began to work with the Electrical/Electronic Systems Engineering department to identify resources that will participate in an increased capacity based on the job families for engineers with project management responsibilities.

The composition of a typical Electrical/Electronic Systems Engineering department project team and the corresponding roles and responsibilities include the following:

- *The project manager.* This person leads the project execution plan development. This also includes gathering the necessary resources, as well as defining the scope, deliverables and time line for the project.
- *The stakeholders.* Usually, they are subject-matter experts who provide feedback about the project deliverables.
- *The governance board.* The board reviews the progress of the project and gives the necessary approval or rejections for recommendations.

In some instances, the stakeholders take the lead role, and the project management office acts as coordinator or facilitator.

Currently, the Electrical/Electronic Systems Engineering department identifies resources and potential leaders using the individual development plan, a tool completed by the department's engineers. Resources identified for advance training take on permanent leadership roles in the organization. Some of these resources will have only part-time responsibilities for project management, and others will be used full-time to manage the project management office.

To maintain the structure necessary for consistent project delivery while allowing for changing circumstances, the project management office and the governance board review projects' status monthly and make any necessary recommendations. Stakeholders also meet monthly for change control of project management tools and processes. This is the formal change control process for any methodology improvements to existing projects. The project scope can be modified as necessary to manage changes to the original project assumptions. The suggested revisions are always reviewed by the top two levels of the project management structure (governance board and stakeholder team), and any revisions are taken from their directions.

Ford used the Berkeley Project Management Maturity Model to quantify the needs assessment results across the project management disciplines and the project life cycle. Level one of the Berkeley Maturity Model is the ad hoc stage, where no formal procedures or plans to execute exist and where project management techniques are applied inconsistently, if at all. Level two is the planned stage, where informal and incomplete processes are used, and planning and management of projects depend primarily on individuals. Level three is the managed stage, where project management processes demonstrate systematic planning and control and where cross-functional teams are becoming integrated. Level four of the model is the integrated stage. Here, project management processes are formal, integrated, and fully implemented. Lastly, level five is the sustained stage, which involves continuous improvement of the project management processes. At the project management office launch in 2003, the Electrical/Electronic Systems Engineering department had a maturity level rating of 1.85, aligning with the average maturity level of most organizations, which is between level one and two.

At the end of 2003, after the implementation of the project management office and the achievement of an organized approach, an informal review of the organization's processes moved the rating to 3.0. To continue increasing its maturity level in 2004, the department's governance board began internalizing the effort to transfer project management knowledge by using technical maturity models, which provide training models, individual development plans, and core training and education online courses in department project management processes. The goal of the department is to internalize competency and to approach project management internally.

Project Management Strategy

The Electrical/Electronic Systems Engineering department has two primary strategies for selecting project management office projects:

1. Base selection on the corporate scorecard objectives for the given calendar year.
2. Base selection on the underlying goal of increasing the department's project management maturity.

The project management strategy aligns with the corporate strategic plan by placing top priority on selecting a project based on its ability to meet the corporate scorecard objectives (i.e., improving the product creation process and engineering disciplines). Other criteria can also include the ability to improve work-related efficiency, standardize reports and processes to improve clarity of data for decision making at the senior level, and realign the organization cross-functionally to increase project synergies.

The department's approach to project management has been used to achieve the strategic objectives of the organization in the following ways:

- The project management office had input into the corporate-level development of the engineering quality operating system. The office also had responsibility for building electrical assessment health charts by system and commodity levels, training the Electrical/Electronic Systems Engineering organization to integrate new corporate reporting tools, implementing an auditing process to ensure proper compliance with procedure, and reporting the efficiency of the organization to senior leadership.
- The department worked toward realigning the sourcing process with the finance department, cataloged issues via the engineering quality

operating system reporting system, and gained the support of the finance department in a joint partnership to improve the supplier sourcing process.

- The department also maintained continuous improvement projects in product development, such as participating in corporate objectives as they pertain to the processes to improve product creation (e.g., improving time to market and the quality of the product launch).

Resource Assignment

Electrical/Electronic Systems Engineering ensures that adequate project resources are devoted to the upfront project phases (project initiation and planning) by defining project execution plans one month prior to the project kick-off. This plan details the scope, timeline, and required resources. Once the governance board approves this plan, it ensures that sufficient organization resources are enabled, and the project management office matches projects to the skill sets of individual project managers.

To effectively manage geographically dispersed or global project teams, the department uses a clearly defined communication plan, including the scope, timeline, resources, and the necessary communication tools that can facilitate a global meeting such as eRoom or Pictel. It is also important to form the project team early and clearly define the objectives, as well as outline regular status-reporting meetings. Cultural differences that might arise during the project are managed by best practices training. For example, the project leader might make recommendations to the team for specific communication plans, the formality of meetings, or conduct, and might negotiate work-related differences and scope disagreements.

Project Management Professionalism/Training

As discussed previously, advanced project managers in the Electrical/Electronic Systems Engineering department are identified through individual development plans as part of the technical maturity model for project management. Resources identified for advanced training will take on permanent leadership roles in the department, which usually consists of managing projects or the project management office.

Training needs for project managers are also identified by comparing the results of the completed individual personal development plans to the technical maturity model for project management. Resources requiring user/expert level skills will be trained by a variety of sources:

- Current professional project managers assigned to train them on project management office operations

- Web-based training or seminar training provided by Ford on core project management disciplines
- Specialized courses developed by the department along with Ford Motor Co. on project management processes, tools, and methodologies

Structuring and Negotiating Project Scope

Professional project managers in the project management office initially prepare the project scope based on a discovery phase approach. The scope is outlined in a project execution plan against the project requirements, timeline, and resources required. Process changes must go through the formal change control process, as outlined earlier, that begins at the monthly stakeholder meeting. Scope changes related to resources are first reviewed with the manager of electrical technology and operations. The governance board must then review the proposed changes before giving its approval or rejection. An adjustment of resources is then made as necessary to meet the approved changes to the scope.

Maintaining Consistency in Project Management Delivery

Overall, the department identifies a number of important ways that it maintains consistency in project management delivery:

- Project management tools, processes, and methods in the department are standardized.
- The project management office institutionalizes approved new processes through training of the organization.
- The project management office audits the correct use of new tools and processes.
- Monthly change control actions are taken to improve gaps.
- Processes are available to the organization through the use of eRoom documentation storage.
- Ongoing organization training and project management pocket cards for engineers are provided.

BUILDING PROJECT PORTFOLIOS BY PRIORITIZING PROJECTS

In Ford's portfolio management approach, projects are ranked based on the priorities identified by the governance board using the corporate scorecard. Initially, the scopes of the various projects are high level, and the project managers review all requested projects and define the scope with the department's business operations manager. In 2004 the organization performed an assessment of this approach and plans to make assessments a biannual process.

Allocation of Resources

As previously outlined, the Electrical/Electronic Systems Engineering department allocates resources to projects based on the project priority, scope, and available resources. If reassignment of resources is necessary because of changes to the project or the personnel, then proposed changes are reviewed and approved by the governance board and department's business office manager. However, the final decision on the prioritization of projects lies with the department's governance board.

The allocation of development funds or resources to different project types, business areas, market sectors, or product lines again depends on the corporate scorecard objectives, areas requiring process improvements, and an increase in the organization's project management maturity level. Organization objectives are cascaded by the governance board to the project management office, which develops high-level project plans that the governance board then reviews for approval. To ensure sufficient resources are available for projects, the governance board conducts monthly reviews to monitor strict adherence to the scope management of projects, as well as manage any over-allocation of resources.

The job of ensuring that low-value projects are terminated before consuming resources is primarily that of the project management office's project manager, governance board, and the Electrical/Electronic Systems Engineering department's business manager. The feedback on value achievement from these sources is provided monthly. Additionally, a periodic formal project management office survey is administered by the Electrical/Electronic Systems Engineering department's business operations planning group to the department to rank the effectiveness and use of project management office tools, processes, and project outcomes. The results of the survey are reviewed with the project management office and the governance board to identify areas of improvement and capture lessons learned.

To enhance ongoing management decisions using the project portfolio, Ford uses the engineering quality operating system reporting system to quantitatively measure the success of program delivery across the North American engineering community, including electrical/electronic commodity and deliverables to the program level. This measurement system is designed to review the history and also present the status of progress across the vehicle programs. The project management office has worked on various projects that have facilitated the communication of these status results in a more streamlined manner to help decision-making capabilities. For example, the Electrical/Electronic Systems Engineering department will prioritize "red issues" and track any red issue closures in a database. These progress reports against the closure of red issues are reviewed as high as the vice president level.

MEASURING PROJECT DELIVERY AND END RESULTS

Ford uses the engineering quality operating system to measure the success of its projects in the engineering community. Its integrator reporting system captures the status of projects and can report these findings up to the system and program levels.

Additionally, the metrics or measures used by the project management office are mostly qualitative and can include completed deliverables assigned to the project or feedback by the user community or other outside sources.

The department's business operations planning department manages all financial aspects of the Electrical/Electronic Systems Engineering project cost. The department's business operations planning manager found the project management approach the most cost effective for managing projects in a large organization. This approach has driven 5 percent efficiency in the operating costs for the electrical area of the company.

The collection of project data is managed by the project management office and can come from various sources, such as the engineering quality operating system health charts (project status reports) or work plans. Data integrity is managed by periodic auditing of the functional engineering team's adherence to the organization's tools and processes. The results of the audit are reported to the chief engineers and also posted in the team's specific eRoom for team feedback. The chief engineers examine the auditing reports to drive 100 percent compliance through the organization.

To make data informational and useful, the organization analyzes various types of data with the following frequency:

- Trend analysis of engineering quality operating system health charts is done twice monthly.
- Timing analysis on work plans is conducted monthly and reported to the electrical program management team.
- Updates to the engineering quality operating system integrator are conducted monthly, but tracking of red issues is conducted on a weekly basis.

Additionally, the following reporting methods or mechanisms are used in the organization:

- The Web-based engineering quality operating system assessment provides red/yellow/green health charts for the commodity and system-level teams.
- Work plans are maintained on eRooms for easy access to project timing data and deliverables.
- Tracking of red issues conducted via a tracking database and a trend analysis is performed on this data.

Decision makers in the organization act on the reported metric data in different ways. The governance board conducts reviews of the engineering quality operating system red-status items across the organization for two hours every week and provides feedback to the managers on action items. The timing reviews are held bimonthly at the system level to review commodity development and testing status. Issues arising from these reviews are elevated to the chief engineers, who actively manage the red issues to green status.

To ensure that the project-related measures add value to the organization, the Electrical/Electronic Systems Engineering department can point to improvements in performance. The quality of the red/yellow/green status at various vehicle program milestones has been steadily improving since 2003. The organization acknowledges an effort to minimize projects in the yellow status. Corrective action plans are for the purpose of changing a commodity status to green, not to merely improve it from red to yellow. The Web-based engineering quality operating system assesses milestone deliverables using the red/yellow/green status and provides managers with immediate issue elevation.

Accountability/Authority

Because project managers execute governance-board-approved projects, team members know they are expected to participate and meet project objectives. Project managers are given the authority to elevate issues or roadblocks that arise during the life of the project to the governance board for any needed feedback or assistance. The overall authority granted to project managers is commensurate with their level of accountability.

The roles and responsibilities for project managers are in the process of being mapped into the Ford Electrical/Electronic Systems Engineering job families. At the manager level, however, achievement and technical excellence is recognized and rewarded by senior management.

In terms of future objectives, the project management office has outlined the following effort to continue to improve the Electrical/Electronic Systems Engineering department's project management maturity:

- Develop a technical maturity model for project management to provide training and organizational structure to transfer project management roles and responsibilities and/or competencies.
- Migrate commodity engineering quality operating system assessment summaries to the integrator and audit/coach/mentor commodity teams on the integrator.
- Continue to expand electrical program management teams and commodity-in-a-box tools and processes.
- Lead electrical work stream development in new product development system.

Greatest Measurement Challenges

The primary measurement challenge for Ford's Electrical/Electronic Systems Engineering department was the length of time it took managers to realize that the project management office approach was necessary for project management processes to improve.

As discussed previously, the current auditing processes used by the project management office to measure project delivery typically address quality issues, whereas the project management change control process in the department allows for ongoing improvement to tools and processes, as well as the management of scope changes. Flexibility in these measurement systems has been important in achieving a higher rate of successful project outcomes. Additionally, process-training surveys are conducted with team members after the rolling out of a new process or tool to gather feedback and to identify areas of improvement.

FINAL COMMENTS AND THOUGHTS

Learning from Project Management Missteps

Even with a strong effort to engage personnel, the objectives of the project management office were not initially clearly understood in the Electrical/Electronic Systems Engineering organization. Because most personnel had not previously experienced a working project management office, incorrect assumptions were sometimes made regarding its scope, roles, and responsibilities. It took the project management office some time to get the entire organization aligned on its value and the most effective method for execution of projects. The participation of the stakeholder board was key to the eventual acceptance of the project management office, along with constant communication.

To summarize, the Electrical/Electronic Systems Engineering department's project manage office's project execution plans were developed and reviewed and then approved by the governance board to clearly define the quarterly project management office objectives, scope, and resource allocation. These plans were made available to the organization via the eRoom and also reviewed at the manager level. Any overextending of project management resources or changes in project scope are routinely reviewed by the governance board at the monthly status review. After one year, the department had developed an effective working relationship with the project management office and had accepted the accompanying project management tools and methodologies.

Excerpted from APQC's Best-practice Report *Project Management,* which is available for purchase at www.apqc.org/pubs. APQC is an international non-profit research organization.

Part 11

PROJECT RISK MANAGEMENT

In today's world of project management, perhaps the single most important skill that a project manager can possess is risk management. This includes identifying the risks, assessing the risks either quantitatively or qualitatively, choosing the appropriate method for handling the risks, and then monitoring and documenting the risks.

Effective risk management requires that the project manager be proactive and demonstrate a willingness to develop contingency plans, actively monitor the project, and be willing to respond quickly when a serious risk event occurs. Time and money is required for effective risk management to take place.

The Space Shuttle
Challenger Disaster

On January 28, 1986, the space shuttle *Challenger* lifted off the launch pad at 11:38 A.M., beginning the flight of mission 51-L.[1] Approximately seventy-four seconds into the flight, the *Challenger* was engulfed in an explosive burn and all communication and telemetry ceased. Seven brave crewmembers lost their lives. On board the *Challenger* were Francis R. (Dick) Scobee (commander), Michael John Smith (pilot), Ellison S. Onizuka (mission specialist one), Judith Arlene Resnik (mission specialist two), Ronald Erwin McNair (mission specialist three), S. Christa McAuliffe (payload specialist one), and Gregory Bruce Jarvis (payload specialist two). A faulty seal, or O-ring, on one of the two solid rocket boosters caused the accident.

Following the accident, significant energy was expended trying to ascertain whether the accident had been predictable. Controversy arose from the desire to assign, or to avoid, blame. Some publications called it a management failure, specifically in risk management, while others called it a technical failure.

Whenever accidents had occurred in the past at the National Aeronautics and Space Administration (NASA), an internal investigation team had been formed.

---

[1]The first digit indicates the fiscal year of the launch (i.e., "5" means 1985). The second number indicates the launch site (i.e., "1" is the Kennedy Space Center in Florida, "2" is Vandenberg Air Force Base in California). The letter represents the mission number (i.e., "C" would be the third mission scheduled). This designation system was implemented after Space Shuttle flights one through nine, which were designated STS-X. STS is the Space Transportation System and X would indicate the flight number.

But in this case, perhaps because of the visibility, the White House took the initiative in appointing an independent commission. There did exist significant justification for the commission. NASA was in a state of disarray, especially in the management ranks. The agency had been without a permanent administrator for almost four months. The turnover rate at the upper echelons of management was significantly high, and there seemed to be a lack of direction from the top down.

Another reason for appointing a Presidential Commission was the visibility of this mission. This mission had been known as the Teacher in Space mission, and Christa McAuliffe, a Concord, New Hampshire, schoolteacher, had been selected from a list of over 10,000 applicants. The nation knew the names of all of the crewmembers on board *Challenger.* The mission had been highly publicized for months, stating that Christa McAuliffe would be teaching students from aboard the *Challenger* on day four of the mission.

The Presidential Commission consisted of the following members:

- **William P. Rogers,** chairman: Former secretary of state under President Nixon and attorney general under President Eisenhower.
- **Neil A. Armstrong,** vice chairman: Former astronaut and spacecraft commander for Apollo 11.
- **David C. Acheson:**   Former senior vice president and general counsel, Communications Satellite Corporation (1967–1974), and a partner in the law firm of Drinker Biddle & Reath.
- **Dr. Eugene E. Covert:**   Professor and head, Department of Aeronautics and Astronautics at Massachusetts Institute of Technology.
- **Dr. Richard P. Feynman:**   Physicist and professor of theoretical physics at California Institute of Technology; Nobel Prize winner in Physics, 1965.
- **Robert B. Hotz:**   Editor-in-chief of *Aviation Week & Space Technology* magazine (1953–1980).
- **Major General Donald J. Kutyna,** USAF: Director of Space Systems and Command, Control, Communications.
- **Dr. Sally K. Ride:**   Astronaut and mission specialist on STS-7, launched on June 18, 1983, making her the first American woman in space. She also flew on mission 41-G, launched October 5, 1984. She holds a Doctorate in Physics from Stanford University (1978) and was still an active astronaut.
- **Robert W. Rummel:**   Vice president of Trans World Airlines and president of Robert W. Rummel Associates, Inc., of Mesa, Arizona.
- **Joseph F. Sutter:**   Executive vice president of the Boeing Commercial Airplane Company.
- **Dr. Arthur B. C. Walker, Jr.:**   Astronomer and professor of Applied Physics; formerly associate dean of the Graduate Division at Stanford

University, and consultant to Aerospace Corporation, Rand Corporation, and the National Science Foundation.

- **Dr. Albert D. Wheelon:**   Executive vice president, Hughes Aircraft Company.
- **Brigadier General Charles Yeager,** USAF (retired):   Former experimental test pilot. He was the first person to break the sound barrier and the first to fly at a speed of more than 1,600 miles an hour.
- **Dr. Alton G. Keel, Jr.,** Executive Director: Detailed to the Commission from his position in the Executive Office of the President, Office of Management and Budget, as associate director for National Security and International Affairs; formerly assistant secretary of the Air Force for Research, Development and Logistics, and Senate Staff.

The Commission interviewed more than 160 individuals, and more than thirty-five formal panel investigative sessions were held generating almost 12,000 pages of transcript. Almost 6,300 documents totaling more than 122,000 pages, along with hundreds of photographs, were examined and made a part of the Commission's permanent database and archives. These sessions and all the data gathered added to the 2,800 pages of hearing transcript generated by the Commission in both closed and open sessions. Unless otherwise stated, all of the quotations and memos in this case study come from the direct testimony cited in the *Report by the Presidential Commission (RPC).*

BACKGROUND TO THE SPACE TRANSPORTATION SYSTEM

During the early 1960s, NASA's strategic plans for post-*Apollo* manned space exploration rested upon a three-legged stool. The first leg was a reusable space transportation system, the space shuttle, which could transport people and equipment to low earth orbits and then return to earth in preparation for the next mission. The second leg was a manned space station that would be resupplied by the space shuttle and serve as a launch platform for space research and planetary exploration. The third leg would be planetary exploration to Mars. But by the late 1960s, the United States was involved in the Vietnam War, which was becoming costly. In addition, confidence in the government was eroding because of civil unrest and assassinations. With limited funding due to budgetary cuts, and with the lunar landing missions coming to an end, prioritization of projects was necessary. With a Democratic Congress continuously attacking the cost of space exploration, and minimal support from President Nixon, the space program was left standing on one leg only, the space shuttle.

President Nixon made it clear that funding all the programs NASA envisioned would be impossible, and that funding for even one program on the order of the *Apollo* Program was likewise not possible. President Nixon seemed to favor the space station concept, but this required the development of a reusable space shuttle. Thus NASA's Space Shuttle Program became the near-term priority.

One of the reasons for the high priority given to the Space Shuttle Program was a 1972 study completed by Dr. Oskar Morgenstern and Dr. Klaus Heiss of the Princeton-based Mathematica organization. The study showed that the space shuttle would be able to orbit payloads for as little as $100 per pound based on sixty launches per year with payloads of 65,000 pounds. This provided tremendous promise for military applications such as reconnaissance and weather satellites, as well as for scientific research.

Unfortunately, the pricing data were somewhat tainted. Much of the cost data were provided by companies who hoped to become NASA contractors and who therefore provided unrealistically low cost estimates in hopes of winning future bids. The actual cost per pound would prove to be more than twenty times the original estimate. Furthermore, the main engines never achieved the 109 percent of thrust that NASA desired, thus limiting the payloads to 47,000 pounds instead of the predicted 65,000 pounds. In addition, the European Space Agency began successfully developing the capability to place satellites into orbit and began competing with NASA for the commercial satellite business.

NASA SUCCUMBS TO POLITICS AND PRESSURE

To retain shuttle funding, NASA was forced to make a series of major concessions. First, facing a highly constrained budget, NASA sacrificed the research and development necessary to produce a truly reusable shuttle, and instead accepted a design that was only partially reusable, eliminating one of the features that had made the shuttle attractive in the first place. Solid rocket boosters (SRBs) were used instead of safer liquid-fueled boosters because they required a much smaller research and development effort. Numerous other design changes were made to reduce the level of research and development required.

Second, to increase its political clout and to guarantee a steady customer base, NASA enlisted the support of the United States Air Force. The Air Force could provide the considerable political clout of the Department of Defense and it used many satellites, which required launching. However, Air Force support did not come without a price. The shuttle payload bay was required to meet Air Force size and shape requirements, which placed key constraints on the ultimate design. Even more important was the Air Force requirement that the shuttle be able to launch from Vandenburg Air Force Base in California. This constraint required a larger cross range than the Florida site, which, in turn, decreased the total allowable vehicle weight. The weight reduction required the elimination of the design's air breathing engines, resulting in a single-pass unpowered landing. This greatly limited the safety and landing versatility of the vehicle.[2]

As the year 1986 began, there was extreme pressure on NASA to "Fly out the Manifest." From its inception, the Space Shuttle Program had been plagued by exaggerated expectations, funding inconsistencies, and political pressure. The ultimate vehicle and mission design were shaped almost as much by politics as by physics. President Kennedy's declaration that the United States would land a man on the moon before the end of the decade (the 1960s) had provided NASA's *Apollo* Program with high visibility, a clear direction, and powerful political backing. The Space Shuttle Program was not as fortunate; it had neither a clear direction nor consistent political backing.

Cost containment became a critical issue for NASA. In order to minimize cost, NASA designed a space shuttle system that utilized both liquid and solid propellants. Liquid propellant engines are more easily controllable than solid propellant engines. Flow of liquid propellant from the storage tanks to the engine can be throttled and even shut down in case of an emergency. Unfortunately, an all-liquid-fuel design was prohibitive because a liquid fuel system is significantly more expensive to maintain than a solid fuel system.

Solid fuel systems are less costly to maintain. However, once a solid propellant system is ignited, it cannot be easily throttled or shut down. Solid propellant rocket motors burn until all of the propellant is consumed. This could have a significant impact on safety, especially during launch, at which time the solid rocket boosters are ignited and have maximum propellant loads. Also, solid rocket boosters can be designed for reusability, whereas liquid engines are generally used only once.

The final design that NASA selected was a compromise of both solid and liquid fuel engines. The space shuttle would be a three-element system composed of the orbiter vehicle, an expendable external liquid fuel tank carrying liquid fuel for the orbiter's engines, and two recoverable solid rocket boosters.[3] The orbiter's engines were liquid fuel because of the necessity for throttle capability. The two solid rocket boosters would provide the added thrust necessary to launch the space shuttle into its orbiting altitude.

In 1972, NASA selected Rockwell as the prime contractor for building the orbiter. Many industry leaders believed that other competitors who had actively participated in the *Apollo* Program had a competitive advantage. Rockwell, however,

---

[2]Kurt Hoover and Wallace T. Fowler (The University of Texas at Austin and The Texas Space Grant Consortium), "Studies in Ethics, Safety and Liability for Engineers" (Web site: http://www.tsgc.utexas. edu/archive/general/ethics/shuttle.html page 2).

[3]The terms *solid rocket booster* (SRB) and *solid rocket motor* (SRM) will be used interchangeably.

was awarded the contract. Rockwell's proposal did not include an escape system. NASA officials decided against the launch escape system since it would have added too much weight to the shuttle at launch and was very expensive. There was also some concern on how effective an escape system would be if an accident occurred during launch while all of the engines were ignited. Thus, the Space Shuttle Program became the first U.S. manned spacecraft without a launch escape system for the crew.

In 1973, NASA went out for competitive bidding for the solid rocket boosters. The competitors were Morton-Thiokol, Inc. (MTI) (henceforth called Thiokol), Aerojet General, Lockheed, and United Technologies. The contract was eventually awarded to Thiokol because of its low cost, $100 million lower than the nearest competitor. Some believed that other competitors, who ranked higher in technical design and safety, should have been given the contract. NASA believed that Thiokol-built solid rocket motors would provide the lowest cost per flight.

THE SOLID ROCKET BOOSTERS

Thiokol's solid rocket boosters had a height of approximately 150 feet and a diameter of 12 feet. The empty weight of each booster was 192,000 pounds and the full weight was 1,300,000 pounds. Once ignited, each booster provided 2.65 million pounds of thrust, which is more than 70 percent of the thrust needed to lift off the launch pad.

Thiokol's design for the boosters was criticized by some of the competitors, and even by some NASA personnel. The boosters were to be manufactured in four segments and then shipped from Utah to the launch site, where the segments would be assembled into a single unit. The Thiokol design was largely based upon the segmented design of the Titan III solid rocket motor produced by United Technologies in the 1950s for Air Force satellite programs. Satellite programs were unmanned efforts.

The four solid rocket sections made up the case of the booster, which essentially encased the rocket fuel and directed the flow of the exhaust gases. This is shown in Exhibit I. The cylindrical shell of the case is protected from the propellant by a layer of insulation. The mating sections of the field joint are called the tang and the clevis. One hundred and seventy-seven pins spaced around the circumference of each joint hold the tang and the clevis together. The joint is sealed in three ways. First, zinc chromate putty is placed in the gap between the mating segments and their insulation. This putty protects the second and third seals, which are rubber-like rings, called O-rings. The first O-ring is called the primary

Exhibit I. **Solid rocket booster (SRB)**

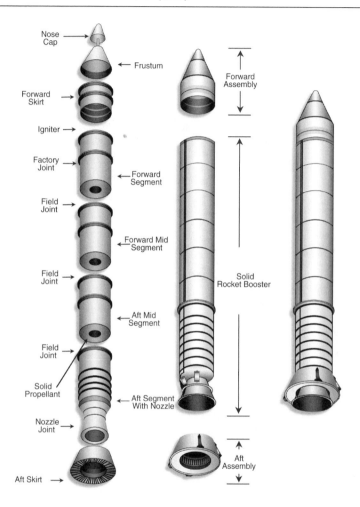

O-ring and is lodged in the gap between the tang and the clevis. The last seal is called the secondary O-ring, which is identical to the primary O-ring except it is positioned further downstream in the gap. Each O-ring is 0.280 inches in diameter. The placement of each O-ring can be seen in Exhibit II. Another component of the field joint is called the leak check port, which is shown in Exhibit III. The leak check port is designed to allow technicians to check the status of the two O-ring seals. Pressurized air is inserted through the leak check port into the gap

Exhibit II. *Location of the O-rings*

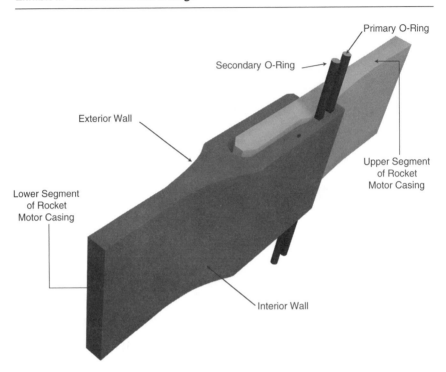

between the two O-rings. If the O-rings maintain the pressure, and do not let the pressurized air past the seal, the technicians know the seal is operating properly.[4]

In the Titan III assembly process, the joints between the segmented sections contained one O-ring. Thiokol's design had two O-rings instead of one. The second O-ring was initially considered as redundant, but included to improve safety. The purpose of the O-rings was to seal the space in the joints such that the hot exhaust gases could not escape and damage the case of the boosters.

Both the Titan III and Shuttle O-rings were made of Viton rubber, which is an elastomeric material. For comparison, rubber is also an elastomer. The elastomeric material used is a fluoroelastomer, which is an elastomer that contains fluorine. This material was chosen because of its resistance to high temperatures and its compatibility with the surrounding materials. The Titan III O-rings were

---

[4]"The *Challenger* Accident: Mechanical Causes of the *Challenger* Accident"; University of Texas (web site: http://www.me.utexas.edu/~uer/challenger/chall2.html pages 1–2).

Exhibit III. ***Cross section showing the leak test port***

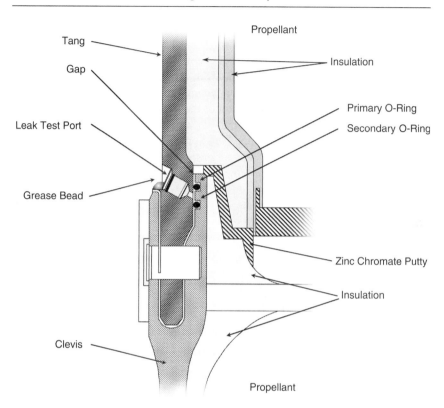

molded in one piece, whereas the shuttle's SRB O-rings would be manufactured in five sections and then glued together. Routinely, repairs would be necessary for inclusions and voids in the rubber received from the material suppliers.

BLOWHOLES

The primary purpose of the zinc chromate putty was to act as a thermal barrier that protected the O-rings from the hot exhaust. As mentioned before, the O-ring seals were tested using the leak check port to pressurize the gap between the seals. During the test, the secondary seal was pushed down into the same, seated position as it occupied during ignition pressurization. However, because the leak check port was between the two O-ring seals, the primary O-ring was pushed up and seated against the putty. The position of the O-rings during flight and their position during the leak check test is shown in Exhibit III.

During early flights, engineers worried that, because the putty above the primary seal could withstand high pressures, the presence of the putty would prevent the leak test from identifying problems with the primary seal. They contended that the putty would seal the gap during testing regardless of the condition of the primary seal. Since the proper operation of the primary seal was essential, engineers decided to increase the pressure used during the test to above the pressure that the putty could withstand. This would ensure that the primary O-ring was properly sealing the gap without the aid of the putty. Unfortunately, during this new procedure, the high-test pressures blew holes through the putty before the primary O-ring could seal the gap.

Since the putty was on the interior of the assembled solid rocket booster, technicians could not mend the blowholes in the putty. As a result, this procedure left small, tunneled holes in the putty. These holes would allow focused exhaust gases to contact a small segment of the primary O-ring during launch. Engineers realized that this was a problem, but decided to test the seals at the high pressure despite the formation of blowholes, rather than risking a launch with a faulty primary seal.

The purpose of the putty was to prevent the hot exhaust gases from reaching the O-rings. For the first nine successful shuttle launches, NASA and Thiokol used asbestos-bearing putty manufactured by the Fuller-O'Brien Company of San Francisco. However, because of the notoriety of products containing asbestos, and the fear of potential lawsuits, Fuller-O'Brien stopped manufacturing the putty that had served the shuttle so well. This created a problem for NASA and Thiokol.

The new putty selected came from Randolph Products of Carlstadt, New Jersey. Unfortunately, with the new putty, blowholes and O-ring erosion were becoming more common to a point where the shuttle engineers became worried. Yet the new putty was still used on the boosters. Following the Challenger disaster, testing showed that, at low temperatures, the Randolph putty became much stiffer than the Fuller-O'Brien putty and lost much of its stickiness.[5]

O-RING EROSION

If the hot exhaust gases penetrated the putty and contacted the primary O-ring, the extreme temperatures would break down the O-ring material. Because

---

[5]Ibid., p. 3.

engineers were aware of the possibility of O-ring erosion, the joints were checked after each flight for evidence of erosion. The amount of O-ring erosion found on flights before the new high-pressure leak check procedure was around 12 percent. After the new high-pressure leak test procedure, the percentage of O-ring erosion was found to increase by 88 percent. High percentages of O-ring erosion in some cases allowed the exhaust gases to pass the primary O-ring and begin eroding the secondary O-ring. Some managers argued that some O-ring erosion was "acceptable" because the O-rings were found to seal the gap even if they were eroded by as much as one-third their original diameter.[6] The engineers believed that the design and operation of the joints were an acceptable risk because a safety margin could be identified quantitatively. This numerical boundary would become an important precedent for future risk assessment.

JOINT ROTATION

During ignition, the internal pressure from the burning fuel applies approximately 1000 pounds per square inch on the case wall, causing the walls to expand. Because the joints are generally stiffer than the case walls, each section tends to bulge out. The swelling of the solid rocket sections causes the tang and the clevis to become misaligned; this misalignment is called joint rotation. A diagram showing a field joint before and after joint rotation is seen in Exhibit IV. The problem with joint rotation is that it increases the gap size near the O-rings. This increase in size is extremely fast, which makes it difficult for the O-rings to follow the increasing gap and keep the seal.[7]

Prior to ignition, the gap between the tang and the clevis is approximately 0.004 inches. At ignition, the gap will enlarge to between 0.042 and 0.060 inches, *but for a maximum of 0.60 second,* and then return to its original position.

O-RING RESILIENCE

The term *O-ring resilience* refers to the ability of the O-ring to return to its original shape after it has been deformed. This property is analogous to the ability of a rubber band to return to its original shape after it has been stretched. As with a rubber band, the resiliency of an O-ring is directly related to its temperature. As the temperature of the O-ring gets lower, the O-ring material becomes stiffer.

---

[6]Ibid., p. 4.

[7]Ibid., p. 4.

Exhibit IV. Field joint rotation

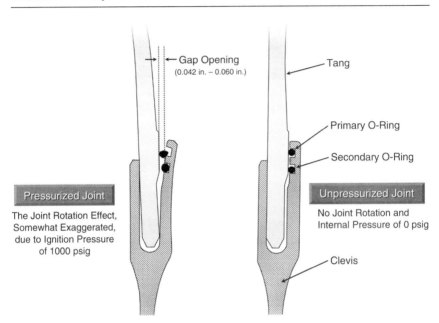

Tests have shown that an O-ring at 75°F is five times more responsive in returning to its original shape than an O-ring at 30°F. This decrease in O-ring resiliency during a cold weather launch would make the O-ring much less likely to follow the increasing gap size during joint rotation. As a result of poor O-ring resiliency, the O-ring would not seal properly.[8]

THE EXTERNAL TANK

The solid rockets are each joined forward and aft to the external liquid fuel tank. They are not connected to the orbiter vehicle. The solid rocket motors are mounted first, and the external liquid fuel tank is put between them and connected. Then the orbiter is mounted to the external tank at two places in the back and one place forward, and those connections carry all of the structural loads for

---

[8]Ibid., pp. 4–5.

the entire system at liftoff and through the ascent phase of flight. Also connected to the orbiter, under the orbiter's wing, are two large propellant lines 17 inches in diameter. The one on the port side carries liquid hydrogen from the hydrogen tank in the back part of the external tank. The line on the right side carries liquid oxygen from the oxygen tank at the forward end, inside the external tank.[9]

The external tank contains about 1.6 million pounds of propellant, or about 526,000 gallons. The orbiter's three engines burn the liquid hydrogen and liquid oxygen at a ratio of 6:1 and at a rate equivalent to emptying out a family swimming pool every 10 seconds! Once ignited, the exhaust gases leave the orbiter's three engines at approximately 6,000 miles per hour. After the fuel is consumed, the external tank separates from the orbiter, falls to earth, and disintegrates in the atmosphere on reentry.

THE SPARE PARTS PROBLEM

In March 1985, NASA's administrator, James Beggs, announced that there would be one shuttle flight per month for all of fiscal year 1985. In actuality, there were only six flights. Repairs became a problem. Continuous repairs were needed on the heat tiles required for reentry, the braking system, and the main engines' hydraulic pumps. Parts were routinely borrowed from other shuttles. The cost of spare parts was excessively high, and NASA was looking for cost containment.

RISK IDENTIFICATION PROCEDURES

The necessity for risk management was apparent right from the start. Prior to the launch of the first shuttle in April of 1981, hazards were analyzed and subjected to a formalized hazard reduction process as described in NASA Handbook, NHB5300.4. The process required that the credibility and probability of the hazards be determined. A Senior Safety Review Board was established for overseeing the risk assessment process. For the most part, the risks assessment process was qualitative. The conclusion reached was that no single hazard or combination of hazards should prevent the launch of the first shuttle *as long as the aggregate risk remained acceptable.*

NASA used a rather simplistic Safety (Risk) Classification System. A quantitative method for risk assessment was not in place at NASA because gathering

---

[9]*RPC,* page 50.

Exhibit V. *Risk classification system*

Level	Description
Criticality 1 (C1)	Loss of life and/or vehicle if the component fails.
Criticality 2 (C2)	Loss of mission if the component fails.
Criticality 3 (C3)	All others.
Criticality 1R (C1R)	Redundant components exist. The failure of both could cause loss of life and/or vehicle.
Criticality 2R (C2R)	Redundant components exist. The failure of both could cause loss of mission.

the data needed to generate statistical models would be expensive and labor-intensive. If the risk identification procedures were overly complex, NASA would have been buried in paperwork due to the number of components on the space shuttle. The risk classification system selected by NASA is shown in Exhibit V.

From 1982 on, the O-ring seal was labeled Criticality 1. By 1985, there were 700 components identified as Criticality 1.

TELECONFERENCING

The Space Shuttle Program involves a vast number of people at both NASA and the contractors. Because of the geographical separation between NASA and the contractors, it became impractical to have continuous meetings. Travel between Thiokol in Utah and the Cape in Florida took one day each way. Therefore, teleconferencing became the primary method of communication and a way of life. Interface meetings were still held, but the emphasis was on teleconferencing. All locations could be linked together in one teleconference and data could be faxed back and forth as needed.

PAPERWORK CONSTRAINTS

With the rather optimistic flight schedule provided to the news media, NASA was under scrutiny and pressure to deliver. For fiscal 1986, the mission manifest called for sixteen flights. The pressure to meet schedule was about to take its toll. Safety problems had to be resolved quickly.

As the number of flights scheduled began to increase, so did the requirements for additional paperwork. The majority of the paperwork had to be completed prior to NASA's Flight Readiness Review (FRR) meetings. Approximately one week, prior to every flight, flight operations and cargo managers were required to endorse the commitment of flight readiness to the NASA associate administrator for space flight at the FRR meeting. The responsible project/element managers would conduct pre-FRR meetings with their contractors, center managers, and the NASA Level II manager. The content of the FRR meetings included the following:

- Determine overall status, as well as establish the baseline in terms of significant changes since the last mission.
- Review significant problems resolved since the last review, and significant anomalies from the previous flight.
- Review all open items and constraints remaining to be resolved before the mission.
- Present all new waivers since the last flight.

NASA personnel were working excessive overtime, including weekends, to fulfill the paperwork requirements and prepare for the required meetings. As the number of space flights increased, so did the paperwork and overtime.

The paperwork constraints were affecting the contractors as well. Additional paperwork requirements existed for problem solving and investigations. On October 1, 1985, an interoffice memo was sent from Scott Stein, space booster project engineer at Thiokol, to Bob Lund, vice president for engineering at Thiokol, and to other selected managers concerning the O-Ring Investigation Task Force:

> We are currently being hog-tied by paperwork every time we try to accomplish anything. I understand that for production programs, the paperwork is necessary. However, for a priority, short schedule investigation, it makes accomplishment of our goals in a timely manner extremely difficult, if not impossible. We need the authority to bypass some of the paperwork jungle. As a representative example of problems and time that could easily be eliminated, consider assembly or disassembly of test hardware by manufacturing personnel. . . . I know the established paperwork procedures can be violated if someone with enough authority dictates it. We did that with the DR system when the FWC hardware "Tiger Team" was established. If changes are not made to allow us to accomplish work in a reasonable amount of time, then the O-ring investigation task force will never have the potency necessary to resolve problems in a timely manner.

Both NASA and the contractors were now feeling the pressure caused by the paperwork constraints.

ISSUING WAIVERS

One quick way of reducing paperwork and meetings was to issue a waiver. Historically, a waiver was a formalized process that allowed an exception to either a rule, a specification, a technical criterion, or a risk. Waivers were ways to reduce excessive paperwork requirements. Project managers and contract administrators had the authority to issue waivers, often with the intent of bypassing standard protocols in order to maintain a schedule. The use of waivers had been in place well before the manned space program even began. What is important here was *not* NASA's use of the waiver, but the *justification* for the waiver given the risks.

NASA had issued waivers on both Criticality 1 status designations and launch constraints. In 1982, the solid rocket boosters were designated C1 by the Marshall Space Flight Center because failure of the O-rings could have caused loss of crew and the shuttle. This meant that the secondary O-rings were not considered redundant. The SRB project manager at Marshall, Larry Malloy, issued a waiver just in time for the next shuttle launch to take place as planned. Later, the O-rings designation went from C1 to C1R (i.e., a redundant process), thus partially avoiding the need for a waiver. The waiver was a necessity to keep the shuttle flying according to the original manifest.

Having a risk identification of C1 was not regarded as a sufficient reason to cancel a launch. It simply meant that component failure could be disastrous. It implied that this might be a potential problem that needed attention. If the risks were acceptable, NASA could still launch. A more serious condition was the issuing of launch constraints. Launch constraints were official NASA designations for situations in which mission safety was a serious enough problem to justify a decision not to launch. But once again, a launch constraint did not imply that the launch should be delayed. It meant that this was an important problem and needed to be addressed.

Following the 1985 mission that showed O-ring erosion and exhaust gas blow-by, a launch constraint was imposed. Yet on each of the next five shuttle missions, NASA's Malloy issued a launch constraint waiver allowing the flights to take place on schedule without any changes to the O-rings.

Were the waivers a violation of serious safety rules just to keep the shuttle flying? The answer is *no*! NASA had protocols such as policies, procedures, and rules for adherence to safety. Waivers were also protocols but for the purpose of deviating from other existing protocols. Larry Malloy, his colleagues at NASA, and the contractors had no intentions of doing evil. Waivers were simply a way of saying that we believe that the risk is an *acceptable risk.*

The lifting of launch constraints and the issuance of waivers became the norm—standard operating procedure. Waivers became a way of life. If waivers were issued and the mission was completed successfully, then the same waivers would exist for the next flight and did not have to be brought up for discussion at the Flight Readiness Review meeting. The justification for the waivers seemed to be the similarity between flight launch conditions, temperature, and so on. Launching under similar conditions seemed to be important for the engineers at NASA and Thiokol because it meant that the forces acting on the O-rings were within their region of experience and could be correlated to existing data. The launch temperature effect on the O-rings was considered predictable, and therefore constituted an acceptable risk to both NASA and Thiokol, thus perhaps eliminating costly program delays that would have resulted from having to redesign the O-rings. The completion of each shuttle mission added another data point to the region of experience, thus guaranteeing the same waivers on the next launch. Flying with acceptable risk became the norm in NASA's culture.

LAUNCH LIFTOFF SEQUENCE PROFILE: POSSIBLE ABORTS

During the countdown to liftoff, the launch team closely monitors weather conditions, not only at the launch site, but also at touchdown sites should the mission need to be prematurely aborted.

Dr. Feynman: "Would you explain why we are so sensitive to the weather?"

Mr. Moore (NASA's deputy administrator for space flight): "Yes, there are several reasons. I mentioned the return to the landing site. We need to have visibility if we get into a situation where we need to return to the landing site after launch, and the pilots and the commanders need to be able to see the runway and so forth. So, you need a ceiling limitation on it [i.e., weather].

"We also need to maintain specifications on wind velocity so we don't exceed crosswinds. Landing on a runway and getting too high of a crosswind may cause us to deviate off of the runway and so forth, so we have a crosswind limit. During ascent, assuming a normal flight, a chief concern is damage to tiles due to rain. We have had experiences in seeing what the effects of a brief shower can do in terms of the tiles. The tiles are thermal insulation blocks, very thick. A lot of them are very thick on the bottom of the orbiter. But if you have a raindrop and you are going at a very high velocity, it tends to erode the tiles, pock the tiles, and that causes us a grave concern regarding the thermal protection.

"In addition to that, you are worried about the turnaround time of the orbiters as well, because with the kind of tile damage that one could get in rain, you have an awful lot of work to do to go back and replace tiles back on the system. So, there are a number of concerns that weather enters into, and it is a major factor in our assessment of whether or not we are ready to launch."[10]

---

[10]Ibid., p. 18.

Approximately six to seven seconds prior to the liftoff, the Shuttle's main engines (liquid fuel) ignite. These engines consume one-half million gallons of liquid fuel. It takes nine hours prior to launch to fill the liquid fuel tanks. At ignition, the engines are throttled up to 104 percent of rated power. Redundancy checks on the engines' systems are then made. The launch site ground complex and the orbiter's onboard computer complex check a large number of details and parameters about the main engines to make sure that everything is proper and that the main engines are performing as planned.

If a malfunction is detected, the system automatically goes into a shutdown sequence, and the mission is scrubbed. The primary concern at this point is to make the vehicle "safe." The crew remains on board and performs a number of functions to get the vehicle into a safe mode. These functions include making sure that all propellant and electrical systems are properly safed. Ground crews at the launch pad begin servicing the launch pad. Once the launch pad is in a safe condition, the hazard and safety teams begin draining the remaining liquid fuel out of the external tank.

If no malfunction is detected during this six-second period of liquid fuel burn, then a signal is sent to ignite the two solid rocket boosters, and liftoff occurs. For the next two minutes, with all engines ignited, the shuttle goes through a Max Q, or high dynamic pressure phase, that exerts maximum pressure loads on the orbiter vehicle. Based upon the launch profile, the main engines may be throttled down slightly during the Max Q phase to lower the loads.

After 128 seconds into the launch sequence, all of the solid fuel is expended and the solid rocket boosters (SRBs) staging occurs. The SRB parachutes are deployed. The SRBs then fall back to earth 162 miles from the launch site and are recovered for examination, cleaning, and reuse on future missions. The main liquid fuel engines are then throttled up to maximum power. After 523 seconds into the liftoff, the external liquid fuel tanks are essentially expended of fuel. The main engines are shut down. Ten to eighteen seconds later, the external tank is separated from the orbiter and disintegrates on reentry into the atmosphere.

From a safety perspective, the most hazardous period is the first 128 seconds when the SRBs are ignited. Here's what Arnold Aldrich, manager of NASA's STS Program, Johnson Space Center, had to say:

Mr. Aldrich: "Once the shuttle system starts off the launch pad, there is no capability in the system to separate these [solid propellant] rockets until they reach burnout. They will burn for two minutes and eight or nine seconds, and the system must stay together. There is not a capability built into the vehicle that would allow these to separate. There is a capability available to the flight crew to separate at this interface the orbiter from the tank, but that is thought to be unacceptable during the first stage when the booster rockets are on and thrusting. So, essentially the first two minutes and a little more of flight, the stack is intended and designed to stay together, and it must stay together to fly successfully."

Exhibit VI. Abort options for shuttle

Type of Abort	Landing Site
Once-around abort	Edwards Air Force Base
Trans-Atlantic abort	DaKar
Trans-Atlantic abort	Casablanca
Return-to-landing-site (RTLS)	Kennedy Space Center

Mr. Hotz: "Mr. Aldrich, why is it unacceptable to separate the orbiter at that stage?"

Mr. Aldrich: "It is unacceptable because of the separation dynamics and the rupture of the propellant lines. You cannot perform the kind of a clean separation required for safety in the proximity of these vehicles at the velocities and the thrust levels they are undergoing, [and] the atmosphere they are flying through. In that regime, it is the design characteristic of the total system."[11]

If an abort is deemed necessary during the first 128 seconds, the actual abort will not begin until *after* SRB staging has occurred, which is after 128 seconds into the launch sequence. Based on the reason and timing of an abort, options include those listed in Exhibit VI.

Arnold Aldrich commented on different abort profiles:

Chairman Rogers: "During the two-minute period, is it possible to abort through the orbiter?"

Mr. Aldrich: "You can abort for certain conditions. You can start an abort, but the vehicle won't do anything yet, and the intended aborts are built around failures in the main engine system, the liquid propellant systems and their controls. If you have a failure of a main engine, it is well detected by the crew and by the ground support, and you can call for a return-to-launch-site abort. That would be logged in the computer. The computer would be set up to execute it, but everything waits until the solids take you to altitude. At that time, the solids will separate in the sequence I described, and then the vehicle flies downrange some 400 miles, maybe 10 to 15 additional minutes, while all of the tank propellant is expelled through these engines.

"As a precursor to setting up the conditions for this return-to-launch-site abort to be successful towards the end of that burn downrange, using the propellants and the thrust of the main engines, the vehicle turns and actually points heads up back towards Florida. When the tank is essentially depleted, automatic signals are sent to close off the [liquid] propellant lines and to separate the orbiter, and the orbiter then does a similar approach to the one we are familiar with with orbit back to the Kennedy Space Center for approach and landing."

---

[11]Ibid., p. 51.

Dr. Walker: "So, the propellant is expelled but not burned?"

Mr. Aldrich: "No, it is burned. You burn the system on two engines all the way down-range until it is gone, and then you turn around and come back because you don't have enough to burn to orbit. That is the return-to-launch-site abort, and it applies during the first 240 seconds of—no, 240 is not right. It is longer than that—the first four minutes, either before or after separation you can set that abort up, but it will occur after the solids separate, and if you have a main engine anomaly after the solids separate, at that time you can start the RTLS, and it will go through that same sequence and come back."

Dr. Ride: "And you can also only do an RTLS if you have lost just one main engine. So if you lose all three main engines, RTLS isn't a viable abort mode."

Mr. Aldrich: "Once you get through the four minutes, there's a period where you now don't have the energy conditions right to come back, and you have a forward abort, and Jesse mentioned the sites in Spain and on the coast of Africa. We have what is called a trans-Atlantic abort, and where you can use a very similar sequence to the one I just described. You still separate the solids, you still burn all the propellant out of the tanks, but you fly across and land across the ocean."

Mr. Hotz: "Mr. Aldrich, could you recapitulate just a bit here? Is what you are telling us that for two minutes of flight, until the solids separate, there is no practical abort mode?"

Mr. Aldrich: "Yes, sir."

Mr. Hotz: "Thank you."

Mr. Aldrich: "A trans-Atlantic abort can cover a range of just a few seconds up to about a minute in the middle where the across-the-ocean sites are effective, and then you reach this abort once-around capability where you go all the way around and land in California or back to Kennedy by going around the earth. And finally, you have abort-to-orbit where you have enough propulsion to make orbit but not enough to achieve the exact orbital parameters that you desire. That is the way that the abort profiles are executed.

"There are many, many nuances of crew procedure and different conditions and combinations of sequences of failures that make it much more complicated than I have described it."[12]

THE O-RING PROBLEM

There were two kinds of joints on the shuttle—field joints that were assembled at the launch site connecting together the SRB's cylindrical cases, and nozzle joints

---

[12]Ibid., pp. 51–52.

that connected the aft end of the case to the nozzle. During the pressure of ignition, the field joints could become bent such that the secondary O-ring could lose contact within an estimated 0.17 to 0.33 seconds after ignition. If the primary O-ring failed to seal properly before the gap within the joints opened up and the secondary seal failed, the results could be disastrous.

When the solid propellant boosters are recovered after separation, they are disassembled and checked for damage. The O-rings could show evidence of coming into contact with heat. Hot gases from the ignition sequence could blow by the primary O-ring briefly before sealing. This "blow-by" phenomenon could last for only a few milliseconds before sealing and result in no heat damage to the O-ring. If the actual sealing process takes longer than expected, then charring and erosion of the O-rings can occur. This would be evidenced by gray or black soot and erosion to the O-rings. The terms used are impingement erosion and "bypass" erosion, with the latter identified also as sooted "blow-by."

Roger Boisjoly of Thiokol describes blow-by erosion and joint rotation as follows:

O-ring material gets removed from the cross section of the O-ring much, much faster than when you have bypass erosion or blow-by, as people have been terming it. We usually use the characteristic blow-by to define gas past it, and we use the other term [bypass erosion] to indicate that we are eroding at the same time. And so you can have blow-by without erosion, [and] you [can] have blow-by with erosion.[13]

At the beginning of the transient cycle [initial ignition rotation, up to 0.17 seconds] . . . [the primary O-ring] is still being attacked by hot gas, and it is eroding at the same time it is trying to seal, and it is a race between, will it erode more than the time allowed to have it seal.[14]

On January 24, 1985, STS 51-C [Flight No. 15] was launched at 51°F, which was the lowest temperature of any launch up to that time. Analyses of the joints showed evidence of damage. Black soot appeared between the primary and secondary O-rings. The engineers concluded that the cold weather had caused the O-rings to harden and move more slowly. This allowed the hot gases to blow by and erode the O-rings. This scorching effect indicated that low temperature launches could be disastrous.

On July 31, 1985, Roger Boisjoly of Thiokol sent an interoffice memo to R. K. Lund, vice president for engineering at Thiokol:

This letter is written to insure that management is fully aware of the seriousness of the current O-ring erosion problem in the SRM joints from an engineering standpoint.

---

[13]Ibid., pp. 784–785.

[14]Ibid., p. 136.

The mistakenly accepted position on the joint problem was to fly without fear of failure and to run a series of design evaluations which would ultimately lead to a solution or at least a significant reduction of the erosion problem. This position is now drastically changed as a result of the SRM 16A nozzle joint erosion which eroded a secondary O-ring with the primary O-ring never sealing.

If the same scenario should occur in a field joint (and it could), then it is a jump ball as to the success or failure of the joint because the secondary O-ring cannot respond to the clevis opening rate and may not be capable of pressurization. The result would be a catastrophe of the highest order—loss of human life.

An unofficial team (a memo defining the team and its purpose was never published) with [a] leader was formed on 19 July 1985 and was tasked with solving the problem for both the short and long term. This unofficial team is essentially nonexistent at this time. In my opinion, the team must be officially given the responsibility and the authority to execute the work that needs to be done on a non-interference basis (full time assignment until completed).

It is my honest and very real fear that if we do not take immediate action to dedicate a team to solve the problem with the field joint having the number one priority, then we stand in jeopardy of losing a flight along with all the launch pad facilities.[15]

On August 9, 1985, a letter was sent from Brian Russell, manager of the SRM Ignition System, to James Thomas at the Marshall Space Flight Center. The memo addressed the following:

Per your request, this letter contains the answers to the two questions you asked at the July Problem Review Board telecon.

1. *Question:* If the field joint secondary seal lifts off the metal mating surfaces during motor pressurization, how soon will it return to a position where contact is re-established?

   Answer: Bench test data indicate that the O-ring resiliency (its capability to follow the metal) is a function of temperature and rate of case expansion. MTI [Thiokol] measured the force of the O-ring against Instron plattens, which simulated the nominal squeeze on the O-ring and approximated the case expansion distance and rate.

   At 100°F, the O-ring maintained contact. At 75°F, the O-ring lost contact for 2.4 seconds. At 50°F, the O-ring did not re-establish contact in 10 minutes at which time the test was terminated.

   The conclusion is that secondary sealing capability in the SRM field joint cannot be guaranteed.

---

[15]Ibid., pp. 691–692.

2. *Question:* If the primary O-ring does not seal, will the secondary seal seat in sufficient time to prevent joint leakage?

Answer: MTI has no reason to suspect that the primary seal would ever fail after pressure equilibrium is reached; i.e., after the ignition transient. If the primary O-ring were to fail from 0 to 170 milliseconds, there is a very high probability that the secondary O-ring would hold pressure since the case has not expanded appreciably at this point. If the primary seal were to fail from 170 to 330 milliseconds, the probability of the secondary seal holding is reduced. From 330 to 600 milliseconds the chance of the secondary seal holding is small. This is a direct result of the O-ring's slow response compared to the metal case segments as the joint rotates.[16]

At NASA, the concern for a solution to the O-ring problem became not only a technical crisis, but also a budgetary crisis. In a July 23, 1985, memorandum from Richard Cook, program analyst, to Michael Mann, chief of the STS Resource Analysis Branch, the impact of the problem was noted:

Earlier this week you asked me to investigate reported problems with the charring of seals between SRB motor segments during flight operations. Discussions with program engineers show this to be a potentially major problem affecting both flight safety and program costs.

Presently three seals between SRB segments use double O-rings sealed with putty. In recent Shuttle flights, charring of these rings has occurred. The O-rings are designed so that if one fails, the other will hold against the pressure of firing. However, at least in the joint between the nozzle and the aft segment, not only has the first O-ring been destroyed, but the second has been partially eaten away.

Engineers have not yet determined the cause of the problem. Candidates include the use of a new type of putty (the putty formerly in use was removed from the market by EPA because it contained asbestos), failure of the second ring to slip into the groove which must engage it for it to work properly, or new, and as yet unidentified, assembly procedures at Thiokol. MSC is trying to identify the cause of the problem, including on-site investigation at Thiokol, and OSF hopes to have some results from their analysis within thirty days. There is little question, however, that flight safety has been and is still being compromised by potential failure of the seals, and it is acknowledged that failure during launch would certainly be catastrophic. There is also indication that staff personnel knew of this problem sometime in advance of management's becoming apprised of what was going on.

The potential impact of the problem depends on the as yet undiscovered cause. If the cause is minor, there should be little or no impact on budget or flight rate. A worst case scenario, however, would lead to the suspension of

---

[16]Ibid., pp. 1568–1569.

Shuttle flights, redesign of the SRB, and scrapping of existing stockpiled hardware. The impact on the FY 1987-8 budget could be immense.

It should be pointed out that Code M management [NASA's associate administrator for space flight] is viewing the situation with the utmost seriousness. From a budgetary standpoint, I would think that any NASA budget submitted this year for FY 1987 and beyond should certainly be based on a reliable judgment as to the cause of the SRB seal problem and a corresponding decision as to budgetary action needed to provide for its solution.[17]

On October 30, 1985, NASA launched Flight STS 61-A [Flight no. 22] at 75°F. This flight also showed signs of sooted blow-by, but the color was significantly blacker. Although there was some heat effect, there was no measurable erosion observed on the secondary O-ring. Since blow-by and erosion had now occurred at a higher launch temperature, the original premise that launches under cold temperatures were a problem was now being questioned. Exhibit VII shows the temperature at launch of all the shuttle flights up to this time and the O-ring damage, if any.

Management at both NASA and Thiokol wanted *concrete* evidence that launch temperature was directly correlated to blow-by and erosion. Other than simply a "gut feel," engineers were now stymied on how to show the direct correlation. NASA was not ready to cancel a launch simply due to an engineer's "gut feel."

William Lucas, director of the Marshall Space Center, made it clear that NASA's manifest for launches would be adhered to. Managers at NASA were pressured to resolve problems internally rather than to escalate them up the chain of command. Managers became afraid to inform anyone higher up that they had problems, even though they knew that one existed.

Richard Feynman, Nobel laureate and member of the Rogers Commission, concluded that a NASA official altered the safety criteria so that flights could be certified on time under pressure imposed by the leadership of William Lucas. Feynman commented:

> . . . They, therefore, fly in a relatively unsafe condition with a chance of failure of the order of one percent. Official management claims to believe that the probability of failure is a thousand times less.

Without concrete evidence of the temperature effect on the O-rings, the secondary O-ring was regarded as a redundant safety constraint and the criticality factor was changed from C1 to C1R. Potentially serious problems were treated as anomalies peculiar to a given flight. Under the guise of anomalies, NASA began

---

[17]Ibid., pp. 391–392.

Exhibit VII.   Erosion and blow-by history (temperature in ascending order from coldest to warmest)

Flight	Date	Temperature (°F)	Erosion Incidents	Blow-by Incidents	Comments
51-C	01/24/85	53	3	2	Most erosion any flight; blow-by; secondary O-rings heated up
41-B	02/03/84	57	1		Deep, extensive erosion
61-C	01/12/86	58	1		O-rings erosion
41-C	04/06/84	63	1		O-rings heated but no damage
1	04/12/81	66			Coolest launch without problems
6	04/04/83	67			
51-A	11/08/84	67			
51-D	04/12/85	67			
5	11/11/82	68			
3	03/22/82	69			
2	11/12/81	70	1		Extent of erosion unknown
9	11/28/83	70			
41-D	08/30/84	70	1		
51-G	06/17/85	70			
7	06/18/83	72			
8	08/30/83	73			
51-B	04/29/85	75			
61-A	10/20/85	75		2	No erosion but soot between O-rings
51-1	08/27/85	76			
61	11/26/85	76			
41-G	10/05/84	78			
51-J	10/03/85	79			
4	06/27/82	80			No data; casing lost at sea
51-F	07/29/85	81			

issuing waivers to maintain the flight schedules. Pressure was placed upon contractors to issue closure reports. On December 24, 1985, L. O. Wear, NASA's SRM Program Office manager, sent a letter to Joe Kilminster, Thiokol's vice president for the Space Booster Program:

> During a recent review of the SRM Problem Review Board open problem list I found that we have 20 open problems, 11 opened during the past 6 months, 13 open over 6 months, 1 three years old, 2 two years old, and 1 closed during the past six months. As you can see our closure record is very poor. You are requested to initiate the required effort to assure more timely closures and the MTI personnel shall coordinate directly with the S&E personnel the contents of the closure reports.[18]

---

[18]Ibid., p. 1554.

PRESSURE, PAPERWORK, AND WAIVERS

To maintain the flight schedule, critical issues such as launch constraints had to be resolved or waived. This would require extensive documentation. During the Rogers Commission investigation, it seemed that there had been a total lack of coordination between NASA's Marshall Space Center and Thiokol prior to the *Challenger* disaster. Joe Kilminster, Thiokol's vice president for the Space Booster Program, testified:

Mr. Kilminster: "Mr. Chairman, if I could, I would like to respond to that. In response to the concern that was expressed—and I had discussions with the team leader, the task force team leader, Mr. Don Kettner, and Mr. Russell and Mr. Ebeling. We held a meeting in my office and that was done in the October time period where we called the people who were in a support role to the task team, as well as the task force members themselves.

"In that discussion, some of the task force members were looking to circumvent some of our established systems. In some cases, that was acceptable; in other cases, it was not. For example, some of the work that they had recommended to be done was involved with full-scale hardware, putting some of these joints together with various putty layup configurations; for instance, taking them apart and finding out what we could from that inspection process."

Dr. Sutter: "Was that one of these things that was outside of the normal work, or was that accepted as a good idea or a bad idea?"

Mr. Kilminster: "A good idea, but outside the normal work, if you will."

Dr. Sutter: "Why not do it?"

Mr. Kilminster: "Well, we were doing it. But the question was, can we circumvent the system, the paper system that requires, for instance, the handling constraints on those flight hardware items? And I said no, we can't do that. We have to maintain our handling system, for instance, so that we don't stand the possibility of injuring or damaging a piece of flight hardware.

"I asked at that time if adding some more people, for instance, a safety engineer—that was one of the things we discussed in there. The consensus was no, we really didn't need a safety engineer. We had the manufacturing engineer in attendance who was in support of that role, and I persuaded him that, typical of the way we normally worked, that he should be calling on the resources from his own organization, that is, in Manufacturing, in order to get this work done and get it done in a timely fashion.

"And I also suggested that if they ran across a problem in doing that, they should bubble that up in their management chain to get help in getting the resources to get that done. Now, after that session, it was my impression that there was improvement based on some of the concerns that had been expressed, and we did get quite a bit of work done. For your evaluation, I would like to talk a little bit about the sequence of events for this task force."

Chairman Rogers: "Can I interrupt? Did you know at that time it was a launch constraint, a formal launch constraint?"

Mr. Kilminster: "Not an overall launch constraint as such. Similar to the words that have been said before, each Flight Readiness Review had to address any anomalies or concerns that were identified at previous launches and in that sense, each of those anomalies or concerns were established in my mind as launch constraints unless they were properly reviewed and agreed upon by all parties."

Chairman Rogers: "You didn't know there was a difference between the launch constraint and just considering it an anomaly? You thought they were the same thing?"

Mr. Kilminster: "No, sir. I did not think they were the same thing."

Chairman Rogers: "My question is: Did you know that this launch constraint was placed on the flights in July 1985?"

Mr. Kilminster: "Until we resolved the O-ring problem on that nozzle joint, yes. We had to resolve that in a fashion for the subsequent flight before we would be okay to fly again."

Chairman Rogers: "So you did know there was a constraint on that?"

Mr. Kilminster: "On a one flight per one flight basis; yes, sir."

Chairman Rogers: "What else would a constraint mean?"

Mr. Kilminster: "Well, I get the feeling that there's a perception here that a launch constraint means all launches, whereas we were addressing each launch through the Flight Readiness Review process as we went."

Chairman Rogers: "No, I don't think—the testimony that we've had is that a launch constraint is put on because it is a very serious problem and the constraint means don't fly unless it's fixed or taken care of, but somebody has the authority to waive it for a particular flight. And in this case, Mr. Mulloy was authorized to waive it, which he did, for a number of flights before 51-L. Just prior to 51-L, the papers showed the launch constraint was closed out, which I guess means no longer existed. And that was done on January 23, 1986. Now, did you know that sequence of events?"

Mr. Kilminster: "Again, my understanding of *closing out,* as the term has been used here, was to close it out on the problem actions list, but not as an overall standard requirement. We had to address these at subsequent Flight Readiness Reviews to ensure that we were all satisfied with the proceeding to launch."

Chairman Rogers: "Did you understand the waiver process, that once a constraint was placed on this kind of a problem, that a flight could not occur unless there was a formal waiver?"

Mr. Kilminster: "Not in the sense of a formal waiver, no, sir."

Chairman Rogers: "Did any of you? Didn't you get the documents saying that?"

Mr. McDonald: "I don't recall seeing any documents for a formal waiver."[19]

MISSION 51-L

On January 25, 1986, questionable weather caused a delay of Mission 51-L to January 27. On January 26, the launch was reconfirmed for 9:37 A.M. on the 27th. However, on the morning of January 27, a malfunction with the hatch, combined with high crosswinds, caused another delay. All preliminary procedures had been completed and the crew had just boarded when the first problem appeared. A microsensor on the hatch indicated that the hatch was not shut securely. It turned out that the hatch was shut securely but the sensor had malfunctioned. Valuable time was lost in determining the problem.

After the hatch was finally closed, the external handle could not be removed. The threads on the connecting bolt were stripped and instead of cleanly disengaging when turned, simply spun around. Attempts to use a portable drill to remove the handle failed. Technicians on the scene asked Mission Control for permission to saw off the bolt. Fearing some form of structural stress to the hatch, engineers made numerous time-consuming calculations before giving the go-ahead to cut off the bolt. The entire process consumed almost two hours before the countdown resumed.

However, the misfortunes continued. During the attempts to verify the integrity of the hatch and remove the handle, the wind had been steadily rising. Chief Astronaut John Young flew a series of approaches in the shuttle training aircraft and confirmed the worst fears of mission control. The crosswinds at the Cape were in excess of the level allowed for the abort contingency. The opportunity had been missed. The mission was then reset to launch the next day, January 28, at 9:38 A.M. Everyone was quite discouraged since extremely cold weather was forecast for Tuesday that could further postpone the launch.[20]

Weather conditions indicated that the temperature at launch could be as low as 26°F. This would be much colder and well below the temperature range that the O-rings were designed to operate in. The components of the solid rocket motors were qualified only to 40°F at the lower limit. Undoubtedly, when the sun

---

[19]Ibid., pp. 1577–1578.

[20]Hoover and Wallace, pp. 3–4.

came up and launch time approached, both the air temperature and vehicle would warm up, but there was still concern. Would the ambient temperature be high enough to meet the launch requirements? NASA's Launch Commit Criteria stated that no launch should occur at temperatures below 31°F. There were also worries over any permanent effects on the shuttle due to the cold overnight temperatures. NASA became concerned and asked Thiokol for their recommendation on whether or not to launch. NASA admitted under testimony that if Thiokol had recommended not launching, then the launch would not have taken place.

At 5:45 P.M. eastern standard time, a teleconference was held between the Kennedy Space Center, Marshall Space Flight Center, and Thiokol. Bob Lund, vice president for engineering, summarized the concerns of the Thiokol engineers that in Thiokol's opinion, the launch should be delayed until noontime or even later such that a launch temperature of at least 53°F could be achieved. Thiokol's engineers were concerned that no data were available for launches at this temperature of 26°F. This was the first time in fourteen years that Thiokol had recommended not to launch.

The design validation tests originally done by Thiokol covered only a narrow temperature range. The temperature data did not include any temperatures below 53°F. The O-rings from Flight 51-C, which had been launched under cold conditions the previous year, showed very significant erosion. These were the only data available on the effects of cold, but all of the Thiokol engineers agreed that the cold weather would decrease the elasticity of the synthetic rubber O-rings, which in turn might cause them to seal slowly and allow hot gases to surge through the joint.[21]

Another teleconference was set up for 8:45 P.M. to invite more parties to be involved in the decision. Meanwhile, Thiokol was asked to fax all relevant and supporting charts to all parties involved in the 8:45 P.M. teleconference.

The following information was included in the pages that were faxed:

Blow-by History:

SRM-15 Worst Blow-by
- Two case joints (80°), (110°) *Arc*
- Much worse visually than SRM-22

SRM-22 Blow-by
- Two case joints (30–40°)

SRM-13A, 15, 16A, 18, 23A, 24A
- Nozzle blow-by

Field Joint Primary Concerns—SRM-25
- A temperature lower than the current database results in changing primary O-ring sealing timing function

---

[21]Ibid., p. 4.

- SRM-15A—80° arc black grease between O-rings
  SRM-15B—110° arc black grease between O-rings
- Lower O-ring squeeze due to lower temp
- Higher O-ring shore hardness
- Thicker grease viscosity
- Higher O-ring pressure activation time
- If actuation time increases, threshold of secondary seal pressurization capability is approached.
- If threshold is reached then secondary seal may not be capable of being pressurized.

Conclusions:

- Temperature of O-ring is not only parameter controlling blow-by:
  SRM-15 with blow-by had an O-ring temp at 53°F.
  SRM-22 with blow-by had an O-ring temp at 75°F.
  Four development motors with no blow-by were tested at O-ring temp of 47° to 52°F.
  Development motors had putty packing which resulted in better performance.
- At about 50°F blow-by could be experienced in case joints.
- Temp for SRM-25 on 1-28-86 launch will be: 29°F 9 A.M.
  38°F 2 P.M.
- Have no data that would indicate SRM-25 is different than SRM-15 other than temp.

Recommendations:

- O-ring temp must be ≥ 53°F at launch.
  Development motors at 47° to 52°F with putty packing had no blow-by.
  SRM-15 (the best simulation) worked at 53°F.
- Project ambient conditions (temp & wind) to determine launch time.

From NASA's perspective, the launch window was from 9:30 A.M. to 12:30 P.M. on January 28. This was based on weather conditions and visibility, not only at the launch site but also at the landing sites should an abort be necessary. An additional consideration was the fact that the temperature might not reach 53°F prior to the launch window closing. Actually, the temperature at the Kennedy Space Center was not expected to reach 50°F until two days later. NASA was hoping that Thiokol would change its mind and recommend launch.

THE SECOND TELECONFERENCE

At the second teleconference, Bob Lund once again asserted Thiokol's recommendation not to launch below 53°F. NASA's Mulloy then burst out over the teleconference network:

My God, Morton Thiokol! When do you want me to launch—next April?

NASA challenged Thiokol's interpretation of the data and argued that Thiokol was inappropriately attempting to establish a new Launch Commit Criterion just prior to launch. NASA asked Thiokol to reevaluate its conclusions. Crediting NASA's comments with some validity, Thiokol then requested a five-minute *off-line* caucus. In the room at Thiokol were fourteen engineers, namely:

1. Jerald Mason, senior vice president, Wasatch Operations
2. Calvin Wiggins, vice president and general manager, Space Division
3. Joe C. Kilminster, vice president, Space Booster Programs
4. Robert K. Lund, vice president, Engineering
5. Larry H. Sayer, director, Engineering and Design
6. William Macbeth, manager, Case Projects, Space Booster Project
7. Donald M. Ketner, supervisor, Gas Dynamics Section and head Seal Task Force
8. Roger Boisjoly, member, Seal Task Force
9. Arnold R. Thompson, supervisor, Rocket Motor Cases
10. Jack R. Kapp, manager, Applied Mechanics Department
11. Jerry Burn, associate engineer, Applied Mechanics
12. Joel Maw, associate scientist, Heat Transfer Section
13. Brian Russell, manager, Special Projects, SRM Project
14. Robert Ebeling, manager, Ignition System and Final Assembly, SRB Project

There were no safety personnel in the room because nobody thought to invite them. The caucus lasted some thirty minutes. Thiokol (specifically Joe Kilminster) then returned to the teleconference stating that they were unable to sustain a valid argument that temperature affects O-ring blow-by and erosion. *Thiokol then reversed its position and was now recommending launch.*

NASA stated that the launch of the *Challenger* would not take place without Thiokol's approval. But when Thiokol reversed its position following the caucus and agreed to launch, NASA interpreted this as an acceptable risk. The launch would now take place.

Mr. McDonald (Thiokol): "The assessment of the data was that the data was not totally conclusive, that the temperature could affect everything relative to the seal. But there was data that indicated that there were things going in the wrong direction, and this was far from our experience base.

"The conclusion being that Thiokol was directed to reassess all the data because the recommendation was not considered acceptable at that time of [waiting for] the 53 degrees [to occur]. NASA asked us for a reassessment and some more data to show that the temperature in itself can cause this to be a more serious concern than we had said it would be. At that time Thiokol in Utah said that they would like to go off-line and caucus for about five minutes and reassess what data they had there or any other additional data.

"And that caucus lasted for, I think, a half hour before they were ready to go back on. When they came back on they said they had reassessed all the data and had come to the conclusions that the temperature influence, based on the data they had available to them, was inconclusive and therefore they recommended a launch."[22]

During the Rogers Commission testimony, NASA's Mulloy stated his thought process in requesting Thiokol to rethink their position:

General Kutyna: "You said the temperature had little effect?"

Mr. Mulloy: "I didn't say that. I said I can't get a correlation between O-ring erosion, blow-by and O-ring, and temperature."

General Kutyna: "51-C was a pretty cool launch. That was January of last year."

Mr. Mulloy: "It was cold before then but it was not that much colder than other launches."

General Kutyna: "So it didn't approximate this particular one?"

Mr. Mulloy: "Unfortunately, that is one you look at and say, aha, is it related to a temperature gradient and the cold. The temperature of the O-ring on 51-C, I believe, was 53 degrees. We have fired motors at 48 degrees."[23]

Mulloy asserted he had not pressured Thiokol into changing their position. Yet, the testimony of Thiokol's engineers stated they believed they were being pressured.

Roger Boisjoly, one of Thiokol's experts on O-rings, was present during the caucus and vehemently opposed the launch. During testimony, Boisjoly described his impressions of what occurred during the caucus:

"The caucus was started by Mr. Mason stating that a management decision was necessary. Those of us who were opposed to the launch continued to speak out, and I am specifically speaking of Mr. Thompson and myself because in my recollection, he and I were the only ones who vigorously continued to oppose the launch. And we were attempting to go back and rereview and try to make clear what we were trying to get across, and we couldn't understand why it was going to be reversed.

---

[22]*RPC,* p. 300.

[23]Ibid., p. 290.

"So, we spoke out and tried to explain again the effects of low temperature. Arnie actually got up from his position which was down the table and walked up the table and put a quad pad down in front of the table, in front of the management folks, and tried to sketch out once again what his concern was with the joint, and when he realized he wasn't getting through, he just stopped.

"I tried one more time with the photos. I grabbed the photos and I went up and discussed the photos once again and tried to make the point that it was my opinion from actual observations that temperature was indeed a discriminator, and we should not ignore the physical evidence that we had observed.

"And again, I brought up the point that SRM-15 had a 110 degree arc of black grease, while SRM-22 had a relatively different amount, which was less and wasn't quite as black. I also stopped when it was apparent that I could not get anybody to listen."

Dr. Walker: "At this point did anyone else [i.e., engineers] speak up in favor of the launch?"

Mr. Boisjoly: "No, sir. No one said anything, in my recollection. Nobody said a word. It was then being discussed amongst the management folks. After Arnie and I had our last say, Mr. Mason said we have to make a management decision. He turned to Bob Lund and asked him to take off his engineering hat and put on his management hat. From this point on, management formulated the points to base their decision on. There was never one comment in favor, as I have said, of launching by any engineer or other nonmanagement person in the room before or after the caucus. I was not even asked to participate in giving any input to the final decision charts.

"I went back on the net with the final charts or final chart, which was the rationale for launching, and that was presented by Mr. Kilminster. It was hand-written on a notepad, and he read from that notepad. I did not agree with some of the statements that were being made to support the decision. I was never asked nor polled, and it was clearly a management decision from that point.

"I must emphasize, I had my say, and I never take any management right to take the input of an engineer and then make a decision based upon that input, and I truly believe that. I have worked at a lot of companies, and that has been done from time to time, and I truly believe that, and so there was no point in me doing anything any further [other] than [what] I had already attempted to do.

"I did not see the final version of the chart until the next day. I just heard it read. I left the room feeling badly defeated, but I felt I really did all I could to stop the launch. I felt personally that management was under a lot of pressure to launch, and they made a very tough decision, but I didn't agree with it.

"One of my colleagues who was in the meeting summed it up best. This was a meeting where the determination was to launch, and it was up to us to prove beyond a shadow of a doubt that it was not safe to do so. This is in total reverse to what the position usually is in a preflight conversation or a Flight Readiness Review. It is usually exactly opposite that."

Dr. Walker: "Do you know the source of the pressure on management that you alluded to?"

Mr. Boisjoly: "Well, the comments made over the net are what I felt. I can't speak for them, but I felt it. I felt the tone of the meeting exactly as I summed up, that we were being put in a position to prove that we should not launch rather than being put in the position and prove that we had enough data to launch."[24]

General Kutyna: "What was the motivation driving those who were trying to overturn your opposition?"

Mr. Boisjoly: "They felt that we had not demonstrated, or I had not demonstrated, because I was the prime mover in SRM-15. Because of my personal observations and involvement in the Flight Readiness Reviews, they felt that I had not conclusively demonstrated that there was a tie-in between temperature and blow-by.

"My main concern was if the timing function changed and that seal took longer to get there, then you might not have any seal left because it might be eroded before it seats. And then, if that timing function is such that it pushes you from the 170 millisecond region into the 330 second region, you might not have a secondary seal to pick up if the primary is gone. That was my major concern.

"I can't quantify it. I just don't know how to quantify that. But I felt that the observations made were telling us that there was a message there telling us that temperature was a discriminator, and I couldn't get that point across. I basically had no direct input into the final recommendation to launch, and I was not polled.

"I think Astronaut Crippin hit the tone of the meeting exactly right on the head when he said that the opposite was true of the way the meetings were normally conducted. We normally have to absolutely prove beyond a shadow of a doubt that we have the ability to fly, and it seemed like we were trying to prove, have proved that we had data to prove that we couldn't fly at this time, instead of the reverse. That was the tone of the meeting, in my opinion."[25]

Jerald Mason, senior vice president at Thiokol's Wasatch Division, directed the caucus at Thiokol. Mason continuously asserted that a management decision was needed and instructed Bob Lund, vice president for engineering, to take off his engineering hat and put on his management hat. During testimony, Mason commented on his interpretation of the data:

Dr. Ride [a member of the Commission]:   "You know, what we've seen in the charts so far is that the data was inconclusive and so you said go ahead."

---

[24]Ibid., pp. 793–794.

[25]Ibid., p. 676.

Mr. Mason: ". . . I hope I didn't convey that. But the reason for the discussion was the fact that we didn't have enough data to quantify the effect of the cold, and that was the heart of our discussion . . . We have had blow-by on earlier flights. We had not had any reason to believe that we couldn't experience it again at any temperature. . . ."[26]

At the end of the second teleconference, NASA's Hardy at Marshall Space Flight Center requested that Thiokol put their recommendation to launch in writing and fax it to both Marshall Space Flight Center and Kennedy Space Center. The memo that follows was signed by Joe Kilminster, vice president for Thiokol's Space Booster Program, and faxed at 11:45 P.M. the night before the launch.

- Calculations show that SRM-25 O-rings will be 20° colder than SRM-15 O-rings.
- Temperature data not conclusive on predicting primary O-ring blow-by.
- Engineering assessment is that:
  - Colder O-rings will have increased effective durometer ("harder").
  - "Harder" O-rings will take longer to "seat."
    - More gas may pass primary O-ring before the primary seal seats (relative to SRM-15).
    - Demonstrated sealing threshold is three times greater than 0.038" erosion experienced on SRM-15.
  - **If the primary seal does not seat, the secondary seal will seat.**
    - **Pressure will get to secondary seal before the metal parts rotate.**
    - **O-ring pressure leak check places secondary seal in outboard position, which minimizes sealing time.**
- **MTI recommends STS-51L launch proceed on 28 January 1986.**
  - **SRM-25 will not be significantly different from SRM-15.**[27]

THE ICE PROBLEM

At 1:30 A.M. on the day of the launch, NASA's Gene Thomas, launch director, ordered a complete inspection of the launch site due to cold weather and severe ice conditions. The prelaunch inspection of the *Challenger* and the launch pad by the ice-team was unusual, to say the least. The ice-team's responsibility was to remove any frost or ice on the vehicle or launch structure. What they found during their inspection looked like something out of a science fiction movie. The

---

[26]Ibid., p. 764.

[27]Ibid., p. 764.

freeze-protection plan implemented by Kennedy personnel had gone very wrong. Hundreds of icicles, some up to 16 inches long, clung to the launch structure. The handrails and walkways near the shuttle entrance were covered in ice, making them extremely dangerous if the crew had to make an emergency evacuation. One solid sheet of ice stretched from the 195 foot level to the 235 foot level on the gantry. However, NASA continued to cling to its calculations that there would be no damage due to flying ice shaken loose during the launch.[28] A decision was then made to delay the launch from 9:38 A.M. to 11:30 A.M. so that the ice on the launch pad could melt. The delay was still within the launch window of 9:30 A.M.–12:30 P.M.

At 8:30 A.M., a second ice inspection was made. Ice was still significantly present at the launch site. Robert Glaysher, vice president for orbital operations at Rockwell, stated that the launch was unsafe. Rockwell's concern was that falling ice could damage the heat tiles on the orbiter. This could have a serious impact during reentry.

At 10:30 A.M., a third ice inspection was made. Though some of the ice was beginning to melt, there was still significant ice on the launch pad. The temperature of the left solid rocket booster was measured at 33°F and the right booster was measured at 19°F. Even though the right booster was 34 degrees colder than Thiokol's original recommendation for a launch temperature (i.e., 53°F), no one seemed alarmed. Rockwell also agreed to launch, even though its earlier statement had been that the launch was unsafe.

Arnold Aldrich, manager of the STS Program at the Johnson Space Center, testified on the concern over the ice problem:

Mr. Aldrich: "Kennedy facility people at that meeting, everyone in that meeting, voted strongly to proceed and said they had no concern, except for Rockwell. The comment to me from Rockwell, which was not written specifically to the exact words, and either recorded or logged, was that they had some concern about the possibility of ice damage to the orbiter. Although it was a minor concern, they felt that we had no experience base launching in this exact configuration before, and therefore they thought we had some additional risk of orbiter damage from ice than we had on previous meetings, or from previous missions."

Chairman Rogers: "Did they sign off on it or not?"

Mr. Aldrich: "We don't have a sign-off at that point. It was not—it was not maybe 20 minutes, but it was close to that. It was within the last hour of launch."

Chairman Rogers: "But they still objected?"

---

[28]Hoover and Wallace, page 5.

Mr. Aldrich: "They issued what I would call a concern, a less than 100 percent concurrence in the launch. They did not say we do not want to launch, and the rest of the team overruled them. They issued a more conservative concern. They did not say don't launch."

General Kutyna: "I can't recall a launch that I have had where there was 100 percent certainty that everything was perfect, and everyone around the table would agree to that. It is the job of the launch director to listen to everyone, and it's our job around the table to listen and say there is this element of risk, and you characterize this as 90 percent, or 95, and then you get a consensus that that risk is an acceptable risk, and then you launch.

"So I think this gentleman is characterizing the degree of risk, and he's honest, and he had to say something."

Dr. Ride: "But one point is that their concern is a specific concern, and they weren't concerned about the overall temperature or damage to the solid rockets or damage to the external tank. They were worried about pieces of ice coming off and denting the tile."[29]

Following the accident, the Rogers Commission identified three major concerns about the ice-on-the-pad issue:

1. An analysis of all of the testimony and interviews established that Rockwell's recommendation on launch was ambiguous. The Commission found it difficult, as did Mr. Aldrich, to conclude that there was a no-launch recommendation. Moreover, all parties were asked specifically to contact Aldrich or Moore about launch objections due to weather. Rockwell made no phone calls or further objections to Aldrich or other NASA officials after the 9:00 A.M. Mission Management Team meeting and subsequent to the resumption of the countdown.
2. The Commission was also concerned about the NASA response to the Rockwell position at the 9:00 A.M. meeting. While it was understood that decisions have to be made in launching a Shuttle, the Commission was not convinced Levels I and II [of NASA's management] appropriately considered Rockwell's concern about the ice. However ambiguous Rockwell's position was, it was clear that they did tell NASA that the ice was an unknown condition. Given the extent of the ice on the pad, the admitted unknown effect of the Solid Rocket Motor and Space Shuttle Main Engines ignition on the ice, as well as the fact that debris striking the orbiter was a potential flight safety hazard, the Commission found the decision to launch questionable under those circumstances. In this situation,

---

[29]Ibid., pp. 237–238.

NASA appeared to be requiring a contractor to prove that it was not safe to launch, rather than proving it was safe. Nevertheless, the Commission had determined that the ice was not a cause of the 51-L accident and does not conclude that NASA's decision to launch specifically overrode a no-launch recommendation by an element contractor.

3. The Commission concluded that the freeze protection plan for launch pad 39B was inadequate. The Commission believed that the severe cold and presence of so much ice on the fixed service structure made it inadvisable to launch on the morning of January 28, and that margins of safety were whittled down too far.

It became obvious that NASA's management knew of the ice problem, but did they know of Thiokol's original recommendation not to launch and then their reversal? Larry Malloy, the SRB Project manager for NASA, and Stanley Reinartz, NASA's manager of the Shuttle Office, both admitted that they told Arnold Aldrich, manager of the STS program, Johnson Space Center, about their concern for the ice problem but there was no discussion about the teleconferences with Thiokol over the O-rings. It appeared that Malloy and Reinartz considered the ice as a potential problem whereas the O-rings constituted an acceptable risk. Therefore, only potential problems went up the chain of command, not the components of the "aggregate acceptable launch risk." It became common practice in Flight Readiness Review documentation to use the term *acceptable risk*. This became the norm at NASA and resulted in insulating senior management from certain potential problems. It was the culture that had developed at NASA that created the flawed decision-making process rather than an intent by individuals to withhold information and jeopardize safety.

THE ACCIDENT

Just after liftoff at 0.678 seconds into the flight, photographic data showed a strong puff of gray smoke spurting from the vicinity of the aft field joint on the right solid rocket booster. The two pad 39B cameras that would have recorded the precise location of the puff were inoperative. Computer graphic analysis of film from other cameras indicated the initial smoke came from the 270- to 310-degree sector of the circumference of the aft field joint of the right solid rocket booster. This area of the solid booster faced the external tank. The vaporized material streaming from the joint indicated there was incomplete sealing action within the joint.

Eight more distinctive puffs of increasingly blacker smoke were recorded between 0.836 and 2.500 seconds. The smoke appeared to puff upward from the joint. While each smoke puff was being left behind by the upward flight of the Shuttle, the next fresh puff could be seen near the level of the joint. The multiple smoke puffs in this sequence occurred about four times per second, approximating the frequency of the structural load dynamics and resultant joint flexing. Computer graphics applied to NASA photos from a variety of cameras in this sequence again placed the smoke puffs' origin in the same 270- to 310-degree sector of the circumference as the original smoke spurt.

As the shuttle *Challenger* increased its upward velocity, it flew past the emerging and expanding smoke puffs. The last smoke was seen above the field joint at 2.733 seconds.

The black color and dense composition of the smoke puffs suggested that the grease, joint insulation, and rubber O-rings in the joint seal were being burned and eroded by the hot propellant gases.

At approximately 37 seconds, *Challenger* encountered the first of several high altitude wind shear conditions that lasted about 64 seconds. The wind shear created forces of relatively large fluctuations on the vehicle itself. These were immediately sensed and countered by the guidance, navigation, and control systems.

The steering system (thrust vector control) of the solid rocket booster responded to all commands and wind shear effects. The wind shear caused the steering system to be more active than on any previous flight.

Both the *Challenger*'s main engines and the solid rockets operated at reduced thrust approaching and passing through the area of maximum dynamic pressure of 720 pounds per square foot. Main engines had been throttled up to 104 percent thrust, and the solid rocket boosters were increasing their thrust when the first flickering flame appeared on the right solid rocket booster in the area of the aft field joint. This first very small flame was detected on image-enhanced film at 58.788 seconds into the flight. It appeared to originate at about 305 degrees around the booster circumference at or near the aft field joint.

One film frame later from the same camera, the flame was visible without image enhancement. It grew into a continuous, well-defined plume at 59.262 seconds. At approximately the same time (60 seconds), telemetry showed a pressure differential between the chamber pressures in the right and left boosters. The right booster chamber pressure was lower, confirming the growing leak in the area of the field joint.

As the flame plume increased in size, it was deflected rearward by the aerodynamic slipstream and circumferentially by the protruding structure of the upper ring attaching the booster to the external tank. These deflections directed the flame plume onto the surface of the external tank. This sequence of flame spreading is confirmed by analysis of the recovered wreckage. The growing flame also impinged on the strut attaching the solid rocket booster to the external tank.

The first visual indication that swirling flame from the right solid rocket booster breached the external tank was at 64.660 seconds, when there was an abrupt change in the shape and color of the plume. This indicated that it was mixing with leaking hydrogen from the external tank. Telemetered changes in the hydrogen tank pressurization confirmed the leak. Within 45 milliseconds of the breach of the external tank, a bright, sustained glow developed on the black tiled underside of the *Challenger* between it and the external tank.

Beginning around 72 seconds, a series of events occurred extremely rapidly that terminated the flight. Telemetered data indicated a wide variety of flight system actions that supported the visual evidence of the photos as the shuttle struggled futilely against the forces that were destroying it.

At about 72.20 seconds, the lower strut linking the solid rocket booster and the external tank was severed or pulled away from the weakened hydrogen tank, permitting the right solid rocket booster to rotate around the upper attachment strut. This rotation was indicated by divergent yaw and pitch rates between the left and right solid rocket boosters.

At 73.124 seconds, a circumferential white vapor pattern was observed blooming from the side of the external tank bottom dome. This was the beginning of the structural failure of the hydrogen tank that culminated in the entire aft dome dropping away. This released massive amounts of liquid hydrogen from the tank and created a sudden forward thrust of about 2.8 million pounds, pushing the hydrogen tank upward into the intertank structure. About the same time, the rotating right solid rocket booster impacted the intertank structure and the lower part of the liquid oxygen tank. These structures failed at 73.137 seconds, as evidenced by the white vapors appearing in the intertank region.

Within milliseconds there was massive, almost explosive, burning of the hydrogen streaming from the failed tank bottom and the liquid oxygen breach in the area of the intertank.

At this point in its trajectory, while traveling at a Mach number of 1.92 at an altitude of 46,000 feet, the *Challenger* was totally enveloped in the explosive burn. The *Challenger*'s reaction control system ruptured, and a hypergolic burn of its propellants occurred, producing the oxygen-hydrogen flames. The reddish brown colors of the hypergolic fuel burn were visible on the edge of the main fireball. The orbiter, under severe aerodynamic loads, broke into several large sections, which emerged from the fireball. Separate sections that can be identified on film include the main engine/tail section with the engines still burning, one wing of the orbiter, and the forward fuselage trailing a mass of umbilical lines pulled loose from the payload bay.

The consensus of the Commission and participating investigative agencies was that the loss of the space shuttle *Challenger* was caused by a failure in the joint between the two lower segments of the right solid rocket motor. The specific failure was the destruction of the seals that were intended to prevent hot gases from leaking through the joint during the propellant burn of the rocket motor. The evidence assembled by the Commission indicates that no other element of the space shuttle system contributed to this failure.

In arriving at this conclusion, the Commission reviewed in detail all available data, reports, and records; directed and supervised numerous tests, analyses, and experiments by NASA, civilian contractors, and various government agencies; and then developed specific failure scenarios and the range of most probably causative factors.

The failure was due to a faulty design unacceptably sensitive to a number of factors. These factors were the effects of temperature, physical dimensions, the character of materials, the effects of reusability, processing, and the reaction of the joint to dynamic loading.

NASA AND THE MEDIA

Following the tragedy, many believed that NASA's decision to launch had been an attempt to minimize further ridicule by the media. Successful shuttle flights were no longer news because they were almost ordinary. However, launch aborts and delayed landings were more newsworthy because they were less common. The *Columbia* launch, which had immediately preceded the *Challenger* mission, had been delayed seven times. The *Challenger* launch had gone through four delays already. News anchor personnel were criticizing NASA. Some believed that NASA felt it had to do something quickly to dispel its poor public image.

The *Challenger* mission had had more media coverage and political ramifications than other recent missions. This would be the launch of the Teacher in Space Project. The original launch date of the *Challenger* had been scheduled just before President Reagan's State of the Union message, that was to be delivered the evening of January 28. Some believed that the president had intended to publicly praise NASA for the Teacher in Space Project and possibly even talk to Ms. McAuliffe live during his address. This would certainly have enhanced NASA's image. Following the tragedy, there were questions as to whether the White House had pressured NASA into launching the Shuttle because of President Reagan's (and NASA's) love of favorable publicity. The commission, however, found no evidence of White House intervention in the decision to launch.

FINDINGS OF THE COMMISSION

Determining the cause of an engineering disaster can take years of investigation. The *Challenger* disaster arose from many factors, including launch conditions, mechanical failure, faulty communication, and poor decision making. In the end, the last-minute decision to launch combined all possible factors into a lethal action.

The Commission concluded that the accident was rooted in history. The space shuttle's solid rocket booster problem began with the faulty design of its joint and increased as both NASA and contractor management first failed to recognize that they had a problem, then failed to fix it, and finally treated it as an acceptable flight risk.

Morton Thiokol, Inc., the contractor, did not accept the implication of tests early in the program that the design had a serious and unanticipated flaw. NASA did not accept the judgment of its engineers that the design was unacceptable, and as the joint problems grew in number and severity, NASA minimized them in management briefings and reports. Thiokol's stated position was that "the condition is not desirable but is acceptable."

Neither Thiokol nor NASA expected the rubber O-rings sealing the joints to be touched by hot gases of motor ignition, much less to be partially burned. However, as tests and then flights confirmed damage to the sealing rings, the reaction by both NASA and Thiokol was to increase the amount of damage considered "acceptable." At no time did management either recommend a redesign of the joint or call for the shuttle's grounding until the problem was solved.

The genesis of the *Challenger* accident—the failure of the joint of the right solid rocket motor—lay in decisions made in the design of the joint and in the failure by both Thiokol and NASA's Solid Rocket Booster project office to understand and respond to facts obtained during testing.

The Commission concluded that neither Thiokol nor NASA had responded adequately to internal warnings about the faulty seal design. Furthermore, Thiokol and NASA did not make a timely attempt to develop and verify a new seal after the initial design was shown to be deficient. Neither organization developed a solution to the unexpected occurrences of O-ring erosion and blow-by, even though this problem was experienced frequently during the shuttle flight history. Instead, Thiokol and NASA management came to accept erosion and blow-by as unavoidable and an acceptable flight risk. Specifically, the Commission found six things:

1. The joint test and certification program was inadequate. There was no requirement to configure the qualifications test motor as it would be in flight, and the motors were static tested in a horizontal position, not in the vertical flight position.
2. Prior to the accident, neither NASA nor Thiokol fully understood the mechanism by which the joint sealing action took place.
3. NASA and Thiokol accepted escalating risk apparently because they "got away with it last time." As Commissioner Feynman observed, the decision-making was:

> A kind of Russian roulette. . . . [The Shuttle] flies [with O-ring erosion] and nothing happens. Then it is suggested, therefore, that the risk is no longer so high for the next flights. We can lower our

standards a little bit because we got away with it last time. . . . You got away with it, but it shouldn't be done over and over again like that.

4. NASA's system for tracking anomalies for Flight Readiness Reviews failed in that, despite a history of persistent O-ring erosion and blow-by, flight was still permitted. It failed again in the strange sequence of six consecutive launch constraint waivers prior to 51-L, permitting it to fly without any record of a waiver, or even of an explicit constraint. Tracking and continuing only anomalies that are outside the database of prior flight allowed major problems to be removed from, and lost by, the reporting system.
5. The O-ring erosion history presented to Level I at NASA Headquarters in August 1985 was sufficiently detailed to require corrective action prior to the next flight.
6. A careful analysis of the flight history of O-ring performance would have revealed the correlation of O-ring damage and low temperature. Neither NASA nor Thiokol carried out such an analysis; consequently, they were unprepared to properly evaluate the risks of launching the 51-L mission in conditions more extreme than they had encountered before.

The Commission also identified a concern for the "silent" safety program. The Commission was surprised to realize after many hours of testimony that NASA's safety staff was never mentioned. No witness related the approval or disapproval of the reliability engineers, and none expressed the satisfaction or dissatisfaction of the quality assurance staff. No one thought to invite a safety representative or a reliability and quality assurance engineer to the January 27, 1986, teleconference between Marshall and Thiokol. Similarly, there was no safety representative on the Mission Management Team that made key decisions during the countdown on January 28, 1986.

The unrelenting pressure to meet the demands of an accelerating flight schedule might have been adequately handled by NASA if it had insisted on the exactingly thorough procedures that had been its hallmark during the *Apollo* program. An extensive and redundant safety program comprising interdependent safety, reliability, and quality assurance functions had existed during the lunar program to discover any potential safety problems. Between that period and 1986, however, the safety program had become ineffective. This loss of effectiveness seriously degraded the checks and balances essential for maintaining flight safety.

On April 3, 1986, Arnold Aldrich, the Space Shuttle Program manager, appeared before the Commission at a public hearing in Washington, D.C. He described five different communication or organization failures that affected the launch decision on January 28, 1986. Four of those failures related directly to faults within the safety program. These faults included a lack of problem reporting requirements, inadequate trend analysis, misrepresentation of criticality, and lack of involvement in critical discussions. A robust safety organization that was properly staffed and supported might well have avoided these faults, and thus eliminated the communication failures.

NASA had a safety program to ensure that the communication failures to which Mr. Aldrich referred did not occur. In the case of mission 51-L, however, that program fell short.

The Commission concluded that there were severe pressures placed on the launch decision-making system to maintain a flight schedule. These pressures caused rational men to make irrational decisions.

With the 1982 completion of the orbital flight test series, NASA began a planned acceleration of the space shuttle launch schedule. One early plan contemplated an eventual rate of a mission a week, but realism forced several downward revisions. In 1985, NASA published a projection calling for an annual rate of twenty-four flights by 1990. Long before the *Challenger* accident, however, it was becoming obvious that even the modified goal of two flights a month was overambitious.

In establishing the schedule, NASA had not provided adequate resources. As a result, the capabilities of the launch decision-making system were strained by the modest nine-mission rate of 1985, and the evidence suggested that NASA would not have been able to accomplish the fifteen flights scheduled for 1986. These were the major conclusions of a Commission examination of the pressures and problems attendant upon the accelerated launch schedule:

1. The capabilities of the launch decision-making system were stretched to the limit to support the flight rate in winter 1985/1986. Projections into the spring and summer of 1986 showed a clear trend; the system, as it existed, would have been unable to deliver crew training software for scheduled flights by the designated dates. The result would have been an unacceptable compression of the time available for the crews to accomplish their required training.

2. Spare parts were in critically short supply. The shuttle program made a conscious decision to postpone spare parts procurements in favor of budget items of perceived higher priority. Lack of spare parts would likely have limited flight operations in 1986.

3. Stated manifesting policies were not enforced. Numerous late manifest changes (after the cargo integration review) had been made to both major payloads and minor payloads throughout the shuttle program
   - Late changes to major payloads or program requirements required extensive resources (money, manpower, facilities) to implement.
   - If many late changes to "minor" payloads occurred, resources were quickly absorbed.
   - Payload specialists frequently were added to a flight well after announced deadlines.

- Late changes to a mission adversely affected the training and development of procedures for subsequent missions.
4. The scheduled flight rate did not accurately reflect the capabilities and resources.
   - The flight rate was not reduced to accommodate periods of adjustment in the capacity of the work force. There was no margin for error in the system to accommodate unforeseen hardware problems.
   - Resources were primarily directed toward supporting the flights and thus not enough were available to improve and expand facilities needed to support a higher flight rate.
5. Training simulators may have been the limiting factor on the flight rate: the two simulators available at that time could not train crews for more than twelve to fifteen flights per year.
6. When flights came in rapid succession, the requirements then current did not ensure that critical anomalies occurring during one flight would be identified and addressed appropriately before the next flight.

CHAIN-OF-COMMAND COMMUNICATION FAILURE

The Commission also identified a communication failure within the reporting structure at both NASA and Thiokol. Part of the problem with the chain of command structure was the idea of the proper reporting channel. Engineers report only to their immediate managers, while those managers report only to their direct supervisors. Engineers and managers believed in the chain of command structure; they felt reluctant to go above their superiors with their concerns. Boisjoly at Thiokol and Powers at Marshall felt that they had done all that they could as far as voicing their concerns. Anything more could have cost them their jobs. When questioned at the Rogers Commission hearing about why he did not voice his concerns to others, Powers replied, "That would not be my reporting channel." The chain of command structure dictated the only path that information could travel at both NASA and Thiokol. If information was modified or silenced at the bottom of the chain, there was not an alternate path for it to take to reach high-level officials at NASA. The Rogers Commission concluded that there was a breakdown in communication between Thiokol engineers and top NASA officials and faulted the management structure for not allowing important information about the SRBs to flow to the people who needed to know it. The Commission reported that the "fundamental problem was poor technical decision-making over a period of several years by top NASA and contractor personnel."

Bad news does not travel well in organizations like NASA and Thiokol. When the early signs of problems with the SRBs appeared, Thiokol managers did not believe that the problems were serious. Thiokol did not want to accept the fact that there could be a problem with its boosters. When Marshall received news of the problems, it considered it Thiokol's problem and did not pass the bad news upward to NASA headquarters. At Thiokol, Boisjoly described his managers as shutting out the bad news. He claims that he argued about the importance of the O-ring seal problems until he was convinced that "no one wanted to hear what he had to say." When Lund finally decided to recommend delay of the launch to Marshall, managers at Marshall rejected the bad news and refused to accept the recommendation not to launch. As with any information going up the chain of command at these two organizations, bad news was often modified so that it had less impact, perhaps skewing its importance.[30]

On January 31, 1986, President Ronald Reagan stated:

> The future is not free: the story of all human progress is one of a struggle against all odds. We learned again that this America, which Abraham Lincoln called the last, best hope of man on Earth, was built on heroism and noble sacrifice. It was built by men and women like our seven star voyagers, who answered a call beyond duty, who gave more than was expected or required and who gave it with little thought of worldly reward.

EPILOGUE

Following the tragic accident, virtually every senior manager that was involved in the space shuttle *Challenger* decision-making processes, at both NASA and Thiokol, accepted early retirement. Whether this was the result of media pressure, peer pressure, fatigue, or stress we can only postulate. The only true failures are the ones from which nothing is learned. Lessons on how to improve the risk management process were learned, unfortunately at the expense of human life.

On January 27, 1967, Astronauts Gus Grissom, Edward White, and Roger Chaffee were killed on board a test on *Apollo-Saturn 204*. James Webb, NASA's Administrator at that time, was allowed by President Johnson to conduct an internal investigation of the cause. The investigation was primarily a technical investigation. NASA was fairly open with the media during the investigation. As a result of the openness, the credibility of the agency was maintained.

With the *Challenger* accident, confusion arose as to whether it had been a technical failure or a management failure. There was no question in anyone's mind that the decision-making process was flawed. NASA and Thiokol acted independently in their response to criticism. Critical information was withheld, at

---

[30]"The *Challenger* Accident: Administrative Causes of the *Challenger* Accident" (Web site: http://www.me.utexas.edu/~uer/challenger/chall3.html pages 8–9).

least temporarily, and this undermined people's confidence in NASA. The media, as might have been expected, began vengeful attacks on NASA and Thiokol.

Following the *Apollo-Saturn 204* fire, there were few changes made in management positions at NASA. Those changes that did occur were the result of a necessity for improvement and where change was definitely warranted. Following the *Challenger* accident, almost every top management position at NASA underwent a change of personnel.

How an organization fares after an accident is often measured by how well it interfaces with the media. Situations such as the Tylenol tragedy (subject of another case study in this volume) and the *Apollo-Saturn 204* fire bore this out.

Following the accident, and after critical data were released, papers were published showing that the O-ring data correlation was indeed possible. In one such paper, Lighthall[31] showed that not only was a correlation possible, but the real problem may be a professional weakness shared by many people, but especially engineers, who have been required to analyze technical data. Lighthall's argument was that engineering curriculums might not provide engineers with strong enough statistical education, especially in covariance analysis. The Rogers Commission also identified this conclusion when they found that there were no engineers at NASA trained in statistical sciences.

Almost all scientific achievements require the taking of risks. The hard part is deciding which risk is worth taking and which is not. Every person who has ever flown in space, whether military or civilian, was a volunteer. They were all risk-takers who understood that safety in space can never be guaranteed with 100 percent accuracy.

QUESTIONS

Following are a series of questions categorized according to the principles of risk management. There may not be any single right or wrong answer to these questions.

Risk Management Plan

1. Does it appear, from the data provided in the case, that a risk management plan was in existence?
2. If such a plan did exist, then why wasn't it followed—or was it followed?

---

[31]Frederick F. Lighthall, "Launching The Space Shuttle Challenger: Disciplinary Deficiencies in the Analysis of Engineering Data," *IEEE Transactions on Engineering Management,* vol. 38, no. 1, (February 1991), pp. 63–74.

3. Is there a difference between a risk management plan, a quality assurance plan, and a safety plan, or are they the same?
4. Would there have been a better way to handle risk management planning at NASA assuming sixteen flights per year, twenty-five flights per year, or as originally planned, sixty flights per year? Why is the number of flights per year critical in designing a formalized risk management plan?

Risk Identification

5. What is the difference between a risk and an anomaly? Who determines the difference?
6. Does there appear to have been a structured process in place for risk identification at either NASA or Thiokol?
7. How should problems with risk identification be resolved if there exist differences of opinion between the customer and the contractors?
8. Should senior management or sponsors be informed about all risks identified or just the overall "aggregate" risk?
9. How should one identify or classify the risks associated with using solid rocket boosters on manned spacecraft rather than the conventional liquid fuel boosters?
10. How should one identify or classify trade-off risks such as trading off safety for political acceptability?
11. How should one identify or classify the risks associated with pressure resulting from making promises that may be hard to keep?
12. Suppose that a risk identification plan had been established at the beginning of the space program when the shuttle was still considered an experimental design. If the shuttle is now considered as an operational vehicle rather than as an experimental design, could that affect the way that risks were identified to the point where the risk identification plan would need to be changed?

Risk Quantification

13. Given the complexity of the Space Shuttle Program, is it feasible and/or practical to develop a methodology for quantifying risks, or should each situation be addressed individually? Can we have both a quantitative and qualitative risk evaluation system in place at the same time?
14. How does one quantify the dangers associated with the ice problem?
15. How should risk quantification problems be resolved if there exist differences of opinion between the customer and the contractors?
16. If a critical risk is discovered, what is the proper way for the project manager to present to senior management the impact of the risk? How do you as a project manager make sure that senior management understand the ramifications?

17. How were the identified risks quantified at NASA? Is the quantification system truly quantitative or is it a qualitative system?
18. Were probabilities assigned to any of the risks? Why or why not?

Risk Response (Risk Handling)

19. How does an organization decide what is or is not an acceptable risk?
20. Who should have final say in deciding upon the appropriate response mechanism for a risk?
21. What methods of risk response were used at NASA?
22. Did it appear that the risk response method selected was dependent on the risk or on other factors?
23. How should an organization decide whether or not to accept a risk and launch if the risks cannot be quantified?
24. What should be the determining factors in deciding which risks are brought upstairs to the executive levels for review before selecting the appropriate risk response mechanism?
25. Why weren't the astronauts involved in the launch decision (i.e., the acceptance of the risk)? Should they have been involved?
26. What risk response mechanism did NASA administrators use when they issued waivers for the Launch Commit Criteria?
27. Are waivers a type of risk response mechanism?
28. Did the need to maintain a flight schedule compromise the risk response mechanism that would otherwise have been taken?
29. What risk response mechanism were managers at Thiokol and NASA using when they ignored the recommendations of their engineers?
30. Did the engineers at Thiokol and NASA do all they could to convince their own management that the wrong risk response mechanism was about to be taken?
31. When NASA pressed its contractors to recommend a launch, did NASA's risk response mechanism violate their responsibility to ensure crew safety?
32. When NASA discounted the effects of the weather, did NASA's risk response mechanism violate their responsibility to ensure crew safety?

Risk Control

33. How much documentation should be necessary for the tracking of a risk management plan? Can this documentation become overexcessive and create decision-making problems?
34. Risk management includes the documentation of lessons-learned. In the case study, was there an audit trail of lessons learned or was that audit trail simply protection memos?

35. How might Thiokol engineers have convinced both their own management and NASA to postpone the launch?

36. Should someone have stopped the *Challenger* launch and, if so, how could this have been accomplished without risking one's job and career?

37. How might an engineer deal with pressure from above to follow a course of action that the engineer knows to be wrong?

38. How could the chains of communication and responsibility for the Space Shuttle Program have been made to function better?

39. Because of the ice problem, Rockwell could not guarantee the shuttle's safety, but did nothing to veto the launch. Is there a better way for situations as this to be handled in the future?

40. What level of risk should have been acceptable for launch?

41. How should we handle situations where people in authority believe that the potential rewards justify what they believe to be relatively minor risks?

42. If you were on a jury attempting to place liability, whom would you say was responsible for the *Challenger* disaster?

The Space Shuttle
Columbia Disaster[1]

Few projects start with an explosion. Even fewer start with a deliberate explosion. Yet every time the space shuttle is launched into space, five tremendous explosions in the rocket engines are needed to hurl the orbiter into orbit around the earth. In just over ten minutes, the orbiter vehicle goes from zero miles an hour to more than 17,500 miles per hour as it circles the Earth.

Shuttle launches are a very dangerous business. The loss of the second shuttle on February 1, 2003, shocked everyone. It is apparent now that some fuel-tank insulation dislodged during liftoff and struck the orbiter during its powered ascent to earth orbit, and that the insulation punched a fatal hole in the leading edge of the left wing. This hole allowed superheated gases, about 10,000°F, to melt the left wing during the re-entry phase of the mission. The loss of the orbiter was the result of the loss of the left wing.

Reading through the results of the disaster, one cannot help but conclude how simple and straightforward the project risks can be that are handled by most project managers. As an example, we can consider the writing of software. Writing and delivering computer software has its challenges, but the risks are not on the same scale of a space shuttle launch. Even the standard risk response

---

strategies (avoidance, transference, mitigation and acceptance) take on new meanings when accelerating to achieve speeds of more than 15,000 mph. For example:

- Avoidance is not possible.
- Acceptance has to be active, not passive.
- Transference is not possible.
- Mitigation entails a lot of work, and under massive constraints.

For the space shuttle, risk analysis is nonlinear, but for most software projects, a simple, linear impact analysis may be sufficient. The equation for linear impact analysis can be written as follows:[2]

$$\text{Risk impact} = (\text{Risk probability}) \times (\text{Risk consequence})$$

For a given risk event, there is a probability of the risk occurring and a consequence expressed in some numerical units of the damage done to the project cost, timeline, or quality. This is a simple linear equation. If one of the factors on the right side of the equation doubles, the risk impact doubles. For a given set of factors on the right, there is one answer, regardless of when the risk occurs. So, based on the equation, impact can be understood and planned for.

Most of the computer software projects have relatively simple functions that either happened or did not happen. The vendor either delivered on time or did not deliver on time. If a particular *risk* event trigger appeared, then there usually existed a time period, usually in days, when the *risk response* could be initiated. There might be dozens of risks, but each one could be defined and explained with only two or three variables.

This linear approach to risk management had several advantages for computer software projects:

- The risks were understandable and could be explained quite easily.
- Management could understand the process from which a probability and a consequence were obtained.
- There was usually one risk impact for a given risk event.
- No one was aware that one risk event may require dozens of strategies to anticipate all the possible consequences.

One valid argument is that the risk of external collisions with the space vehicle as it accelerates to make orbital speed results in a multivariant, multidimensional,

---

[2]Kerzner, H., *Project Management: A Systems Approach to Planning, Scheduling and Controlling*, 8th ed. (New York: John Wiley & Sons, 2003), p. 653.

nonlinear risk function that is very difficult to comprehend, much less manage. This is orders of magnitude more complex than the project risks encountered when managing computer software development projects.

RISK DEFINITIONS AND SOME TERMS

For this case study, risks and related terms will be defined according to the Project Management Institute's PMBOK® Guide (*Project Management Body of Knowledge*).

- *Risk:* An uncertain event or condition that, if it occurs, has a positive or negative effect on a project's objectives.

For this discussion, the focus will be on negative risks. This family of negative risks can have detrimental consequences to the successful completion of the project. These risks may not happen, but if they do, we know the consequences will make it difficult to complete the project successfully. The consequences may range from a minor change in the timeline to total project failure. The key here is that for each risk, two variables are needed: probability of occurrence and a measurement.

- *Risk triggers:* These are indicators that a risk event has happened or is about to happen.
- *Risk consequence(s):* What could happen if the risk is triggered? Are we going to lose a few dollars, lose our job, or lose an entire business?

To analyze these standard terms, additional terms can be included. These terms are needed to adequately support managing risks that are multivariant, multidimensional, and nonlinear risk functions:

- *Risk scope:* What parts of the project are affected if the risk is triggered? Does this risk jeopardize a task, a phase, or the entire project? Is the risk confined to one project or an entire portfolio of projects?
- *Risk response rules:* Given that the event occurred, and based on available information, what is the best response? Can we derive rules to make intelligent decisions based on the information acquired when the risk event triggers or even the risk events occur?
- *Risk response levels:* Based on the variables and the response rules, the level of concern may range from *not a problem* to *total destruction*.
- *Risk timeline:* If the risk event or risk trigger occurs, how much time is available to make a decision about the best response to the risk? Are there two days to make a decision, or two seconds?

All we know is that if the risk event is "triggered" or occurs, bad things can and will happen. Our goal is to minimize the consequences. Our plan is that by early identification and rigorous analysis of the risks, we will have time to develop a portfolio of responses to minimize the consequences from a risk event.

BACKGROUND TO THE SPACE SHUTTLE LAUNCH

The three liquid fuel motors consume an amazing quantity of super cooled fuel. The main fuel tank is insulated to ensure that the fuel stays hundreds of degrees below the freezing point of water. It is this insulation that had a history of coming off the fuel tank and hitting the orbiter. It most cases, it caused very minor damage to the orbiter because the foam was usually the size of popcorn. In one or two previous launches, the foam was able to knock a tile off the orbiter. Fortunately, the orbiter was able to return safely. So for most of the launch team, the news that *Columbia* had been struck by foam was of minor concern.

After all, if the risk was not a major problem in one hundred previous launches, then it could not be a problem in this launch. Reviewing, our linear impact equation:

$$\text{Risk impact} = (\text{Risk probability}) \times (\text{Risk consequence})$$

The risk probability was very high, but the consequences were always acceptable. Therefore, the conclusion was that it would always be an acceptable risk. This is what happens when there is only one risk consequence for the life of the risk event. People want to believe that the future is just the same history waiting to happen.

DESCRIPTION OF WHAT HAPPENS AS THE SHUTTLE RE-ENTERS THE ATMOSPHERE

If getting the orbiter into space is one problem, then getting the orbiter back is another problem. Re-entry is a complex set of computer-guided maneuvers to change the speed of the vehicle into heat. And as the heat grows, the speed decreases. Since the metal components of the shuttle melt around 2,000°F, the leading edges of the orbiter are covered in ceramic tiles that melt at about 3,000°F. The tiles keep the 10,000°F re-entry heat from penetrating the vehicle. If all goes well, the computers bring the orbiter to a slow enough speed that a human being can land the vehicle.

In *Columbia*'s launch, the foam knocked several of the tiles off the leading edge of the left wing and created a hole where the tiles had been attached. Upon re-entry, the hot gases entered *Columbia*'s left wing and melted the internal structure. When enough of the wing melted, the wing collapsed and the orbiter blew apart.

THE RISK FUNCTION

What are some of the variables needed to understand the risk of foreign objects colliding with the vehicle from the time the rocket engines start until the rocket engines are jettisoned from the orbiter some ten minutes later?

Since the linear risk-impact equation may not be applicable, what kind of questions should we ask if we are to find a risk impact equation that could work?

Exhibit I examines what you need to measure and/or track if an object strikes the shuttle:

Exhibit I.   *Concerns if an object strikes a space shuttle*

1. What are the attributes of the foreign object?
   - What was it that you collided with?
   - What is the length, width, thickness?
   - What is the mass of the object?
   - What is the density of the object?
   - How hard is the object?
   - How is the mass of the object distributed?
   - Is it like a cannon ball, or dumbbells, or sheet of paper?
2. What are the attributes of the collision?
   - Where did it hit?
   - Were there multiple impact points?
   - How much damage was done?
   - Can the damage be verified and examined?
   - Is this an isolated event, or the first of many?
   - What was the angle of the collision? Was it a glancing blow or a direct contact?
   - Did the object hit and leave the area, or is it imbedded in the vehicle?
   - Why did the shuttle collide with it? Are you off course? Is something coming apart?
3. What are the attributes of the vehicle?
   - How fast was it going at the time of the collision?
   - Was it in the middle of a complex maneuver?
   - Did the collision damage a component needed in the current phase of the mission?
   - Did the collision damage a component needed later in the mission?

This is certainly not an exhaustive list, but it is already orders of magnitude more complex compared to most project managers' experiences in risk management. Unfortunately, the problem is even more complex.

The acceleration of the vehicles adds another dimension to the risk function. A collision with an object at 100 mph is not the same as that which occurs when the vehicle is going 200 mph. The damage will not be twice as much as with a linear equation (i.e., if you are going twice as fast, then there will be twice the damage). These risk functions have now become nonlinear. The damage caused when the speed doubles may be sixteen times more, not just twice as much. This has a significant impact on how often you track and record the ongoing events.

Time is also a critical issue. Time is not on your side in a project that moves this fast. It is not just the fact that the risks are nonlinear, but the response envelope is constantly changing. In a vehicle going from 0 to 15,000 mph, a lot can happen in a very short time.

Now let's look at what happens to the simple risk–impact equation:

$$\text{Risk impact} = (\text{Risk probability}) \times (\text{Risk consequence})$$

One probability for a risk event may be sufficient, but the risk consequences are now a function of many variables that have to be measured before an impact can be computed. Also, the risk consequence may be a non-linear function. This is a much more complex problem than trying to identify one probability and one consequence per risk event.

CONCLUSIONS

It may be necessary to compress the risk consequence function into some relatively simple equations and then combine the simple equations into a much more complex mathematical statement. For example, consider the variables of dimensions, weight, and speed. What type of rules can we define to make the risk impact easily derived and of value in making responses to the risk? We might apply the following parameters:

Rule 1: If the sum of the three dimensions (length + width + height) is less than 30, then the risk level is "10."
If the sum of the three dimensions (length + width + height) is more than 30, the risk level is "20."
Rule 2: If the weight is more than 500 grams, then the Risk-Level is multiplied by 1.5.
Rule 3: For every 5 seconds of flight, the risk level doubles.

This process can be continued for all relevant variables.

Risk response level (RRL) is the sum of the individual risk levels computed. If the RRL is less than 50, the event is taken as noncritical. If the RRL is less than 100, procedures A, B, and C should be initiated, and so on.

This exercise provides us with "rules" to initiate action. There is no discussion or guessing as to the proper response to a hazardous event. There is no necessity to contact management for approval to start further actions. There are no stare downs with management to minimize the event for political or other considerations.

The more complicated things get, the more important rules and preplanned responses become to successfully managing project risk.

LESSONS LEARNED

In reviewing articles on the space shuttle events before and after its destruction, several things were learned:

- Debris had hit the shuttle during its powered ascent in previous launches. Management believed that because there were few problems in the past, the risk impact was known and would not change in the future.

  The lesson learned is not to make the same mistake.
- Risks can be very complex.

  The lesson learned is to study more about risk and how to document the impact so even managers unfamiliar with risk management concepts can grasp complex impact functions.
- The shuttle crew never knew the spacecraft was doomed. By the time they were aware of the danger, the shuttle disintegrated.

  The lesson learned is that life is like that, and probably more often than you realize.

REFERENCES

1. Peter Sprent, *Taking Risks—The Science of Uncertainty* (Penguin Books, 1988).
2. Daniel Kehrer, *The Art of Taking Intelligent Risks* (Times Books, 1989).
3. William Langewiesche, "Columbia's Last Flight," *The Atlantic Monthly* (November, 2003).

Packer Telecom

BACKGROUND

The rapid growth of the telecom industry made it apparent to Packer's executives that risk management must be performed on all development projects. If Packer were late in the introduction of a new product, then market share would be lost. Furthermore, Packer could lose valuable opportunities to "partner" with other companies if Packer were regarded as being behind the learning curve with regard to new product development.

Another problem facing Packer was the amount of money being committed to R&D. Typical companies spend 8 to 10 percent of earnings on R&D, whereas in the telecom industry, the number may be as high as 15 to 18 percent. Packer was spending 20 percent on R&D, and only a small percentage of the projects that started out in the conceptual phase ever reached the commercialization phase, where Packer could expect to recover its R&D costs. Management attributed the problem to a lack of effective risk management.

THE MEETING

PM: "I have spent a great deal of time trying to benchmark best practices in risk management. I was amazed to find that most companies are in the same boat as us, with very little knowledge in risk management. From the limited results I have found from other companies, I have been able to develop a risk management template for us to use."

Sponsor: "I've read over your report and looked at your templates. You have words and expressions in the templates that we don't use here at Packer. This concerns me greatly. Do we have to change the way we manage projects to use these templates? Are we expected to make major changes to our existing project management methodology?"

PM: "I was hoping we could use these templates in their existing format. If the other companies are using these templates, then we should also. These templates also have the same probability distributions that other companies are using. I consider these facts equivalent to a validation of the templates."

Sponsor: "Shouldn't the templates be tailored to our methodology for managing projects and our life cycle phases? These templates may have undergone validation, but not at Packer. The probability distributions are also based upon someone else's history, not our history. I cannot see anything in your report that talks about the justification of the probabilities.

"The final problem I have is that the templates are based upon history. It is my understanding that risk management should be forward looking, with an attempt at predicting the possible future outcomes. I cannot see any of this in your templates."

PM: "I understand your concerns, but I don't believe they are a problem. I would prefer to use the next project as a 'breakthrough project' using these templates. This will give us a good basis to validate the templates."

Sponsor: "I will need to think about your request. I am not sure that we can use these templates without some type of risk management training for our employees."

QUESTIONS

1. Can templates be transferred from one company to another, or should tailoring be mandatory?
2. Can probability distributions be transferred from one company to another? If not, then how do we develop a probability distribution?
3. How do you validate a risk management template?
4. Should a risk management template be forward looking?
5. Can employees begin using a risk management template without some form of specialized training?

Luxor Technologies

Between 1992 and 1996, Luxor Technologies had seen their business almost quadruple in the wireless communications area. Luxor's success was attributed largely to the strength of its technical community, which was regarded as second to none. The technical community was paid very well and given the freedom to innovate. Even though Luxor's revenue came from manufacturing, Luxor was regarded by Wall Street as being a technology-driven company.

The majority of Luxor's products were based upon low cost, high quality applications of the state-of-the-art technology, rather than advanced state-of-the-art technological breakthroughs. Applications engineering and process improvement were major strengths at Luxor. Luxor possessed patents in technology breakthrough, applications engineering, and even process improvement. Luxor refused to license their technology to other firms, even if the applicant was not a major competitor.

Patent protection and design secrecy were of paramount importance to Luxor. In this regard, Luxor became vertically integrated, manufacturing and assembling all components of their products internally. Only off-the-shelf components were purchased. Luxor believed that if they were to use outside vendors for sensitive component procurement, they would have to release critical and proprietary data to the vendors. Since these vendors most likely also serviced Luxor's competitors, Luxor maintained the approach of vertical integration to maintain secrecy.

Being the market leader technically afforded Luxor certain luxuries. Luxor saw no need for expertise in technical risk management. In cases where the technical

Exhibit I. Likelihood of a technical risk

Event	Likelihood Rating
• State-of-the-art advance needed	0.95
• Scientific research required (without advancements)	0.80
• Concept formulation	0.40
• Prototype development	0.20
• Prototype testing	0.15
• Critical performance demonstrated	0.10

community was only able to achieve 75–80 percent of the desired specification limit, the product was released as it stood, accompanied by an announcement that there would be an upgrade the following year to achieve the remaining 20–25 percent of the specification limit, together with other features. Enhancements and upgrades were made on a yearly basis.

By the fall of 1996, however, Luxor's fortunes were diminishing. The competition was catching up quickly, thanks to major technological breakthroughs. Marketing estimated that by 1998, Luxor would be a "follower" rather than a market leader. Luxor realized that something must be done, and quickly.

In January 1999, Luxor hired an expert in risk analysis and risk management to help Luxor assess the potential damage to the firm and to assist in development of a mitigation plan. The consultant reviewed project histories and lessons learned on all projects undertaken from 1992 through 1998. The consultant concluded that the major risk to Luxor would be the technical risk and prepared Exhibits I and II.

Exhibit I shows the likelihood of a technical risk event occurring. The consultant identified the six most common technical risk events that could occur at

Exhibit II. Impact of a technical risk event

Event	Impact Rating	
	With State-of-the-Art Changes	Without State-of-the-Art Changes
• Product performance not at 100 percent of specification	0.95	0.80
• Product performance not at 75–80 percent of specification	0.75	0.30
• Abandonment of project	0.70	0.10
• Need for further enhancements	0.60	0.25
• Reduced profit margins	0.45	0.10
• Potential systems performance degradation	0.20	0.05

Luxor over the next several years, based upon the extrapolation of past and present data into the future. Exhibit II shows the impact that a technical risk event could have on each project. Because of the high probability of state-of-the-art advancements needed in the future (i.e., 95 percent from Exhibit I), the consultant identified the impact probabilities in Exhibit II for both with and without state-of-the-art advancement needed.

Exhibits I and II confirmed management's fear that Luxor was in trouble. A strategic decision had to be made concerning the technical risks identified in Exhibit I, specifically the first two risks. The competition had caught up to Luxor in applications engineering and was now surpassing Luxor in patents involving state-of-the-art advancements. From 1992 to 1998, time was considered as a luxury for the technical community at Luxor. Now time was a serious constraint.

The strategic decision facing management was whether Luxor should struggle to remain a technical leader in wireless communications technology or simply console itself with a future as a "follower." Marketing was given the task of determining the potential impact of a change in strategy from a market leader to a market follower. The following list was prepared and presented to management by marketing:

1. The company's future growth rate will be limited.
2. Luxor will still remain strong in applications engineering but will need to outsource state-of-the-art development work.
3. Luxor will be required to provide outside vendors with proprietary information.
4. Luxor may no longer be vertically integrated (i.e., have backward integration).
5. Final product costs may be heavily influenced by the costs of subcontractors.
6. Luxor may not be able to remain a low cost supplier.
7. Layoffs will be inevitable, but perhaps not in the near term.
8. The marketing and selling of products may need to change. Can Luxor still market products as a low-cost, high quality, state-of-the-art manufacturer?
9. Price-cutting by Luxor's competitors could have a serious impact on Luxor's future ability to survive.

The list presented by marketing demonstrated that there was a serious threat to Luxor's growth and even survival. Engineering then prepared a list of alternative courses of action that would enable Luxor to maintain its technical leadership position:

1. Luxor could hire (away from the competition) more staff personnel with pure and applied R&D skills. This would be a costly effort.
2. Luxor could slowly retrain part of its existing labor force using existing, experienced R&D personnel to conduct the training.
3. Luxor could fund seminars and university courses on general R&D methods, as well as R&D methods for telecommunications projects. These programs were available locally.

4. Luxor could use tuition reimbursement funds to pay for distance learning courses (conducted over the Internet). These were full semester programs.
5. Luxor could outsource technical development.
6. Luxor could purchase or license technology from other firms, including competitors. This assumed that competitors would agree to this at a reasonable price.
7. Luxor could develop joint ventures/mergers with other companies which, in turn, would probably require Luxor to disclose much of its proprietary knowledge.

With marketing's and engineering's lists before them, Luxor's management had to decide which path would be best for the long term.

QUESTIONS

1. Can the impact of one specific risk event, such as a technical risk event, create additional risks, which may or may not be technical risks? Can risk events be interrelated?
2. Does the list provided by marketing demonstrate the likelihood of a risk event or the impact of a risk event?
3. How does one assign probabilities to the marketing list?
4. The seven items in the list provided by engineering are all ways of mitigating certain risk events. If the company follows these suggestions, is it adopting a risk response mode of avoidance, assumption, reduction, or deflection?
5. Would you side with marketing or engineering? What should Luxor do at this point?

Altex Corporation

BACKGROUND

Following World War II, the United States entered into a Cold War with Russia. To win this Cold War, the United States had to develop sophisticated weapon systems with such destructive power that any aggressor knew that the retaliatory capability of the United States could and would inflict vast destruction.

Hundreds of millions of dollars were committed to ideas concerning technology that had not been developed as yet. Aerospace and defense contractors were growing without bounds, thanks to cost-plus-percentage-of-cost contract awards. Speed and technological capability were judged to be significantly more important than cost. To make matters worse, contracts were often awarded to the second or third most qualified bidder for the sole purpose of maintaining competition and maximizing the total number of defense contractors.

CONTRACT AWARD

During this period Altex Corporation was elated when it learned that it had just been awarded the R&D phase of the Advanced Tactical Missile Program (ATMP). The terms of the contract specified that Altex had to submit to the Army, within 60 days after contract award, a formal project plan for the two-year ATMP effort. Contracts at that time did not require that a risk management plan be developed. A meeting was held with the project manager of R&D to assess the risks in the ATMP effort.

PM: "I'm in the process of developing the project plan. Should I also develop a risk management plan as part of the project plan?"

Sponsor: "Absolutely not! Most new weapon systems requirements are established by military personnel who have no sense of reality about what it takes to develop a weapon system based upon technology that doesn't even exist yet. We'll be lucky if we can deliver 60–70 percent of the specification imposed upon us."

PM: "But that's not what we stated in our proposal. I wasn't brought on board until after we won the award, so I wasn't privileged to know the thought process that went into the proposal. The proposal even went so far as to imply that we might be able to exceed the specification limits, and now you're saying that we should be happy with 60–70 percent."

Sponsor: "We say what we have to say to win the bid. Everyone does it. It is common practice. Whoever wins the R&D portion of the contract will also be first in line for the manufacturing effort and that's where the megabucks come from! If we can achieve 60–70 percent of specifications, it should placate the Army enough to give us a follow-on contract. If we told the Army the true cost of developing the technology to meet the specification limits, we would never get the contract. The program might even be canceled. The military people want this weapon system. They're not stupid! They know what is happening and they do not want to go to their superiors for more money until later on, downstream, after approval by DoD and project kickoff. The government wants the lowest cost and we want long-term, follow-on production contracts, which can generate huge profits."

PM: "Aren't we simply telling lies in our proposal?"

Sponsor: "My engineers and scientists are highly optimistic and believe they can do the impossible. This is how technological breakthroughs are made. I prefer to call it 'over-optimism of technical capability' rather than 'telling lies.' If my engineers and scientists have to develop a risk management plan, they may become pessimistic, and that's not good for us!"

PM: "The problem with letting your engineers and scientists be optimistic is that they become reactive rather than proactive thinkers. Without proactive thinkers, we end up with virtually no risk management or contingency plans. When problems surface that require significantly more in the way of resources than we budgeted for, we will be forced to accept crisis management as a way of life. Our costs will increase and that's not going to make the Army happy."

Sponsor: "But the Army won't penalize us for failing to meet cost or for allowing the schedule to slip. If we fail to meet at least 60–70 percent of the specification limits, however, then we may well be in trouble. The Army knows there will be a follow-on contract request if we cannot meet specification limits. I consider 60–70 percent of the specifications to be the minimum acceptable limits for the Army. The Army wants the program kicked off right now.

"Another important point is that long-term contracts and follow-on production contracts allow us to build up a good working relationship with the Army. This is critical. Once we get the initial contract, as we did, the Army will always work with us for follow-on efforts. Whoever gets the R&D effort will almost always get the lucrative production contract. Military officers are under pressure to work with us because their careers may be in jeopardy if they have to tell their superiors that millions of dollars were awarded to the wrong defense contractor. From a career standpoint, the military officers are better off allowing us to downgrade the requirements than admitting that a mistake was made."

PM: "I'm just a little nervous managing a project that is so optimistic that major advances in the state of the art must occur to meet specifications. This is why I want to prepare a risk management plan."

Sponsor: "You don't need a risk management plan when you know you can spend as much as you want and also let the schedule slip. If you prepare a risk management plan, you will end up exposing a multitude of risks, especially technical risks. The Army might not know about many of these risks, so why expose them and open up Pandora's box? Personally, I believe that the Army does already know many of these risks, but does not want them publicized to their superiors.

"If you want to develop a risk management plan, then do it by yourself, and I really mean by yourself. Past experience has shown that our employees will be talking informally to Army personnel at least two to three times a week. I don't want anyone telling the customer that we have a risk management plan. The customer will obviously want to see it, and that's not good for us.

"If you are so incensed that you feel obligated to tell the customer what you're doing, then wait about a year and a half. By that time, the Army will have made a considerable investment in both us and the project, and they'll be locked into us for follow-on work. Because of the strategic timing and additional costs, they will never want to qualify a second supplier so late in the game. Just keep the risk management plan to yourself for now.

"If it looks like the Army might cancel the program, then we'll show them the risk management plan, and perhaps that will keep the program alive."

QUESTIONS

1. Why was a risk management plan considered unnecessary?
2. Should risk management planning be performed in the proposal stage or after contract award, assuming that it must be done?
3. Does the customer have the right to expect the contractor to perform risk analysis and develop a risk management plan if it is not called out as part of the contractual statement of work?
4. Would Altex have been more interested in developing a risk management plan if the project were funded entirely from within?
5. How effective will the risk management plan be if developed by the project manager in seclusion?
6. Should the customer be allowed to participate in or assist the contractor in developing a risk management plan?
7. How might the Army have responded if it were presented with a risk management plan early during the R&D activities?
8. How effective is a risk management plan if cost overruns and schedule slippages are always allowed?
9. How can severe optimism or severe pessimism influence the development of a risk management plan?
10. How does one develop a risk management plan predicated upon needed advances in the state of the art?
11. Can the sudden disclosure of a risk management plan be used as a stopgap measure to prevent termination of a potentially failing project?
12. Can risk management planning be justified on almost all programs and projects?

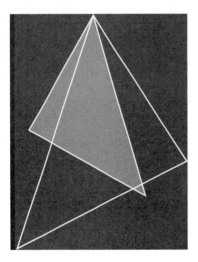

Acme Corporation

BACKGROUND

Acme Corporation embarked upon an optimistic project to develop a new product for the marketplace. Acme's scientific community made a technical breakthrough and now the project appears to be in the development stage, more than being pure or applied research.

The product is considered to be high tech. If the product can be launched within the next four months, Acme expects to dominate the market for at least a year or so until the competition catches up. Marketing has stated that the product must sell for not more than $150 to $160 per unit to be the cost-focused market leader.

Acme uses a project management methodology for all multifunctional projects. The methodology has six life cycle phases:

1. Preliminary planning
2. Detailed planning
3. Execution/design selection
4. Prototyping
5. Testing/buyoff
6. Production

At the end of each life cycle phase a gate/phase review meeting is held with the project sponsor and other appropriate stakeholders. Gate review meetings are formal meetings. The company has demonstrated success following this methodology for managing projects.

At the end of the second life cycle stage of this project, detailed planning, a meeting is held with just the project manager and the project sponsor. The purpose of the meeting is to review the detailed plan and identify any future problem areas that will require involvement by the project sponsor.

THE MEETING

Sponsor: "I simply do not understand this document you sent me entitled 'Risk Management Plan.' All I see is a work breakdown structure with work packages at level 5 of the WBS accompanied by almost 100 risk events. Why am I looking at more than 100 risk events? Furthermore, they're not categorized in any manner. Doesn't our project management methodology provide any guidance on how to do this?"

PM: "All of these risk events can and will impact the design of the final product. We must be sure we select the right design at the lowest risk. Unfortunately, our project management methodology does not include any provisions or guidance on how to develop a risk management plan. Perhaps it should."

Sponsor: "I see no reason for an in-depth analysis of 100 or so risk events. That's too many. Where are the probabilities and expected outcomes or damages?"

PM: "My team will not be assigning probabilities or damages until we get closer to prototype development. Some of these risk events may go away altogether."

Sponsor: "Why spend all of this time and money on risk identification if the risks can go away next month? You've spent too much money doing this. If you spend the same amount of money on all of the risk management steps, then we'll be way over budget."

PM: "We haven't looked at the other risk management steps yet, but I believe all of the remaining steps will require less than 10 percent of the budget we used for risk identification. We'll stay on budget."

QUESTIONS

1. Was the document given to the sponsor a risk management plan?
2. Did the project manager actually perform effective risk management?

3. Was the appropriate amount of time and money spent identifying the risk events?

4. Should one step be allowed to "dominate" the entire risk management process?

5. Are there any significant benefits to the amount of work already done for risk identification?

6. Should the 100 or so risk events identified have been categorized? If so, how?

7. Can probabilities of occurrence and expected outcomes (i.e., damage) be accurately assigned to 100 risk events?

8. Should a project management methodology provide guidance for the development of a risk management plan?

9. Given the life cycle phases in the case study, in which phase would it be appropriate to identify the risk management plan?

10. What are your feelings on the project manager's comments that he must wait until the prototyping phase to assign probabilities and outcomes?

Part 12

CONFLICT MANAGEMENT

Conflicts can occur anywhere in the project and with anyone. Some conflicts are severe, while others are easily solvable. In the past, we avoided conflicts when possible. Today, we believe that conflicts can produce beneficial results if the conflicts are managed correctly.

There are numerous methods available to project managers for the resolution of conflicts. The methods selected may vary depending on the severity of the conflict, the person with whom the conflict exists and his/her level of authority, the life-cycle phase of the project, the priority of the project, and the relative importance of the project as seen by senior management.

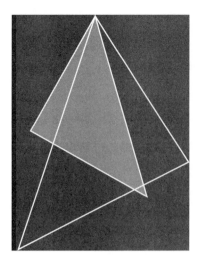

Facilities Scheduling at Mayer Manufacturing

Eddie Turner was elated with the good news that he was being promoted to section supervisor in charge of scheduling all activities in the new engineering research laboratory. The new laboratory was a necessity for Mayer Manufacturing. The engineering, manufacturing, and quality control directorates were all in desperate need of a new testing facility. Upper-level management felt that this new facility would alleviate many of the problems that previously existed.

The new organizational structure (as shown in Exhibit I) required a change in policy over use of the laboratory. The new section supervisor, on approval from his department manager, would have full authority for establishing priorities for the use of the new facility. The new policy change was a necessity because upper-level management felt that there would be inevitable conflict between manufacturing, engineering, and quality control.

After one month of operations, Eddie Turner was finding his job impossible, so Eddie has a meeting with Gary Whitehead, his department manager.

Eddie: "I'm having a hell of a time trying to satisfy all of the department managers. If I give engineering prime-time use of the facility, then quality control and manufacturing say that I'm playing favorites. Imagine that! Even my own people say that I'm playing favorites with other directorates. I just can't satisfy everyone."

Exhibit I.   *Mayer Manufacturing organizational structure*

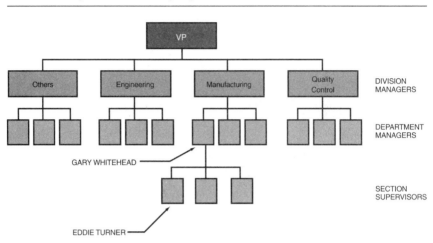

Gary: "Well, Eddie, you know that this problem comes with the job. You'll get the job done."

Eddie: "The problem is that I'm a section supervisor and have to work with department managers. These department managers look down on me like I'm their servant. If I were a department manager, then they'd show me some respect. What I'm really trying to say is that I would like you to send out the weekly memos to these department managers telling them of the new priorities. They wouldn't argue with you like they do with me. I can supply you with all the necessary information. All you'll have to do is to sign your name."

Gary: "Determining the priorities and scheduling the facilities is your job, not mine. This is a new position and I want you to handle it. I know you can because I selected you. I do not intend to interfere."

During the next two weeks, the conflicts got progressively worse. Eddie felt that he was unable to cope with the situation by himself. The department managers did not respect the authority delegated to him by his superiors. For the next two weeks, Eddie sent memos to Gary in the early part of the week asking whether Gary agreed with the priority list. There was no response to the two memos. Eddie then met with Gary to discuss the deteriorating situation.

Eddie: "Gary, I've sent you two memos to see if I'm doing anything wrong in establishing the weekly priorities and schedules. Did you get my memos?"

Gary: "Yes, I received your memos. But as I told you before, I have enough problems to worry about without doing your job for you. If you can't handle the work, let me know and I'll find someone who can."

Eddie returned to his desk and contemplated his situation. Finally, he made a decision. Next week he was going to put a signature block under his for Gary to sign, with carbon copies for all division managers. "Now, let's see what happens," remarked Eddie.

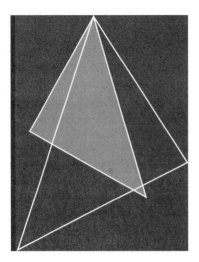

Scheduling the Safety Lab

"Now see here, Tom, I understand your problem well," remarked Dr. Polly, director of the Research Laboratories. "I pay you a good salary to run the safety labs. That salary also includes doing the necessary scheduling to match our priorities. Now, if you can't handle the job, I'll get someone who can."

Tom: "Every Friday morning your secretary hands me a sheet with the listing of priorities for the following week. Once, just once, I'd like to sit in on the director's meeting and tell you people what you do to us in the safety lab when you continually shuffle around the priorities from week to week.

"On Friday afternoons, my people and I meet with representatives from each project to establish the following week's schedules."

Dr. Polly: "Can't you people come to an agreement?"

Tom: "I don't think you appreciate my problem. Two months ago, we all sat down to work out the lab schedule. Project X-13 had signed up to use the lab last week. Now, mind you, they had been scheduled for the past two months. But the Friday before they were to use it, your new priority list forced them to reschedule the lab at a later date, so that we could give the use of the lab to a higher-priority project. We're paying an awful lot of money for idle time and the redoing of network schedules. Only the project managers on the top-priority projects end up smiling after our Friday meetings."

Dr. Polly: "As I see your problem, you can't match long-range planning with the current priority list. I agree that it does create conflicts for you. But you have to remember that we, upstairs, have many other conflicts to resolve. I want that one solved at your level, not mine."

Tom: "Every project we have requires use of the safety lab. This is the basis for our problem. Would you consider letting us modify your priority list with regard to the safety lab?"

Dr. Polly: "Yes, but you had better have the agreement of all of the project managers. I don't want them coming to see me about your scheduling problems."

Tom: "How about if I let people do long-range scheduling for the lab, for three out of four weeks each month? The fourth week will be for the priority projects."

Dr. Polly: "That might work. You had better make sure that each project manager informs you immediately of any schedule slippages so that you can reschedule accordingly. From what I've heard, some of the project managers don't let you know until the last minute."

Tom: "That has been part of the problem. Just to give you an example, Project VX-161 was a top-priority effort and had the lab scheduled for the first week in March. I was never informed that they had accelerated their schedule by two weeks. They walked into my office and demanded use of the lab for the third week in February. Since they had the top priority, I had to grant them their request. However, Project BP-3 was planning on using the lab during that week and was bumped back three weeks. That cost them a pile of bucks in idle time pay and, of course, they're blaming me."

Dr. Polly: "Well Tom, I'm sure you'll find a solution to your problem."

Telestar
International

On November 15, 1998, the Department of Energy Resources awarded Telestar a $475,000 contract for the developing and testing of two waste treatment plants. Telestar had spent the better part of the last two years developing waste treatment technology under its own R&D activities. This new contract would give Telestar the opportunity to "break into a new field"—that of waste treatment.

The contract was negotiated at a firm-fixed price. Any cost overruns would have to be incurred by Telestar. The original bid was priced out at $847,000. Telestar's management, however, wanted to win this one. The decision was made that Telestar would "buy in" at $475,000 so that they could at least get their foot into the new marketplace.

The original estimate of $847,000 was very "rough" because Telestar did not have any good man-hour standards, in the area of waste treatment, on which to base their man-hour projections. Corporate management was willing to spend up to $400,000 of their own funds in order to compensate the bid of $475,000.

By February 15, 1999, costs were increasing to such a point where overrun would be occurring well ahead of schedule. Anticipated costs to completion were now $943,000. The project manager decided to stop all activities in certain functional departments, one of which was structural analysis. The manager of the structural analysis department strongly opposed the closing out of the work order prior to the testing of the first plant's high-pressure pneumatic and electrical systems.

Structures manager: "You're running a risk if you close out this work order. How will you know if the hardware can withstand the stresses that will be imposed during the test? After all, the test is scheduled for next month and I can probably finish the analysis by then."

Project manager: "I understand your concern, but I cannot risk a cost overrun. My boss expects me to do the work within cost. The plant design is similar to one that we have tested before, without any structural problems being detected. On this basis I consider your analysis unnecessary."

Structures manager: "Just because two plants are similar does not mean that they will be identical in performance. There can be major structural deficiencies."

Project manager: "I guess the risk is mine."

Structures manager: "Yes, but I get concerned when a failure can reflect on the integrity of my department. You know, we're performing on schedule and within the time and money budgeted. You're setting a bad example by cutting off our budget without any real justification."

Project manager: "I understand your concern, but we must pull out all the stops when overrun costs are inevitable."

Structures manager: "There's no question in my mind that this analysis should be completed. However, I'm not going to complete it on my overhead budget. I'll reassign my people tomorrow. Incidentally, you had better be careful; my people are not very happy to work for a project that can be canceled immediately. I may have trouble getting volunteers next time."

Project manager: "Well, I'm sure you'll be able to adequately handle any future work. I'll report to my boss that I have issued a work stoppage order to your department."

During the next month's test, the plant exploded. Postanalysis indicated that the failure was due to a structural deficiency.

QUESTIONS

1. Who is at fault?
2. Should the structures manager have been dedicated enough to continue the work on his own?
3. Can a functional manager, who considers his organization as strictly support, still be dedicated to total project success?

The Problem
with Priorities

For the past several years, Kent Corporation had achieved remarkable success in winning R&D contracts. The customers were pleased with the analytical capabilities of the R&D staff at Kent Corporation. Theoretical and experimental results were usually within 95 percent agreement. But many customers still felt that 95 percent was too low. They wanted 98–99 percent.

In 1989, Kent updated their computer facility by purchasing a large computer. The increased performance with the new computer encouraged the R&D group to attempt to convert from two-dimensional to three-dimensional solutions to their theoretical problems. Almost everyone except the director of R&D thought that this would give better comparison between experimental and theoretical data.

Kent Corporation had tried to develop the computer program for three-dimensional solutions with their own internal R&D programs, but the cost was too great. Finally, after a year of writing proposals, Kent Corporation convinced the federal government to sponsor the project. The project was estimated at $750,000, to begin January 2, 1991, and to be completed by December 20, 1991. Dan McCord was selected as project manager. Dan had worked with the EDP department on other projects and knew the people and the man-hour standards.

Kent Corporation was big enough to support 100 simultaneous projects. With so many projects in existence at one time, continual reshuffling of resources was necessary. The corporation directors met every Monday morning to establish project priorities. Priorities were not enforced unless project and functional managers could not agree on the allocation and distribution of resources.

Because of the R&D director's persistence, the computer project was given a low priority. This posed a problem for Dan McCord. The computer department manager refused to staff the project with his best people. As a result, Dan had severe skepticism about the success of the project.

In July, two other project managers held a meeting with Dan to discuss the availability of the new computer model.

"We have two proposals that we're favored to win, providing that we can state in our proposal that we have this new computer model available for use," remarked one of the project managers.

"We have a low priority and, even if we finish the job on time, I'm not sure of the quality of work because of the people we have assigned," said Dan.

"How do you propose we improve our position?" asked a project manager.

"Let's try to get in to see the director of R&D," asserted Dan.

"And what are we going to say in our defense?" asked one of the project managers.

Part 13

MORALITY AND ETHICS

When the survival of the firm is at stake, workers often make decisions that may violate moral and ethical principles. Some may view an action as a violation whereas others may view it as an acceptable practice. Every day, people are placed in situations that may require a moral or ethical decision.

Some companies have found a solution to this problem by creating a standard practice manual or corporate credo that provides guidelines for how these decisions should be made. The guidelines identify the order in which certain stakeholders' interests should be satisfied.

The Tylenol
Tragedies

ABSTRACT

Both the academic community that teaches project management and practitioners of project management appear to be in agreement that the most critical phase of any project is planning. But what if a major crisis happens, especially one that could have an extremely serious consequence upon the financial health, image or reputation of the company? Based upon the seriousness of the crisis, there may not be sufficient time to prepare a statement of work, work breakdown structure, detailed schedules, a budget, or even any semblance of a project plan. Yet action must be taken as quickly as possible.

During the last week of September and the first week of October in 1982, seven people died ingesting Extra-Strength Tylenol capsules laced with cyanide. Four years later in 1986, the same situation of product tampering of Tylenol occurred again, this time with the death of only one person. During both crises, Johnson & Johnson set the standard on how crises should be managed. Academia has been teaching the Tylenol case study for over nineteen years as an example of morality and ethics in business and what constitutes effective corporate responsibility.

This case study focuses more on the project management decisions than on the business decisions. The case also identifies the lessons learned in project management and crisis project management.

UNDERSTANDING CRISIS MANAGEMENT

For some time, corporations in specific industries have found it necessary to simulate and analyze worst-case scenarios for their products and services. These worst-case scenarios have been referred to as contingency plans, emergency plans, or disaster plans. These scenarios are designed around "known unknowns" where at least partial information exists on what events could happen.

Crisis management focuses on the "unknown unknowns," which are tragedies without precedent. Crisis management requires a heads-up approach with a very quick reaction time combined with a concerted effort on the part of possibly all employees. In crisis management, decisions have to be made often without even partial information and perhaps before the full extent of the damage is known.

In a crisis, events happen so quickly and so unpredictably that it may be impossible to perform any kind of planning. Statements of work, work breakdown structures, and detailed scheduling are nonexistent. Roles and responsibilities of key individuals may change on a daily basis. There may be very active involvement by a majority of the stakeholders. Company survival could rest entirely on how well a company manages the crisis.

THE HISTORY OF TYLENOL

In 1982, Johnson & Johnson was a health care giant with annual sales of over $5.4 billion. Johnson & Johnson owned 150 companies, one of which was McNeil Consumer Products, the maker of Tylenol.

Beginning in 1960, McNeil had carefully promoted Tylenol to physicians, hospitals, and pharmacies as an alternative pain reliever for people who suffered from the side effects of aspirin. By 1976, McNeil began aggressively advertising Tylenol to the general public, building on its reputation as a "professional product."

With a massive advertising budget, Tylenol's market share dominated other nonprescription painkillers such as Anacin, Bayer aspirin, Bufferin, and Excedrin. By 1982, Tylenol commanded an astounding 37 percent share of the $1 billion plus analgesic market. So much advertising and marketing money was poured into Tylenol that none of the other makers of acetaminophen pain relievers could threaten its dominance. Surprisingly enough, acetaminophen is the only active ingredient in Tylenol, and any drug company could produce the product.

Tylenol accounted for 7 percent of Johnson & Johnson's worldwide sales and 15 to 20 percent of its 1981 profits. Even though the U.S. economy was in the midst of a recession in 1981, Johnson & Johnson's earnings were up 16.7 percent, and 1982 looked even better. McNeil executives were predicting that Tylenol could achieve a 50 percent market share within the next few years. Within the previous years (during the recession), the stock had climbed from the low 20s to 46 the night before the poisonings. McNeil and Johnson & Johnson were certainly in an enviable position.

THE TYLENOL POISONINGS

In September 1982, seven people died after taking Extra-Strength Tylenol laced with cyanide. All of the victims were relatively young. These deaths were the first ever to result from what came to be known as product tampering. All seven individuals died within a one-week time period. The symptoms of cyanide poisoning are rapid collapse and coma and are difficult to treat.

On the morning of September 30, 1982, reporters began calling the headquarters of Johnson & Johnson asking about information on Tylenol and Johnson & Johnson's reaction to the deaths. This was the first that Johnson & Johnson had heard about the deaths and the possible link to Tylenol.

The news quickly spread to the fifth floor of Johnson & Johnson's headquarters building in New Brunswick, New Jersey. The chairman of Johnson & Johnson was James Burke, 57, a thirty-year veteran of Johnson & Johnson. The news came as a shock to Chairman Burke. Despite the company's size, the news could have a huge, damaging impact on earnings.

Chairman Burke assigned David Collins, 48, to take charge of coordinating the company's response to the Tylenol crisis. Collins was a former general counsel and company group chairman. A month earlier he had been named to Johnson & Johnson's twelve-man executive committee and given the additional job of chairman of McNeil Consumer Products.

In addition to the personal qualifications of David Collins, there were several reasons why Burke asked him to take charge. First, Burke was a strong proponent of decentralized decision-making. Second, Tylenol was a McNeil product and Burke was hoping to insulate the parent company, Johnson & Johnson, from bad publicity. Third, Collins was the chairman of McNeil and, therefore, had the authority to commit McNeil resources to the crisis.

> ***Lessons Learned:*** The project manager assigned to manage the crisis must be high enough in the organization to possess the authority for the immediate commitment of corporate resources. Approval processes that must follow the chain of command can rob the project manager of valuable time and prolong the crisis.

Collins was asked by Burke to take a lawyer, a public relations aide, and a security person and fly immediately to McNeil, 60 miles away in Fort Washington, Pennsylvania, to handle things from there. It was important to Burke to isolate the parent company as much as possible from the potentially bad news that could possibly affect other Johnson & Johnson products. Burke was hoping that the news media would view Tylenol as a McNeil product, rather than a Johnson & Johnson product. Having the crisis managed from McNeil rather than Johnson & Johnson corporate certainly seemed the right thing to do at the time.

With very little information available at that time, and very little time to act, the crisis project was managed using three phases. The first phase was discovery, which included the gathering of any and all information from every possible source. The full complexity of the problem had to be known, as well as the associated risks. The second phase was the assessment and quantification of the risks and the containment of potential damage. The third phase was the establishment of a recovery plan and risk mitigation. Unlike traditional "life-cycle" phases, which could be months or years in duration, these phases would be in hours or days.

> *Lessons Learned:* Because of the potential lack of information available at the beginning of the crisis, an abbreviated life-cycle phase approach is often more appropriate to use. This provides at least some initial guidance for crisis management. It is highly unlikely that during crisis management, sufficient time will exist for formal planning, scheduling, and WBS construction.

By the time Collins arrived at McNeil, the switchboards were lighting up at both McNeil and Johnson & Johnson. At first the calls came from the newspapers, TV, and radio stations, some as far away as Honolulu and Ireland. But as the story started to break, even more calls began to pour in from pharmacies, doctors, hospitals, poison control centers, and hundreds of panicked consumers, many asking for clarifications (which Johnson & Johnson couldn't give), and many others making what turned out to be false reports of possible poisonings.

"It looked like the plague," remarked Collins. "We had no idea where it would end. And the only information we had was that we didn't know what was going on." Collins's first move was to call an old roommate of his from Notre Dame, a lawyer who handled some of Johnson & Johnson's business in Chicago, and ask him to get down to the Cook County Medical Examiner's Office, find out as much as he could, and call him back at McNeil. "I needed my own eyes and ears on the scene," he said.

However serene the impeccably landscaped McNeil plant grounds appeared to Collins as his helicopter touched down on the pad, inside, the natural order of things was in turmoil. Harried managers were running back and forth between telephone banks and the office of McNeil President Joseph Chiesa, bringing in new reports of fatalities and other supposed poisonings. Each bit of information was scribbled with a laundry marker on drawing paper held by a big easel. As the reports accumulated, the sheets were ripped from the easel and pinned up on the walls. Soon the room was papered with a confusing mass of information with arrows drawn between them: victims, causes of deaths, lot numbers on the poisoned Tylenol bottles, the outlets where they had been purchased, dates when they had been manufactured, and the route they had taken through the distribution system— all the way back to the fourteen stainless steel machines in Fort Washington that encapsulated and spewed out pills at the rate of over 1,000 a minute.

From the start, the company found itself entering a closer relationship with the press than it was accustomed to. Johnson & Johnson bitterly recalled an incident nine years earlier in which the media had circulated a misleading report suggesting that some baby powder had been contaminated by asbestos. But in the Tylenol case, Johnson & Johnson opened its doors. For one thing, the company was getting some of its most accurate and up-to-date information about what was going on around the country from the reporters calling in for comment. For another, Johnson & Johnson needed the media to get out as much information to the public as quickly as possible and prevent a panic.

The dangers of trying to manage the news were firmly in mind when the company had to reverse itself on whether any cyanide was used on the premises. It was Collins's first question to McNeil executives when he got off the helicopter. He was told no, but later in the day he learned to his dismay that cyanide was in fact used in the quality assurance facility next to the manufacturing plant to test the purity of raw materials. The public relations department released this startling bit of information to the press the next morning. While the reversal embarrassed the company briefly, Johnson & Johnson's openness made up for any damage to its credibility— the last thing the company could afford to lose under the circumstances.

By the end of the first day, a Thursday spent largely sorting out facts from false alarms, Collins and the other McNeil executives felt strongly that the poisonings did not occur at their plant, either accidentally or intentionally. If someone had dumped a dose of cyanide small enough to escape detection into one of the drug mixing machines, the mixture would have been so diluted as to be nearly harmless, and the contaminated pills would have ended up all around the country, not simply on Chicago's West Side. Moreover, all the samples tested from the lot reported to have poisoned the first five Chicago victims turned out to be normal.

Regardless, the company couldn't take the chance that the whole lot had not been poisoned and recalled all 93,000 bottles scattered across the country, an expensive process for which the telegrams to doctors, hospitals, and distributors alone cost a half million dollars. McNeil also suspended all advertising for Tylenol.

AN IMPORTANT DISCOVERY

The first phase of the crisis ended early Friday morning when the company learned that the sixth victim had been poisoned with Tylenol capsules from a lot manufactured at McNeil's other plant in Round Rock, Texas. That proved the tampering had to have taken place in Chicago and not in the manufacturing process, because poisoning at both plants would have been almost impossible. The discovery was important for the company because it signaled the end of its initial helter-skelter involvement with fact gathering and the beginning of its effort to assess the impact the poisonings would have on its product. Also, Johnson & Johnson had to figure out what to do about it. But for Collins, who had gone to bed exhausted at a nearby motel at 2 A.M. only to be reawakened an hour later by a phone call reporting the Round Rock development, its significance— like so much else that first day—was not immediately apparent. "The fact the second batch came from Round Rock didn't say a damn thing to me," he admitted, "except that, oh Jesus, now I've got two lots to recall instead of one."[1] This was both bad and good news for Collins. The bad news, obviously, was that two lots had to be recalled. The good news, however, was that it now seemed unlikely that the tampering had occurred at McNeil.

Had the incident not been so extraordinary, Johnson & Johnson, ardent in its commitment to decentralization, would have expected McNeil Consumer Products to cope with the problem on its own. Reassuring as it was to have the resources of Johnson & Johnson at its disposal, McNeil executives didn't seem altogether thrilled by the new scrutiny they were getting from above. "Managing a crisis is one thing," said McNeil President Chiesa, "but managing all the helpful advice is another."

In Johnson & Johnson's eyes, the Tylenol crisis was a major public health problem—and a major threat to the company. Johnson & Johnson carefully restricts the company name to relatively few items, such as baby products and Band-Aids. "One of the things that was bothering me," said Burke, "is the extent to which Johnson & Johnson was becoming deeply involved in the affair. The public was learning that Tylenol was a Johnson & Johnson product, and the dilemma was how to protect the name and not incite whoever did this to attack other Johnson & Johnson products." According to company surveys, less than 1 percent of consumers knew before the poisonings that Johnson & Johnson was the parent company behind Tylenol; now more than 47 percent were aware of that fact.[2]

On the weekend of October 9th, Lawrence G. Foster, Johnson & Johnson's vice president for corporate public relations, told the chairman that the crisis had

---

[1] Adapted from Thomas Moore, "The Fight to Save Tylenol," *Fortune,* November 29, 1982, pages 44–49, ©1982 Time Inc. All rights reserved.

[2] "The Fight to Save Tylenol," pages 44–49.

unquestionably become a Johnson & Johnson problem. The company was at risk. Financially it wouldn't topple—Tylenol accounted for only a small fraction of the profits—but there was an international reputation at stake.

On October 10th, Mr. Burke made up his mind. All 150 sister companies would pitch in; there would be no new name or second "fighting" brand. Tylenol would fight under its own flag. It was hell or high water.

The next morning, Monday, Mr. Burke and Mr. Clare, Johnson & Johnson's president, huddled for three hours and concurred on the strategy. Other top executives were notified that afternoon. "It would almost be an admission of some kind of guilt in my opinion to walk away from that name," Mr. Nelson agreed. "We'd be very foolish. And even if a third of this business never came back, we'd still have the top-selling pain reliever in the world. . . . It's better than a sharp stick in the eye."

On Tuesday morning, October 12th, at a meeting of fifty company presidents and corporate staffers, Mr. Burke declared, "This is an unequivocal decision."

Suddenly the corporation had some direction. Bleary eyes got a little brighter; snippy impatience was leavened with occasional humor. The executives even knitted themselves together with shared bromides, such as, "We're the guy who got hit by the truck," or, "Is there more upside than downside if we make this move?"[3]

Burke quickly decided to elevate the management of the crisis to the corporate level, personally taking charge of the company's response and delegating responsibility for running the rest of the company to other members of the executive committee.

The members of the executive committee responsible for developing strategies for the crisis project, as well as crisis decision making, were:

- James Burke, Johnson & Johnson chairman
- David Clare, Johnson & Johnson president
- George Frazza, general counsel
- Lawrence Foster, vice president of public relations
- David Collins, McNeil Consumer Products chairman
- Wayne Nelson, group chairman
- Arthur Quilty, executive committee member

Lessons Learned: Based on the seriousness of the crisis, there could be multiple committees with the overseeing or strategy committee made up entirely of senior corporate executives.

---

[3] Adapted from Rick Atkinson, "The Tylenol Nightmare: How a Corporate Giant Fought Back," *The Kansas City Times,* November 12, 1982, page 3.

There were several reasons why Burke decided to take control of the situation himself. First, Burke believed that the crisis could become a national crisis with the future of self-medication at stake. Second, Burke recognized that the reputation of Johnson & Johnson was now at stake, even though all of the spokespeople up to this point had carefully been labeled as McNeil employees. Third, and perhaps the toughest decision of all, was Burke's belief that McNeil may not be able to battle the crisis alone.

The fourth reason was the need for a Johnson & Johnson corporate spokesperson. James Burke was about to become that corporate spokesperson. This was one of the few times that a CEO had appeared on television. Burke's first decision was to completely cooperate with the news media. The general public, medical community, and Food and Drug Administration were immediately notified.

There was some concern that pulling the capsules off of the shelves would provide instant gratification to the killer, resulting in the tampering of other Johnson & Johnson products. Also, there could be a whole series of "copycat" tamperings that could affect the entire industry.

There was also a discussion over offering a reward for information leading to prosecution of the killer. At first, they settled for a $1 million reward leading to the culprit's conviction. However, the FBI feared that this would result in more blind leads than the agency could handle. The reward was then reduced to $100,000 and announced at a news conference.

Lessons Learned: When managing a crisis project, especially during the early phases, effective communication is critical. All communication channels must remain open, free of political intervention, and hopefully based upon trust and honesty. Failing to do this could result in "burning bridges" with information sources such that repairs cannot be made prior to the closure of the crisis. The project manager assigned to the project must possess strong communication skills and foster a culture of trust with all of the stakeholders.

Lessons Learned: The company spokesperson must be a professional communicator who understands how to represent both the crisis project and the company.

Instead of providing incomplete information or only the most critical pieces and stonewalling the media, Burke provided all information available. He quickly and honestly answered all questions from anyone. This was the first time that a corporate CEO had become so visible to the media and the public. James Burke spoke with an aura of trust.

Tylenol quickly captured the nation's attention. Queries from the press on the Tylenol story exceeded 2,500. Two news clipping services generated in excess of 125,000 clippings. One of them said the Tylenol story had resulted in the widest domestic coverage of any story since the assassination of President John F. Kennedy. Associated Press and United Press International gave it second place as the impact story of 1982—only coverage of the nation's economy ranked higher. The television and radio coverage was staggering.[4]

Before the first week came to an end, more than 100 state and federal agents were spread across the Chicago area in a painstaking effort to reconstruct the route of the poisoned capsules. The route of the contaminated Tylenol capsules was as follows:

- The capsules were manufactured in Fort Washington, Pennsylvania, and Round Rock, Texas.
- From the plants, McNeil then shipped the capsules to thirty-five states, including Illinois.
- In Chicago, McNeil delivered the capsules to almost a hundred wholesalers, some of whom would keep them in the warehouse for a few days.
- Sometime in August, the wholesalers sold the Tylenol capsules to retail outlets.

The investigators believed that the tampering occurred after the capsules reached Illinois. This was based upon the theory that potassium cyanide is corrosive and would eventually destroy the gelatin shell. Investigators began experimenting with potassium cyanide and its decomposition to see if they could pinpoint the precise point in time when the tampering occurred.

Other investigative teams were focusing on possible disgruntled workers or former employees in one of a number of companies that physically handled the product along its route. The initial conclusion was that the poisoning was a willful act and not a manufacturing accident.

While the first phase had been one of problem identification and containment, the second phase was one of communication. Burke allotted the next week to

---

[4]Lawrence G. Foster, "The Johnson & Johnson Credo and The Tylenol Crisis," *New Jersey Bell Journal,* Vol. 6, No. 1 (Spring 1983), page 3.

establishing a good working relationship between the company and the police and health authorities investigating the crime. On Monday, he went to Washington to meet with the FBI and the Food and Drug Administration. Burke had begun to advocate a recall of all Extra-Strength Tylenol capsules but—in a surprising role reversal—both the FBI and the FDA counseled him against recalling the drug precipitously. "The FBI didn't want us to do it," explained Burke, "because it would say to whoever did this: Hey, I'm winning. I can bring a major corporation to its knees.' And the FDA argued a recall might cause more public anxiety than it would relieve."

On Tuesday, however, following what appeared to be a copycat strychnine poisoning with Tylenol capsules in California, the FDA agreed with Burke that he had to recall all Tylenol capsules—31 million bottles with a retail value of over $100 million. "Often our society rails against bigness," Burke said, "but this has been an example where size helps. If Tylenol had been a separate company, the decisions would have been much tougher. As it was, it was hard to convince the McNeil people that we didn't care what it cost to fix the problem."[5]

There were several options available to Burke and the strategy committee. Some of the options included:

- Tell the Johnson & Johnson story in hopes that the public would be sympathetic and Tylenol could recover quickly.
- Take aggressive action in a search for the killer, placing all blame elsewhere.
- Replace the capsules with another type of product (i.e., caplets).
- Recall only those batches that were contaminated.
- Recall all Extra-Strength Tylenol capsules.
- After recall, relaunch the product under the same name but different packaging.
- After recall, relaunch the product under a different name and different packaging.

Deciding which option to select would certainly be a difficult decision. Looking over Burke's shoulder were the stakeholders who would be affected by Johnson & Johnson's decision. Among the stakeholders were stockholders, lending institutions, employees, managers, suppliers, government agencies, and the consumers.

Consumers: The consumers had the greatest stake in the crisis because their lives were on the line. The consumers must have confidence in the products they purchase and believe that they are safe to use as directed.

---

[5]"The Fight to Save Tylenol," page 48.

Stockholders: The stockholders had a financial interest in the selling price of the stock and the dividends. If the cost of removal and replacement, or in the worst-case scenario of product redesign, were substantial, it could lead to a financial hardship for some investors who were relying on the income.

Lending institutions: Lending institutions provide loans and lines of credit. If the present and/or future revenue stream is impaired, then the funds available might be reduced and the interest rate charge could increase. The future revenue stream of its products could affect the quality rating of its debt.

Government: The primary concern of the government was in protecting public health. In this regard, government law enforcement agencies were committed to apprehending the murderer. Other government agencies would provide assistance in promoting and designing tamper-resistant packages in an effort to restore consumer confidence.

Management: Company management had the responsibility to protect the image of the company, as well as its profitability. To do this, management must convince the public that management will take whatever steps are necessary to protect the consumer.

Employees: Employees have the same concerns as management but are also somewhat worried about possible loss of income, or even employment.

Whatever decision Johnson & Johnson selected was certain to displease at least some of the stakeholders. Therefore, how does a company decide which stakeholders' needs are more important? How does a company prioritize stakeholders?

For Jim Burke and the entire strategy committee, the decision was not very difficult—just follow the corporate credo.

For more than forty-five years, Johnson & Johnson had a corporate credo, shown in Appendix A, which clearly stated that the company's first priority is to the users of Johnson & Johnson's products and services. Everyone knew the credo, what it stood for, and the fact that it must be followed. The corporate credo guided the decision-making process, and everyone knew it without having to be told.

When the crisis had ended, Burke recalled that no meeting had been convened for the first critical decision: to be open with the press and put the consumer's interest first. "Every one of us knew what we had to do," Mr. Burke

---

Lessons Learned: Some sort of structured decision-making process should be in place during crisis management. Whatever process is used should be readily understood and acceptable to all parties involved in the crisis. Corporate credos or corporate standard practice manuals can make the decision-making process easier.

commented. "There was no need to meet. We had the credo philosophy to guide us."

By mid-week, an extortion note threatening a second wave of poisonings turned up at McNeil. "Imagine our reaction," explained Collins. "We get this note that says send $1 million to a bank account number at Continental Bank in Illinois. We had to laugh. This guy's gotta be an idiot. We're still not convinced he did it."

Through advertisements promising to exchange capsules for tablets, through thousands of letters to the trade, and through statements to the media, the company was hoping to demystify the incident. "There was a lot of noise out there, most of it associating Tylenol with death," said Chiesa. "We wanted to clear up any misunderstanding, to make sure everyone had all the facts we did, that the problem was limited to one area of the country, and only a few bottles of Tylenol capsules were contaminated."[6]

Advice was pouring in by the sackful. A Pittsburgh man offered some new names for Tylenol—perhaps Lespane or Apamin or Painex. There was a suggestion from Atlantic City that Tylenol be canned like chili. An Ontario couple sent $10—which was returned—as a contribution to the reward.

A psychic in Schenectady, New York, breezily notified the company that the killer was a Chicago pharmacist "dressed in a white smock, buttoned up to the neck, and white trousers." A boy sent a $5 billion extortion note, carefully composed of letters clipped from the newspaper, and included his home address. The Colorado School of Mines offered to extract cyanide from any contaminated capsules so thoroughly that they could still be marketed. Thanks anyway, the company politely replied.

Mr. Burke argued that:

> There are some very real problems with all the suggestions put forth. "Tylenol II: Change the color from red to green because red is stop, green is go; change the name.
>
> The public collectively just isn't easy to fool. When the public is watching carefully, they make incredibly smart decisions. They're just so much smarter collectively than we are individually."[7]

WORST-CASE SCENARIO

From the start, Burke squelched one obvious option: abandoning Tylenol and reintroducing the pain reliever under a new name. Despite the long odds many outside marketing experts gave against a complete comeback, and despite the fact

---

[6]"The Fight to Save Tylenol," page 48.

[7]"The Tylenol Nightmare," page 3.

that sales of Tylenol products initially dropped 80 percent, company executives said they never had any question about whether to bring back Tylenol. Said Wayne Nelson, Collins's predecessor at McNeil, who was now a company group chairman, "Even in our worst-case scenarios where we get back only half the base we had before, it would still be the market leader."

By the second weekend, Burke had moved on to the third phase: rebuilding the brand. "We were still in a state of shock," explained Burke. "It's like going through a death in the family. But the urgency of bringing about Tylenol's recovery makes it important we move out of the mourning stage faster than usual."

GOOD NEWS, BAD NEWS

It seemed clear that the company would have to come up with a new tamper-resistant package, as would the rest of the drug industry. But how consumers ultimately felt about the product—and what conflicts the poisonings posed in their minds—would be the determining factor in the comeback. Burke called in Young & Rubicam, Johnson & Johnson's oldest advertising agency, to begin polling consumer attitudes. Initially he wanted to know how the public was reacting to the crisis, but he also knew the surveys would be indispensable in building a database for what was obviously going to be, as he put it, "a very complicated communications problem."

One of the more astonishing things learned from the first surveys was that an overwhelming number of people—94 percent of the consumers surveyed—were aware that Tylenol had been involved with the poisonings. The implications of that figure, when coupled with other data, were both good and bad.

The good news, said Burke, was that 87 percent of the Tylenol users surveyed said they realized the maker of Tylenol was not responsible for the deaths. The bad news was that although a high percentage didn't blame Tylenol, 61 percent still said they were not likely to buy Extra-Strength capsules in the future. Worse, 50 percent felt that way about Tylenol tablets as well as capsules. In short, many consumers knew it wasn't Tylenol's fault but said they were not going to buy it anyway, revealing a fear associated with the name that was not likely to dissipate soon.

The most heartening piece of information in the surveys—and the one on which the company based its comeback strategy—was that the frequent Tylenol user seemed much more inclined to go back to the product than the infrequent user. The message: the company can forget about making new converts in the next year or two. Instead, it would concentrate on bringing back to the fold the loyal customers of the past.

"People forget how we built up such a big and important franchise," said Burke. "It was based on trust. People started taking Tylenol in hospitals or because their doctors recommended it. In other words, they were not well and in a highly emotional state." The contrary view, it can be argued, is that those same people who originally bought Tylenol because they didn't want to take a chance on aspirin's side effects are the last people who would want to take a chance now with the emotionally charged brand name.

The competition was not standing idly by. American Home Products had increased production of its acetaminophen, Anacin-3, at both its plants from two to three shifts on a round-the-clock basis. Bristol-Myers did not discuss any marketing plans it had for its acetaminophen, Datril, except to say that demand was up considerably and the company was looking into new packaging for all its analgesic products.[8]

THE RACE BEGINS

It would not be easy designing a new tamper-resistant package. There were a thousand tasks, and each time one was completed, two others seemed to spring up in its stead. Nothing was more crucial than the new packaging, and the chairman headed the task force on tamper resistance himself. Everyone in the industry realized that the first product on the market to be sheathed in some kind of anti-tampering protection would reap enormous psychological benefits. And that meant dollars—each share of the analgesic market was worth $15 million to retail sales.

A small team was formed at McNeil, quickly christened Machiavelli & Co., which tried to outwit the dozen or so tamper-resistant methods available. "Tylenol had been on TV right alongside a skull and crossbones." Mr. Clare said, "so we knew whatever the rest of industry was going to do, we had to do more."

After a mad scramble among the drug companies for machines and material, McNeil settled on a triple seal: A glued carton; a shrink sleeve on the bottle neck; and foil covering beneath the cap. It was decided that the first "put ups," or shipments, would be Extra-Strength Tylenol capsules in bottles of 50, the most popular size.

By this time, the engineers and executives were beginning to think of the whole ordeal as a kind of over-the-counter space race. Production went to three shifts, seven days a week. There were agonizing logistical roadblocks—the shrink sleeve, for instance, had to be mounted by hand, despite a search through Asia and Europe for enough machinery.

Carton machinery had to be reconfigured to glue rather than fold boxes. New graphics had to be designed. They wanted to call the anti-tampering device the

---

[8]"The Fight to Save Tylenol," page 49.

Tylenol "safety seal," but first they had to hunt down the man who owned the trademark to that phrase and license it from him for $2,000 a year.[9]

> *Lessons Learned:* Under the pressure of the crisis, very little time existed for planning. The only viable way to plan effectively, if at all possible, is with rolling wave planning. Hopefully, this can be achieved with minimal risk and minimal scope changes.

David R. Clare, president of Johnson & Johnson and chairman of its executive committee had these comments:

> There probably are as many emergency plans worked out and ready to go within the Johnson & Johnson organization as there are in any other company that tries to prepare for unforeseen emergencies. But the events surrounding the Tylenol crisis were so atypical that we found ourselves improvising every step of the way.
>
> I doubt that even now we could devise a plan of action to deal with all aspects of the Tylenol situation. Events happened so quickly and so unpredictably that it would be impossible to anticipate the critical decisions that had to be made.[10]

Estimates on the extra cost ranged from a penny to 4 cents a bottle. But somehow no one had thought about whether to pass it along to the customer until Joe Chiesa, the McNeil president, suddenly announced on a television talk show late one night that the company would eat the price increase. The Johnson & Johnson executives were stunned. Of course! Brilliant!

"If he hadn't made that judgment, I think we would have horsewhipped him," Mr. Burke quipped.

The packaging frenzy came to a climax on Thursday afternoon, November 4th. Mr. Burke, Mr. Frazza, and Mr. Collins swooped down to Washington in the company chopper for an emergency meeting with Richard Schweiker, secretary of Health and Human Services, and Dr. Hayes of the Food and Drug Administration (FDA). In mid-meeting, the chairman pulled out samples of the newly packaged capsules.

Both Mr. Schweiker and Dr. Hayes had trouble getting the bottles open. The secretary turned to Mr. Burke and said, "Everybody else is going to have a package. You've got an armored tank."

---

[9]"The Tylenol Nightmare," page 4.

[10]"The Johnson & Johnson Credo and the Tylenol Crisis," page 2.

Amid this small triumph there were dozens of other critical decisions to make. No mine field was more hazardous than advertising. If the product wasn't aggressively peddled, the market would wither away permanently. But if the company was perceived as pushy or manipulative, consumers would balk at returning to the fold.

Mr. Burke and his executives watched nearly six hours of taped consumer reactions to Tylenol. They commissioned survey after survey of public sentiment. The polls showed that one American in five still didn't know the tampering had occurred outside the plant, but they also showed that millions would be willing to take Tylenol capsules again if the bottles were made tamper-proof.

The campaign began to take shape. Three commercials were filmed, all geared at luring back former users since the surveys indicated that consumers who had never taken Tylenol before were a hopeless cause for now. A print ad was drafted for release November 21st that said, "The Makers of Tylenol want to say 'THANK YOU, AMERICA' for your continuing confidence and support."[11]

THE MARKETING WAR

The makers of Tylenol embarked on an extremely delicate mission of psychological warfare. Timing was crucial. If Johnson & Johnson brought Tylenol back before the hysteria had subsided, the product could die on the shelves. If the company waited too long, the competition could gain an enormous lead.

Sophisticated though the consumer surveys were, they didn't give a clear answer on timing. "The problem with consumer research," said an impatient Joe Chiesa, "is that it reflects attitudes and not behavior. The best way to know what consumers are really going to do is put the product back on the shelves and let them vote with their hands."

With carefully measured public-service-like ads vowing to regain consumer trust, the company had set a discreet tone for its new campaign. Said Collins, "We're coming back against a tragedy, so there's no way we can come riding in on elephants, blowing horns and saying here we are."

Perhaps even more important was the effort that most of the public would never see. At the end of October, a month to the day after the crisis began, Burke mobilized 2,259 salespeople from all of Johnson & Johnson's domestic subsidiaries to persuade doctors and pharmacists to begin recommending Tylenol tablets to patients and customers. It was the same road the makers of Tylenol had taken when they began marketing the product twenty-two years earlier. Tylenol was not out of the game, but it was back at square one.[12]

---

[11]"The Tylenol Nightmare," page 4.

[12]"The Fight to Save Tylenol," page 49.

Everything was in place. Return to the marketplace would be announced in a New York news conference on Thursday, November 11th, and transmitted by satellite to thirty cities. Then Mr. Burke would fly to San Diego for a "high road" speech to a convention of editors, stressing that "We don't want to let the bastard (the killer) win."

The rest would be up to the consumer. And the courts would decide whether Johnson & Johnson should have foreseen the lunatic in Chicago. The question persisted: Should the packages have been tamper-resistant?[13]

From the day the deaths were linked to the poisoned Tylenol until the recall on Thursday, November 11th, Johnson & Johnson had succeeded in portraying itself to the public as a company willing to do what was right, regardless of cost.

Serving the public interest had simultaneously saved the company's reputation. That lesson in public responsibility—the public relations—would survive at Johnson & Johnson, regardless of what happened to Tylenol.

THE TYLENOL RECOVERY

From crisis to comeback, the following is just a partial list of activities undertaken by the company:

- McNeil established toll-free consumer hot lines in the first week of the crisis to respond to inquiries related to the safety of Tylenol. Through November, more than 30,000 phone calls were handled through this medium.
- A full-page ad was placed in major newspapers across the country on October 12th, offering consumers the opportunity to exchange capsules for tablets.
- In October, Johnson & Johnson communicated by letter on two separate occasions with its domestic employees and retirees, keeping them updated on important information and expressing thanks for continued support and assistance. In part, the communication urged employees and friends of Johnson & Johnson to request that Tylenol tablets be returned to those local drug stores and retail outlets where they had been removed.
- A sixty-second spot was broadcast in October and November featuring Dr. Thomas Gates, medical director for McNeil, alerting consumers to the impending return of the Tylenol capsules in tamper-resistant packaging. An estimated 85 percent of all TV households in the United States saw the commercial an average of 2.5 times during the first week of airing.
- Members of the Corporate Relations Department of Johnson & Johnson visited more than 160 congressional offices in Washington to accomplish

---

[13]"The Tylenol Nightmare," page 4.

a number of goals related to the Tylenol comeback. These included voicing support for federal criminal legislation making product tampering a felony and endorsing public service announcements by the Food & Drug Administration on tamper-resistant packaging. Resolutions were under consideration in Congress to commend the FDA, the industry, and Johnson & Johnson for the prompt and effective response to the Tylenol crisis.

- Johnson & Johnson executives made personal appearances or were interviewed for such print and video feature presentations as *Fortune, The Wall Street Journal, 60 Minutes, The Phil Donahue Show, ABC Nightline,* and *Live at Five* (New York). A number of additional executives were briefed for interviews on Tylenol that were being requested by TV and radio talk shows.

- Four videotaped special reports on the Tylenol crisis and comeback were prepared and distributed or shown to employees and retirees. The tapes, which lasted more than three hours, covered all important aspects of the evolving Tylenol story and treated at length the November 11 teleconference and the appearance of James E. Burke, chairman of the board, on *The Phil Donahue Show.*

- The Johnson & Johnson quarterly report in October informed stockholders of the impact of the Tylenol capsule withdrawal.

- A four-minute videotape was prepared for use by television programs covering tamper-resistant packaging. The footage depicted the production of Tylenol in the new tamper-resistant packaging and established the important triple-seal features as the standard for the industry.

- As a matter of policy, all letters directed to Johnson & Johnson and McNeil on Tylenol from consumers were answered. By late November, the company had responded to more than 3,000 inquiries and letters of support.[14]

By Christmas week, 1982, Tylenol had recovered 67 percent of its original market. The product was coming back faster and stronger than the company had anticipated.

Among the key components of the McNeil/Johnson & Johnson Tylenol comeback campaign were the following:

- Tylenol capsules were reintroduced in November in triple-seal, tamper-resistant packaging, with the new packages beginning to appear on retail shelves in December. Despite the unsettled conditions at McNeil caused

---

[14]"The Tylenol Comeback," A Special Report from the Editors of *Worldwide,* A Publication of Johnson & Johnson Corporate Public Relations, undated. Reprinted from Johnson & Johnson *Worldwide,* Vol. 17, No. 5 (December 1982).

by the withdrawal of the Tylenol capsules in October, the company, with its new triple-sealed package, was the first in the industry to respond to the national mandate for tamper-resistant packaging and the new regulations from the Food & Drug Administration.

- In an effort to encourage the American consumer to become reaccustomed to using Tylenol, McNeil Consumer Products Company provided the opportunity of obtaining free $2.50-off coupons good toward the purchase of any Tylenol product. Consumers simply phoned a special toll-free number to be placed on the list of those receiving the coupons. The same offer was made on two separate occasions in November and December through high-circulation newspapers containing the $2.50 coupon. McNeil estimated that these two programs would stimulate millions of trials of Tylenol before the end of the year.

- McNeil sales people were working to recover former stock levels for Tylenol by implementing an off-invoice pricing program that provided the buyer with discounts linked to wholesale purchasing patterns established prior to October. Discounts went as high as 25 percent.[15]

Three years later, by 1985, the company had recovered virtually its entire market share, outselling the next four analgesics combined. The company had spent more than $175 million to survive and conquer what was potentially one of the biggest disasters in the drug industry. This included more than $60 million in one year's advertising to reintroduce its new Tylenol. The Tylenol disaster had far-reaching effects in that virtually all nonprescription drugs, as well as many other products, are now packaged in tamper-resistant packages.

Some people believed that James Burke almost single-handedly saved Tylenol, especially when Wall Street believed that the Tylenol name was dead. Burke courageously made some decisions against the advice of government agents and some of his own colleagues. He appeared on a variety of talk shows, such as *The Phil Donahue Show* and *60 Minutes.* His open and honest approach to the crisis convinced people that Johnson & Johnson was also a victim. According to Johnson & Johnson spokesman, Bob Andrews, "The American public saw this company was also the victim of an unfortunate incident and gave us our market back."

Both Johnson & Johnson and James Burke received nothing but accolades and support from the media and general public for the way the crisis was handled. A sampling of opinion from newspapers across the United States includes:

- *Wall Street Journal:* "Johnson & Johnson, the parent company that makes Tylenol, set the pattern of industry response. Without being asked,

---

[15]"The Tylenol Comeback."

it quickly withdrew Extra-Strength Tylenol from the market at a very considerable expense . . . the company chose to take a large loss rather than expose anyone to further risk. The anti-corporation movement may have trouble squaring that with the devil theories it purveys."

- *Washington Post:* "Though the hysteria and frustration generated by random murder have often obscured the company's actions, Johnson & Johnson has effectively demonstrated how a major business ought to handle a disaster. From the day the deaths were linked to the poisoned Tylenol . . . Johnson & Johnson has succeeded in portraying itself to the public as a company willing to do what's right regardless of cost."

- *Express and News* (San Antonio, Texas): "In spite of the $100 million loss it was facing, the company . . . never put its interests ahead of solving the murders and protecting the public. Such corporate responsibility deserves support."

- *Evening Independent* (St. Petersburg, Florida): "The company has been straightforward and honest since the first news of the possible Tylenol link in the Chicago-area deaths. Some firms would have tried to cover up, lie or say 'no comment.' Johnson & Johnson knows better. Its first concern was to safeguard the public from further contamination, and the best way to do that was to let people know what had occurred by speaking frankly with the news media."

- *Morning News* (Savannah, Georgia): "Tylenol's makers deserve applause for their valiant attempt to recover from the terrible blow they have suffered. . . ."[16]

Federal and local law enforcement agencies had logged into the computer some 20,000 names of potential Tylenol killers, and 400 of them were scrutinized extensively. Two of them were eventually convicted. James Lewis, forty years old, admitted writing a letter to Johnson & Johnson threatening more deaths unless he was paid $1 million. He was found guilty of extortion and received a ten-year sentence in federal prison.

Six weeks after the murders, after receiving a tip, police questioned Roger Arnold, forty-eight years old, a laborer in the Jewel Food Stores warehouse. He was released due to lack of evidence. The next June, according to the police, he shot and killed a man he believed to be a tipster. He was convicted of second-degree murder. Neither Lewis, Arnold, nor anyone else was convicted for the original Tylenol deaths.

---

[16]"The Tylenol Comeback."

THE CORPORATE CULTURE

The speed with which a company can react to a crisis is often dependent upon the corporate culture. If the culture of the firm promotes individualism and internal competition, employees will feel threatened by the crisis, become nonsupportive, and refuse to help even if it is in their own best interest. Such was not the case at Johnson & Johnson. The culture at Johnson & Johnson was one of cooperation. Employees were volunteering to assist in any way possible to help Johnson & Johnson and McNeil out of the crisis.

On November 11, 1982, Mr. Burke announced that Johnson & Johnson would give consumers a free $2.50 coupon good toward the purchase of any Tylenol product. The free coupon could be obtained by calling an 800 number, which Mr. Burke gave out at a teleconference for the media, portions of which were rebroadcast on local TV and radio news shows.

Within minutes after the conference closed, a telephone center that had been set up at McNeil headquarters in Fort Washington, Pennsylvania, was swamped with calls—and would continue to be inundated for the next two weeks.

In fact, the number of calls proved to be much more than McNeil management had anticipated. The day after the teleconference, Peter Scarperi, vice president of finance and a member of the management board, appealed to McNeil employees to pitch in and handle the phones on a volunteer basis—on the weekend. How fast was the response? "Within an hour we had all the people we needed for nearly the entire weekend," said Mr. Scarperi.[17] During the eleven-day period following the announcement, McNeil received 136,000 calls. By the first week of December, there were over 210,000 calls by consumers.

Emblematic of the responsiveness, determination, and spirit characterizing the manner in which the crisis and the recovery program had been handled—both inside and outside the company—is the story of McNeil's employee buttons. The following is reprinted from *TYLINE,* a McNeil employee publication:

> Like everyone, (McNeil employee) Tony McGeorge wanted to be of help any way he could. "It was so frustrating not to be able to do more," he said. "So I headed up an ad hoc employee morale committee (with Logan P. Hottle, director, professional sales staff), and we came up with a suggestion for the button."
>
> The button, of course, is the thumbs-up "We're coming back" badge worn all over McNeil. The committee received approval for the idea late one Friday afternoon. They called an outside manufacturer and described how the button should look. . . . The manufacturer called his production facility in California at 5:30 P.M. Eastern time.

---

[17]"The Tylenol Comeback."

The finished buttons were delivered to McNeil by that Sunday noon—free of charge, a gift from the manufacturer.[18]

Lessons Learned: Project management works exceptionally well when the organization has a cooperative culture. Decision-making is rapid and full organizational support exists. Employees make decisions in the best interest of the company (and consumers) rather than their own self-interest.

FOUR YEARS LATER: THE SECOND TYLENOL TRAGEDY[19]

On Monday afternoon, February 10, 1986, Johnson & Johnson was informed that Diane Elsroth, a young woman in Westchester County, New York, had died of cyanide poisoning after ingesting Extra-Strength Tylenol capsules. Johnson & Johnson immediately sent representatives to Yonkers to attempt to learn more and to assist in the investigation. Johnson & Johnson also began conferring by telephone with the FDA and the FBI, both in Washington and at the respective field offices.

Johnson & Johnson endorsed the recommendation of the FDA and local authorities that people in the Bronxville/Yonkers, New York area not take any of these capsules until the investigation was completed. Although from the outset Johnson & Johnson had no reason to believe that this was more than an isolated event, Johnson & Johnson concurred with the FDA recommendation that nationally no one take any capsules from the affected lot number ADF 916 until further notice. The consumers were asked to return products from this lot for credit or exchange. The tainted bottle was part of a batch of 200,000 packages shipped to retailers during the previous August, 95 percent of which had already been sold to consumers. Johnson & Johnson believed other people would have reported problems months ago if the batch had been tainted either at the manufacturing plant or at distribution sites.

Once again Johnson & Johnson made its CEO, James Burke, its lead spokesperson. And once again the media treated Johnson & Johnson fairly because of the openness and availability of James Burke and other Johnson & Johnson personnel. Johnson & Johnson responded rapidly and honestly to all

---

[18]"The Tylenol Comeback."

[19]This entire section has been adapted from James E. Burke's speech to the U.S. Senate Committee on Labor and Human Resources, February 28, 1986.

information requests by the media, a lesson remembered from the 1982 Tylenol tragedy.

Johnson & Johnson responded to hundreds of news media inquiries, providing up-to-the minute information, and also held three major press conferences in a week's time at its headquarters. It established a free, 800 telephone hotline through which consumers and retailers could contact McNeil Consumer Products Company to receive the latest information and advisories. And it suspended all television advertisements as of the evening of February 10th.

On February 11th, Johnson & Johnson initiated a pick-up of the entire Tylenol capsule inventory from all retail outlets in a three-mile radius of the store where it is believed the bottle of poisoned capsules had been purchased. Laboratories of the FDA and McNeil Consumer Products Company began chemical analyses of capsules retrieved from the Westchester County area. Additionally, as they did in 1982, Johnson & Johnson began polling consumers through independent research organizations to track their awareness and understanding of the extent of the problem and attitudes toward their products.

On the afternoon of February 11th, Johnson & Johnson called a major news conference to aid in communicating accurate information to the public and to make certain that the public was aware of the nature of the problem. The FDA also picked up and analyzed thousands of Tylenol capsules from the Westchester County area, which led to the finding of a second bottle that contained five capsules of cyanide on February 13th.

The second bottle confirmed to be tainted came from the shelves of a Woolworth's store in the New York City suburb of Bronxville, about two blocks from an A&P Food Store that was the source of the capsules taken by Diane Elsroth. The second bottle apparently was manufactured at a different plant than the first.

Within minutes of notification that the FDA had found a second bottle of contaminated capsules, Johnson & Johnson issued a nationwide release to the news media urging consumers not to use Tylenol capsules until further notice. On the evening of the 13th, Johnson & Johnson initiated a withdrawal of all capsule put-ups from wholesalers, retailers, and consumers in the Westchester County area. Johnson & Johnson also began urging the trade nationwide to remove all capsule put-ups from their shelves.

At the second news conference on February 14th, Johnson & Johnson again requested retailers to remove Tylenol capsule products from their shelves. All consumers were advised to return those products to Johnson & Johnson. A reward of $100,000 was offered for information leading to the identification of the person or persons responsible for the poisoned Tylenol capsules in Westchester County.

During exhaustive deliberations over the weekend of February 15th and 16th, Johnson & Johnson concluded that the company could no longer guarantee the safety of capsules to a degree consistent with Johnson & Johnson's standards of responsibility to its consumers. The packaged capsules were tamper-resistant, not tamper-proof. Thus on Monday, February 17th, Johnson & Johnson announced at the third press conference that Johnson & Johnson would no longer manufacture or sell any capsule products made directly available to the consumer, and that they have no plans to reenter this business in the foreseeable future. "We were aided in this judgment by the knowledge that we had the caplet available, a solid dosage form that all our research showed was an acceptable alternative to the majority of our capsule users," remarked Burke. "We also announced on February 17th that we would replace all capsules in the hands of consumers and the trade with caplets or tablets. We issued a toll-free 800 number consumers could call for a product exchange or a full refund, and we announced a Tylenol capsule exchange address."

TAMPER-RESISTANT PACKAGING

After the tamperings in the Chicago area in the fall of 1982, the over-the-counter medications and the packaging industries examined a broad range of technologies designed to protect the consumer against product tamperings. Johnson & Johnson had conducted an exhaustive review of virtually every viable technology. Ultimately, they had selected for Tylenol capsules a triple-safety-sealed system that consisted of glued flaps on the outer box, a tight printed plastic seal over the cap and neck of the bottle, and a strong foil seal over the mouth of the bottle.

The decision to reintroduce capsules had been based on marketing research done during that time. This research indicated that the capsules remained the dosage form of choice for many consumers. Many felt they were easier to swallow, and some felt that they provided more potent pain relief. While there was no basis in fact for the latter perception—tablets, caplets, and capsules are all equally effective—it was not an irrelevant consideration. "To the extent that some people think a pain reliever may be more powerful, a better result can often be achieved," said Burke. "This is due to a placebo effect but is nonetheless beneficial to the consumer."

Given these findings and Johnson & Johnson's confidence in the new triple-seal packaging system, the decision had been made to reintroduce Extra-Strength Tylenol pain relievers in capsule form.

WITHDRAWAL FROM THE DIRECT-TO-CONSUMER CAPSULE BUSINESS

Clearly, circumstances were different in 1986. In light of the Westchester events, Johnson & Johnson no longer believed that it could provide an adequate level of assurance of the safety of hollow capsule products sold direct to consumers. For this reason, management made the decision to go out of that business. "We take this action with great reluctance and a heavy heart," Burke said in announcing the action. "But we cannot control random tampering with capsules after they leave our plant. Therefore, we feel our obligation to consumers is to remove capsules from the market to protect the public." The company, Burke said, had "fought our way back" after the deaths of seven people who ingested cyanide-laced Tylenol capsules in 1982. He went on to say:

> We will do it again. We will encourage consumers to use either the solid tablets or caplets. The caplet is especially well-suited to serve the needs of capsule users. It is oval-shaped like a capsule, 35 percent smaller and coated to facilitate swallowing. It is also hard like a tablet and thus extremely difficult to violate without leaving clearly visible signs. We developed this dosage form as an alternative to the capsule. Since its introduction in 1984, it has become the analgesic dosage form of choice for many consumers.

Burke noted further that, "while this decision is a financial burden to us, it does not begin to compare to the loss suffered by the family and friends of Diane Elsroth." His voice quavered as he referred to the woman who died. He said he had expressed, on behalf of the company, "our heartfelt sympathy to Diane's family and loved ones."

In abandoning the capsule business, Johnson & Johnson had taken the boldest option open to it in dealing with an attack on its prized Tylenol line. At the same time, the company used the publicity generated by the latest Tylenol scare as an opportunity to promote other forms of the drug, particularly its so-called caplets. Johnson & Johnson initially balked at leaving the Tylenol capsule business, which represented about one-third of the $525 million in Tylenol sales in 1985. Mr. Burke said he was loath to take such a step, noting that if "we get out of the capsule business, others will get into it." Also, he said, pulling the capsules would be a "victory for terrorism."

In Washington, Food and Drug Administration Commissioner Frank Young said the agency respected Johnson & Johnson's decision. "This is a matter of Johnson & Johnson's own business judgment and represents responsible actions under tough circumstances," the agency said in a statement, adding that it "didn't suggest, direct or pressure Johnson & Johnson into this action." Meanwhile, other over-the-counter drug makers said Johnson & Johnson's decision was premature, and some indicated that they wouldn't abandon capsules.

People had thought that the 1982 tampering incident would mean an end to Tylenol. Now the same people believed that Tylenol would survive because it had already demonstrated once that it could do so. If survival would be in caplets, then industry would eventually follow. The FDA was planning to meet with industry officials to discuss what technological changes might be necessary to respond effectively to this problem. At stake was a re-examination of over-the-counter capsules, which included dozens of products ranging from Contac decongestant to Dexatrim diet formula.

CONCLUSION[20]

During congressional testimony, Burke said,

> I have been deeply impressed by the commitment and performance of government agencies, especially the FDA and the FBI. I cannot imagine how any organizations could have been more professional, more energetic or more rational in exercising their responsibilities to the American public.
>
> In addition, the media performed a critical role in telling the public what it needed to know in order to provide for its own protection. In the vast majority of instances, this was accomplished in a timely and accurate fashion. Within the first week following the Westchester incidents, polling revealed nearly 100 percent of consumers in the New York area were aware of the problem. I believe this is an example of how a responsible press can serve the public well being.
>
> The wide availability of self-medications through the open market in the United States is part of our unparalleled health care delivery system. We must work together to maintain the advantages of this system. For industry and government, this means doing what is appropriate to provide for the safety of consumer products. At the same time, we must maximize the freedom and accessibility needed for maintaining full advantage of an open marketplace—not the least of which is enlightened product selection. For the consumer, this means continued education and awareness. As has often been noted, no system yet designed can absolutely guarantee freedom from product tampering; the careful consumer must always be the most important single element in providing for his or her own safety.

ACCOLADES AND SUPPORT

Once again, Johnson & Johnson received high marks, this time for the way it handled the second Tylenol tragedy. This was evident from the remarks of Mr. Simonds, McNeil's vice president of marketing. "There's a real unity of purpose and positive attitude here at McNeil. Any reservations people had about the effect of removing capsules from the company's product line are long since resolved. We realized there was no good alternative" to the action announced by Johnson & Johnson Chairman James E. Burke because, Mr. Simonds went on to say, "as we simply could not guarantee the safety of the capsule form."

---

[20] Adapted from Burke, 1986.

As in the aftermath of the 1982 tragedies, Mr. Simonds adds, everybody at McNeil "pulled together." More than 300 McNeil employees manned the consumer phone lines, with four of them even fielding calls in Spanish.

The employees were buoyed by unsolicited testimonials from consumers. One man in Savannah, Georgia, wrote Mr. Burke that "You and your people deserve the right to walk with pride!" Members of a fifth-grade class in Manchester, Missouri, wrote that they "will continue to support and buy your products."

Many of the correspondents did not know what they could do to help but clearly wanted to do something. A man in Yonkers, New York (near where the tampering occurred), for example, wrote, "In my small way I am donating a check in the amount of $10 to offset some of the cost of this kind of terrorism. I would like to donate it in the names of my two children, Candice and Jennifer. Now it (the capsule recall and discontinuation) will only cost Johnson & Johnson $149,999,990." (The check was returned with thanks.)

Nor were the testimonials and words of thanks just from consumers. The *New York Times* said that in dealing with a public crisis "in a forthright way and with his decision to stop selling Tylenol in capsule form, Mr. Burke is receiving praise from analysts, marketing experts and from consumers themselves."

The *Cleveland Plain Dealer* said "the decision to withdraw all over-the-counter capsule medications, in the face of growing public concern about the vulnerability of such products, was sadly but wisely arrived at."

The magazine *U.S. News and World Report* wrote, "No company likes bad news, and too few prepare for it. For dealing with the unexpected, they could take lessons from Johnson & Johnson."

And columnist Tom Blackburn in the *Miami News* put it this way: "Johnson & Johnson is in business to make money. It has done that very well. But when the going gets tough, the corporation gets human, and that makes it something special in the bloodless business world."

Perhaps most significantly, President Reagan, opening a meeting of the Business Council (comprised of corporate chief executives) in Washington, D.C., said, "Let me congratulate one of your members, someone who in recent days has lived up to the highest ideals of corporate responsibility and grace under pressure. Jim Burke of Johnson & Johnson, you have our deepest admiration."

That kind of support from many sources has spurred McNeil, and the results are evident. "There is absolutely no doubt about it. We're coming back—again!"[21]

---

[21]"Tylenol Begins Making a Solid Recovery," *Worldwide,* A Publication of Johnson & Johnson Corporate Public Relations, 1986.

APPENDIX A*

JOHNSON & JOHNSON CREDO

We believe our first responsibility is to the doctors, nurses and patients, to mothers and all others who use our products and services. In meeting their needs everything we do must be of high quality. We must constantly strive to reduce our costs in order to maintain reasonable prices. Customers' orders must be serviced promptly and accurately. Our suppliers and distributors must have an opportunity to make a fair profit.

We are responsible to our employees, the men and women who work with us throughout the world. Everyone must be considered as an individual. We must respect their dignity and recognize their merit. They must have a sense of security in their jobs. Compensation must be fair and adequate, and working conditions clean, orderly and safe. Employees must feel free to make suggestions and complaints. There must be equal opportunity for employment, development and advancement for those qualified. We must provide competent management, and their actions must be just and ethical.

We are responsible to the communities in which we live and work and to the world community as well. We must be good citizens—support good works and charities and bear our fair share of taxes. We must encourage civic improvements and better health and education. We must maintain in good order the property we are privileged to use, protecting the environment and natural resources.

Our final responsibility is to our stockholders. Business must make a sound profit. We must experiment with new ideas. Research must be carried on, innovative programs developed and mistakes paid for. New equipment must be purchased, new facilities provided and new products launched. Reserves must be created to provide for adverse times. When we operate according to these principles, the stockholders should realize a fair return.

---

*Reproduced by permission of Johnson & Johnson. This is the 1979–1989 version of the Credo. Small changes were made to the Credo in 1989.

Part 14

MANAGING SCOPE CHANGES

Scope changes on a project can occur regardless of how well the project is planned or executed. Scope changes can be the result of something that was omitted during the planning stage, because the customer's requirements have changed, or because changes in technology have taken place.

The two most commonly used methods for scope change control are (1) allowing continuous scope changes to occur but under the guidance of the configuration management process and (2) clustering all scope changes together to be accomplished later as an enhancement project. In each of these two methods there are risks and rewards. The decision of when to select one over the other is not always black or white, but more so a gray area.

Denver International Airport (DIA)

BACKGROUND

How does one convert a $1.2 billion project into a $5.0 billion project? It's easy. Just build a new airport in Denver. The decision to replace Denver's Stapleton Airport with Denver International Airport (DIA) was made by well-intentioned city officials. The city of Denver would need a new airport eventually, and it seemed like the right time to build an airport that would satisfy Denver's needs for at least fifty to sixty years. DIA could become the benchmark for other airports to follow.

A summary of the critical events is listed below:

1985: Denver Mayor Federico Pena and Adams County officials agree to build a replacement for Stapleton International Airport.
Project estimate: $1.2 billion

1986: Peat Marwick, a consulting firm, is hired to perform a feasibility study including projected traffic. Their results indicate that, depending on the season, as many as 50 percent of the passengers would change planes. The new airport would have to handle this smoothly. United and Continental object to the idea of building a new airport, fearing the added cost burden.

May 1989: Denver voters pass an airport referendum.
Project estimate: $1.7 billion

March 1993: Denver Mayor Wellington Webb announces the first delay. Opening day would be postponed from October, 1993 to December 1993. (Federico Pena becomes Secretary of Transportation under Clinton). Project estimate: $2.7 billion

October 1993: Opening day is to be delayed to March 1994. There are problems with the fire and security systems in addition to the inoperable baggage handling system. Project estimate: $3.1 billion

December 1993: The airport is ready to open, but without an operational baggage handling system. Another delay is announced.

February 1994: Opening day is to be delayed to May 15, 1994 because of baggage handling system.

May 1994: Airport misses the fourth deadline.

August 1994: DIA finances a backup baggage handling system. Opening day is delayed indefinitely. Project estimate: $4 billion plus.

December 1994: Denver announces that DIA was built on top of an old Native American burial ground. An agreement is reached to lift the curse.

AIRPORTS AND AIRLINE DEREGULATION

Prior to the Airline Deregulation Act of 1978, airline routes and airfare were established by the Civil Aeronautics Board (CAB). Airlines were allowed to charge whatever they wanted for airfare, based on CAB approval. The cost of additional aircraft was eventually passed on to the consumer. Initially, the high cost for airfare restricted travel to the businessperson and the elite who could afford it.

Increases in passenger travel were moderate. Most airports were already underutilized and growth was achieved by adding terminals or runways on existing airport sites. The need for new airports was not deemed critical for the near term.

Following deregulation, the airline industry had to prepare for open market competition. This meant that airfares were expected to decrease dramatically. Airlines began purchasing hoards of planes, and most routes were "free game." Airlines had to purchase more planes and fly more routes in order to remain profitable. The increase in passenger traffic was expected to come from the average person who could finally afford air travel.

Deregulation made it clear that airport expansion would be necessary. While airport management conducted feasibility studies, the recession of 1979–1983 occurred. Several airlines, such as Braniff, filed for bankruptcy protection under Chapter 11 and the airline industry headed for consolidation through mergers and leveraged buyouts.

Cities took a wait-and-see attitude rather than risk billions in new airport development. Noise abatement policies, environmental protection acts, and land acquisition were viewed as headaches. The only major airport built in the last twenty years was Dallas–Ft. Worth, which was completed in 1974.

DOES DENVER NEED A NEW AIRPORT?

In 1974, even prior to deregulation, Denver's Stapleton Airport was experiencing such rapid growth that Denver's Regional Council of Governments concluded that Stapleton would not be able to handle the necessary traffic expected by the year 2000. Modernization of Stapleton could have extended the inevitable problem to 2005. But were the headaches with Stapleton better cured through modernization or

Exhibit I. Current service characteristics: United Airlines and Continental Airlines, December 1993 and April 1994

	Enplaned Passengers[a]	Scheduled Seats[b]	Boarding Load Factor	Scheduled Departures[b]	Average Seats per Departure
December 1993					
United Airlines	641,209	1,080,210	59%	7,734	140
United Express	57,867	108,554	53%	3,582	30
Continental Airlines	355,667	624,325	57%	4,376	143
Continental Express	52,680	105,800	50%	3,190	33
Other	236,751	357,214	66%	2,851	125
Total	1,344,174	2,276,103	59%	21,733	105
April 1994					
United Airlines	717,093	1,049,613	68%	7,743	136
United Express	44,451	92,880	48%	3,395	27
Continental Airlines	275,948	461,168	60%	3,127	147
Continental Express	24,809	92,733	27%	2,838	33
Other	234,091	354,950	66%	2,833	125
Total	1,296,392	2,051,344	63%	19,936	103

[a] Airport management records.
[b] Official Airline Guides, Inc. (on-line database), for periods noted.

Exhibit II. Airlines serving Denver, June 1994

Major/National Airlines	Regional/Commuter Airlines
America West Airlines	Air Wisconsin (United Express)[b]
American Airlines	Continental Express
Continental Airlines	GP Express Airlines
Delta Air Lines	Great Lakes Aviation (United Express)
Markair	Mesa Airlines (United Express)
Midway Airlines	Midwest Express[b]
Morris Air[a]	
Northwest Airlines	*Cargo Airlines*
TransWorld Airlines	
United Airlines	Airborne Express
USAir	Air Vantage
	Alpine Air
Charter Airlines	American International Airways
	Ameriflight
Aero Mexico	Bighorn Airways
American Trans Air	Burlington Air Express
Casino Express	Casper Air
Express One	Corporate Air
Great American	DHL Worldwide Express
Private Jet	Emery Worldwide
Sun Country Airlines	Evergreen International Airlines
	EWW Airline/Air Train
Foreign Flag Airlines (scheduled)	Federal Express
	Kitty Hawk
Martinair Holland	Majestic Airlines
Mexicana de Aviacion	Reliant Airlines
	United Parcel Service
	Western Aviators

[a] Morris Air was purchased by Southwest Airlines in December 1993. The airline announced that it would no longer serve Denver as of October 3, 1994.

[b] Air Wisconsin and Midwest Express have both achieved the level of operating revenues needed to qualify as a national airline as defined by the FAA. However, for purposes of this report, these airlines are referred to as regional airlines.

Source: Airport management, June 1994.

by building a new airport? There was no question that insufficient airport capacity would cause Denver to lose valuable business. Being 500 miles from other major cities placed enormous pressure upon the need for air travel in and out of Denver.

In 1988, Denver's Stapleton International Airport ranked as the fifth busiest in the country, with 30 million passengers. The busiest airports were Chicago, Atlanta, Los Angeles, and Dallas–Ft. Worth. By the year 2000, Denver anticipated 66 million passengers, just below Dallas–Ft. Worth's 70 million and Chicago's 83 million estimates.

Delays at Denver's Stapleton Airport caused major delays at all other airports. By one estimate, bad weather in Denver caused up to $100 million in lost income to the airlines each year because of delays, rerouting, canceled flights, putting travelers into hotels overnight, employee overtime pay, and passengers switching to other airlines.

Exhibit III. U.S. airports served nonstop from Denver

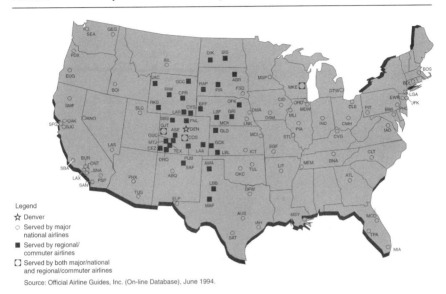

Legend

☆ Denver

○ Served by major
   national airlines

■ Served by regional/
   commuter airlines

☐ Served by both major/national
   and regional/commuter airlines

Source: Official Airline Guides, Inc. (On-line Database), June 1994.

Denver's United Airlines and Continental comprised 80 percent of all flights in and out of Denver. Exhibit I shows the service characteristics of United and Continental between December 1993 and April 1994. Exhibit II shows all of the airlines serving Denver as of June 1994. Exhibit III shows the cities that are serviced from Denver. It should be obvious that delays in Denver could cause delays in each of these cities. Exhibit IV shows the top ten domestic passenger origin-destination markets from Denver Stapleton.

Stapleton was ranked as one of the ten worst air traffic bottlenecks in the United States. Even low clouds at Denver Stapleton could bring delays of 30 to 60 minutes.

Stapleton has two parallel north-south runways that are close together. During bad weather where instrument landing conditions exist, the two runways are considered as only one. This drastically reduces the takeoffs and landings each hour.

The new airport would have three north-south runways initially with a master plan calling for eight eventually. This would triple or quadruple instrument flights occurring at the same time to 104 aircraft per hour. Currently, Stapleton can handle only thirty landings per hour under instrument conditions with a *maximum* of eighty aircraft per hour during clear weather.

The runway master plan called for ten 12,000 foot and two 16,000 foot runways. By opening day, three north-south and one east-west 12,000 foot runways would be in operation and one of the 16,000 foot north-south runways would be operational shortly thereafter.

Exhibit IV.   Top ten domestic passenger origin-destination markets and airline service, Stapleton International Airport (for the 12 months ended September 30, 1993)

City of Orgin or Destination[a]	Air Miles from Denver	Percentage of Certificated Airline Passengers	Average Daily Nonstop Departures[b]
1. Los Angeles[c]	849	6.8	34
2. New York[d]	1,630	6.2	19
3. Chicago[e]	908	5.6	26
4. San Francisco[f]	957	5.6	29
5. Washington, D.C.[g]	1,476	4.9	12
6. Dallas–Forth Worth	644	3.5	26
7. Houston[h]	864	3.2	15
8. Phoenix	589	3.1	19
9. Seattle	1,019	2.6	14
10. Minneapolis	693	2.3	16
Cities listed		43.8	210
All others		56.2	241
Total		100.0	451

[a] Top ten cities based on total inbound and outbound passengers (on large certificated airlines) at Stapleton International Airport in 10 percent sample for the 12 months ended September 30, 1993.
[b] Official Airline Guides, Inc. (on-line database), April 1994. Includes domestic flights operated at least four days per week by major/national airlines and excludes the activity of foreign-flag and commuter/regional airlines.
[c] Los Angeles International, Burbank–Glendale–Pasadena, John Wayne (Orange County), Ontario International, and Long Beach Municipal Airports.
[d] John F. Kennedy International, LaGuardia, and Newark International Airports.
[e] Chicago-O'Hare International and Midway Airports.
[f] San Franciscio, Metropolitan Oakland, and San Jose International Airports.
[g] Washington Dulles International, Washington National, and Baltimore/Washington International Airports.
[h] Houston Intercontinental and William P. Hobby Airports.
Sources: U.S. Department of Transportation/Air Transport Association of America, "Origin-Destination Survey of Airline Passenger Traffic, Domestic," third quarter 1993, except as noted.

The airfield facilities also included a 327-foot FAA air traffic control tower (the nation's tallest) and base building structures. The tower's height allowed controllers to visually monitor runway thresholds as much as three miles away. The runway/taxiway lighting system, with lights imbedded in the concrete pavement to form centerlines and stopbars at intersections, would allow air traffic controllers to signal pilots to wait on taxiways and cross active runways, and to lead them through the airfield in poor visibility.

Due to shifting winds, runway operations were shifted from one direction to another. At the new airport, the changeover would require four minutes as opposed to the 45 minutes at Stapleton.

Sufficient spacing was provided for in the concourse design such that two FAA Class 6 aircraft (i.e. 747-XX) could operate back-to-back without impeding each other.

Even when two aircraft (one from each concourse) have pushed back at the same time, there could still exist room for a third FAA Class 6 aircraft to pass between them. City officials believed that Denver's location, being equidistant from Japan and Germany, would allow twin-engine, extended range transports to reach both countries nonstop. The international opportunities were there. Between late 1990 and early 1991, Denver was entertaining four groups of leaders per month from Pacific Rim countries to look at DIA's planned capabilities.

In the long term, Denver saw the new airport as a potential hub for Northwest or USAir. This would certainly bring more business to Denver. Very few airports in the world can boast of multiple hubs.

THE ENPLANED PASSENGER MARKET

Perhaps the most critical parameter that illustrates the necessity for a new airport is the enplaned passenger market. (An enplaned passenger is one who gets on a flight, either an origination flight or connecting flight.)

Exhibit V identifies the enplaned passengers for individual airlines servicing Denver Stapleton for 1992 and 1993.

Exhibit V. Enplaned passengers by airline, 1992–1993, Stapleton International Airport

Enplaned Passengers	1992	1993
United	6,887,936	7,793,246
United Express[a]	470,841	578,619
	7,358,777	8,371,865
Continental	5,162,812	4,870,861
Continental Express	514,293	532,046
	5,677,105	5,402,907
American Airlines	599,705	563,119
America West Airlines	176,963	156,032
Delta Air Lines	643,644	634,341
MarkAir	2,739	93,648
Northwest Airlines	317,507	320,527
TransWorld Airlines	203,096	182,502
USAir	201,949	197,095
Other	256,226	398,436
	2,401,829	2,545,700
Total	15,437,711	16,320,472

[a] Includes Mesa Airlines, Air Wisconsin, Great Lakes Aviation, and Westair Airlines.
Source: Department of Aviation management records.

Connecting passengers were forecast to decrease about 1 million between 1993 and 1995 before returning to a steady 3.0 percent per year growth, totaling 8,285,500 in 2000. As a result, the number of connecting passengers is forecast to represent a smaller share (46 percent) of total enplaned passengers at the Airport in 2000 than in 1993 (50 percent). Total enplaned passengers at Denver are forecast to increase from 16,320,472 in 1993 to 18,161,000 in 2000—an average increase of 1.5 percent per year (decreasing slightly from 1993 through 1995, then increasing 2.7 percent per year after 1995).

The increase in enplaned passengers will necessitate an increase in the number of aircraft departures. Since landing fees are based upon aircraft landed weight, more parrivals and departures will generate more landing fee revenue. Since airport revenue is derived from cargo operations as well as passenger activities, it is important to recognize that enplaned cargo is also expected to increase.

LAND SELECTION[1]

The site selected was a 53-square-mile area 18 miles northeast of Denver's business district. The site would be larger than the Chicago O'Hare and Dallas–Ft. Worth airports combined. Unfortunately, a state law took effect prohibiting political entities from annexing land without the consent of its residents. The land was in Adams County. Before the vote was taken, Adams County and Denver negotiated an agreement limiting noise and requiring the creation of a buffer zone to protect surrounding residents. The agreement also included continuous noise monitoring, as well as limits on such businesses as airport hotels that could be in direct competition with existing services provided in Adams County. The final part of the agreement limited DIA to such businesses as airline maintenance, cargo, small package delivery, and other such airport-related activities.

With those agreements in place, Denver annexed 45 square miles and purchased an additional 8 square miles for noise buffer zones. Denver rezoned the buffer area to prohibit residential development within a 65 LDN (Level Day/Night) noise level. LDN is a weighted noise measurement intended to determine perceived noise in both day and night conditions. Adams County enacted even stiffer zoning regulations, calling for no residential development with an LDN noise level of 60.

Most of the airport land embodied two ranches. About 550 people were relocated. The site had overhead power lines and gas wells, which were relocated or

---

[1]Adapted from David A. Brown, "Denver Aims for Global Hub Status with New Airport Under Construction," *Aviation Week and Space Technology,* March 11, 1991, p. 44.

abandoned. The site lacked infrastructure development and there were no facilities for providing water, power, sewage disposal, or other such services.

FRONT RANGE AIRPORT

Located 2.5 miles southeast of DIA is Front Range Airport, which had been developed to relieve Denver's Stapleton Airport of most nonairline traffic operations. As a satellite airport to DIA, Front Range Airport had been offering six aviation business services by 1991:

- Air cargo and air freight, including small package services. (This is direct competition for DIA.)
- Aircraft manufacturing.
- Aircraft repair. (This is direct competition for DIA.)
- Fixed base operators to service general (and corporate) aviation.
- Flight training.
- Military maintenance and training.

The airport was located on a 4,800-acre site and was surrounded by a 12,000-acre industrial park. The airport was owned and operated by Adams County, which had completely different ownership than DIA. By 1991, Front Range Airport had two east-west runways: a 700-foot runway for general aviation use and an 8,000-foot runway to be extended to 10,000 feet. By 1992, the general plans called for two more runways to be built, both north-south. The first runway would be 10,000 feet initially with expansion capability to 16,000 feet to support wide body aircraft. The second runway would be 7,000 feet to service general aviation.

Opponents of DIA contended that Front Range Airport could be enlarged significantly, thus reducing pressure on Denver's Stapleton Airport, and that DIA would not be necessary at that time. Proponents of DIA argued that Front Range should be used to relieve pressure on DIA if and when DIA became a major international airport as all expected. Both sides were in agreement that initially, Front Range Airport would be a competitor to DIA.

AIRPORT DESIGN

The Denver International Airport was based upon a "Home-on-the-Range" design. The city wanted a wide open entry point for visitors. In spring of 1991, the city began soliciting bids.

To maintain a distinctive look that would be easily identified by travelers, a translucent tent-like roof was selected. The roof was made of two thicknesses of translucent, Teflon-coated glass fiber material suspended from steel cables hanging from the structural supports. The original plans for the roof called for a conventional design using 800,000 tons of structural steel. The glass fiber roof would require only 30,000 tons of structural steel, thus providing substantial savings on construction costs. The entire roof would permit about 10 percent of the sunlight to shine through, thus providing an open, outdoors-like atmosphere.

The master plan for the airport called for four concourses, each with a maximum of sixty gates. However, only three concourses would be built initially, and none would be full size. The first, Concourse A, would have thirty-two airline gates and six commuter gates. This concourse would be shared by Continental and any future international carriers. Continental had agreed to give up certain gate positions if requested to do so in order to accommodate future international operations. Continental was the only long-haul international carrier, with one daily flight to London. Shorter international flights were to Canada and Mexico.

Concourses B and C would each have twenty gates initially for airline use plus six commuter gates. Concourse B would be the United Concourse. Concourse C would be for all carriers other than Continental or United.

All three concourses would provide a total of seventy-two airline gates and eighteen commuter gates. This would be substantially less than what the original master plan called for.

Although the master plan identified sixty departure gates for each concourse, cost became an issue. The first set of plans identified 106 departure gates (not counting commuter gates) and was then scaled down to 72 gates. United Airlines originally wanted forty-five departure gates, but settled for twenty. The recession was having its effect.

The original plans called for a train running through a tunnel beneath the terminal building and the concourses. The train would carry 6,000 passengers per hour. Road construction on and adjacent to the airport was planned to take one year. Runway construction was planned to take one year but was deliberately scheduled for two years in order to save on construction costs.

The principal benefits of the new airport compared to Stapleton were:

- A significantly *improved airfield configuration* that allowed for triple simultaneous instrument landings in all weather conditions, improved efficiency and safety of airfield operations, and reduced taxiway congestion
- *Improved efficiency in the operation of the regional airspace,* which, coupled with the increased capacity of the airfield, was supposed to significantly reduce aircraft delays and airline operating costs both at Denver and system-wide
- *Reduced noise impacts* resulting from a large site that was situated in a relatively unpopulated area

- *A more efficient terminal/concourse/apron layout* that minimized passenger walking distance, maximized the exposure of concessions to passenger flows, provided significantly greater curbside capacity, and allowed for the efficient maneuvering of aircraft in and out of gates
- *Improved international facilities* including longer runway lengths for improved stage length capability for international flights and larger Federal Inspection Services (FIS) facilities for greater passenger processing capability
- *Significant expansion capability* of each major functional element of the airport
- *Enhanced efficiency of airline operations* as a result of new baggage handling, communications, deicing, fueling, mail sorting, and other specialty systems

One of the problems with the airport design related to the high wind shears that would exist where the runways were placed. This could eventually become a serious issue.

PROJECT MANAGEMENT

The city of Denver selected two companies to assist in the project management process. The first was Greiner Engineering, an engineering, architecture, and airport planning firm. The second company was Morrison-Knudsen Engineering (MKE) which is a design-construct firm. The city of Denver and Greiner/MKE would function as the project management team (PMT) responsible for schedule coordination, cost control, information management, and administration of approximately 100 design contracts, 160 general contractors, and more than 2000 subcontractors.

In the selection of architects, it became obvious that there would be a split between those who would operate the airport and the city's aspirations. Airport personnel were more interested in an "easy-to-clean" airport and convinced the city to hire a New Orleans-based architectural firm with whom Stapleton personnel had worked previously. The city wanted a "thing of beauty" rather than an easy-to-clean venture.

In an unusual split of responsibilities, the New Orleans firm was contracted to create standards that would unify the entire airport and to take the design of the main terminal only through schematics and design development, at which point it would be handed off to another firm. This sharing of the wealth with several firms would later prove more detrimental than beneficial.

The New Orleans architectural firm complained that the direction given by airport personnel focused on operational issues rather than aesthetic values.

Furthermore, almost all decisions seemed to be made in reaction to maintenance or technical issues. This created a problem for the design team because the project's requirements specified that the design reflect a signature image for the airport, one that would capture the uniqueness of Denver and Colorado.

The New Orleans team designed a stepped-roof profile supported by an exposed truss system over a large central atrium, thus resembling the structure of train sheds. The intent was to bring the image of railroading, which was responsible for Denver's early growth, into the jet age.

The mayor, city council, and others were concerned that the design did not express a $2 billion project. A blue-ribbon commission was formed to study the matter. The city council eventually approved the design.

Financial analysis of the terminal indicated that the roof design would increase the cost of the project by $48 million and would push the project off schedule. A second architectural firm was hired. The final design was a peaked roof with Teflon-coated fabric designed to bring out the image of the Rocky Mountains. The second architectural firm had the additional responsibility to take the project from design development through to construction. The cost savings from the new design was so substantial that the city upgraded the floor finish in the terminal and doubled the size of the parking structure to 12,000 spaces.

The effectiveness of the project management team was being questioned. The PMT failed to sort out the differences between the city's aspirations and the maintenance orientation of the operators. It failed to detect the cost and constructability issues with the first design even though both PMT partners had vast in-house expertise. The burden of responsibility was falling on the shoulders of the architects. The PMT also did not appear to be aware that the first design may not have met the project's standards.

Throughout the design battle, no one heard from the airlines. Continental and United controlled 80 percent of the flights at Stapleton. Yet the airlines refused to participate in the design effort, hoping the project would be canceled. The city ordered the design teams to proceed for bids without any formal input from the users.

With a recession looming in the wings and Contentinal fighting for survival, the city needed the airlines to sign on. To entice the airlines to participate, the city agreed to a stunning range of design changes while assuring the bond rating agencies that the 1993 opening date would be kept. Continental convinced Denver to move the international gates away from the north side of the main terminal to terminal A, and to build a bridge from the main terminal to terminal A. This duplicated the function of a below-ground people-mover system. A basement was added the full length of the concourses. Service cores, located between gates, received a second level.

United's changes were more significant. It widened concourse B by 8 feet to accommodate two moving walkways in each direction. It added a second level of service cores, and had the roof redesigned to provide a clerestory of natural light.

Most important, United wanted a destination-coded vehicle (DCV) baggage handling system where bags could be transferred between gates in less than 10 minutes, thus supporting short turnaround times. The DCV was to be on Concourse B (United) only. Within a few weeks thereafter, DIA proposed that the baggage handling system be extended to the entire airport. Yet even with these changes in place, United and Continental *still* did not sign a firm agreement with DIA, thus keeping bond interest expense at a higher than anticipated level. Some people contended that United and Continental were holding DIA hostage.

From a project management perspective, there was no question that disaster was on the horizon. Nobody knew what to do about the DCV system. The risks were unknown. Nobody realized the complexity of the system, especially the software requirements. By one account, the launch date should have been delayed by at least two years. The contract for DCV hadn't been awarded yet, and terminal construction was already under way. Everyone wanted to know why the design (and construction) was not delayed until after the airlines had signed on. How could DIA install and maintain the terminal's baseline design without having a design for the baggage handling system? Everyone felt that what they were now building would have to be ripped apart.

There were going to be massive scope changes. DIA management persisted in its belief that the airport would open on time. Work in process was now $130 million per month. Acceleration costs, because of the scope changes, would be $30–$40 million. Three shifts were running at DIA with massive overtime. People were getting burned out to the point where they couldn't continue.

To reduce paperwork and maintain the schedule, architects became heavily involved during the construction phase, which was highly unusual. The PMT seemed to be abdicating control to the architects who would be responsible for coordination. The trust that had developed during the early phases began evaporating.

Even the car rental companies got into the act. They balked at the fees for their in-terminal location and said that servicing within the parking structures was inconvenient. They demanded and finally received a separate campus. Passengers would now be forced to take shuttle buses out of the terminal complex to rent or return vehicles.

THE BAGGAGE HANDLING SYSTEM

DIA's $200 million baggage handling system was designed to be state of the art. Conventional baggage handling systems are manual. Each airline operates its own system. DIA opted to buy a single system and lease it back to the airlines. In effect, it would be a one-baggage-system-fits-all configuration.

The system would contain 100 computers, 56 laser scanners, conveyor belts, and thousands of motors. As designed, the system would contain 400 fiberglass carts, each carrying a single suitcase through 22 miles of steel tracks. Operating at 20 miles per hour, the system could deliver 60,000 bags per hour from dozens of gates. United was worried that passengers would have to wait for luggage since several of their gates were more than a mile from the main terminal. The system design was for the luggage to go from the plane to the carousel in 8–10 minutes. The luggage would reach the carousel before the passengers.

The baggage handling system would be centered on track-mounted cars propelled by linear induction motors. The cars slow down, but don't stop, as a conveyor ejects bags onto their platform. During the induction process, a scanner reads the bar-coded label and transmits the data through a programmable logic controller to a radio frequency identification tag on a passing car. At this point, the car knows the destination of the bag it is carrying, as does the computer software that routes the car to its destination. To illustrate the complexity of the situation, consider 4,000 taxicabs in a major city, all without drivers, being controlled by a computer through the streets of a city.

EARLY RISK ANALYSIS

Construction began in 1989 without a signed agreement from Continental and United. By March 1991, the bidding process was in full swing for the main terminal, concourses, and tunnel. Preliminary risk analysis involved three areas: cost, human resources, and weather.

- *Cost:* The grading of the terminal area was completed at about $5 million under budget and the grading of the first runway was completed at about $1.8 million under budget. This led management to believe that the original construction cost estimates were accurate. Also, many of the construction bids being received were below the city's own estimates.
- *Human resources:* The economic recession hit Denver a lot harder than the rest of the nation. DIA was at that time employing about 500 construction workers. By late 1992, it was anticipated that 6,000 construction workers would be needed. Although more than 3,000 applications were on file, there remained the question of available, qualified labor. If the recession were to be prolonged, then the lack of qualified suppliers could be an issue as well.
- *Bad weather:* Bad weather, particularly in the winter, was considered as the greatest risk to the schedule. Fortunately, the winters of 1989–1990 and 1990–1991 were relatively mild, which gave promise to future mild winters. Actually, more time was lost due to bad weather in the summer of 1990 than in either of the two previous winters.

MARCH 1991

By early March 1991, Denver had already issued more than $900 million in bonds to begin construction of the new airport. Denver planned to issue another $500 million in bonds the following month. Standard & Poor's Corporation lowered the rating on the DIA bonds from BBB to BBB−, just a notch above the junk grade rating. This could prove to be extremely costly to DIA because any downgrading in bond quality ratings would force DIA to offer higher yields on their new bond offerings, thus increasing their yearly interest expense.

Denver was in the midst of an upcoming mayoral race. Candidates were calling for the postponement of the construction, not only because of the lower ratings, but also because Denver *still* did not have a firm agreement with either Continental or United Airlines that they would use the new airport. The situation became more intense because three months earlier, in December of 1990, Continental had filed for bankruptcy protection under Chapter 11. Fears existed that Continental might drastically reduce the size of its hub at DIA or even pull out altogether.

Denver estimated that cancelation or postponement of the new airport would be costly. The city had $521 million in contracts that could not be canceled. Approximately $22 million had been spent in debt service for the land, and $38 million in interest on the $470 million in bond money was already spent. The city would have to default on more than $900 million in bonds if it could not collect landing fees from the new airport. The study also showed that a two year delay would increase the total cost by $2 billion to $3 billion and increase debt service to $340 million per year. It now appeared that the point of no return was at hand.

Fortunately for DIA, Moody's Investors Service, Inc. did *not* lower their rating on the $1 billion outstanding of airport bonds. Moody's confirmed their conditional Baa1 rating, which was slightly higher than the S & P rating of BBB−. Moody's believed that the DIA effort was a strong one and that even at depressed airline traffic levels, DIA would be able to service its debt for the scaled-back airport. Had both Moody's and S & P lowered their ratings together, DIA's future might have been in jeopardy.

APRIL 1991

Denver issued $500 million in serial revenue bonds with a maximum yield of 9.185 percent for bonds maturing in 2023. A report by Fitch Investors Service estimated that the airport was ahead of schedule and 7 percent below budget. The concerns of the investor community seemed to have been tempered despite the bankruptcy filing of Continental Airlines. However, there was still concern that no formal agreement existed between DIA and either United Airlines or Continental Airlines.

MAY 1991

The city of Denver and United Airlines finally reached a tentative agreement. United would use 45 of the potential 90–100 gates at Concourse B. This would be a substantial increase from the 26 gates DIA had originally thought that United would require. The 50 percent increase in gates would also add 2,000 reservations jobs. United also expressed an interest in building a $1 billion maintenance facility at DIA employing 6,000 people.

United stated later that the agreement did not constitute a firm commitment but was contingent upon legislative approval of a tax incentive package of $360 million over 30 years plus $185 million in financing and $23 million in tax exemptions. United would decide by the summer in which city the maintenance facility would be located. United reserved the right to renegotiate the hub agreement if DIA was not chosen as the site for the maintenance facility.

Some people believed that United had delayed signing a formal agreement until it was in a strong bargaining position. With Continental in bankruptcy and DIA beyond the point of no return, United was in a favorable position to demand tax incentives of $200 million in order to keep its hub in Denver and build a maintenance facility. The state legislature would have to be involved in approving the incentives. United Airlines ultimately located the $1 billion maintenance facility at the Indianapolis Airport.

AUGUST 1991

Hotel developers expressed concern about building at DIA, which is 26 miles from downtown compared to 8 miles from Stapleton to downtown Denver. DIA officials initially planned for a 1,000-room hotel attached to the airport terminal, with another 300–500 rooms adjacent to the terminal. The 1,000-room hotel had been scaled back to 500–700 rooms and was not likely to be ready when the airport was scheduled to open in October 1993. Developers had expressed resistance to building close to DIA unless industrial and office parks were also built near the airport. Even though ample land existed, developers were putting hotel development on the back burner until after 1993.

NOVEMBER 1991

Federal Express and United Parcel Service (UPS) planned to move cargo operations to the smaller Front Range Airport rather than to DIA. The master plan for DIA called for cargo operations to be at the northern edge of DIA, thus increasing the time and cost for deliveries to Denver. Shifting operations to Front Range Airport would certainly have been closer to Denver but would have alienated northern Adams County cities that counted on an economic boost in their areas. Moving cargo operations would have been in violation of the original agreement between Adams County and Denver for the annexation of the land for DIA.

The cost of renting at DIA was estimated at $0.75 per square foot, compared to $0.25 per square foot at Front Range. DIA would have higher landing fees of $2.68 per 1000 pounds compared to $2.15 for Front Range. UPS demanded a cap on landing fees at DIA if another carrier were to go out of business. Under the UPS proposal, area landholders and businesses would set up a fund to compensate DIA if landing fees were to exceed the cap. Cargo carriers at Stapleton were currently paying $2 million in landing fees and rental of facilities per year.

As the "dog fight" over cargo operations continued, the Federal Aviation Administration (FAA) issued a report calling for cargo operations to be collocated with passenger operations at the busier metropolitan airports. This included both full cargo carriers as well as passenger cargo (i.e., "belly cargo") carriers. Proponents of Front Range argued that the report didn't preclude the use of Front Range because of its proximity to DIA.

DECEMBER 1991

United Airlines formally agreed to a 30-year lease for forty-five gates at Concourse B. With the firm agreement in place, the DIA revenue bonds shot up in price almost $30 per $1000 bond. Earlier in the year, Continental signed a five-year lease agreement.

Other airlines also agreed to service DIA. Exhibit VI sets forth the airlines that either executed use and lease agreements for, or indicated an interest in leasing, the 20 gates on Concourse C on a first-preferential-use basis.

JANUARY 1992

BAE was selected to design and build the baggage handling system. The airport had been under construction for three years before BAE was brought on board. BAE agreed to do eight years of work in two years to meet the October 1993 opening date.

Exhibit VI. Airline agreements

Airline	Term (Years)	Number of Gates
American Airlines	5	3
Delta Air Lines[a]	5	4
Frontier Airlines	10	2
MarkAir	10	5
Northwest Airlines	10	2
TransWorld Airlines	10	2
USAir[a]	5	2
Total		20

[a] The city has entered into Use and Lease Agreements with these airlines. The USAir lease is for one gate on Concourse C and USAir has indicated its interest in leasing a second gate on Concourse C.

JUNE 1992

DIA officials awarded a $24.4 million conract for the new airport's telephone services to U.S. West Communication Services. The officials of DIA had considered controlling its own operations through shared tenant service, which would allow the airport to act as its own telephone company. All calls would be routed through an airport-owned computer switch. By grouping tenants together into a single shared entity, the airport would be in a position to negotiate discounts with long distance providers, thus enabling cost savings to be passed on to the tenants.

By one estimate, the city would generate $3 million to $8 million annually in new, nontax net revenue by owning and operating its own telecommunication network. Unfortunately, DIA officials did not feel that sufficient time existed for them to operate their own system. The city of Denver was unhappy over this lost income.

SEPTEMBER 1992

By September 1992, the city had received $501 million in Federal Aviation Administration grants and $2.3 billion in bonds with interest rates of 9.0–9.5 percent in the first issue to 6 percent in the latest issue. The decrease in interest rates due to the recession was helpful to DIA. The rating agencies also increased the city's bond rating one notch.

The FAA permitted Denver to charge a $3 departure tax at Stapleton with the income earmarked for construction of DIA. Denver officials estimated that over 34 years, the tax would generate $2.3 billion.

The cities bordering the northern edge of DIA (where the cargo operations were to be located) teamed up with Adams County to file lawsuits against DIA in its attempt to relocate cargo operations to the southern perimeter of DIA. This relocation would appease the cargo carriers and hopefully end the year-long battle with Front Range Airport. The Adams County Commissioner contended that relocation would violate the Clean Air Act and the National Environmental Policy Act and would be a major deviation from the original airport plan approved by the FAA.

OCTOBER 1992

The city issued $261 million of Airport Revenue Bonds for the construction of facilities for United Airlines. (See Appendix A at the end of this case.)

MARCH 1993

The city of Denver announced that the launch date for DIA would be pushed back to December 18 rather than the original October 30 date in order to install and test all of the new equipment. The city wanted to delay the opening until late in the first quarter of 1994 but deemed it too costly because the airport's debt would have to be paid without an adequate stream of revenue. The interest on the bond debt was now at $500,000 per day.

The delay to December 18 angered the cargo carriers. This would be their busiest time of the year, usually twice their normal cargo levels, and a complete revamping of their delivery service would be needed. The Washington-based Air Freight Association urged the city to allow the cargo carriers to fly out of Stapleton through the holiday period.

By March 1993, Federal Express, Airborne Express, and UPS (reluctantly) had agreed to house operations at DIA after the city pledged to build facilities for them at the south end of the airport. Negotiations were also underway with Emery Worldwide and Burlington Air Express. The "belly" carriers, Continental and United, had already signed on.

UPS had wanted to create a hub at Front Range Airport. If Front Range Airport were a cargo-only facility, it would free up UPS from competing with passenger traffic for runway access even though both Front Range and DIA were in the same air traffic control pattern. UPS stated that it would not locate a regional hub at DIA. This would mean the loss of a major development project that would have attracted other businesses that relied on UPS delivery.

For UPS to build a regional hub at Front Range would have required the construction of a control tower and enlargement of the runways, both requiring federal funds. The FAA refused to free up funds for Front Range, largely due to a lawsuit by United Airlines and environmental groups.

United's lawsuit had an ulterior motive. Adams County officials repeatedly stated that they had no intention of building passenger terminals at Front Range. However, once federal funds were given to Front Range, a commercial passenger plane could not be prevented from setting up shop in Front Range. The threat to United was the low-cost carriers such as Southwest Airlines. Because costs were fixed, fewer passengers traveling through DIA meant less profits for the airlines. United simply did not want any airline activities removed from DIA!

AUGUST 1993

Plans for a train to connect downtown Denver to DIA were underway. A $450,000 feasibility study and federal environmental assessment were being conducted, with the results due November 30, 1993. Union Pacific had spent $350,000 preparing a design for the new track, which could be constructed in thirteen to sixteen months.

The major hurdle would be the financing, which was estimated between $70 million and $120 million, based upon hourly trips or twenty-minute trips. The more frequent the trips, the higher the cost.

The feasibility study also considered the possibility of baggage check-in at each of the stops. This would require financial support and management assistance from the airlines.

SEPTEMBER 1993

Denver officials disclosed plans for transfering airport facilities and personnel from Stapleton to DIA. The move would be stage-managed by Larry Sweat, a retired military officer who had coordinated troop movements for Operation Desert Shield. Bechtel Corporation would be responsible for directing the transport and setup of machinery, computer systems, furniture, and service equipment, all of which had to be accomplished overnight since the airport had to be operational again in the morning.

OCTOBER 1993

DIA, which was already $1.1 billion over budget, was to be delayed again. The new opening date would be March 1994. The city blamed the airlines for the delays, citing the numerous scope changes required. Even the fire safety system hadn't been completed.

Financial estimates became troublesome. Airlines would have to charge a $15 per person tax, the largest in the nation. Fees and rent charged the airlines would triple from $74 million at Stapleton to $247 million at DIA.

JANUARY 1994

Front Range Airport and DIA were considering the idea of being designated as one system by the FAA. Front Range could legally be limited to cargo only. This would also prevent low-cost carriers from paying lower landing fees and rental space at Front Range.

FEBRUARY 1994

Southwest Airlines, being a low-cost no-frills carrier, said that it would not service DIA. Southwest wanted to keep its airport fees below $3 a passenger. Current projections indicated that DIA would have to charge between $15 and $20 per passenger in order to service its debt. This was based on a March 9 opening day.

Continental announced that it would provide a limited number of low-frill service flights in and out of Denver. Furthermore, Continental said that because of the high landing fees, it would cancel 23 percent of its flights through Denver and relocate some of its maintenance facilities.

United Airlines expected its operating cost to be $100 million more per year at DIA than at Stapleton. With the low-cost carriers either pulling out or reducing service to Denver, United was under less pressure to lower airfares.

MARCH 1994

The city of Denver announced the fourth delay in opening DIA, from March 9 to May 15. The cost of the delay, $100 million, would be paid mostly by United and Continental. As of March, only Concourse C, which housed the carriers other than United and Continental, was granted a temporary certificate of occupancy (TCO) by the city.

As the finger-pointing began, blame for this delay was given to the baggage handling system, which was experiencing late changes, restricted access flow, and a slowdown in installation and testing. A test by Continental Airlines indicated that only 39 percent of baggage was delivered to the correct location. Other problems also existed. As of December 31, 1993, there were 2,100 design changes. The city of Denver had taken out insurance for construction errors and omissions. The city's insurance claims cited failure to coordinate design of the ductwork with ceiling and structure, failure to properly design the storm draining systems for the terminal to prevent freezing, failure to coordinate mechanical and structural designs of the terminal, and failure to design an adequate subfloor support system.

Consultants began identifying potential estimating errors in DIA's operations. The runways at DIA were six times longer than the runways at Stapleton, but DIA had purchased only 25 percent more equipment. DIA's cost projections would be $280 million for debt service and $130 million for operating costs, for a total of $410 million per year. The total cost at Stapleton was $120 million per year.

APRIL 1994

Denver International Airport began having personnel problems. According to DIA's personnel officer, Linda Rubin Royer, moving seventeen miles away from its present site was creating serious problems. One of the biggest issues was the additional twenty-minute drive that employees had to bear. To resolve this problem, she proposed a car/van pooling scheme and tried to get the city bus company to transport people to and from the new airport. There was also the problem of transfering employees to similar jobs elsewhere if they truly disliked working at DIA. The scarcity of applicants wanting to work at DIA was creating a problem as well.

MAY 1994

Standard and Poor's Corporation lowered the rating on DIA's outstanding debt to the noninvestment grade of BB, citing the problems with the baggage handling system and no immediate cure in sight. Denver was currently paying $33.3 million per month to service debt. Stapleton was generating $17 million per month and United Airlines had agreed to pay $8.8 million in cash for the next three months only. That left a current shortfall of $7.5 million each month that the city would have to fund. Beginning in August 1994, the city would be burdened with $16.3 million each month.

BAE Automated Systems personnel began to complain that they were pressured into doing the impossible. The only other system of this type in the world was in Frankfurt, Germany. That system required six years to install and two years to debug. BAE was asked to do it all in two years.

BAE underestimated the complexity of the routing problems. During trials, cars crashed into one another, luggage was dropped at the wrong location, cars that were needed to carry luggage were routed to empty waiting pens, and some cars traveled in the wrong direction. Sensors became coated with dirt, throwing the system out of alignment, and luggage was dumped prematurely because of faulty latches, jamming cars against the side of a tunnel. By the end of May, BAE was conducting a worldwide search for consultants who could determine what was going wrong and how long it would take to repair the system.

BAE conducted an end-of-month test with 600 bags. Outbound (terminal to plane), the sort accuracy was 94 percent and inbound the accuracy was 98 percent. The system had a zero downtime for both inbound and outbound testing. The specification requirements called for 99.5 percent accuracy.

BAE hired three technicians from Germany's Logplan, which helped solve similar problems with the automated system at Frankfurt, Germany. With no opening date set, DIA contemplated opening the east side of the airport for general aviation and air cargo flights. That would begin generating at least some revenue.

JUNE 1994

The cost for DIA was now approaching $3.7 billion and the jokes about DIA appeared everywhere. One common joke as that when you fly to Denver, you will have to stop in Chicago to pick up your luggage. Other common jokes included the abbreviation, DIA. Appendix B provides a listing of some 152 of the jokes.

The people who did not appear to be laughing at these jokes were the concessionaires, including about fifty food service operators, who had been forced to rehire, retrain, and reequip, at considerable expense. Several small businesses were forced to call it quits because of the eight-month delay. Red ink was flowing despite the fact that the $45-a-square-foot rent would not have to be paid until DIA officially opened. Several of the concessionaires had requested that the rent be cut by $10 a square foot for the first six months or so, after the airport opened. A merchant's association was formed at DIA to fight for financial compensation.

THE PROJECT'S WORK BREAKDOWN STRUCTURE (WBS)

The city had managed the design and construction of the project by grouping design and construction activities into seven categories, or *areas:*

    Area #0   Program management/preliminary design
    Area #1   Site development
    Area #2   Roadways and on-grade parking
    Area #3   Airfield
    Area #4   Terminal complex

Area #5   Utilites and specialty systems
Area #6   Other

Since the fall of 1992, the project budget had increased by $224 million (from $2,700 million to $2,924 million), principally as a result of scope changes.

- Structural modifications to the terminal buildings (primarily in the Landside Terminal and Concourse B) to accommodate the automated baggage system
- Changes in the interior configuration of Concourse B
- Increases in the scope of various airline tenant finished, equipment, and systems, particularly in Concourse B
- Grading, drainage, utilities, and access costs associated with the relocation of air cargo facilities to the south side of the airport
- Increases in the scope and costs of communication and control systems, particularly premises wiring
- Increases in the costs of runway, taxiway, and apron paving and change orders as a result of changing specifications for the runway lighting system
- Increased program management costs because of schedule delays

Yet even with all of these design changes, the airport was ready to open except for the baggage handling system.

JULY 1994

The Securities and Exchange Commission (SEC) disclosed that DIA was one of thirty municipal bond issuers that were under investigation for improper contributions to the political campaigns of Pena and his successor, Mayor Wellington Webb. Citing public records, Pena was said to have received $13,900 and Webb's campaign fund increased by $96,000. The SEC said that the contributions may have been in exchange for the right to underwrite DIA's muncipal bond offerings. Those under investigation included Merrill Lynch, Goldman Sachs & Co., and Lehman Brothers, Inc.

AUGUST 1994

Continental confirmed that as of November 1, 1994, it would reduce its flights out of Denver from eighty to twenty-three. At one time, Continental had 200 flights out of Denver.

Denver announced that it expected to sell $200 million in new bonds. Approximately $150 million would be used to cover future interest payments on existing DIA debt and to replenish interest and other money paid due to the delayed opening.

Approximately $50 million would be used to fund the construction of an interim baggage handling system of the more conventional tug-and-conveyor type. The interim system would require 500 to 600 people rather than the 150 to 160 people needed for the computerized system. Early estimates said that the conveyor belt/tug-and-cart system would be at least as fast as the system at Stapleton and would be built using proven technology and off-the-shelf parts. However, modifications would have to be made to both the terminal and the concourses.

United Airlines asked for a thirty-day delay in approving the interim system for fear that it would not be able to satisfy their requirements. The original lease agreement with DIA and United stipulated that on opening day there would be a fully operational automated baggage handling system in place. United had 284 flights a day out of Denver and had to be certain that the interim system would support a twenty-five-minute turnaround time for passenger aircraft.

The city's District Attorney's Office said it was investigating accusations of falsified test data and shoddy workmanship at DIA. Reports had come in regarding fraudulent construction and contracting practices. No charges were filed at that time.

DIA began repairing cracks, holes, and fissures that had emerged in the runways, ramps, and taxiways. Officials said that the cracks were part of the normal settling problems and might require maintenance for years to come.

United Airlines agreed to invest $20 million and act as the project manager to the baggage handling system at Concourse B. DIA picked February 28, 1995, as the new opening date as long as either the primary or secondary baggage handling systems was operational.

UNITED BENEFITS FROM CONTINENTAL'S DOWNSIZING

United had been building up its Denver hub since 1991, increasing its total departures 9 percent in 1992, 22 percent in 1993, and 9 percent in the first six months of 1994. Stapleton is United's second largest connecting hub after Chicago O'Hare (ORD), ahead of San Francisco (SFO), Los Angeles (LAX), and Washington Dulles (IAD) International Airports, as shown in Exhibit VII.

In response to the downsizing by Continental, United is expected to absorb a significant portion of Continental's Denver traffic by means of increased load factors and increased service (i.e. capacity), particularly in larger markets where significant voids in service might be left by Continental. United served twenty-four of the twenty-eight cities served by Continental from Stapleton in June, 1994, with about 79 percent more total available seats to those cities—23,937 seats

**Exhibit VII.   Comparative United Airlines service at hub airports,
June 1983 and June 1994**

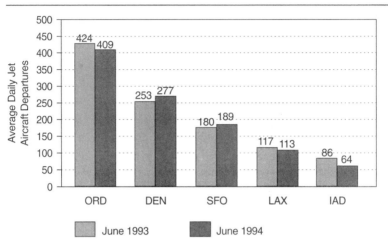

Note: Does not include activity by United Express.

Source: Official Airline Guides, Inc.
(On-line Database), for periods shown.

provided by United compared with 13,400 seats provided by Continental. During 1993, United's average load factor from Denver was 63 percent, indicating that, with its existing service and available capacity, United had the ability to absorb many of the passengers abandoned by Continental. In addition, United had announced plans to increase service at Denver to 300 daily flights by the end of the calendar year.

As a result of its downsizing in Denver, Continental was forecasted to lose more than 3.9 million enplaned passengers from 1993 to 1995—a total decrease of 80 percent. However, this decrease was expected to be largely offset by the forecasted 2.2 million increase in enplaned passengers by United and 1.0 million by the other airlines, resulting in a total of 15,877,000 enplaned passengers at Denver in 1995. As discussed earlier, it was assumed that, in addition to a continuation of historical growth, United and the other airlines would pick up much of the traffic abandoned by Continental through a combination of added service, larger average aircraft size, and increased load factors.

From 1995 to 2000, the increase in total enplaned passengers is based on growth rates of 2.5 percent per year in originating passengers and 3.0 percent per year in connecting passengers. Between 1995 and 2000, United's emerging dominance at the airport (with almost twice the number of passengers of all other

Exhibit VIII. Enplaned passenger market shares at Denver Airports

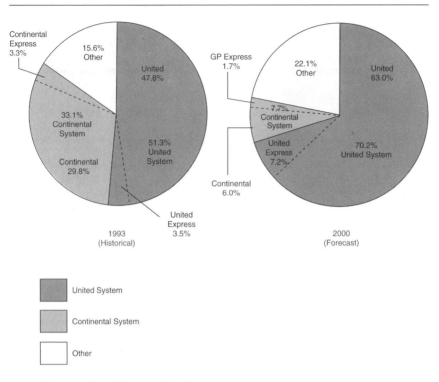

United System

Continental System

Other

Source: 1993: Airport Management Records.

airlines combined) should result in somewhat higher fare levels in the Denver markets, and therefore may dampen traffic growth. As shown in Exhibit VIII, of the 18.2 million forecasted enplaned passengers in 2000, United and United Express together are forecasted to account for 70 percent of total passengers at the airport—up from about 51 percent in 1993—while Continental's share, including GP Express, is forecasted to be less than 8 percent—down from about 33 percent in 1993.

Total connecting passengers at Stapleton increased from about 6.1 million in 1990 to about 8.2 million in 1993—an average increase of about 10 percent per year. The number of connecting passengers was forecast to decrease in 1994 and 1995, as a result of the downsizing by Continental, and then return to steady growth of 3.0 percent per year through 2000, reflecting expected growth in passenger traffic nationally and a stable market share by United in Denver. Airline market share of connecting passengers in 1993 and 1995 are shown in Exhibit IX.

Exhibit IX. Connecting passenger market shares at Denver Airports

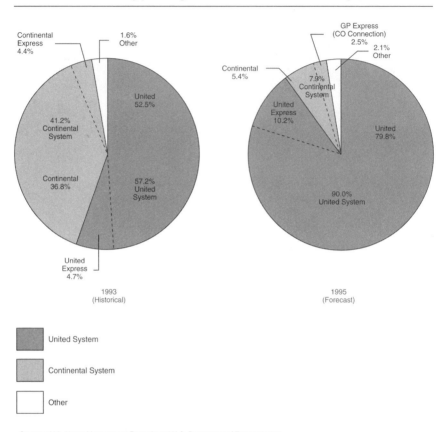

Continental Express 4.4%

1.6% Other

Continental 5.4%

GP Express (CO Connection) 2.5%

2.1% Other

United 52.5%

41.2% Continental System

Continental 36.8%

57.2% United System

United Express 4.7%

1993 (Historical)

7.9% Continental System

United Express 10.2%

United 79.8%

90.0% United System

1995 (Forecast)

United System

Continental System

Other

Source: 1993: Airport Management Records and U.S. Department of Transportation

SEPTEMBER 1994

Denver began discussions with cash-strapped MarkAir of Alaska to begin service at DIA. For an undercapitalized carrier, the prospects of tax breaks, favorable rents, and a $30 million guaranteed city loan were enticing.

DIA officials estimated an $18 per person charge on opening day. Plans to allow only cargo carriers and general aviation to begin operations at DIA were canceled.

Total construction cost for the main terminal exceeded $455 million (including the parking structure and the airport office building). See Exhibit X.

Exhibit X. *Total construction costs for Denver Airport*

General site expenses, commission	$ 38,667,967
Sitework, building excavations	15,064,817
Concrete	89,238,296
Masonry	5,501,608
Metals	40,889,411
Carpentry	3,727,408
Thermal, moisture protection	8,120,907
Doors and windows	13,829,336
Finishes	37,025,019
Specialties	2,312,691
Building equipment	227,720
Furnishings	3,283,852
Special construction	39,370,072
Conveying systems	23,741,336
Mechanical	60,836,566
Electrical	73,436,575
Total	$455,273,581

OCTOBER 1994

A federal grand jury convened to investigate faulty workmanship and falsified records at DIA. The faulty workmanship had resulted in falling ceilings, buckling walls, and collapsing floors.

NOVEMBER 1994

The baggage handling system was working, but only in segments. Frustration still existed in not being able to get the whole system to work at the same time. The problem appeared to be with the software required to get computers to talk to computers. The fact that a mere software failure could hold up Denver's new airport for more than a year put in question the project's risk management program.

Jerry Waddles was the risk manager for Denver. He left that post to become risk manager for the State of Colorado. Eventually the city found an acting risk manager, Molly Austin Flaherty, to replace Mr. Waddles, but for the most part, DIA construction over the past several months had continued without a full-time risk manager.

The failure of the baggage handling system had propelled DIA into newspaper headlines around the country. The U.S. Securities and Exchange Commission had launched a probe into whether Denver officials had deliberately deceived bondholders about how equipment malfunctions would affect the December 19, 1993 opening. The allegations were made by Denver's KCNC-TV. Internal memos indicated that in the summer of 1993 city engineers believed it would take at least until March, 1994 to get the system working. However, Mayor Wellington Webb did not announce the delayed opening until October 1993. The SEC was investigating whether the last postponement misled investors holding $3 billion in airport bonds.

Under a new agreement, the city agreed to pay BAE an additional $35 million for modifications *if* the system was working for United Airlines by February 28, 1995. BAE would then have until August 1995 to complete the rest of the system for the other tenants. If the system was not operational by February 28, the city could withhold payment of the $35 million.

BAE lodged a $40 million claim against the city, alleging that the city caused the delay by changing the system's baseline configuration after the April 1, 1992, deadline. The city filed a $90 million counterclaim, blaming BAE for the delays.

The lawsuits were settled out of court when BAE agreed to pay $12,000 a day in liquidated damages dating from December 19, 1993, to February 28, 1995, or approximately $5 million. The city agreed to pay BAE $6.5 million to cover some invoices submitted by BAE for work already done to repair the system.

Under its DIA construction contract, BAE's risks were limited. BAE's liability for consequential damages resulting from its failure to complete the baggage handling system on time was capped at $5 million. BAE had no intention of being held liable for changes to the system. The system as it was at the time was not the system that BAE had been hired to install.

Additional insurance policies also existed. Builder's risk policies generally pay damages caused by defective parts or materials, but so far none of the parts used to construct the system had been defective. BAE was also covered for design errors or omissions. The unknown risk at that point was who would be responsible if the system worked for Concourse B (i.e., United) but then failed when it was expanded to cover all concourses.

A study was underway to determine the source of respiratory problems suffered by workers at the construction site. The biggest culprit appeared to be the use of concrete in a confined space.

The city and DIA were also protected from claims filed by vendors whose businesses were put on hold because of the delays under a hold-harmless agreement in the contracts. However, the city had offered to permit the concessionaires to charge higher fees and also to extend their leases for no charge to make up for lost income due to the delays.

DECEMBER 1994

The designer of the baggage handling system was asked to reexamine the number of bags per minute that the BAE system was required to accommodate as per the specifications. The contract called for departing luggage to Concourse A to be delivered at a peak rate of 90 bags per minute. The designer estimated peak demand at 25 bags per minute. Luggage from Concourse A was contracted for at 223 bags per minute but again, the designer calculated peak demand at a lower rate of 44 bags per minute.

AIRPORT DEBT

By December 1994, DIA was more than $3.4 billion in debt, as shown in Exhibit XI.

AIRPORT REVENUE

Airports generally have two types of contracts with their tenants. The first type is the residual contract where the carriers guarantee that the airport will remain solvent. Under this contract, the carriers absorb the majority of the risk. The airport maintains the right to increase rents and landing fees to cover operating expenses and debt coverage. The second type of contract is the compensatory contract where the airport is at risk. DIA has a residual contract with its carriers.

Airports generate revenue from several sources. The most common breakdown includes landing fees and rent from the following entities: airline carriers, passenger

Exhibit XI. Outstanding debt at Denver Airport

Series 1984 Bonds	$ 103,875,000
Series 1985 Bonds	175,930,000
Series 1990A Bonds	700,003,843
Series 1991A Bonds	500,003,523
Series 1991D Bonds	600,001,391
Series 1992A Bonds	253,180,000
Series 1992B Bonds	315,000,000
Series 1992C Bonds	392,160,000
Series 1992D–G Bonds	135,000,000
Series 1994A Bonds	257,000,000
	$3,432,153,757

facilities, rental car agencies, concessionary stores, food and beverage services, retail shops, and parking garages. Retail shops and other concessionary stores also pay a percent of sales.

AIRLINE COSTS PER ENPLANED PASSENGER

Revenues derived from the airlines are often expressed on a per enplaned passenger basis. The average airline cost per enplaned passenger at Stapleton in 1993 was $5.02. However, this amount excludes costs related to major investments in terminal facilities made by United Airlines in the mid-1980s and, therefore, understates the true historical airline cost per passenger.

Average airline costs per enplaned passenger at the airport in 1995 and 2000 are forecast to be as shown in Exhibit XII.

The forecasted airline costs per enplaned passenger at the airport are considerably higher than costs at Stapleton today and the highest of any major airport in the United States. (The cost per enplaned passenger at Cleveland Hopkins is $7.50). The relatively high airline cost per passenger is attributable, in part, to (1) the unusually large amount of tenant finishes, equipment, and systems costs being financed as part of the project relative to other airport projects and (2) delayed costs incurred since the original opening date for purposes of the Plan of Financing (January 1, 1994).

The City estimates that, as a result of the increased capacity and efficiency of the airfield, operation of the airport will result in annual delay savings to the airlines of $50 million to $100 million per year (equivalent to about $3 to $6 per enplaned passenger), and that other advanced technology and systems incorporated into the design of the airport will result in further operational savings. In the final analysis, the cost effectiveness of operating at the airport is a judgment that must be made by the individual airlines in deciding to serve the Denver market.

It is assumed for the purposes of this analysis that the city and the airlines will resolve the current disputes regarding cost allocation procedures and responsibility for delay costs, and that the airlines will pay rates generally in accordance with the procedures of the use and lease agreements as followed by the city and as summarized in the accompanying exhibits.

Exhibit XII. *Total average airline costs per enplaned passenger*

Year	Current Dollars	1990 Dollars
1995	$18.15	$14.92
2000	17.20	11.62

FEBRUARY 28, 1995

The airline opened as planned on February 28, 1995. However, several problems became apparent. First, the baggage handling system did have "bad days." Passengers traveling to and from Denver felt more comfortable carrying bags than having them transfered by the computerized baggage handling system. Large queues began to form at the end of the escalators in the main terminal going down to the concourse trains. The trains were not running frequently enough, and the number of cars in each train did not appear to be sufficient to handle the necessary passenger traffic.

The author flew from Dallas–Ft. Worth to Denver in one hour and 45 minutes. It then took one hour and 40 minutes to catch the airport shuttles (which stop at all the hotels) and arrive at the appropriate hotel in downtown Denver. Passengers began to balk at the discomfort of the remote rental car facilities, the additional $3 tax per day for each rental car, and the fact that the nearest gas station was fifteen miles away. How does one return a rental car with a full tank of gas?

Departing passengers estimated it would take two hours to drive to the airport from downtown Denver, unload luggage, park their automobile, check in, and take the train to the concourse.

Faults in the concourse construction were becoming apparent. Tiles that were supposed to be 5/8 inches thick were found to be 1/2 inch thick. Tiles began to crack. During rainy weather, rain began seeping in through the ceiling.

Appendix A*
Municipal Bond Prospectus

$261,415,000
City and County Of Denver, Colorado
6.875% Special Facilities Airport Revenue Bonds
(United Airlines Project)
Series 1992A
Date: October 1, 1992
Due: October 1, 2032
Rating: Standard & Poor's BBB–
Moody's Baa2

INTRODUCTION

This official statement is provided to furnish information in connection with the sale by the City and County of Denver, Colorado (the "City") of 6.875% Special Facilities Airport Revenue Bonds (United Airlines Project) series 1992A in the aggregate

---

*Only excerpts from the prospectus are included here.

principle amount of $261,415,000 (the "Bonds"). The bonds will be dated, mature, bear interest, and be subject to redemption prior to maturity as described herein.

The Bonds will be issued pursuant to an Ordinance of the City and County of Denver, Colorado (the "Ordinance").

The proceeds received by the City from the sale of the Bonds will be used to acquire, construct, equip, or improve (or a reimbursement of payments for the acquisition, construction, equipping, or improvement of) certain terminals, Concourse B, aircraft maintenance, ground equipment maintenance, flight kitchen, and air freight facilities (the "Facilities") at the new Denver International Airport (the "New Airport").

The City will cause such proceeds to be deposited, distributed, and applied in accordance with the terms of a Special Facilities and Ground Lease, dated as of October 1, 1992 (the "Lease") between United Airlines and the City. Under the Lease, United has agreed to make payments sufficient to pay the principal, premium, if any, and interest on the Bonds. Neither the Facilities nor the ground rental payments under the Lease are pledged as security for the payment of principal, premium, if any, and interest on the bonds.

AGREEMENT BETWEEN UNITED AND THE CITY

On June 26, 1991, United and the City entered into an agreement followed by a second agreement on December 12, 1991, which, among other things, collectively provide for the use and lease by United of certain premises and facilities at the New Airport. In the United Agreement, United agrees among other things, to (1) support the construction of the New Airport, (2) relocate its present air carrier operations from Stapleton to the New Airport, (3) occupy and lease certain facilities at the New Airport, including no less than 45 gates on Concourse B within two years of the date of beneficial occupancy as described in the United Agreement, and (4) construct prior to the date of beneficial occupancy, a regional reservation center at a site at Stapleton.

In conjunction with the execution of the United Agreement, United also executes a 30-year use and lease agreement. United has agreed to lease, on a preferential use basis, Concourse B, which is expected to support 42 jet aircraft with up to 24 commuter aircraft parking positions at the date of beneficial occupancy, and, on an exclusive use basis, certain ticket counters and other areas in the terminal complex of the New Airport.

THE FACILITIES

The proceeds of the bonds will be used to finance the acquisition, construction, and equipping of the Facilities, as provided under the Lease. The Facilities will be located on approximately 100 acres of improved land located within the New Airport, which United will lease from the City. The Facilities will include an aircraft maintenance facility capable of housing ten jet aircraft, a ground equipment support facility with 26 maintenance bays, an approximately 55,500-square-foot air freight facility, and an approximately 155,000-square-foot flight kitchen. Additionally, the proceeds of the Bonds will be used to furnish, equip, and install certain facilities to be used by United in Concourse B and in the terminal of the New Airport.

REDEMPTION OF BONDS

The Bonds will be subject to optional and mandatory redemption prior to maturity in the amounts, at the times, at the prices, and in the manner as provided in the Ordinance. If less than all of the Bonds are to be redeemed, the particular Bonds to be called for redemption will be selected by lot by the Paying Agent in any manner deemed fair and reasonable by the Paying Agent.

The bonds are subject to redemption prior to maturity by the City at the request of United, in whole or in part, by lot, on any date on or after October 1, 2002, from an account created pursuant to the Ordinance used to pay the principal, premium, if any, and interest on the Bonds (the "Bond Fund") and from monies otherwise available for such purpose. Such redemptions are to be made at the applicable redemption price shown below as a percentage of the principal amount thereof, plus interest accrued to the redemption date:

Redemption Period	Optional Redemption Price
October 1, 2002 through September 30, 2003	102%
October 1, 2003 through September 30, 2004	101%
October 1, 2004 and thereafter	100%

The Bonds are subject to optional redemption prior to maturity, in whole or in part by lot, on any date, upon the exercise by United of its option to prepay Facilities Rentals under the Lease at a redemption price equal to 100% of the principal amount thereof plus interest accrued to the redemption date, if one or more of the following events occurs with respect to one or more of the units of the Leased Property:

(a) the damage or destruction of all or substantially all of such unit or units of the Leased Property to such extent that, in the reasonable opinion of United, repair and restoration would not be economical and United elects not to restore or replace such unit or units of the Leased Property; or,

(b) the condemnation of any part, use, or control of so much of such unit or units of the Leased Property that such unit or units cannot be reasonably used by United for carrying on, at substantially the same level or scope, the business theretofore conducted by United on such unit or units.

In the event of a partial extraordinary redemption, the amount of the Bonds to be redeemed for any unit of the Leased Property with respect to which such prepayment is made shall be determined as set forth below (expressed as a percentage of the original principal amount of the Bonds) plus accrued interest on the Bonds to be redeemed to the redemption date of such Bonds provided that the amount of Bonds to be redeemed may be reduced by the aggregate principal amount (valued at par) of any Bonds purchased by or on behalf of United and delivered to the Paying Agent for cancelation:

Terminal Concourse B Facility	Aircraft Maintenance Facility	Ground Equipment Maintenance Facility	Flight Kitchen	Air Freight Facility
20%	50%	10%	15%	5%

The Bonds shall be subject to mandatory redemption in whole prior to maturity, on October 1, 2023, at a redemption price equal to 100% of the principal amount thereof, plus accrued interest to the redemption date if the term of the Lease is not extended to October 1, 2032, in accordance with the provisions of the Lease and subject to the conditions in the Ordinance.

LIMITATIONS

Pursuant to the United Use and Lease Agreement, if costs at the New Airport exceed $20 per revenue enplaned passenger, in 1990 dollars, for the preceding calendar year, calculated in accordance with such agreement, United can elect to terminate its Use and Lease Agreement. Such termination by United would not, however, be an event of default under the Lease.

If United causes an event of default under the Lease and the City exercises its remedies thereunder and accelerates Facilities Rentals, the City is not obligated to relet the Facilities. If the City relets the Facilities, it is not obligated to use any of the payments received to pay principal, premium, if any, or interest on the Bonds.

APPLICATION OF THE BOND PROCEEDS

It is estimated that the proceeds of the sale of the Bonds will be applied as follows:

Cost of Construction	$226,002,433
Interest on Bonds During Construction	22,319,740
Cost of Issuance Including Underwriters' Discount	1,980,075
Original Issue Discount	11,112,742
Principal Amount of the Bonds	$261,415,000

TAX COVENANT

Under the terms of the lease, United has agreed that it will not take or omit to take any action with respect to the Facilities or the proceeds of the bonds (including any investment earnings thereon), insurance, condemnation, or any other proceeds derived in connection with the Facilities, which would cause the interest on the Bonds to become included in the gross income of the Bondholder for federal income tax purposes.

OTHER MATERIAL COVENANTS

United has agreed to acquire, construct, and install the Facilities to completion pursuant to the terms of the Lease. If monies in the Construction Fund are insufficient to pay the cost of such acquisition, construction, and installation in full, then United shall pay the excess cost without reimbursement from the City, the Paying Agent, or any Bondholder.

United has agreed to indemnify the City and the Paying Agent for damages incurred in connection with the occurrence of certain events, including without limitation, the construction of the Facilities, occupancy by United of the land on which the Facilities are located, and violation by United of any of the terms of the Lease or other agreements related to the Leased Property.

During the Lease Term, United has agreed to maintain its corporate existence and its qualifications to do business in the state. United will not dissolve or otherwise dispose of its assets and will not consolidate with or merge into another corporation provided, however, that United may, without violating the Lease, consolidate or merge into another corporation.

ADDITIONAL BONDS

At the request of United, the City may, at its option, issue additional bonds to finance the cost of special Facilities for United upon the terms and conditions in the Lease and the Ordinance.

THE GUARANTY

Under the Guaranty, United will unconditionally guarantee to the Paying Agent, for the benefit of the Bondholders, the full and prompt payment of the principal, premium, if any, and interest on the Bonds, when and as the same shall become due whether at the stated maturity, by redemption, acceleration, or otherwise. The obligations of United under the Guaranty are unsecured, but are stated to be absolute and unconditional, and the Guaranty will remain in effect until the entire principal, premium, if any, and interest on the Bonds has been paid in full or provision for the payment thereof has been made in accordance with the Ordinance.

Appendix B
Jokes about the Abbreviation DIA

DENVER—The Denver International Airport, whose opening has been delayed indefinitely because of snafus, has borne the brunt of joke writers

Punsters in the aviation and travel community have done their share of work on one particular genre, coming up with new variations on the theme of DIA, the star-crossed airport's new and as-yet-unused city code.

Here's what's making the rounds on electronic bulletin boards; it originated in the May 15 issue of the *Boulder* (Colo.) *Camera* newspaper.

1. Dis Is Awful
2. Doing It Again
3. Dumbest International Airport
4. Dinosaur In Action
5. Debt In Arrival
6. Denver's Intense Adventure
7. Darn It All
8. Dollar Investment Astounding
9. Delay It Again
10. Denver International Antique
11. Date Is AWOL
12. Denver Intellects Awry
13. Dance Is Autumn
14. Dopes In Authority

15. Don't Ice Attendance
16. Drop In Asylum
17. Don't Immediately Assume
18. Don't Ignore Aspirin
19. Dittohead Idle Again
20. Doubtful If Atall
21. Denver In Action
22. Deces, l'Inaugural Arrivage (means "dead on arrival" in French)
23. Dummies In Action
24. Dexterity In Action
25. Display In Arrogance
26. Denver Incomplete Act
27. D'luggage Is A'coming
28. Defect In Automation
29. Dysfunctional Itinerary Apparatus
30. Dis Is Absurd
31. Delays In Abundance
32. Did It Arrive?
33. Denver's Infamous Air-or-port (sounds like "error")
34. Dopes In Action
35. Doubtful Intermittent Access
36. Don't Intend Atall
37. Damned Inconvenient Airport
38. Duped In Anticipation
39. Delay In Action
40. Delirious In Accounting
41. Date Indeterminate, Ah?
42. Denver's Indisposed

Access
43. Detained Interphase Ahead
44. Denver's Interminably Aground
45. Deceit In Action
46. Delay Institute America
47. Denver's Intractable Airport
48. Delayed Indefinitely Again
49. Delayed Introduction Again
50. Disaster In Arrears
51. Denver International Amusementpark
52. Debacle In Action
53. Deadline (of) Incomprehensible Attainment
54. Duffel Improbable Arrival
55. Delay In America
56. Dying In Anticipation
57. Dazzling Inaccessible Absurdity
58. Damned Intractable Automation
59. Da Infamous Annoyance
60. Dare I Ask?
61. Done In Arrears
62. Done In Ancestral
63. Denver International Accident
64. Dumb Idea Anyway
65. Diversion In Accounting

Appendix B (continued).
Jokes about the Abbreviation DIA

66. Doesn't Include Airlines
67. Disparate Instruments in Action
68. Delay International Airport
69. Dumb Idea Askew
70. Delayed Indefinitely Airport
71. Delays In Arrival
72. Deja In Absentee
73. Done In Aminute
74. Done In August
75. Denver's Inordinate Airport
76. Denver's Imaginary Airport
77. Debentures In Arrears
78. Denver Isn't Airborne
79. Descend Into Abyss
80. Done In April 2000
81. Disaster In Aviation
82. Denver's Interminable Airport
83. Denver In Arrears
84. Dallying Is Aggravating
85. Don't In Angst
86. Distress Is Acute
87. Development Is Arrested
88. Darned Inevitable Atrocity
89. Debt In Airport
90. Devastation In Aviation
91. Debacle in Automation
92. Denver's Inconstructable Airport
93. Denver Is Awaitin'
94. DIsAster
95. Denver's Inoperable Airport

96. Delay, Impede, Await
97. Date Isn't Available
98. Delayed International Airport
99. Denver Irrational Airport
100. Denver Irate Association
101. Denver's Ignominious Atrocity
102. Daytrippers Invitational Airport
103. Delay Is Anticipated
104. Doofis, Interruptness, Accidentalis
105. Denver International Arrival
106. Denver's Interminable Apparition
107. Distance Is Astronomical
108. Doubtful It's Able
109. Dreadfully Ineffective Automation
110. Do It Again
111. Did it, Installed it, Ate it
112. Drowned In Apoplexy
113. Dodo International Airport (the dodo is an extinct, flightless bird)
114. Dead In the Air
115. Denouncement In Ambiguity
116. Deserted, Inactive Airport
117. Definitely Incapable of Activation
118. Democracy In Action
119. Dysfunction Imitating Art
120. Design In Alabaster
121. Desperately In Arrears
122. Dazzling, If Anything
123. Delays In Aeternum

124. Delighted If Actualized
125. Destination: Imagine Arabia
126. Dumb Idea: Abandoned?
127. Deem It Apiary
128. Dollars In Action
129. Definitely Iffy Achievement
130. Dreadfully Incompetent Architects
131. Denver International Ain't
132. Delayed In Automation
133. Dragging Its Ass
134. Driving Is Advantageous
135. Dang It All
136. Druggies Installing Automation
137. Dumb Idea Approved
138. Didn't Invite Airplanes
139. Died In April
140. Deplane In Albuquerque
141. Departure Is Agonizing
142. Denver's Infuriating Abscess
143. Denver's Ill-fated Airport
144. Domestic International Aggravation
145. Duffels In Anchorage
146. Denver's Indeterminate Abomination
147. Damn It All
148. Darn Idiotic Airport
149. Delay Is Acceptable
150. Denver's Idle Airport
151. Does It Arrive?
152. Damned Inconvenient Anyway

Source: Reprinted from *Boulder* (Colorado) *Camera* newspaper (May 15, 1991).

REFERENCES (IN CHRONOLOGICAL ORDER)

David A. Brown, "Denver Aims for Global Hub Status with New Airport Under Construction," *Aviation Week & Space Technology* (March 11, 1991), pp. 42–45.

"Satellite Airport to Handle Corporate, General Aviation for Denver Area," *Aviation Week & Space Technology* (March 11, 1991), pp. 44–45.

"Denver to Seek Bids This Spring for Wide-Open Terminal Building," *Aviation Week & Space Technology* (March 11, 1991), p. 50.

"Denver City Council Supports Airport Despite Downgrade," *The Wall Street Journal* (March 20, 1991), p. A1D.

"Denver Airport Bonds' Rating Is Confirmed by Moody's Investors," *The Wall Street Journal* (March 22, 1991), p. C14.

"Bonds for Denver Airport Priced to Yield up to 9.185%," *New York Times* (April 10, 1991), p. D16.

Marj Charlier, "Denver Reports a Tentative Agreement with United over Hub at New Airport," *The Wall Street Journal* (May 3, 1991), p. B2.

Brad Smith, "New Airport Has Its Ups and Downs," *Los Angeles Times* (July 9, 1991), p. A5.

Christopher Wood, "Hotel Development at New Airport Not Likely Until After '93," *Denver Business Journal* (August 2, 1991), p. 8S.

Christopher Wood, "FAA: Link Air Cargo, Passengers," *Denver Business Journal* (November 1–7, 1991), p. 3.

Christopher Wood, "Airport May Move Cargo Operations, Offer Reserve Funds," *Denver Business Journal* (December 6–12, 1991), pp. 1, 34.

"UAL in Accord on Denver," *The New York Times* (December 7, 1991), p. 39L.

Thomas Fisher, "Projects Flights of Fantasy," *Progressive Architecture* (March 1992), p. 103.

Tom Locke, "Disconnected," *Denver Business Journal* (June 12–18, 1992), p. 19.

"Big Ain't Hardly the Word for It," *ENR* (September 7, 1992), pp. 28–29.

Christopher Wood, "Adams Seeks Action," *Denver Business Journal* (September 4–10, 1992), pp. 1, 13.

"Denver Airport Rises under Gossamer Roof," *The Wall Street Journal* (November 17, 1992), p. B1.

Mark B. Solomon, "Denver Airport Delay Angers Cargo Carriers," *Journal of Commerce* (March 17, 1993), p. 3B.

"Denver Airport Opening Delayed Until December," *Aviation Week & Space Technology* (May 10, 1993), p. 39.

Aldo Svaldi, "DIA Air Train Gathering Steam as Planners Shift Possible Route," *Denver Business Journal* (August 27–September 2, 1993), p. 74.

Dirk Johnson, "Opening of New Denver Airport is Delayed Again," *The New York Times* (October 26, 1993), p. A19.

"Denver's Mayor Webb Postpones Opening International Airport," *The Wall Street Journal* (October 26, 1993), p. A9.

"An Airport Comes to Denver," *Skiing* (December 1993), p. 66.

Ellis Booker, "Airport Prepares for Takeoff," *Computerworld* (January 10, 1994).

Aldo Svaldi, "Front Range, DIA Weigh Merging Airport Systems," *Denver Business Journal* (January 21–27, 1994), p. 3.

Don Phillips, "$3.1 Billion Airport at Denver Preparing for a Rough Takeoff," *The Washington Post* (February 13, 1994), p. A10.

"New Denver Airport Combines Several State-of-the-Art Systems," *Travel Weekly* (February 21, 1994) p. 20.

Steve Munford, "Options in Hard Surface Flooring," *Buildings* (March 1994), p. 58.

Mars Charles, "Denver's New Airport, Already Mixed in Controversy, Won't Open Next Week," *The Wall Street Journal* (March 2, 1994), pp. B1, B7.

"Denver Grounded for Third Time," *ENR* (March 7, 1994), p. 6.

Shannon Peters, "Denver's New Airport Creates HR Challenges," *Personnel Journal* (April 1994), p. 21.

Laura Del Rosso, "Denver Airport Delayed Indefinitely," *Travel Weekly* (May 5, 1994), p. 37.

"DIA Bond Rating Cut," *Aviation Week & Space Technology* (May 16, 1994), p. 33.

Robert Scheler, "Software Snafu Grounds Denver's High-Tech Airport," *PC Week* (May 16, 1994), p. 1.

John Dodge, "Architects Take a Page from Book on Denver Airport-Bag System," *PC Week* (May 16, 1994), p. 3.

Jean S. Bozman, "Denver Airport Hits Systems Layover," *Computerworld* (May 16, 1994), p. 30.

Richard Woodbury, "The Bag Stops Here," *Time* (May 16, 1994), p. 52.

"Consultants Review Denver Baggage Problems," *Aviation Week & Space Technology* (June 6, 1994), p. 38.

"Doesn't It Amaze? The Delay that Launched a Thousand Gags," *Travel Weekly* (June 6, 1994), p. 16.

Michael Romano, "This Delay Is Costing Business a Lot of Money," *Restaurant Business* (June 10, 1994), p. 26.

Scott Armstrong, "Denver Builds New Airport, Asks 'Will Planes Come?'" *The Christian Science Monitor* (June 21, 1994), p. 1.

Benjamin Weiser, "SEC Turns Investigation to Denver Airport Financing," *The Washington Post* (July 13, 1994), p. D1.

Bernie Knill, "Flying Blind at Denver International Airport," *Material Handling Engineering* (July 1994), p. 47.

Keith Dubay, "Denver Airport Seeks Compromise on Baggage Handling," *American Banker Washington Watch* (July 25, 1994), p. 10.

Dirk Johnson, "Denver May Open Airport in Spite of Glitches," *The New York Times* (July 27, 1994), p. A14.

Jeffrey Leib, "Investors Want a Plan," *The Denver Post* (August 2, 1994), p. A1.

Marj Charlier, "Denver Plans Backup Baggage System for Airport's Troubled Automated One," *The Wall Street Journal* (August 5, 1994), p. B2.

Louis Sahagun, "Denver Airport to Bypass Balky Baggage Mover," *Los Angeles Times* (August 5, 1994), p. A1.

Len Morgan, "Airports Have Growing Pains," *Flying* (August 1994), p. 104.

Adam Bryant, "Denver Goes Back to Basics for Baggage," *The New York Times* (August 6, 1994), pp. 5N, 6L.

"Prosecutors Scrutinize New Denver Airport," *The New York Times* (August 21, 1994), p. 36L.

Kevin Flynn, "Panic Drove New DIA Plan," *Rocky Mountain News* (August 7, 1994), p. 5A.

David Hughes, "Denver Airport Still Months from Opening," *Aviation Week & Space Technology* (August 8, 1994), p. 30.

"Airport May Open in Early '95," *Travel Weekly* (August 8, 1994), p. 57.

Michael Meyer, and Daniel Glick, "Still Late for Arrival," *Newsweek* (August 22, 1994), p. 38.

Andrew Bary, "A $3 Billion Joke," *Barron's* (August 22, 1994), p. MW10.

Jean Bozman, "Baggage System Woes Costing Denver Airport Millions," *Computerworld* (August 22, 1994), p. 28.

Edward Phillips, "Denver, United Agree on Baggage System Fixes," *Aviation Week & Space Technology* (August 29, 1994).

Glenn Rifkin, "What Really Happened at Denver's Airport," *Forbes* (August 29, 1994), p. 110.

Andrew Bary, "New Denver Airport Bond Issue Could Face Turbulence from Investors," *Barron's* (August 29, 1994), p. MW9.

Andrew Bary, "Denver Airport Bonds Take Off as Investors Line Up for Higher Yields," *Barron's* (August 29, 1994), p. MW9.

Susan Carey, "Alaska's Cash-Strapped MarkAir Is Wooed by Denver," *The Wall Street Journal* (September 1, 1994), p. B6.

Dana K. Henderson, "It's in the Bag(s)," *Air Transport World* (September 1994), p. 54.

Dirk Johnson, "Late Already, Denver Airport Faces More Delays," *The New York Times* (September 25, 1994), p. 26L.

Gordon Wright, "Denver Builds a Field of Dreams," *Building Design and Construction* (September 1994), p. 52.

Alan Jabez, "Airport of the Future Stays Grounded," *Sunday Times* (October 9, 1994), Features Section.

Jean Bozman, "United to Simplify Denver's Troubled Baggage Project," *Computerworld* (October 10, 1994), p. 76.

"Denver Aide Tells of Laxity in Airport Job," *The New York Times* (October 17, 1994), p. A12.

Brendan Murray, "In the Bags: Local Company to Rescue Befuddled Denver Airport," *Marietta Daily Journal* (October 21, 1994), p. C1.

Joanne Wojcik, "Airport in Holding Pattern, Project Is Insured, but Denver to Retain Brunt of Delay Costs," *Business Insurance* (November 7, 1994), p. 1.
James S. Russell, "Is This Any Way to Build an Airport?," *Architectural Record* (November 1994), p. 30.

QUESTIONS

1. Is the decision to build a new airport at Denver strategically a sound decision?
2. Perform an analysis for strengths, weaknesses, opportunities, and threats (SWOT) on the decision to build DIA.
3. Who are the stakeholders and what are their interests or objectives?
4. Did the airlines support the decision to build DIA?
5. Why was United Opposed to expansion at Front Range Airport?
6. Why was the new baggage handling system so important to United?
7. Is DIA a good strategic fit for Continental?
8. What appears to be the single greatest risk in the decision to build DIA?
9. United is a corporation in business to make money. How can United issue tax-free municipal bonds?
10. What impact do the rating agencies (i.e., Moody's and Standard & Poor's) have in the financing of the airport?
11. According to the prospectus, the DIA bonds were rated as BBB− by Standard & Poor's Corporation. Yet, at the same time, the City of Denver was given a rating of AA. How can this be?
12. On October 1, 1992, the United bonds were issued at an interest rate of 6.875 percent. Was this an appropriate coupon for the bonds?
13. There are numerous scenarios that can occur once the airport opens. The following questions are "what if" exercises and may not have a right or wrong answer. The questions are used to stimulate classroom discussion. The students must use the prospectus excerpts in the exhibit at the end of the case study. For each situation, what will be the possible outcome and what impact is there upon the bondholders?
14. Assume that DIA finally opens and with a debt of $3 billion. Is the revenue stream sufficient to pay interest each year *and* pay the principal at maturity?
15. What options are available to DIA if the coverage falls below 100 percent?
16. If the debt coverage were actually this good, why would the ratings on the bonds be BB?
17. One of the critical parameters that airlines use is the cost per enplaned passenger. Using Exhibit V, determine whether the cost per enplaned passenger can be lowered.
18. Is there additional revenue space available (i.e., unused capacity)?

19. What is the function of the project management team (PMT) and why were two companies involved?
20. When did the effectiveness of the project management team begin to be questioned?
21. Did it sound as though the statement of work/specifications provided by the city to the PMT was "vague" for the design phase?
22. During the design phase, contractors were submitting reestimates for work, 30 days after their original estimates, and the new estimates were up to $50 million larger than the prior estimate. Does this reflect upon the capabilities of the PMT?
23. Should the PMT be qualified to perform risk analyses?
24. Why were the architects coordinating the changes at the construction site?
25. Should the PMT have been replaced?
26. Do scope changes reflect upon the ineffectiveness of a project management team?
27. Why did United Airlines decide to act as the project manager for the baggage handling system on Concourse B?

Part 15

WAGE AND SALARY ADMINISTRATION

It is very difficult for the true benefits of project management to be realized unless project management is integrated into the wage and salary administration program. Some companies view project management as a career path position while others view it simply as a part time profession.

The situation becomes even more complex when dealing with functional employees who report to multiple bosses. When employees are notified that they are being assigned to a new project, their first concern is what is in it for them? How will they be evaluated? How will their boss know whether or not they did a good job? Project managers must have either a formal or informal input into the employee's performance review.

Photolite
Corporation (A)

Photolite Corporation is engaged in the sale and manufacture of cameras and photographic accessories. The company was founded in Baltimore in 1980 by John Benet. After a few rough years, the company began to flourish, with the majority of its sales coming from the military. By 1985, sales had risen to $5 million.

By 1995, sales had increased to almost $55 million. However, in 1996 competition from larger manufacturers and from some Japanese and German imports made itself felt on Photolite's sales. The company did what it could to improve its product line, but due to lack of funds, it could not meet the competition head-on. The company was slowly losing its market share and was approached by several larger manufacturers as to the possibility of a merger or acquisition. Each offer was turned down.

During this time period, several meetings took place with department heads and product managers regarding the financial health of Photolite. At one of the more recent meetings, John Benet expressed his feelings in this manner:

> I have been offered some very attractive buyouts, but frankly the companies that want to acquire us are just after our patents and processes. We have a good business, even though we are experiencing some tough times. I want our new camera lens project intensified. The new lens is just about complete, and I want it in full-scale production as soon as possible! Harry Munson will be in charge of this project as of today, and I expect everyone's full cooperation. This may be our last chance for survival.

With that, the meeting was adjourned.

PROJECT INFORMATION

The new lens project was an innovation that was sure to succeed if followed through properly. The innovation was a lens that could be used in connection with sophisticated camera equipment. It was more intense than the wide-angle lens and had no distortion. The lens was to be manufactured in three different sizes, enabling the lens to be used with the top selling cameras already on the market. The lens would not only be operable with the camera equipment manufactured by Photolite, but also that of their competitors.

Management was certain that if the manufactured lens proved to be as precise as the prototypes, the CIA and possibly government satellite manufacturers would be their largest potential customers.

THE PROJECT OFFICE

Harry Munson was a young project manager, twenty-nine years of age, who had both sales and engineering experience, in addition to an MBA degree. He had handled relatively small projects in the past and realized that this was the most critical, not only to his career but also for the company's future.

Project management was still relatively new at Photolite, having been initiated only fifteen months earlier. Some of the older department heads were very much against letting go of their subordinates for any length of time, even though it was only a sharing arrangement. This was especially true of Herb Wallace, head of the manufacturing division. He felt his division would suffer in the long run if any of his people were to spend much time on projects and reporting to another manager or project leader.

Harry Munson went directly to the personnel office to review the personnel files of available people from the manufacturing division. There were nine folders available for review. Harry had expected to see at least twenty folders, but decided to make the best of the situation. Harry was afraid that it was Herb Wallace's influence that had reduced the number of files down to nine.

Harry Munson had several decisions to make before looking at the folders. He felt that it was important to have a manufacturing project engineer assigned full-time to the project, rather than having to negotiate for part-time specialists who would have to be shared with other projects. The ideal manufacturing project engineer would have to coordinate activity in production scheduling, quality control, manufacturing engineering, procurement, and inventory control. Because project management had only recently been adopted, there were no individuals qualified for this position. This project would have to become the training ground for development of a manufacturing project engineer.

Due to the critical nature of the project, Harry realized that he must have the most competent people on his team. He could always obtain specialists on a part-time basis, but his choice for the project engineering slot would have to be not only the best person available, but someone who would be willing to give as much extra time as the project demanded for at least the next 18 months. After all, the project engineer would also be the assistant project manager since only the project manager and project engineer would be working full-time on the project. Now, Harry Munson was faced with the problem of trying to select the individual who would be best qualified for this slot. Harry decided to interview each of the potential candidates, in addition to analyzing their personnel files.

QUESTIONS

1. What would be the ideal qualifications for the project engineering slot?
2. What information should Harry look for in the personnel files?
3. Harry decided to interview potential candidates after reviewing the files. This is usually a good idea, because the files may not address all of Harry's concerns. What questions should Harry ask during the interviews? Why is Harry interviewing candidates? What critical information may not appear in the personnel files?

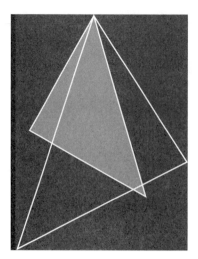

Photolite Corporation (B)

On October 3, 1998, a meeting was held between Jesse Jaimeson, the director of personnel, and Ronald Ward, the wage and salary administrator. The purpose of the meeting was to discuss the grievances by the functional employees that Photolite's present employee evaluation procedures are inadequate for an organization that supports a project management structure.

Jesse Jaimeson: "Ron, we're having a lot of trouble with our functional employees over their evaluation procedures. The majority of the complaints stem from situations where the functional employee works closely with the project manager. If the functional manager does not track the work of this employee closely, then the functional manager must rely heavily upon the project manager for information during employee evaluation."

Ron Ward: "There aren't enough hours in a day for a functional manager to keep close tabs on all of his or her people, especially if those people are working in a project environment. Therefore, the functional manager will ask the project manager for evaluation information. This poses several problems. First, there are always situations where functional and project management disagree as to either direction or quality of work. The functional employee has a tendency of bending toward the individual who signs his or her promotion and evaluation form. This can alienate the project manager into recommending a poor evaluation regardless of how well the functional employee performs."

588

In the second situation, the functional employee will spend most of this time working by her or himself, with very little contact with the project manager. In this case, the project manager tends to give an average evaluation, even if the employee's performance is superb. This could result from a situation where the employee has perhaps only a one to two week effort on a given project. This doesn't give that employee enough time to get to know anybody.

In the third situation, the project manager allows personal feelings to influence his or her decision. A project manager who knows an employee personally might be tempted to give a strong or weak recommendation, regardless of the performance. When personalities influence the evaluation procedure, chaos usually results.

Jaimeson: "There's also a problem if the project manager makes an overly good recommendation to a functional manager. If the employee knows that he or she has received a good appraisal for work done on a given project, that employee feels that he or she should be given an above average pay increase or possibly a promotion. Many times this puts severe pressure upon the functional manager. We have one functional manager here at Photolite who gives only average salary increases to employees who work a great deal of time on one project, perhaps away from view of the functional manager. In this case, the functional manager claims that he cannot give the individual an above average evaluation because he hasn't seen him enough. Of course, this is the responsibility of the functional manager.

"We have another manager who refuses to give employees adequate compensation if they are attached to a project that could eventually grow into a product line. His rationale is that if the project grows big enough to become a product line, then the project will have its own cost center account and the employee will then be transferred to the new cost center. The functional manager thus reserves the best salary increases for those employees who he feels will stay in his department and make him look good."

Ward: "Last year we had a major confrontation on the Coral Project. The Coral Project Manager took a grade 5 employee and gave him the responsibilities of a grade 7 employee. The grade 5 employee did an outstanding job and naturally expected a large salary increase or even a promotion. Unfortunately, the functional manager gave the employee an average evaluation and argued that the project manager had no right to give the employee this added responsibility without first checking with the functional manager. We're still trying to work this problem out. It could very easily happen again."

Jaimeson: "Ron, we have to develop a good procedure for evaluating our employees. I'm not sure if our present evaluation form is sufficient. Can we develop multiple evaluation forms, one for project personnel and another one for nonproject personnel?"

Ward: "That might really get us in trouble. Suppose we let each project manager fill out a project evaluation form for each functional employee who works more than, say, 60 hours on a given project. The forms are then given to the functional manager. Should the project manager fill out these forms at project termination or when the employee is up for evaluation?"

Jaimeson: "It would have to be at project termination. If the evaluation were made when the employee is up for promotion and the employee is not promoted, then that employee might slack off on the job if he or she felt that the project manager rated him or her down. Of course, we could always show the employee the project evaluation sheets, but I'm not sure that this would be the wise thing to do. This could easily lead into a situation where every project manager would want to see these forms before staffing a project. Perhaps these forms should be solely for the functional manager's use."

Ward: "There are several problems with this form of evaluation. First, some of our functional employees work on three or four projects at the same time. This could be a problem if some of the evaluations are good while others are not. Some functional people are working on departmental projects and, therefore, would receive only one type of evaluation. And, of course, we have the people who charge to our overhead structure. They also would have one evaluation form."

Jaimeson: "You know, Ron, we have both exempt and nonexempt people charging to our projects. Should we have different evaluation forms for these people?"

Ward: "Probably so. Unfortunately, we're now using just one form for our exempt, nonexempt, technical, and managerial personnel. We're definitely going to have to change. The question is how to do it without disrupting the organization."

Jaimeson: "I'm dumping this problem into your lap, Ron. I want you to develop an equitable way of evaluating our people here at Photolite Corporation, and I want you to develop the appropriate evaluation forms. Just remember one thing—I do not want to open Pandora's Box. We're having enough personnel problems as it is."

QUESTIONS

1. Can a company effectively utilize multiple performance evaluation forms within an organization? What are the advantages and disadvantages?
2. If we use only one form, what information should be evaluated so as to be equitable to everyone?
3. If multiple evaluation forms are used, what information should go into the form filled out by the project manager?
4. What information can and cannot a project manager effectively evaluate? Could it depend upon the project manager's educational background and experience?

Photolite Corporation (C)

On December 11, 1998, after more than two months of effort, Ron Ward (the wage and salary administrator for Photolite Corporation) was ready to present his findings on the most equitable means of evaluating personnel who are required to perform in a project management organizational structure. Jesse Jaimeson (the director of personnel) was eagerly awaiting the results.

Ron Ward: "Well, Jesse, after two months of research and analysis we've come to some reasonable possibilities. My staff looked at the nine basic performance appraisal techniques. They are:

1. Essay appraisal
2. Graphic rating scale
3. Field review
4. Forced choice rating
5. Critical incident appraisal
6. Management by objectives
7. Work-standards approach
8. Ranking methods
9. Assessment centers

(Exhibit I contains a brief description of each technique.)

We tried to look at each technique objectively. Unfortunately, many of my people are not familiar with project management and, therefore, had some difficulties. We had no so-called 'standards of performance' against which we could evaluate each technique. We, therefore, listed the advantages and disadvantages that each technique would have if utilized in a project management structure."

Jesse Jaimeson: "I'm not sure of what value your results are in this case because they might not directly apply to our project management organization."

Ward: "In order to select the technique most applicable to a project management structure, I met with several functional and project managers as to the establishment of a selection criteria. The functional managers felt that conflicts were predominant in a project organization, and that these conflicts could be used as a comparison. I, therefore, decided to compare each of the appraisal techniques to the seven most commonly mentioned conflicts that exist in project management organizational forms. The comparison is shown in Exhibit II.

"Analysis of Exhibit II shows the management by objectives technique to be the most applicable system. Factors supporting this conclusion are as follows:

Essay Appraisal: This technique appears in most performance appraisals and is characterized by a lack of standards. As a result, it tend to be subjective and inconsistent.

Graphic Rating Scale: This technique is marked by checking boxes and does not have the flexibility required by the constantly changing dynamic structure required in project management.

Field Review: This system would probably account for the majority of performance appraisal problems. However, it is costly and provides for another management overlay, as well as an additional cost (time) factor.

Forced-Choice Rating: This technique has the same problems as the essay technique with the added problem of being inflexible.

Critical Incident Appraisal: This technique centers on the individual's performance and does not take into account decisions made by one's superiors or the problems beyond the individual's control. Again, it is time-consuming.

Management by Objectives (MBO): This technique allows all parties, the project manager, the functional manager, and the employee, to share and to participate in the appraisal. It epitomizes the systems approach since it allows for objectives modification without undue or undeserved penalty to the employee. Finally, it uses objective data and downplays subjective data.

Work-Standards Approach: This technique lends itself easily to technical projects. Though not usually recognized formally, it is probably the most common project management performance appraisal technique. However, it is not flexible and downplays the effect of personality conflicts with little employee input.

Ranking Method: This method allows for little individual input. Most conflict possibilities are maximized with this technique.

Assessment Centers: This method is not utilizable on site and is very costly. It is probably most applicable (if not the best technique) for selecting project management human resources.

"In summary, MBO appears to be the best technique for performance appraisal in a project management organization."

Jaimeson: "Your conclusions lead me to believe that the MBO appraisal technique is applicable to all project management appraisal situations and should be recommended. However, I do have a few reservations. A key point is that the MBO approach does not eliminate, or even minimize, the problems inherent in project and matrix management organizations. MBO provides the technique through which human resources can be fairly appraised (and, of course, rewarded and punished). MBO has the weakness that it prohibits individual input and systems that employ poorly trained appraisers and faulty follow-up techniques. Of course, such weaknesses would kill any performance appraisal system. The MBO technique most exemplifies the systems approach and, even with its inherent weaknesses, should be considered when the systems approach to management is being employed."

Ward: "There is another major weakness that you have omitted. What about those situations where the employee has no say in setting the objectives? I'm sure we have project managers, as well as functional managers, who will do all of the objective-setting themselves."

Jaimeson: "I'm sure this situation either exists now or will eventually exist. But that's not what worries me. If we go to an MBO approach, how will it affect our present evaluation forms? We began this study to determine the best appraisal method for our organization. I've yet to see any kind of MBO evaluation form that can be used in a project management environment. This should be our next milestone."

Exhibit I. Basic appraisal techniques

Essay Appraisal
This technique asks raters to write a short statement covering a particular employee's strengths, weaknesses, areas for improvement, potential, and so on. This method is often used in the selection of employees when written recommendations are solicited from former employers, teachers, or supervisors. The major problem with this type of appraisal is the extreme variability in length and content, which makes comparisons difficult.

Graphic Rating Scale

A typical graphic rating scale assesses a person on the quality and quantity of his or her work and on a variety of other factors that vary with the specific job. Usually included are personal traits such as flexibility, cooperation, level of self-motivation, and organizational ability. The graphic rating scale results in more consistent and quantifiable data, though it does not provide the depth of the essay appraisal.

Field Review

As a check on reliability of the standards used among raters, a systematic review process may be utilized. A member of the personnel or central administrative staff meets with small groups of raters from each supervisory unit to go over ratings for each employee to identify areas of dispute and to arrive at an agreement on the standards to be utilized. This group judgment technique tends to be more fair and valid than individual ratings, but is considerably more time-consuming.

Forced-Choice Rating

There are many variations of this method, but the most common version asks raters to choose from among groups of statements those that best fit the person being evaluated and those that least fit. The statements are then weighted and scored in much the same way psychological tests are scored. The theory behind this type of appraisal is that since the rater does not know what the scoring weight of each statement is, he or she cannot play favorites.

Critical Incident Appraisal

Supervisors are asked to keep a record on each employee and to record actual incidents of positive and negative behavior. While this method is beneficial in that it deals with actual behavior rather than abstractions, it is time-consuming for the supervisor, and the standards of recording are set by the supervisor.

Management by Objectives

In this approach, employees are asked to set, or help set, their own performance goals. This approach has considerable merit in its involvement of the individual in setting the standards by which he or she will be judged, and the emphasis on results rather than on abstract personality characteristics.

Work-Standards Approach

Instead of asking each employee to set his or her own performance standards, many organizations set measured daily work standards. The work-standards technique establishes work and staffing targets aimed at increasing productivity. When realistically used and when standards are fair and visible, it can be an effective type of performance appraisal. The most serious problem is that of comparability. With different standards for different people, it is difficult to make comparisons for the purposes of promotion.

Ranking Methods

For purposes of comparing people in different units, the best approach appears to be a ranking technique involving pooled judgment. The two most effective ranking methods include alternation-ranking and paired-comparison ranking. Essentially, supervisors are asked to rank who is "most valuable."

Assessment Centers

Assessment centers are coming into use more for the prediction and assessment of future potential. Typically, individuals from different areas are brought together to spend two or three days working on individual and group assignments. The pooled judgment of observers leads to an order-of-merit ranking of participants. The greatest drawback to this system is that it is very time-consuming and costly.

QUESTIONS

1. Do you agree with the results in Exhibit II? Why or why not? Defend your answers.
2. Are there any other techniques that may be better?

Exhibit II. *Rating evaluation techniques against types of conflict*

Type of Conflict	Essay Appraisal	Graphic Rating Scale	Field Review	Forced-Choice Review	Critical Incident Appraisal	Management by Objectives	Work Standards Approach	Ranking Medthods	Assessment Center
Conflict over schedules	●	●		●	●		●	●	
Conflict over priorities	●	●		●	●		●	●	
Conflict over technical issues	●			●			●		
Conflict over administration	●	●	●	●			●	●	●
Personality conflict	●	●		●			●		
Conflict over cost	●		●	●	●		●	●	●

Note: Shaded circles indicate areas of difficulty.

Photolite Corporation (D)

On June 12, 1999, Ron Ward (the wage and salary administrator for Photolite Corporation) met with Jesse Jaimeson (the director of personnel) to discuss their presentation to senior management for new evaluation techniques in the recently established matrix organization.

Jesse Jaimeson: "I've read your handout on what you're planning to present to senior management, and I feel a brief introduction should also be included (see Exhibit I). Some of these guys have been divorced from lower-level appraisals for over 20 years. How do you propose to convince these guys?"

Ron Ward: "We do have guidelines for employee evaluation and appraisal. These include:

    A. To record an individual's *specific* accomplishments for a given period of time.
    B. To formally communicate to the individual on four basic issues:
        1. What is expected of him/her (in specifics).
        2. How he/she is performing (in specifics).
        3. What his/her manager thinks of his/her performance (in specifics).
        4. Where he/she could progress within the present framework.
    C. To improve performance.
    D. To serve as a basis for salary determination.
    E. To provide a constructive channel for upward communication.

"Linked to the objectives of the performance appraisal, we must also consider some of the possible negative influences impacting on a manager involved in this process. Some of these factors could be:

- A manager's inability to control the work climate.
- A normal dislike to criticize a subordinate.
- A lack of communication skills needed to handle the employee interview.
- A dislike for the general mode in the operation of the business.
- A mistrust of the validity of the appraisal instrument.

"To determine the magnitude of management problems inherent in the appraisal of employees working under the matrix concept, the above-mentioned factors could be increased four or five times, the multiplier effect being caused by the fact that an employee working under the project/matrix concept could be working on as many as four or five projects during the appraisal period, thereby requiring all the project managers and the functional manager to input their evaluation regarding a subordinate's performance and the appraisal system itself."

Jaimeson: "Of course, managers cannot escape making judgments about subordinates. Without these evaluations, Photolite would be unable to adequately administer its promotion and salary policies. But in no instance can a performance appraisal be a simple accept or reject concept involving individuals. Unlike the quality appraisal systems used in accepting or rejecting manufactured units, our personnel appraisal systems must include a human factor. This human factor must take us beyond the scope of job objectives into the values of an individual's worth, human personality, and dignity. It is in this vein that any effective personnel appraisal system must allow the subordinate to participate fully in the appraisal activities."

Ward: "Prior to 1998, this was a major problem within Photolite. Up to that time, all appraisals were based on the manager or managers assessing an individual's progress toward goals that had been established and passed on to subordinates. Although an employee meeting was held to discuss the outcome of an employee's appraisal, in many instances it was one-sided, without meaningful participation by the person being reviewed. Because of such a system, many employees began to view the appraisal concept as inconsistent and without true concern for the development of the individual. This also led many to believe that promotions and salary increases were based on favoritism rather than merit.

"Problems inherent in these situations are compounded in the matrix organization when an individual is assigned to several projects with varying degrees of importance placed on each project, but knowing that each project manager will contribute to the performance appraisal based on the success of their individual projects. Such dilemmas can only be overcome when the individual is considered as the primary participating party in the appraisal process and the functional manager coordinates and places prime responsibility of the subordinate contributor in the project for which prime interest has been focused by the company. Other project contributions are then considered, but on a secondary basis."

Jaimeson: "Although we have discussed problems that are inherent in a matrix organization and can be compounded by the multiple performance determination, a number of positives can also be drawn from such a work environment. It is obvious, based on its design, that a project/matrix organization demands new attitudes, behavior, knowledge, and skills. This in turn has substantial implications for employee selection, development, and career progression. The ultimate success of the individual and the project depends largely on the ability of the organization to help people learn how to function in new ways.

"The matrix organization provides an opportunity for people to develop and grow in ways and rates not normally possible in the more traditional functional organizational setting. Although the project/matrix organization is considered to be high tension in nature, it places greater demands on people but offers greater development and career opportunities than does the functional organization.

"Because of the interdependencies of projects in a matrix, increased communications and contact between people is necessary. This does not mean that in a functional organization interdependency and communication are not necessary. What it does say, however, is that in a functional setting, roles are structured so that individuals can usually resolve conflicting demands by talking to their functional manager. In a matrix, such differences would be resolved by people from different functions who have different attitudes and orientations."

Ward: "From the very outset, organizations such as Photolite ran into conflict between projects involving such items as:

- Assignment of personnel to projects
- Manpower costs
- Project priority
- Project management status (as related to functional managers)
- Overlap of authority and power in the matrix

If not adequately planned for in advance, these factors could be significant factors in the performance appraisal of matrix/project members. However, where procedures exist to resolve authority and evaluation conflicts, a more equitable performance appraisal climate exists. Unfortunately, such a climate rarely exists in any functioning organization.

"With the hope of alleviating such problems, my group has redefined its approach to Exempt Performance Appraisals (see Exhibits I and II). This approach is based on the management by objectives technique. This approach allows both management and employees to work together in establishing performance goals.

Beyond this point of involvement, employees also perform a self-evaluation of their performance, which is considered a vital portion of the performance appraisal. Utilization of this system also opens up communication between management and the employee, thereby allowing two-way communication to become a natural item. Although it is hoped that differences can be reconciled, if this cannot occur, the parties involved have at least established firm grounds on which to disagree. These grounds are not hidden to either and the employee knows exactly how his/her performance appraisal was determined."

Jaimeson: "O.K. I'm convinced we're talking the same language. We won't have any problem convincing these people of what we're trying to do."

Exhibit I. Recommended approach

I. Prework
- Employee and manager record work to be done using goals, work plans, position guide.
- Employee and manager record measurements to be used.

Note: This may not be possible at this time since we are in the middle of a cycle. For 1999 only, the process will start with the employees submitting a list of their key tasks (i.e., job description) as they see it. Manager will review that list with the employee.

II. Self-Appraisal
- Employee submits self-appraisal for key tasks.
- It becomes part of the record.

III. Managerial Appraisal
- Manager evaluates each task.
- Manager evaluates total effort.
- Skills displayed are recorded.
- Development effort required is identified.

Note: Appraisals should describe what happened, both good and bad.

IV. Objective Review
- Employee relations reviews the appraisal.
  - Assure consistent application of ratings.
  - Assist in preparation, if needed.
  - Be a sounding board.

V. One-over-One Review
- Managerial perspective is obtained.
- A consistent point of view should be presented.

VI. Appraisal Discussion
- Discussion should be participative.
- Differences should be reconciled. If this is not possible, participants must agree to disagree.
- Work plans are recycled.
- Career discussion is teed-up.
- Employee and manager commit to development actions.

VII. Follow-up
- Checkpoints on development plan allow for this follow-up.

Exhibit II. Performance summary

When writing the overall statement of performance:
- Consider the degree of difficulty of the work package undertaken in addition to the actual results.
- Reinforce performance outcomes that you would like to see in the future by highlighting them here.
- Communicate importance of missed targets by listing them here.
- Let employees know the direction that performance is taking so that they can make decisions about effort levels, skill training emphasis, future placement possibilities, and so on.

When determining the overall rating number:
- Choose the paragraph that best describes performance in total, then choose the number that shades the direction it leans.
- Use the individual task measurements plus some weighting factor—realistically some projects are worth more than others and should carry more weight.
- Again, consider the degree of difficulty of the work package undertaken.

Strong points are:
- Demonstrated in the accomplishment of the work.
- Found in the completion of more than one project.
- Relevant—avoid trivia.
- Usually not heard well by employees.
- Good subjects for sharpening and growing.

Areas requiring improvement usually:
- Show up in more than one project.
- Are known by subordinate.
- Limit employee effectiveness.
- Can be improved to some degree.

Areas of disagreement:
- Can be manager or subordinate initiated.
- Need not be prepared in advance.
- Require some effort on both parts before recording.
- Are designed to keep problems from hiding beneath the surface.

Your review of the self-appraisal may surface some disagreement. Discuss this with the employee before formally committing it to writing.

QUESTIONS

1. If you were an executive attending this briefing, how would you react?
2. Are there any additional questions that need to be addressed?

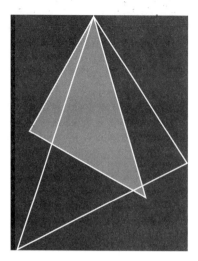

First Security Bank
of Cleveland

The growth rate of First Security of Cleveland had caused several executives to do some serious thinking about whether the present organizational structure was adequate for future operations. The big question was whether the banking community could adapt to a project management structure.

Tom Hood had been the president of First Security for the past ten years. He had been a pioneer in bringing computer technology into the banking industry. Unfortunately, the size and complexity of the new computer project created severe integration problems, problems with which the present traditional organization was unable to cope. What was needed was a project manager who could drive the project to success and handle the integration of work across functional lines.

Tom Hood met with Ray Dallas, one of the bank's vice presidents, to discuss possible organizational restructuring:

Tom Hood: "I've looked at the size and complexity of some twenty projects that First Security did last year. Over 50 percent of these projects required interaction between four or more departments."

Ray Dallas: "What's wrong with that? We're growing and our problems are likewise becoming more complex."

Hood: "It's the other 50 percent that worry me. We can change our organizational structure to adapt to complex problem-solving and integration. But what happens when we have a project that stays in one functional department? Who's going to drive it home? I don't see how we can tell a functional manager that he or she is a support group in one organizational form and a project manager in the other and have both organizational forms going on at the same time.

"We can have either large, complex projects or small ones. The small ones will be the problem. They can exist in one department or be special projects assigned to one person or a task force team. This means that if we incorporate project management, we'll have to live with a variety of structures. This can become a bad situation. I'm not sure that our people will be able to adapt to this changing environment."

Dallas: "I don't think it will be as bad as you make it. As long as we clearly define each person's authority and responsibility, we'll be all right. Other industries have done this successfully. Why can't we?"

Hood: "There are several questions that need answering. Should each project head be called a project manager, even if the project requires only one person? I can see our people suddenly becoming title-oriented. Should all project managers report to the same boss, even if one manager has thirty people working on the project and the other manager has none? This could lead to power struggles. I want to avoid that because it can easily disrupt our organization."

Dallas: "The problem you mentioned earlier concerns me. If we have a project that belongs in one functional department, the ideal solution is to let the department manager wear two hats, the second one being project manager. Disregarding for the moment the problem that this manager will have in determining priorities, to whom should he or she report to as to the status of the work? Obviously, not to the director of project management."

Hood: "I think the solution must be that all project managers report to one person. Therefore, even if the project stays in one functional department, we'll still have to assign a project manager Under project management organizational forms, functional managers become synonymous with resource managers. It is very dangerous to permit a resource manager to act also as a project manager. The resource manager might consider the project as being so important that he or she will commit all the department's best people to it and make it into a success at the expense of all the department's other work. That would be like winning a battle but losing the war."

Dallas: "You realize that we'll need to revamp our wage and salary administration program if we go to project management. Evaluating project managers might prove difficult. Regardless of what policies we establish, there are still going to be project managers who try to build empires, thinking that their progress is dependent upon the number of people they control. Project management will definitely give some people the opportunity to build a empire. We'll have to watch that closely."

Hood: "Ray, I'm a little worried that we might not be able to get good project managers. We can't compete with the salaries the project managers get in other industries such as engineering, construction, or computers. Project management cannot be successful unless we have good managers at the controls. What's your feeling on this?"

Dallas: "We'll have to promote from within. That's the only viable solution. If we try to make project management salaries overly attractive, we'll end up throwing the organization into chaos. We must maintain an adequate salary structure so that people feel that they have the same opportunities in both project management and the functional organization. Of course, we'll still have some people who will be more title-oriented than money-oriented, but at least each person will have the same opportunity for salary advancement."

Hood: "See if you can get some information from our personnel people on how we could modify our salary structure and what salary levels we can pay our project managers. Also, check with other banks and see what they're paying their project managers. I don't want to go into this blind and then find out that we're setting the trend for project management salaries. Everyone would hate us. I'd rather be a follower than a leader in this regard."

QUESTIONS

1. What are the major problems identified in the case?
2. What are your solutions to the above question and problems?

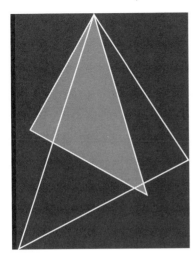

Jackson
Industries

"I wish the hell that they had never invented computers," remarked Tom Ford, president of Jackson Industries. "This damn computer has been nothing but a thorn in our side for the past ten years. We're gonna resolve this problem now. I'm through watching our people fight with one another. We must find a solution to this problem."

In 1982, Jackson Industries decided to purchase a mainframe computer, primarily to handle the large, repetitive tasks found in the accounting and finance functions of the organization. It was only fitting, therefore, that control of the computer came under the director of finance, Al Moody. For two years, operations went smoothly. In 1984, the computer department was reorganized in three sections: scientific computer programming, business computer programming, and systems programming. The reorganization was necessary because the computer department had grown into the fifth largest department, employing some thirty people, and was experiencing some severe problems working with other departments.

After the reorganization, Ralph Gregg, the computer department manager, made the following remarks in a memo distributed to all personnel:

The Computer Department has found it increasingly difficult to work with engineering and operations functional departments, which continue to permit their personnel to write and document their own computer programs. In order to maintain some degree of consistency, the Computer Department will now assume the responsibility for writing all computer programs. All requests should be directed to the department manager. My people are under explicit instructions that they are to provide absolutely no assistance to any functional personnel attempting to write their own programs without authorization from me. Company directives in this regard will be forthcoming.

The memo caused concern among the functional departments. If engineering wanted a computer program written, they would now have to submit a formal request and then have the person requesting the program spend a great deal of time explaining the problem to the scientific programmer assigned to this effort. The department managers were reluctant to have their people "waste time" in training the scientific programmers to be engineers. The computer department manager countered this argument by stating that once the programmer was fully familiar with the engineering problem, then the engineer's time could be spent more fruitfully on other activities until the computer program was ready for implementation.

This same problem generated more concern by department managers when they were involved in computer projects that required integration among several departments. Although Jackson Industries operated on a traditional structure, the new directive implied that the computer department would be responsible for managing all projects involving computer programming even if they crossed into other departments. Many people looked on this as a "baby" project management structure within the traditional organization.

In June 1992, Al Moody and Ralph Gregg met to discuss the deterioration of working relationships between the computer department and other organizations.

Al Moody: "I'm getting complaints from the engineering and operations departments that they can't get any priorities established on the work to be done in your group. What can we do about it?"

Ralph Gregg: "I set the priorities as I see fit, for what's best for the company. Those guys in the engineering and operations have absolutely no idea how long it takes to write, debug, and document a computer program. Then they keep feeding me this crap about how their projects will slip if this computer program isn't ready on time. I've told them what problems I have, and yet they still refuse to let me participate in the planning phase of their activities."

Al Moody: "Well, you may have a valid gripe there. I'm more concerned about this closed shop you've developed for your department. You've built a little empire down there and it looks like your people are unionized where the rest of us are not. Furthermore, I've noticed that your people have their own informal organization and tend to avoid socializing with the other employees. We're supposed to be one big, happy family, you know. Can't you do something about that?"

Ralph Gregg: "The problem belongs to you and Tom Ford. For the last three years, the average salary increase for the entire company has been 7.5 percent and our department has averaged a mere 5 percent because you people upstairs do not feel as though we contribute anything to company profits. My scientific programmers feel that they're doing engineering work and that they're making the same contribution to profits as is the engineer. Therefore, they should be on the engineering pay structure and receive an 8 percent salary increase."

Al Moody: "You could have given your scientific programmers more money. You had a budget for salary increases, the same as everyone else."

Ralph Gregg: "Sure I did. But my budget was less than everyone else's. I could have given the scientific people 7 percent and everyone else 3 percent. That would be an easy way to tell people that we think they should look for another job. My people do good work and do, in fact, contribute to profits. If Tom Ford doesn't change his impression of us, then I expect to lose some of my key people. Maybe you should tell him that."

Al Moody: "Between you and me, all of your comments are correct. I agree with your concerns. But my hands are tied, as you know.

"We are contemplating the installation of a management information system for all departments and, especially, for executive decision making. Tom is contemplating creating a new position, Director of Information Services. This would move the computer out of a department under finance and up to the directorate level. I'm sure this would have an impact on yearly salary increases for your people.

"The problem that we're facing involves the managing of projects under the new directorate. It looks like we'll have to create a project management organization just for this new directorate. Tom likes the traditional structure and wants to leave all other directorates intact. We know that this new directorate will have to integrate the new computer projects across multiple departments and divisions. Once we solve the organizational structure problem, we'll begin looking at implementation. Got any good ideas about the organizational structure?"

Ralph Gregg: "You bet I do. Make me director and I'll see that the work gets done."

Part 16

TIME MANAGEMENT

Managing projects within time, cost, and performance is easier said than done. The project management environment is extremely turbulent and is composed of numerous meetings, report writing, conflict resolution, continuous planning and replanning, customer communications, and crisis management.

To manage all of these activities requires that the project manager and team members effectively manage their time each day. Some people are morning people and soon learn they are more productive in the morning than afternoon. Others are afternoon people. Knowing your own energy cycle is important. Also, good project managers realize that not all of the activities that they are asked to do are their responsibility.

Time Management Exercise

Effective time management is one of the most difficult chores facing even the most experienced managers. For a manager who manages well-planned repetitive tasks, effective time management can be accomplished without very much pain. But for a project manager who must plan, schedule, and control resources and activities on unique, one-of-a-kind projects or tasks, effective time management may not be possible because of the continuous stream of unexpected problems that develop.

This exercise is designed to make you aware of the difficulties of time management both in a traditional organization and in a project environment. Before beginning the exercise, you must make the following assumptions concerning the nature of the project:

- You are the project manager on a project for an outside customer.
- The project is estimated at $3.5 million with a time span of two years.
- The two-year time span is broken down into three phases: Phase I—one year, beginning February 1; Phase II—six months; Phase III—six months. You are now at the end of Phase I. (Phases I and II overlap by approximately two weeks. You are now in the Monday of the next to the last week of Phase I.) Almost all of the work has been completed.
- Your project employs thirty-five to sixty people, depending on the phase that you are in.

- You, as the project manager, have three full-time assistant project managers that report directly to you in the project office; an assistant project manager each for engineering, cost control, and manufacturing. (Material procurement is included as part of the responsibilities of the manufacturing assistant project manager.)
- Phase I appears to be proceeding within the time, cost, and performance constraints.
- You have a scheduled team meeting for each Wednesday from 10 A.M. to 12 noon. The meeting will be attended by all project office team members and the functional team members from all participating line organizations. Line managers are not team members and therefore do not show up at team meetings. It would be impossible for them to show up at the team meetings for all projects and still be able to function as a line manager. Even when requested, they may not show up at the team meeting because it is not effective time management for them to show up for a two-hour meeting simply to discuss ten minutes of business. (Disregard the possibility that a team meeting agenda could resolve this problem.)

It is now Monday morning and you are home eating breakfast, waiting for your car pool to pick you up. As soon as you enter your office, you will be informed about problems, situations, tasks, and activities that have to be investigated. Your problem will be to accomplish effective time management for this entire week based on the problems and situations that occur.

You will take each day one at a time. You will be given ten problems and/or situations that will occur for each day, and the time necessary for resolution. You must try to optimize your time for each of the next five days and get the maximum amount of productive work accomplished. Obviously, the word "productive" can take on several meanings. You must determine what is meant by productive work. For the sake of simplicity, let us assume that your energy cycle is such that you can do eight hours of productive work in an eight-hour day. You do not have to schedule idle time, except for lunch. However, you must be aware that in a project environment, the project manager occasionally becomes the catchall for all work that line managers, line personnel, and even executives do not feel like accomplishing.

Following the ten tasks for each day, you will find a worksheet that breaks down each day into half-hour blocks between 9:00 A.M. and 5:00 P.M. Your job will be to determine which of the tasks you wish to accomplish during each half-hour block. The following assumptions are made in scheduling work:

- Because of car pool requirements, overtime is not permitted.
- Family commitments for the next week prevent work at home. Therefore, you will not schedule any work after 5:00 P.M.
- The project manager is advised of the ten tasks as soon as he arrives at work.

The first step in the solution to the exercise is to establish the priorities for each activity based on:

- *Priority A:* This activity is urgent and must be completed today. (However, some A priorities can be withheld until the team meeting.)
- *Priority B:* This activity is important but not necessarily urgent.
- *Priority C:* This activity can be delayed, perhaps indefinitely.

Fill in the space after each activity as to the appropriate priority. Next, you must determine which of the activities you have time to accomplish for this day. You have either seven or seven and one-half hours to use for effective time management, depending on whether you want a half-hour or a full hour for lunch.

You have choices as to how to accomplish each of the activities. These choices are shown below:

- You can do the activity *yourself* (Symbol = Y).
- You can *delegate* the responsibility to one of your assistant project managers (Symbol = D). If you use this technique, you can delegate only one hour's worth of *your* work to each of your assistants without incurring a penalty. The key word here is that you are delegating *your* work. If the task that you wish to delegate is one that the assistant project manager would normally perform, then it does *not* count toward the one hour's worth of your work. This type of work is transmittal work and will be discussed below. For example, if you wish to delegate five hours of work to one of your assistant project managers and four of those hours are activities that would normally be his responsibility, then no penalty will be assessed. You are actually transmitting four hours and delegating one. You may assume that whatever work you assign to an assistant project manager will be completed on the day it is assigned, regardless of the priority.
- Many times, the project manager and his team are asked to perform work that is normally the responsibility of someone else, say, an executive or a line manager. As an example, a line employee states that he doesn't have sufficient time to write a report and he wants you to do it, since you are the project manager. These types of requests can be returned to the requestor since they normally do not fall within the project manager's responsibilities. You may, therefore, select one of the following four choices:
  - You can *return* the activity request back to the originator, whether line manager, executive, or subordinate, since it is not your responsibility (Symbol = R). Of course, you might want to do this activity, if you have time, in order to build up good will with the requestor.
  - Many times, work that should be requested of an assistant project manager is automatically sent to the project manager. In this case, the project

manager will automatically *transmit* this work to the appropriate assistant project manager (Symbol = T). As before, if the project manager feels that he has sufficient time available or if his assistants are burdened, he may wish to do the work himself. Work that is normally the responsibility of an assistant project manager is transmitted, not delegated. Thus the project manager can transmit four hours of work (T) and still delegate one hour of work (D) to the same assistant project manager without incurring any penalty.

- You can *postpone* work from one day to the next (Symbol = P). As an example, you decide that you want to accomplish a given Monday activity but do not have sufficient time. You can postpone the activity until Tuesday. If you do not have sufficient time on Tuesday, you may then decide to transmit (T) the activity to one of your assistants, delegate (D) the activity to one of your assistants, return (R) the activity to the requestor, or postpone (P) the activity another day. Postponing activities can be a trap. On Monday you decide to postpone a category B priority. On Tuesday, the activity may become a category A priority and you have no time to accomplish it. If you make a decision to postpone an activity from Monday to Tuesday and find that you have made a mistake by not performing this activity on Monday, you *cannot* go back in time and correct the situation.

- You can simply consider the activity as unnecessary and *avoid* doing it (Symbol = A).

After you have decided which activities you will perform each day, place them in the appropriate time slot based on your own energy cycle. Later we will discuss energy cycles and the order of the activities accomplished each day. You will find one worksheet for each day. The worksheets follow the ten daily situations and/or problems.

Repeat the procedure for each of the five days. Remember to keep track of the activities that are carried over from the previous days. Several of the problems can be resolved by more than one method. If you are thoroughly trapped between two or more choices on setting priorities or modes of resolution, then write a note or two to justify your answer in space beneath each activity.

SCORING SYSTEM

Briefly look at the work plan for one of the days. Under the column labeled "priority," the ten activities for each day are listed. You must first identify the priorities for each activity. Next, under the column labeled "method," you must select the method of accomplishment according to the legend at the bottom of the page. At the same time, you must fill in the activities you wish to perform yourself under the "accomplishment" column in the appropriate time slot because your method for accomplishment may be dependent on whether you have sufficient time to accomplish the activity.

Notice that there is a space provided for you to keep track of activities that have been carried over. This means that if you have three activities on Monday's list that you wish to carry over until Tuesday, then you must turn to Tuesday's work plan and record these activities so that you will not forget.

You will not score any points until you complete Friday's work plan. Using the scoring sheets that follow Friday's work plan, you can return to the daily work plans and fill in the appropriate points. You will receive either positive points or negative points for each decision that you make. Negative points should be subtracted when calculating totals.

After completing the work plans for all five days, fill in the summary work plan that follows and be prepared to answer the summary questions.

You will not be told at this time how the scoring points will be awarded because it may affect your answers.

Monday's Activities

Activity	*Description*	*Priority*
1.	The detailed schedules for Phase II must be updated prior to Thursday's meeting with the customer. (Time = 1 hr)	_____
2.	The manufacturing manager calls you and states that he cannot find a certain piece of equipment for tomorrow's production run test. (Time = ½ hr)	_____
3.	The local university has a monthly distinguished lecturer series scheduled for 3–5 P.M. today. You have been directed by the vice president to attend and hear the lecture. The company will give you a car. Driving time to the university is one hour. (Time = 3 hrs)	_____
4.	A manufacturer's representative wants to call on you today to show you why his product is superior to the one that you are now using. (Time = ½ hr)	_____
5.	You must write a two-page weekly status report for the vice president. Report is due on his desk by 1:00 P.M. Wednesday. (Time = 1 hr)	_____
6.	A vice president calls you and suggests that you contact one of the other project managers about obtaining a uniform structure for the weekly progress reports. (Time = ½ hr)	_____
7.	A functional manager calls to inform you that, due to a	_____

WORK PLAN

Day ___Monday___

Activity	Priority	Points	Method of Accomplishment	Points	Time	Activity	Points
	Priority		**Method**			**Accomplishment**	
1					9:00–9:30		
2					9:30–10:00		
3					10:00–10:30		
4					10:30–11:00		
5					11:00–11:30		
6					11:30–12:00		
7					12:00–12:30		
8					12:30–1:00		
9					1:00–1:30		
10					1:30–2:00		
Total			Total		2:00–2:30		
					2:30–3:00		
					3:00–3:30		
					3:30–4:00		
					4:00–4:30		
					4:30–5:00		
					Total		

Activities Postponed Until Today	Today's Priority

Points	
Priority Points	
Method Points	
Accomplishment Points	
Today's Points	

Legend

Method of Accomplishment:

Y = you
D = delegate
T = transmit
R = return
A = avoid
P = postpone

Activity	*Description*	*Priority*
	schedule slippage on another project, your beginning milestones on Phase II may slip to the right because his people will not be available. He wants to know if you can look at the detailed schedules and modify them. (Time = 2 hr)	
8.	The director of personnel wants to know if you have reviewed the three resumes that he sent you last week. He would like your written comments by quitting time today. (Time = 1 hr)	_____
9.	One of your assistant project managers asks you to review a detailed Phase III schedule that appears to have errors. (Time = 1 hr)	_____
10.	The procurement department calls with a request that you tell them approximately how much money you plan to spend on raw materials for Phase III. (Time = ½ hr)	_____

Tuesday's Activities

Activity	*Description*	*Priority*
11.	A functional manager calls you wanting to know if his people should be scheduled for overtime next week. (Time = ½ hr)	_____
12.	You have a safety board meeting today from 1–3 P.M. and must review the agenda. (Time = 2½ hrs)	_____
13.	Because of an impending company cash flow problem, your boss has asked you for the detailed monthly labor expenses for the next three months. (Time = 2 hrs)	_____
14.	The vice president has just called to inform you that two congressmen will be visiting the plant today and you are requested to conduct the tour of the facility from 3–5 P.M. (Time = 2 hrs)	_____
15.	You have developed a new policy for controlling overtime costs on Phase II. You must inform your people either by memo, phone, or team meeting. (Time = ½ hr)	_____
16.	You must sign and review twenty-five purchase order requisitions for Phase III raw materials. It is company policy that the project manager sign all forms. Almost all of the items require a three-month lead time. (Time = 1 hr)	_____
17.	The engineering division manager has asked you to assist one of his people this afternoon in the solution of a technical problem. You are not required to do this. It would be as a personal favor for the engineering	_____

WORK PLAN Day Tuesday

Activity	Priority	Points	Method of Accomplishment	Points	Time	Activity	Points
	Priority		Method		Accomplishment		
11					9:00–9:30		
12					9:30–10:00		
13					10:00–10:30		
14					10:30–11:00		
15					11:00–11:30		
16					11:30–12:00		
17					12:00–12:30		
18					12:30–1:00		
19					1:00–1:30		
20					1:30–2:00		
Total			Total		2:00–2:30		
					2:30–3:00		
					3:00–3:30		
					3:30–4:00		
					4:00–4:30		
					4:30–5:00		
					Total		

Activities Postponed Until Today	Today's Priority

Points	
Priority Points	
Method Points	
Accomplishment Points	
Today's Points	

Legend

Method of Accomplishment:

Y = you
D = delegate
T = transmit
R = return
A = avoid
P = postpone

Activity	*Description*	*Priority*
	manager, a man to whom you reported for the six years that you were an engineering functional manager. (Time = 2 hrs)	
18.	The data processing department manager informs you that the company is trying to eliminate unnecessary reports. He would like you to tell him which reports you can do without. (Time = ½ hr)	_____
19.	The assistant project manager for cost informs you that he does not know how to fill out the revised corporate project review form. (Time = ½ hr)	_____
20.	One of the functional managers wants an immediate explanation of why the scope of effort for Phase II was changed this late into the project and why he wasn't informed. (Time = 1 hr)	_____

Wednesday's Activities

Activity	*Description*	*Priority*
21.	A vice president calls you stating that he has just read the rough draft of your Phase I report and wants to discuss some of the conclusions with you before the report is submitted to the customer on Thursday. (Time = 2 hrs)	_____
22.	The reproduction department informs you that they are expecting the final version of the in-house quarterly report for your project by noon today. The report is on your desk waiting for final review. (Time = 1 hr)	_____
23.	The manufacturing department manager calls to say that they may have to do more work than initially defined in Phase II. A meeting is requested. (Time = 1 hr)	_____
24.	Quality control sends you a memo stating that, unless changes are made, they will not be able to work with the engineering specifications developed for Phase III. A meeting will be required with all assistant project managers in attendance. (Time = 1 hr)	_____
25.	A functional manager calls to tell you that the raw data from yesterday's tests are terrific and invites you to come up to the laboratory and see the results yourself. (Time = 1 hr)	_____
26.	Your assistant project manager is having trouble resolving a technical problem. The functional manager wants to deal with you directly. This problem must be resolved by Friday or else a major Phase II milestone might slip. (Time = 1 hr)	_____

WORK PLAN   Day _____Wednesday_____

Priority			Method			Accomplishment		
Activity	Priority	Points	Method of Accomplishment	Points		Time	Activity	Points
21						9:00–9:30		
22						9:30–10:00		
23						10:00–10:30		
24						10:30–11:00		
25						11:00–11:30		
26						11:30–12:00		
27						12:00–12:30		
28						12:30–1:00		
29						1:00–1:30		
30						1:30–2:00		
Total			Total			2:00–2:30		
						2:30–3:00		
						3:00–3:30		
						3:30–4:00		
						4:00–4:30		
						4:30–5:00		
						Total		

Activities Postponed Until Today	Today's Priority

Points	
Priority Points	
Method Points	
Accomplishment Points	
Today's Points	

Legend
Method of Accomplishment:
Y = you
D = delegate
T = transmit
R = return
A = avoid
P = postpone

Activity	*Description*	*Priority*
27.	You have a technical interchange meeting with the customer scheduled for 1–3 P.M. on Thursday, and must review the handout before it goes to publication. The reproduction department has requested at least twelve hours' notice. (Time = 1 hr)	——
28.	You have a weekly team meeting from 10 A.M. to 12 noon (Time = 2 hrs)	——
29.	You must dictate minutes to your secretary concerning your weekly team meeting which is held on Wednesday 10 A.M. to 12 noon (Time = ½ hr)	——
30.	A new project problem has occurred in the manufacturing area and your manufacturing functional team members are reluctant to make a decision. (Time = 1 hr)	——

Thursday's Activities

Activity	*Description*	*Priority*
31	The electrical engineering department informs you that they have completed some Phase II activities ahead of schedule and want to know if you wish to push any other activities to the left. (Time = 1 hr)	——
32.	The assistant project manager for cost informs you that the corporate overhead rate is increasing faster than anticipated. If this continues, severe cost overruns will occur in Phases II and III. A schedule and cost review is necessary. (Time = 2 hrs)	——
33.	Your insurance man is calling to see if you wish to increase your life insurance. (Time = ½ hr)	——
34.	You cannot find one of last week's manufacturing line manager's technical reports as to departmental project status. You'll need it for the customer technical interchange meeting. (Time = ½ hr)	——
35.	One of your car pool members wants to talk to you concerning next Saturday's golf tournament. (Time = ½ hr)	——
36.	A functional manager calls to inform you that, due to a change in his division's workload priorities, people with the necessary technical expertise may not be available for next week's Phase II tasks. (Time = 2 hrs)	——
37.	An employee calls you stating that he is receiving conflicting instructions from one of your assistant project managers and his line manager. (Time = 1 hr)	——

WORK PLAN

Day Thursday

Priority			Method	
Activity	Priority	Points	Method of Accomplishment	Points
31				
32				
33				
34				
35				
36				
37				
38				
39				
40				
	Total		Total	

Accomplishment		
Time	Activity	Points
9:00–9:30		
9:30–10:00		
10:00–10:30		
10:30–11:00		
11:00–11:30		
11:30–12:00		
12:00–12:30		
12:30–1:00		
1:00–1:30		
1:30–2:00		
2:00–2:30		
2:30–3:00		
3:00–3:30		
3:30–4:00		
4:00–4:30		
4:30–5:00		
	Total	

Activities Postponed Until Today	Today's Priority

Points	
Priority Points	
Method Points	
Accomplishment Points	
Today's Points	

Legend
Method of Accomplishment:
Y = you D = delegate T = transmit R = return A = avoid P = postpone

Activity	*Description*	*Priority*
38.	The customer has requested bimonthly instead of monthly team meetings for Phase II. You must decide whether to add an additional project office team member to support the added workload. (Time = ½ hr)	_____
39.	Your secretary reminds you that you must make a presentation to the Rotary Club tonight on how your project will affect the local economy. You must prepare your speech. (Time = 2 hrs)	_____
40.	The bank has just called you concerning your personal loan. The information is urgent to get loan approval in time. (Time = ½ hr)	_____

Friday's Activities

Activity	*Description*	*Priority*
41.	An assistant project manager has asked for your solution to a recurring problem. (Time = ½ hr)	_____
42.	A functional employee is up for a merit review. You must fill out a brief checklist form and discuss it with the employee. The form must be on the functional manager's desk by next Tuesday. (Time = ½ hr)	_____
43.	The personnel department wants you to review the summer vacation schedule for your project office personnel. (Time = ½ hr)	_____
44.	The vice president calls you into his office stating that he has seen the excellent test results from this week's work, and feels that a follow-on contract should be considered. He wants to know if you can develop reasonable justification for requesting a follow-on contract at this early date. (Time = 1 hr)	_____
45.	The travel department says that you'll have to make your own travel arrangements for next month's trip to one of the customers, since you are taking a planned vacation trip in conjunction with the customer visit. (Time = ½ hr)	_____
46.	The personnel manager has asked if you would be willing to conduct a screening interview for an applicant who wants to be an assistant project manager. The applicant will be available this afternoon 1–2 P.M. (Time = 1 hr)	_____
47.	Your assistant project manager wants to know why you haven't approved his request to take MBA courses this quarter. (Time = ½ hr)	_____

WORK PLAN　　　　　Day ___Friday___

Priority			Method		Accomplishment		
Activity	Priority	Points	Method of Accomplishment	Points	Time	Activity	Points
41					9:00–9:30		
42					9:30–10:00		
43					10:00–10:30		
44					10:30–11:00		
45					11:00–11:30		
46					11:30–12:00		
47					12:00–12:30		
48					12:30–1:00		
49					1:00–1:30		
50					1:30–2:00		
Total			Total		2:00–2:30		
					2:30–3:00		
					3:00–3:30		
					3:30–4:00		
					4:00–4:30		
					4:30–5:00		
					Total		

Activities Postponed Until Today	Today's Priority

Points	
Priority Points	
Method Points	
Accomplishment Points	
Today's Points	

Legend
Method of Accomplishment:
Y = you
D = delegate
T = transmit
R = return
A = avoid
P = postpone

Activity	*Description*	*Priority*
48.	Your assistant project manager wants to know if he has the authority to visit vendors without informing procurement. (Time = ½ hr)	_____
49.	You have just received your copy of *Engineering Review Quarterly* and would like to look it over. (Time = ½ hr)	_____
50.	You have been asked to make a statement before the grievance committee (this Friday, 10 A.M. to 12 noon) because one of the functional employees has complained about working overtime on Sunday mornings. You'll have to be in attendance for the entire meeting. (Time = 2 hrs)	_____

RATIONALE AND POINT AWARDS

In the answers that follow, your recommendations may differ from those of the author because of the type of industry or the nature of the project. You will be given the opportunity to defend your answers at a later time.

a. If you selected the correct priority according to the table on pages 375–376, then the following system should be employed for awarding points:

Priority	*Points*
A	10
B	5
C	3

b. If you selected the correct accomplishment mode according to the table on pages 375–376, then the following system should be employed for assigning points:

Method of Accomplishment	*Points*
Y	10
T	10
P	8
D	8
A	6

c. You will receive 10 bonus points for each correctly postponed or delayed activity accomplished during the team meeting.

d. You will receive 5 points for each half-hour time slot in which you perform a priority A activity (one that is correctly identified as priority A).

e. You will receive a 10-point penalty for any activity that is split.

f. You will receive a 20-point penalty for each priority A or B activity not accomplished by you or your team by Friday at 5:00 P.M.

Activity	Rationale
1.	The updating of schedules, especially for Phase II, should be of prime importance because of the impact on functional resources. These schedules can be delegated to assistant project managers. However, with a team meeting scheduled for Wednesday, it should be an easy task to update the schedules when all of the players are present. The updating of the schedules should *not* be delayed until Thursday. Sufficient time must be allocated for close analysis and reproduction services.
2.	This must be done immediately. Your assistant project manager for manufacturing should be able to handle this activity.
3.	You must handle this yourself.
4.	Here, we assume that the representative is available only today. The assistant project managers can handle this activity. This activity may be important if you were unaware of this vendor's product.
5.	This could be delegated to your assistants provided that you allow sufficient time for personal review on Wednesday.
6.	Delaying this activity for one more week should not cause any problems. This activity can be delegated.
7.	You must take charge at once.
8.	Even though your main concern is the project, you still must fulfill your company's administrative requirements.
9.	This can be delayed until Wednesday's team meeting, especially since these are Phase III schedules. However, there is no guarantee that line people will be ready or knowledgeable to discuss Phase III this early. You will probably have to do this yourself.
10.	The procurement request must be answered. Your assistant project manager for manufacturing should have this information available.
11.	This is urgent and should *not* be postponed until the team meeting. Good project managers will give functional managers as much information as possible as early as possible for resource control. This task can be delegated to the assistant project managers, but it is not recommended.
12.	This belongs to the project manager. The agenda review and the meeting can be split, but it is not recommended.
13.	This must be done immediately. The results could severely limit your resources (especially if overtime would normally be required). Although your assistant project managers will probably be involved, the majority of the work is yours.

Activity	*Rationale*
14.	Most project managers hate a request like this but know that situations such as this are inevitable.
15.	Project policies should be told by the project manager himself. Policy changes should be announced as early as possible. Team meetings are appropriate for such actions.
16.	Obviously, the project manager must do this task himself. Fortunately, there is sufficient time if the lead times are accurate.
17.	The priority of this activity is actually your choice, but an A priority is preferred if you have time. This activity cannot be delegated.
18.	This activity must be done, but the question is when. Parts of this task can be delegated, but the final decision must be made by the project manager.
19.	Obviously you must do this yourself. Your priority, of course, depends on the deadline on the corporate project review form.
20.	The project manager must perform this activity immediately.
21.	Top-level executives from both the customer and contractor often communicate project status among themselves. Therefore, since the conclusions in the report reflect corporate policy, this activity should be accomplished immediately.
22.	The reproduction department considers each job as a project and therefore you should try not to violate their milestones. This activity can be delegated, depending on the nature of the report.
23.	This could have a severe impact on your program. Although you could delegate this to one of your assistants, you should do this yourself because of the ramifications.
24.	This must be done, and the team meeting is the ideal place.
25.	You, personally, should give the functional manager the courtesy of showing you his outstanding results. However, it is not a high priority and could even be delegated or postponed since you'll see the data eventually.
26.	The question here is the importance of the problem. The problem must be resolved by Thursday in case an executive meeting needs to be scheduled to establish company direction. Waiting until the last minute can be catastrophic here.
27.	The project manager should personally review all data presented to the customer. Check Thursday's schedule. Did you forget the interchange meeting?
28.	This is your show.
29.	This should be done immediately. Nonparticipants need to know the project status. The longer you wait, the greater the risk that you will neglect something important. This activity can be delegated, but it is not recommended.

Activity	*Rationale*
30.	You may have to solve this yourself even though you have an assistant project manager for manufacturing. The decision may affect the schedule and miletones.
31.	Activities such as this do not happen very often. But when they do, the project manager should make the most of them, as fast as he can. These are gold mine activities. They can be delegated, but not postponed.
32.	If this activity is not accomplished immediately, the results can be catastrophic. Regardless of the project manager's first inclination to delegate, this activity should be done by the project manager himself.
33.	This activity can be postponed or even avoided, if necessary.
34.	Obviously, if the report is that important, then your assistant project managers should have copies of the report and the activity can be delegated.
35.	This activity should be discussed in the car pool, not on company time.
36.	This is extremely serious. The line manager would probably prefer to work directly with the project manager on this problem.
37.	This is an activity that you should handle. Transmitting this to one of your assistants may aggravate the situation further. Although it is possible that this activity could be postponed, it is highly unlikely that time would smooth out the conflict.
38.	This is a decision for the project manager. Extreme urgency may not be necessary.
39.	Project managers also have a social responsibility.
40.	The solution to this activity is up for grabs. Most companies realize that employees occasionally need company time to complete personal business.
41.	Why is he asking you about a recurring problem? How did he solve it last time? Let him do it again.
42.	You must do this personally, but it can wait until Monday.
43.	This activity is not urgent and can be accomplished by your assistant project managers.
44.	This could be your lucky day.
45.	Although most managers would prefer to delegate this activity to their secretaries, it is really the responsibility of the project manager since it involves personal business.
46.	This is an example of an administrative responsibility that is required of all personnel regardless of the job title or management level. This activity must be accomplished today, if time permits.

Activity	Monday Prior.	Monday Accom.	Tuesday Prior.	Tuesday Accom.	Wednesday Prior.	Wednesday Accom.	Thursday Prior.	Thursday Accom.	Friday Prior.	Friday Accom.
1	B	D,Y,T,P	B	D,Y,T,P	A	D,Y,T				
2	A	D,Y,T								
3	A	Y								
4	A/B	D,Y,T								
5	B	D,Y,P	B	D,Y,P	A	D,Y				
6	B	D,Y,P	B	D,Y,P	B	D,Y,P	B	D,Y,P	B	D,Y,P
7	A	Y								
8	A	Y								
9	B	Y,P	B	Y,P	A	Y				
10	B	Y,T,P	B	Y,T,P	B	Y,T,P	B	Y,T,P	B	Y,T,P
11			A	D,Y,T						
12			A	Y						
13			A	Y,P						
14			A	Y						
15			B	P,Y	A	Y				
16			B	Y,P	B	Y,P	B	Y,P	B	Y,P
17			C	A,Y						
18			B/C	D,Y,P	B	D,Y,P	B	D,Y,P	B	D,Y,P

(continues)

PRIORITY/ACCOMPLISHMENT MODE (Continued)

No.	A/B	Y,P	A/B	Y,P	A/B	Y,P	A/B	Y,P
19								
20	A	Y						
21			A	Y				
22			A	D,Y				
23			A	D,Y,T				
24			A	Y				
25			B	Y,T,P,D	B	Y,T,P,D	B	Y,T,P,R
26			B	Y	A	Y		
27			A	Y				
28			A	Y				
29			A	Y,D				
30			A	Y,T				
31					A	Y,D		
32					A	Y		
33					C	Y,P	C	Y,P
34					A	Y,T		
35					C	A,P	C	A,P
36					A	Y,T		
37					A	Y		
38					B	Y,P	B	Y,P
39					A	Y		

PRIORITY/ACCOMPLISHMENT MODE (*Continued*)

						A	Y		
40									
41								A/B	R
42								B	Y,P
43								B	Y,P,D
44								A	Y
45								B	Y,P
46								A	Y,T,D
47								A	Y
48								B	Y,T,P,D,R
49								C	Y,P,A
50								A	Y

631

47.   Although you might consider this as a B priority or one that can be postponed, you must remember that your assistant project manager considers this as an A priority and would like an answer today. You are morally obligated to give him the answer today.

48.   Why can't he get the answer himself? Whether or not you handle this activity might depend on the priority and how much time you have available.

49.   How important is it for you to review the publication?

50.   This is mandatory attendance on your behalf. You have total responsibility for all overtime scheduled on your project. You may wish to bring one of your assistant project managers with you for moral support.

Now take the total points for each day and complete the following table:

Summary Work Plan	
Day	**Points**
Monday	
Tuesday	
Wednesday	
Thursday	
Friday	
Total	

CONCLUSIONS AND SUMMARY QUESTIONS

1. Project managers have a tendency to want to carry the load themselves, even if it means working sixty hours a week. You were told to do everything within your normal working day. But, as a potentially good project manager, you probably have the natural tendency of wanting to postpone some work until a later date so that you can do it yourself. Doing the activities, when they occur, even through transmittal or delegation, is probably the best policy. You might wish to do the same again at a later time and see if you can beat your present score. Only this time, try to do as many tasks as possible on each day, even if it means delegation.

2. Several of the activities were company, not project, requests. Project managers have a tendency to avoid administrative responsibilities unless it deals directly with their project. This process of project management "tunnel vision" can lead to antagonism and conflicts if the proper attitude is not developed on the part of the project manager. This can easily carry down to his assistants as well.
3. Several of the activities could have been returned to the requestor. However, in a project environment where the project manager cannot be successful without the functional manager's support, most project managers would never turn away a line employee's request for assistance.
4. Make a list of the activities where your answers differ from those of the answer key and where you feel that there exists sufficient justification for your interpretation.
5. Quite often self-productivity can be increased by knowing one's own energy cycle. Are your more important meetings in the mornings or afternoons? What time of day do you perform your most productive work? When do you do your best writing? Does your energy cycle vary according to the day of the week?

Part 17

INDUSTRY SPECIFIC: CONSTRUCTION

Many project management situations or problems are somewhat complex and involve many interacting factors, all contributing to a common situation. For example, poor planning on a project may appear on the surface to be a planning issue, whereas the real problem may be the corporate culture, lack of line management support, or poor employee morale. The case studies in this chapter involve interacting factors.

Robert L. Frank Construction Company

It was Friday afternoon, a late November day in 2003, and Ron Katz, a purchasing agent for Robert L. Frank Construction, poured over the latest earned value measurement reports. The results kept pointing out the same fact; the Lewis project was seriously over budget. Man-hours expended to date were running 30 percent over the projection and, despite this fact, the project was not progressing sufficiently to satisfy the customer. Material deliveries had experienced several slippages, and the unofficial indication from the project scheduler was that, due to delivery delays on several of the project's key items, the completion date of the coal liquefaction pilot plant was no longer possible.

Katz was completely baffled. Each day for the past few months as he reviewed the daily printout of project time charges, he would note that the purchasing and expediting departments were working on the Lewis project, even though it was not an unusually large project, dollarwise, for Frank. Two years earlier, Frank was working on a $300 million contract, a $100 million contract and a $50 million contract concurrently with the Frank Chicago purchasing department responsible for all the purchasing, inspection, and expediting on all three contracts. The Lewis project was the largest project in house and was valued at only $90 million. What made this project so different from previous contracts and caused such problems? There was little Katz felt that he could do to correct the situation. All that could be done was to understand what had occurred in an effort to prevent a recurrence. He began to write his man-hour report for requested by the project manager the next day.

COMPANY BACKGROUND

Robert L. Frank Construction Company was an engineering and construction firm serving the petroleum, petrochemical, chemical, iron and steel, mining, pharmaceutical, and food-processing industries from its corporate headquarters in Chicago, Illinois, and its worldwide offices. Its services include engineering, purchasing, inspection, expediting, construction, and consultation.

Frank's history began in 1947 when Robert L. Frank opened his office. In 1955, a corporation was formed, and by 1960 the company had completed contracts for the majority of the American producers of iron and steel. In 1962, an event occurred that was to have a large impact on Frank's future. This was the merger of Wilson Engineering Company, a successful refinery concern, with Robert L. Frank, now a highly successful iron and steel concern. This merger greatly expanded Frank's scope of operations and brought with it a strong period of growth. Several offices were opened in the United States in an effort to better handle the increase in business. Future expansions and mergers enlarged the Frank organization to the point where it had fifteen offices or subsidiaries located throughout the United States and twenty offices worldwide. Through its first twenty years of operations, Frank had more than 2,500 contracts for projects having an erected value of over $1 billion.

Frank's organizational structure has been well suited to the type of work undertaken. The projects Frank contracted for typically had a time constraint, a budget constraint, and a performance constraint. They all involved an outside customer such as a major petroleum company or a steel manufacturer. Upon acceptance of a project, a project manager was chosen (and usually identified in the proposal). The project manager would head up the project office, typically consisting of the project manager, one to three project engineers, a project control manager, and the project secretaries. The project team also included the necessary functional personnel from the engineering, purchasing, estimating, cost control, and scheduling areas. Exhibit I is a simplified depiction. Of the functional areas, the purchasing department is somewhat unique in its organizational structure. The purchasing department is organized on a project management basis much as the project as a whole would be organized. Within the purchasing department, each project had a project office that included a project purchasing agent, one or more project expeditors and a project purchasing secretary. Within the purchasing department the project purchasing agent had line authority over only the project expeditor(s) and project secretary. However, for the project purchasing agent to accomplish his goals, the various functions within the purchasing

Exhibit I. Frank organization

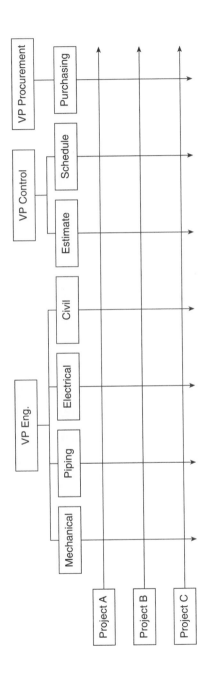

Exhibit II. *Frank purchasing organization*

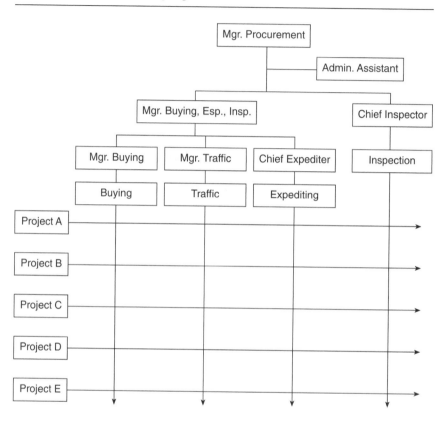

department had to commit sufficient resources. Exhibit II illustrates the organization within the purchasing department.

HISTORY OF THE LEWIS PROJECT

Since 1998, the work backlog at Frank has been steadily declining. The Rovery Project, valued at $600 million, had increased company employment sharply since its inception in 1997. In fact, the engineering on the Rovery project was such a large undertaking that in addition to the Chicago office's participation, two other U.S. offices, the Canadian office, and the Italian subsidiary were heavily involved. However, since the Rovery project completion in 2001, not enough new work was received to support the work force thus necessitating recent lay-offs of engineers, including a few project engineers.

Company officials were very disturbed with the situation. Frank's company policy was to "maintain an efficient organization of sufficient size and resources, and staffed by people with the necessary qualifications, to execute projects in any location for the industries served by Frank." However, the recent down-turn in business meant that there was not enough work even with the reduction in employees. Further cutbacks would jeopardize Frank's prospects of obtaining future large projects as prospective clients look to contractors with a sufficient staff of qualified people to accomplish their work. By contrast, supporting employees out of overhead was not the way to do business, either. It became increasingly important to "cut the fat out" of the proposals being submitted for possible projects. Despite this, new projects were few and far between, and the projects that were received were small in scope and dollar value and therefore did not provide work for very many employees.

When rumors of a possible construction project for a new coal liquefaction pilot plant started circulating, Frank officials were extremely interested in bidding for the work. It was an excellent prospect for two reasons. Besides Frank's desperate need for work, the Lewis chemical process used in the pilot plant would benefit Frank in the long run by the development of state-of-the-art technology. If the pilot plant project could be successfully executed, when it came time to construct the full-scale facility, Frank would have the inside track as they had already worked with the technology. The full-scale facility offered prospects exceeding the Rovery project, Frank's largest project to date. Top priority was therefore put on obtaining the Lewis project. It was felt that Frank had a slight edge due to successful completion of a Lewis project six years ago. The proposal submitted to Lewis contained estimates for material costs, man-hours, and the fee. Any changes in scope after contract award would be handled by change order to the contract. Both Lewis and Frank had excellent scope change control processes as part of their configuration management plans. The functional department affected would submit an estimate of extra man-hours involved to the project manager, who would review the request and submit it to the client for approval. Frank's preference was for cost-plus-fixed-fee contracts.

One of the unique aspects stated in the Lewis proposal was the requirement for participation by both of Frank Chicago's operating divisions. Previous Frank contracts were well suited to either Frank's Petroleum and Chemical Division (P & C) or the Iron and Steel Division (I & S). However, due to the unusual chemical process, one that starts with coal and ends up with a liquid energy form, one of the plant's three units was well suited to the P & C Division and one was well suited to the I & S Division. The third unit was an off-site unit and was not of particular engineering significance.

The award of the contract six weeks later led to expectations by most Frank personnel that the company's future was back on track again. The project began inauspiciously. The project manager was a well-liked, easy-going sort who had been manager of several Frank projects. The project office included three of Frank's most qualified project engineers.

In the purchasing department, the project purchasing agent (PPA) assigned to the project was Frank's most experienced PPA. Bill Hall had just completed his assignment on the Rovery Project and had done well, considering the magnitude of the job. The project had its problems, but they were small in comparison to the achievements. He had alienated some of the departments slightly but that was to be expected. Purchasing upper management was somewhat dissatisfied with him in that, due to the size of the project, he didn't always use the normal Frank purchasing methods; rather, he used whatever method he felt was in the best interest of the project. Also, after the Rovery project, a purchasing upper management reshuffling left him in the same position but with less power and authority rather than receiving a promotion he had felt he had earned. As a result, he began to subtly criticize the purchasing management. This action caused upper management to hold him in less than high regard but, at the time of the Lewis Project, Hall was the best person available.

Due to the lack of float in the schedule and the early field start date, it was necessary to *fast start* the Lewis Project. All major equipment was to be purchased within the first three months. This, with few exceptions, was accomplished. The usual problems occurred such as late receipt of requisition from engineering and late receipt of bids from suppliers.

One of the unique aspects of the Lewis project was the requirement for purchase order award meetings with vendors. Typically, Frank would hold award meetings with vendors of major equipment such as reactors, compressors, large process towers, or large pumps. However, almost each time Lewis approved purchase of a mechanical item or vessel, it requested that the vendor come in for a meeting. Even if the order was for an on-the-shelf stock pump or small drum or tank, a meeting was held. Initially, the purchasing department meeting attendees included the project purchasing agent, the buyer, the manager of the traffic department, the chief expediter, and the chief Inspector. Engineering representatives included the responsible engineer and one or two of the project engineers. Other Frank attendees were the project control manager and the scheduler. Quite often, these meetings would accomplish nothing except the reiteration of what had been included in the proposal or what could have been resolved with a phone call or even e-mail. The project purchasing agent was responsible for issuing meeting notes after each meeting.

One day at the end of the first three-month period, the top-ranking Lewis representative met with Larry Broyles, the Frank project manager.

Lewis rep: Larry, the project is progressing but I'm a little concerned. We don't feel like we have our finger on the pulse of the project. The information we are getting is sketchy and untimely. What we would like to do is meet with Frank every Wednesday to review progress and resolve problems.

Larry: I'd be more than happy to meet with any of the Lewis people because I think your request has a lot of merit.

Lewis rep: Well, Larry, what I had in mind was a meeting between all the Lewis people, yourself, your project office, the project purchasing agent, his assistant, and your scheduling and cost control people.

Larry: This sounds like a pretty involved meeting. We're going to tie up a lot of our people for one full day a week. I'd like to scale this thing down. Our proposal took into consideration meetings, but not to the magnitude we're talking about.

Lewis rep: Larry, I'm sorry but we're footing the bill on this project and we've got to know what's going on.

Larry: I'll set it up for this coming Wednesday.

Lewis rep: Good.

The required personnel were informed by the project manager that effective immediately, meetings with the client would be held weekly. However, Lewis was dissatisfied with the results of the meetings, so the Frank project manager informed his people that a premeeting would be held each Tuesday to prepare the Frank portion of the Wednesday meeting. All of the Wednesday participants attended the Tuesday premeetings.

Lewis requests for additional special reports from the purchasing department were given into without comment. The project purchasing agent and his assistants (project started with one and expanded to four) were devoting a great majority of their time to special reports and putting out fires instead of being able to track progress and prevent problems. For example, recommended spare parts lists were normally required from vendors on all Frank projects. Lewis was no exception. However, after the project began, Lewis decided it wanted the spare parts recommendations early into the job. Usually, spare parts lists are left for the end of an order. For example, on a pump with fifteen-week delivery, normally Frank would pursue the recommended spare parts list three to four weeks prior to shipment, as it would tend to be more accurate. This improved accuracy was due to the fact that at this point in the order, all changes probably had been made. In the case of the Lewis project, spare parts recommendations had to be expedited from the day the material was released for fabrication. Changes could still be made that could dramatically affect the design of the pump. Thus, a change in the pump after receipt of the spare parts list would necessitate a new spare parts list. The time involved in this method of expediting the spare parts list was much greater than the time involved in the normal Frank method. Added to this situation was Lewis's request for a fairly involved biweekly report on the status of spare parts lists on all the orders. In addition, a full time spare parts coordinator was assigned to the project.

The initial lines of communication between Frank and Lewis were initially well defined. The seven in-house Lewis representatives occupied the area adjacent to the Frank project office (see Exhibit III). Initially, all communications from Lewis were channeled through the Frank project office to the applicable functional employee. In the case of the purchasing department, the Frank project office would channel Lewis requests through the purchasing project office. Responses or return communications followed the reverse route. Soon the volume of communications increased to the point where response time was becoming unacceptable. In several special cases, an effort was made to cut this response time. Larry Broyles told the Lewis team members to call or go see the functional person (i.e., buyer or engineer) for the answer. However, this practice soon became the rule rather than the exception. Initially, the project office was kept informed

Exhibit III. Floor plan—Lewis project teams

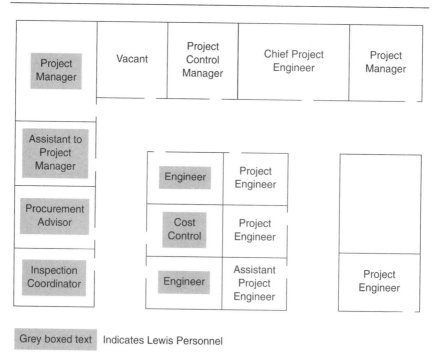

Grey boxed text   Indicates Lewis Personnel

Normal text   Indicates Frank Personnel

of these conversations, but this soon stopped. The Lewis personnel had integrated themselves into the Frank organization to the point where they became part of the organization.

The project continued on, and numerous problems cropped up. Vendors' material delays occurred, companies with Frank purchase orders went bankrupt, and progress was not to Lewis's satisfaction. Upper management soon became aware the problems on this project due to its sensitive nature, and the Lewis project was now receiving much more intense involvement by senior management than it had previously. Upper management sat in on the weekly meetings in an attempt to pacify Lewis. Further problems plagued the project. Purchasing management, in an attempt to placate Lewis, replaced the project purchasing agent. Ron Katz, a promising young MBA graduate, had five years of experience as an assistant to several of the project purchasing agents. He was most recently a project purchasing agent on a fairly small project that had been very successful. It was thought by purchasing upper management that this move was a good one, for two reasons. First, it would remove Bill Hall from the project as PPA. Second, by appointing Ron Katz, Lewis would be pacified, as Katz was a promising talent with a successful project under his belt.

However, the project under direction of Katz still experienced problems in the purchasing area. Revisions by engineering to material already on order caused serious delivery delays. Recently requisitioned material could not be located with an acceptable delivery promise. Katz and purchasing upper management, in an attempt to improve the situation, assigned more personnel to the project, personnel that were more qualified than the positions dictated. Buyers and upper-level purchasing officials were sent on trips to vendors' facilities that were normally handled by traveling expediters. In the last week the Lewis representative met with the project manager, Broyles:

Lewis rep: Larry, I've been reviewing these man-hour expenditures, and I'm disturbed by them.

Larry: Why's that?

Lewis rep: The man-hour expenditures are far outrunning project progress. Three months ago, you reported that the project completion percentage was 30 percent, but according to my calculations, we've used 47 percent of the man hours. Last month you reported 40 percent project completion and I show a 60 percent expenditure of man-hours.

Larry: Well, as you know, due to problems with vendors' deliveries, we've really had to expedite intensively to try to bring them back in line.

Lewis rep: Larry, I'm being closely watched by my people on this project, and a cost or schedule overrun not only makes Frank look bad, it makes me look bad.

Larry: Where do we go from here?

Lewis rep: What I want is an estimate from your people on what is left, man-hour wise. Then I can sit down with my people and see where we are.

Larry: I'll have something for you the day after tomorrow.

Lewis rep: Good.

The functional areas were requested to provide this information, which was reviewed and combined by the project manager and submitted to Lewis for approval. Lewis's reaction was unpleasant, to say the least. The estimated man-hours in the proposal were now insufficient. The revised estimate was for almost 40 percent over the proposal. The Lewis representative immediately demanded an extensive report on the requested increase. In response to this, the project manager requested man-hour breakdowns from the functional areas. Purchasing was told to do a purchase order by purchase order breakdown of expediting and inspection man-hours. The buying section had to break down the estimate of the man-hours needed to purchase each requisition, many of which were not even issued. Things appeared to have gone from bad to worse.

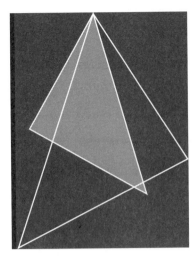

The Lyle Construction Project

At 6:00 P.M. on Thursday in late October 1998, Don Jung, an Atlay Company project manager (assigned to the Lyle contract) sat in his office thinking about the comments brought up during a meeting with his immediate superior earlier that afternoon. During that meeting Fred Franks, the supervisor of project managers, criticized Don for not promoting a cooperative attitude between him and the functional managers. Fred Franks had a high-level meeting with the vice presidents in charge of the various functional departments (i.e., engineering, construction, cost control, scheduling, and purchasing) earlier that day. One of these vice presidents, John Mabby (head of the purchasing department) had indicated that his department, according to his latest projections, would overrun their man-hour allocation by 6,000 hours. This fact had been relayed to Don by Bob Stewart (the project purchasing agent assigned to the Lyle Project) twice in the past, but Don had not seriously considered the request because some of the purchasing was now going to be done by the subcontractor at the job site (who had enough man-hours to cover this additional work). John Mabby complained that, even though the subcontractor was doing some of the purchasing in the field, his department still would overrun its man-hour allocation. He also indicated to Fred Franks that Don Jung had better do something about this man-hour problem now. At this point in the meeting, the vice president of engineering, Harold Mont, stated that he had experienced the same problem in that Don Jung seemed to ignore their requests for additional man-hours. Also at this meeting the various vice presidents indicated that Don Jung had not been operating within the established standard company procedures. In an effort to make up for time lost due to initial delays that occurred in the process development stage of this project, Don and his project team had been getting the various functional people working on the contract to "cut corners" and in many cases to buck the standard operating procedures of their respective functional departments in an effort to save time. His actions and the actions of his project team were alienating the vice presidents in charge of the functional departments. During this meeting, Fred Franks received a good deal of criticism due to this fact. He was also told that Don Jung had better shape up, because it was the consensus opinion of these vice presidents that his method of operating might seriously hamper the project's ability to finish on time and within budget. It was very important that this job be completed in accordance with the Lyle requirements since they would be building two more similar plants within the next ten years. A good effort on this job could further enhance Atlay's chances for being awarded the next two jobs.

Fred Franks related these comments and a few of his own to Don Jung. Fred seriously questioned Don's ability to manage the project effectively and told him so. However, Fred was willing to allow Don to remain on the job if he would begin to operate in accordance with the various functional departments' standard operating procedures and if he would listen and be more attentive to the comments from the various functional departments and do his best to cooperate with them in the best interests of the company and the project itself.

INCEPTION OF THE LYLE PROJECT

In April of 1978, Bob Briggs, Atlay's vice president of sales, was notified by Lyle's vice president of operations (Fred Wilson) that Atlay had been awarded the $600 million contract to design, engineer, and construct a polypropylene plant in Louisiana. Bob Briggs immediately notified Atlay's president and other high-level officials in the organization (see Exhibit I). He then contacted Fred Franks in order to finalize the members of the project team. Briggs wanted George Fitz, who was involved in developing the initial proposal, to be the project manager. However, Fitz was in the hospital and would be essentially out of action for another three months. Atlay then had to scramble to appoint a project manager, since Lyle wanted to conduct a kickoff meeting in a week with all the principals present.

One of the persons most available for the position of project manager was Don Jung. Don had been with the company for about fifteen years. He had started with the company as a project engineer, and then was promoted to the position of manager of computer services. He was in charge of computer services for six months until he had a confrontation with Atlay's upper management regarding

Exhibit I. Atlay and Company organization chart

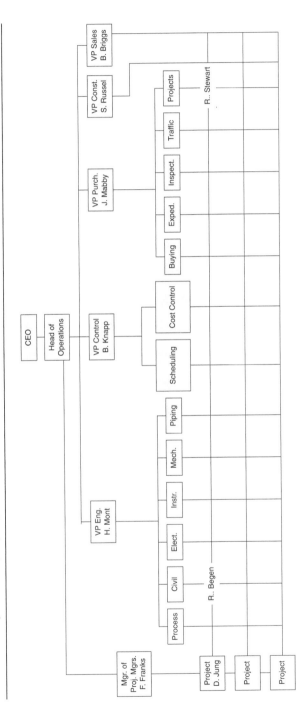

the policies under which the computer department was operating. He had served the company in two other functions since—the most recent position, that of being a senior project engineer on a small project that was handled out of the Houston office. One big plus was the fact that Don knew Lyle's Fred Wilson personally since they belonged to many of the same community organizations. It was decided that Don Jung would be the project manager and John Neber (an experienced project engineer) would be assigned as the senior project engineer. The next week was spent advising Don Jung regarding the contents of the proposal and determining the rest of the members to be assigned to the project team.

A week later, Lyle's contingent arrived at Atlay's headquarters (see Exhibit II). Atlay was informed that Steve Zorn would be the assistant project manager on this job for Lyle. The position of project manager would be left vacant for the time being. The rest of Lyle's project team was then introduced. Lyle's project team consisted of individuals from various Lyle divisions around the country, including Texas, West Virginia, and Philadelphia. Many of the Lyle project team members had met each other for the first time only two weeks ago.

During this initial meeting, Fred Wilson emphasized that it was essential that this plant be completed on time since their competitor was also in the process of preparing to build a similar facility in the same general location. The first plant finished would most likely be the one that would establish control over the southwestern United States market for polypropylene material. Mr. Wilson felt that Lyle had a six-week head start over its competitor at the moment and would like

Exhibit II. *Lyle project team organizational chart*

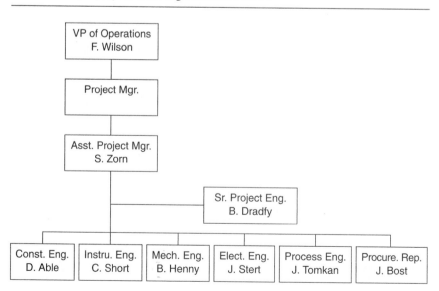

to increase that difference, if at all possible. He then introduced Lyle's assistant project manager who completed the rest of the presentation.

At this initial meeting the design package was handed over to Atlay's Don Jung so that the process engineering stage of this project could begin. This package was, according to their inquiry letter, so complete that all material requirements for this job could be placed within three months after project award (since very little additional design work was required by Atlay on this project). Two weeks later, Don contacted the lead process engineer on the project, Raphael Begen. He wanted to get Raphael's opinion regarding the condition of the design package.

Begen: Don, I think you have been sold a bill of goods. This package is in bad shape.

Jung: What do you mean this package is in bad shape? Lyle told us that we would be able to have all the material on order within three months since this package was in such good shape.

Begen: Well in my opinion, it will take at least six weeks to straighten out the design package. Within three months from that point you will be able to have all the material on order.

Jung: What you are telling me then is that I am faced with a six-week schedule delay right off the bat due to the condition of the package.

Begen: Exactly.

Don Jung went back to his office after his conversation with the lead process engineer. He thought about the status of his project. He felt that Begen was being overly pessimistic and that the package wasn't really all that bad. Besides, a month shouldn't be too hard to make up if the engineering section would do its work quicker than normal and if purchasing would cut down on the amount of time it takes to purchase materials and equipment needed for this plant.

CONDUCT OF THE PROJECT

The project began on a high note. Two months after contract award, Lyle sent in a contingent of their representatives. These representatives would be located at Atlay's headquarters for the next eight to ten months. Don Jung had arranged to have the Lyle offices set up on the other side of the building away from his project team. At first there were complaints from Lyle's assistant project manager regarding the physical distance that separated Lyle's project team and Atlay's project team. However, Don Jung assured him that there just wasn't any available space that was closer to the Atlay project team other than the one they were now occupying.

The Atlay project team operating within a matrix organizational structure plunged right into the project (see Exhibit III). They were made aware of the delay that was incurred at the onset of the job (due to the poor design package) by Don Jung. His instructions to them were to cut corners whenever doing so might result in time savings. They were also to suggest to members of the functional departments that were working on this project methods that could possibly result in quicker turnaround of the work required of them. The project team coerced the various engineering departments into operating outside of their normal procedures due to the special circumstances surrounding this job. For example, the civil engineering section prepared a special preliminary structural steel package, and the piping engineering section prepared preliminary piping packages so that the purchasing department could go out on inquiry immediately. Normally, the purchasing department would have to wait for formal take-offs from both of these departments before they could send out inquiries to potential vendors. Operating in

Exhibit III. Atlay Company procurement department organizational chart

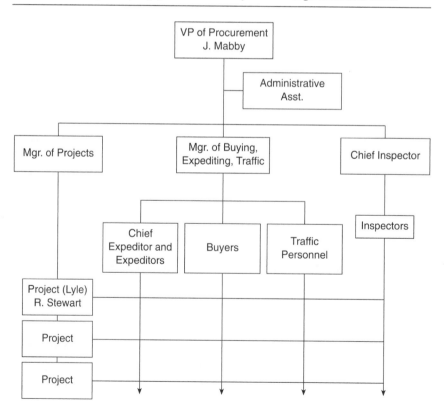

this manner could result in some problems, however. For example, the purchasing department might arrange for discounts from the vendors based on the quantity of structural steel estimated during the preliminary take-off. After the formal take-off has been done by the civil engineering section (which would take about a month), they might find out that they underestimated the quantity of structural steel required on the project by 50 tons. This was damaging, because knowing that there was an additional 50 tons of structural steel might have aided the purchasing department in securing an additional discount of $.20 per pound (or $160,000 discount for 400 tons of steel).

In an effort to make up for lost time, the project team convinced the functional engineering departments to use catalog drawings or quotation information whenever they lacked engineering data on a particular piece of equipment. The engineering section leaders pointed out that this procedure could be very dangerous and could result in additional work and further delays to the project. If, for example, the dimensions for the scale model being built are based on this project on preliminary information without the benefit of having certified vendor drawings in house, then the scale for that section of the model might be off. When the certified data prints are later received and it is apparent that the dimensions are incorrect, that portion of the model might have to be rebuilt entirely. This would further delay the project. However, if the information does *not* change substantially, the company could save approximately a month in engineering time. Lyle was advised in regards to the risks and potential benefits involved when Atlay operates outside of their normal operating procedure. Steve Zorn informed Don Jung that Lyle was willing to take these risks in an effort to make up for lost time. The Atlay project team then proceeded accordingly.

The method that the project team was utilizing appeared to be working. It seemed as if the work was being accomplished at a much quicker rate than what was initially anticipated. The only snag in this operation occurred when Lyle had to review/approve something. Drawings, engineering requisitions, and purchase orders would sit in the Lyle area for about two weeks before Lyle personnel would review them. Half of the time these documents were returned two weeks later with a request for additional information or with changes noted by some of Lyle's engineers. Then the Atlay project team would have to review the comments/changes, incorporate them into the documents, and resubmit them to Lyle for review/approval. They would then sit for another week in that area before finally being reviewed and eventually returned to Atlay with final approval. It should be pointed out that the contract procedures stated that Lyle would have only five days to review/approve the various documents being submitted to it. Don Jung felt that part of the reason for this delay had to do with the fact that all the Lyle team members went back to their homes for the weekends. Their routine was to leave around 10:00 A.M. on Friday and return around 3:00 P.M. on the following Monday. Therefore, essentially two days of work by the Lyle project team

out of the week were lost. Don reminded Steve Zorn that according to the contract, Lyle was to return documents that needed approval within five days after
receiving them. He also suggested that if the Lyle project team would work a full
day on Monday and Friday, it would probably increase the speed at which documents were being returned. However, neither corrective action was undertaken by
Lyle's assistant project manager, and the situation failed to improve. All the time
the project team had saved by cutting corners was now being wasted, and further
project delays seemed inevitable. In addition, other problems were being encountered during the interface process between the Lyle and Atlay project team
members. It seems that the Lyle project team members (who were on temporary
loan to Steve Zorn from various functional departments within the Lyle organization) were more concerned with producing a perfect end product. They did not
seem to realize that their actions, as well as the actions of the Atlay project team,
had a significant impact on this particular project. They did not seem to be aware
of the fact that they were also constrained by time and cost, as well as performance. Instead, they had a very relaxed and informal operating procedure. Many
of the changes made by Lyle were given to Atlay verbally. They explained to
the Atlay project team members that written confirmation of the changes were
unnecessary because "we are all working on the same team." Many significant
changes in the project were made when a Lyle engineer was talking directly
to an Atlay engineer. The Atlay engineer would then incorporate the changes into
the drawings he was working on, and sometimes failed to inform his project engineer about the changes. Because of this informal way of operating, there were
instances in which Lyle was dissatisfied with Atlay because changes were not
being incorporated or were not made in strict accordance with their requests.
Steve Zorn called Don Jung into his office to discuss this problem:

Steve: Don, I've received complaints from my personnel regarding your
teams inability to follow through and incorporate Lyle's comments/changes
accurately into the P & ID drawings.

Don: Steve, I think my staff has been doing a fairly good job of incorporating your team's comments/changes. You know the whole process would work a
lot better, however, if you would send us a letter detailing each change.
Sometimes my engineers are given two different instructions regarding the scope
of the change recommended by your people. For example, one of your people will
tell our process engineer to add a check valve to a specific process line and
another would tell him that check valves are not required in that service.

Steve: Don, you know that if we documented everything that was discussed
between our two project teams we would be buried in paperwork. Nothing would
ever get accomplished. Now, if you get two different instructions from my project team, you should advise me accordingly so that I can resolve the discrepancy.
I've decided that since we seem to have a communication problem regarding

engineering changes, I want to set up a weekly engineering meeting for every Thursday. These meetings should help to cut down on the misunderstandings, as well as keeping us advised of your progress in the engineering area of this contract without the need of a formal status report. I would like all members of your project staff present at these meetings.

Don: Will this meeting be in addition to our overall progress meetings that are held on Wednesdays?

Steve: Yes. We will now have two joint Atlay/Lyle meetings a week—one discussing overall progress on the job and one specifically aimed at engineering.

On the way back to his office Don thought about the request for an additional meeting. That meeting will be a waste of time, he thought, just as the Wednesday meeting currently is. It will just take away another day from the Lyle project team's available time for approving drawings, engineering, requisitions, and purchase orders. Now there are three days during the week where at least a good part of the day is taken up by meetings, in addition to a meeting with his project team on Mondays in order to freely discuss the progress and problems of the job without intervention by Lyle personnel. A good part of his project team's time, therefore, was now being spent preparing for and attending meetings during the course of the week. "Well," Don rationalized, "they are the client, and if they desire a meeting, then I have no alternative but to accommodate them."

JUNG'S CONFRONTATION

When Don returned to his desk he saw a message stating that John Mabby (vice-president of procurement) had called. Don returned his call and found out that John requested a meeting. A meeting was set up for the following day. At 9:00 A.M. the next day Don was in Mabby's office. Mabby was concerned about the unusual procedures that were being utilized on this project. It seems as though he had a rather lengthy discussion with Bob Stewart, the project purchasing agent assigned to the Lyle project. During the course of that conversation it became very apparent that this particular project was not operating within the normal procedures established for the purchasing department. This deviation from normal procedures was the result of instructions given by Don Jung to Bob Stewart. This upset John Mabby, since he felt that Don Jung should have discussed these deviations with him prior to his instructing Bob Stewart to proceed in this manner:

Mabby: Don, I understand that you advised my project purchasing agent to work around the procedures that I established for this department so that you could possibly save time on your project.

Jung: That's right, John. We ran into a little trouble early in the project and started running behind schedule, but by cutting corners here and there we've been able to make up some of the time.

Mabby: Well I wish you had contacted me first regarding this situation. I have to tell you, however, that if I had known about some of these actions I would never have allowed Bob Stewart to proceed. I've instructed Stewart that from now on he is to check with me prior to going against our standard operating procedure.

Jung: But John Stewart has been assigned to me for this project. Therefore, I feel that he should operate in accordance with my requests, whether they are within your procedures or not.

Mabby: That's not true. Stewart is in my department and works for me. I am the one who reviews him, approves the size of his raise, and decides if and when he gets a promotion. I have made that fact very clear to Stewart, and I hope I've made it very clear to you, also. In addition, I hear that Stewart has been predicting a 6,000 man-hour overrun for the purchasing department on your project. Why haven't you submitted an additional change request to the client?

Jung: Well, if what Stewart tells me is true the main reason that your department is short man-hours is because the project manager who was handling the initial proposal (George Fitz) underestimated your requirements by 7,000 man-hours. Therefore, from the very beginning you were short man-hours. Why should I be the one that goes to the client and tells him that we blew our estimate when I wasn't even involved in the proposal stage of this contract? Besides, we are taking away some of your duties on this job, and I personally feel that you won't even need those additional 6,000 man-hours.

Mabby: Well, I have to attend a meeting with your boss Fred Franks tomorrow, and I think I'll talk to him about these matters.

Jung: Go right ahead. I'm sure you'll find out that Fred stands behind me 100 percent.

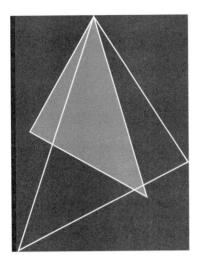

Ivey

Richard Ivey School of Business
The University of Western Ontario

Hong Kong and Shanghai Banking Corporation Limited: Hong Kong Bank Headquarters (A)

Margot Huddart prepared this case under the supervision of Professor Jean-Louis Schaan solely to provide material for class discussion. The authors do not intend to illustrate either effective or ineffective handling of a managerial situation. The authors may have disguised certain names and other identifying information to protect confidentiality.

Roy Munden, assistant general manager of the Hong Kong and Shanghai Bank's (HSBC) Management Services, considered his options for managing the design and construction of the Hong Kong Bank's new headquarters in Central District, Hong Kong. Munden had been appointed by the chairman of the bank, Michael Sandberg, to resolve the problem of providing accommodation for the bank in the future. Two months ago, in November 1979, in his capacity as project director for the development of the new headquarters, Munden had formally engaged the team of building designers for the project. A formal meeting of the board was scheduled two weeks hence on Wednesday, January 22, 1980. Munden needed to assess the options for managing the project and make a recommendation to the board of the bank.

THE HONG KONG AND SHANGHAI BANKING CORPORATION LIMITED

The HSBC, founded in 1865, was one of the first British overseas banks to be opened in the Far East. The merchants of the British colony of Hong Kong founded the bank to finance their trade in China, Japan, and the Philippines. Over time, the western *hongs,* British trade and warehousing companies established in the early days of Hong Kong colonialism, grew into major commercial conglomerates with interests in cargo handling, manufacturing, real estate, and retailing. The powers that ran business in Hong Kong resided in the *hongs;* they were the engine of Hong Kong's economy. Historically, the board of directors of the bank was made up of representatives of the *hongs.*

By 1900, the Hong Kong and Shanghai Bank had become the leading foreign bank in Asia. By 1945, it was the most powerful banking organization of foreign interests in China. In addition to having strong relationships with each of the major *hongs,* the HSBC was the principal banker to the government of Hong Kong. The bank acted for the government in the foreign exchange and money markets and issued 80 percent of the local currency notes. In Hong Kong society, the bank was extremely important: the power structure in Hong Kong was described as the Hong Kong Bank, the Jockey Club, and the Hong Kong government, in that order. Between 1949 and 1979, the profits of the bank rose from HK$17.8 million to HK$2.49 billion, a compounded growth rate of approximately 15 percent per year (see Exhibit I).

In 1977, Michael Sandberg, an Australian by birth, became the bank's chief executive officer and chairman of the board. His aspiration was to lift the bank from its regional roots to be a more multinational and global operation. In 1979, Sandberg was aware of a change in the politics of China, especially with regard to Deng Xiaoping's introduction of new economic policies. To Sandberg, China's move to a market economy meant that the Chinese of Hong Kong would play an increasingly important role in the business affairs of the colony. The bank needed to acquire new allies and associates accessible to Beijing, an avenue to Chinese clients in the People's Republic of China. To this end, the HSBC sold a majority share position in one of the pre-opium war *hongs,* Hutchison Whampoa, to Li Ka-shing in 1979. Li Ka-shing became the third Chinese on the

Exhibit I. **Profit and less statement, HSBC**

The Hong Kong and Shanghai Banking Corporation
Consolidated Profit and Loss Account
for the Years Ending 31 December, 1977, 1978 and 1979

	1977 HK$000s	1978 HK$000s	1979 HK$000s
Combined Net Profit of The Hong Kong and Shanghai Banking Corporation and its subsidiaries	$582,021	$810,225	$1,130,572
Deduct: Profit attributable to outside shareholders of The Hong Kong and Shanghai Banking Corp.	59,977	81,779	116,866
Profit attributable to shareholders of The Hong Kong and Shanghai Banking Corp.	$522,044	$728,446	$1,013,706
Deduct: Transfers to reserves and dividends paid and proposed to parent company and subsidiary companies	415,063	571,111	790,259
	$106,981	$157,335	$223,447
Balance brought forward from previous year less transfer to reserve fund	128,945	137,895	117,801
Retained Profits			
Parent Company	$157,343	$280,186	$309,548
Subsidiary Companies	(78,583)	15,044	31,700
Total: Retained Profits	$235,926	$295,230	$341,248

board, following shipping magnate YK Pao and the chief executive of the Hang Seng Bank, QW Lee.

The Hong Kong and Shanghai Banking Corporation Limited: Symbolism

The image used to denote the bank since 1955 was the bank's headquarters, a monumental building located on the prestigious Statue Square. The square was owned by the bank and dedicated to public use. The square enhanced the bank's visibility and presence in the heart of Hong Kong. In addition, the building's profile was featured on the back of banknotes and on the sign of every branch bank in the colony (see Exhibit II).

Exhibit II. 1979 Hong Kong and Shanghai Bank logo

(a)

(b)

January 1978 to November 1978: The Need for New Headquarters for the Bank

By early 1978, it was clear that the bank had outgrown the existing headquarters. Projections for future growth were at least equal to the growth experienced by the bank in the previous five years. Munden estimated that in five to ten years the bank would need at least another 4,185 square meters (45,000 square feet) of working space.

To accommodate the recent growth of the bank, individual departments had been located throughout central Hong Kong. The total amount of office space outside of the main building amounted to one-third of the floor area of the headquarters. The bank's board of directors had decided that the ad hoc dispersal of offices and staff could no longer continue, especially in view of the planned major international expansion.

Complicating the problem created by the sheer volume of space required by the bank was the need to update the original building's systems and fire safety provisions. The existing building required extensive renovations to provide electrical service for telecommunication systems and to accommodate modern developments in fire precautions.

The board was aware that moving away from Statue Square, while simplifying the problem of office consolidation, could appear to threaten the stability and prosperity of the whole colony by upsetting the *fung shui*[1] of the bank. The *fung shui* of the existing building was considered excellent, and early studies on relocating the bank had been unable to come up with a site that matched Statue Square. To maintain the prestigious location on Statue Square, the board decided to redevelop the existing building site.

"Best Bank Building in the World"

Michael Sandberg wanted the new building to be an architectural landmark in the tradition of fine buildings built for the bank. In addition, the new building would be required to support the bank in carrying out its business into the twenty-first century: It would need to accommodate the rapidly growing bank, be built of the very best materials, and be composed of the most up-to-date systems.

Sandberg had entrusted the management of the project to Roy Munden in addition to his function as assistant general manager of the bank's Management Services. Munden had worked his way up in the bank, after a brief career in the army. During his banking career, Munden had maintained a relationship with Michael Sandberg, who shared with Munden an army background. Both Munden and Sandberg were committed to developing a fine building for the bank. In addition to commissioning a building of high architectural quality, the project was to

---

[1] Fung shui is the ancient Chinese method of interpreting the effect that the surrounding landscape and predominant physical features might have on the fortunes of those who lived in the vicinity.

be five years in advance of any building built at the time and provide twenty-first century technological capacities to enable the bank to leapfrog its competition.

Defining the Problem

Munden hired PA management consultants to refine space calculations for the bank's future needs and to develop a solution for a temporary headquarters. In the fall of 1978, PA produced a 220-page report, which concluded that a phased redevelopment of the bank's existing site was the option the bank should investigate in greater detail.

Subsequent to the presentation of PA management consultants' report, the board initiated a HK$1.2 million feasibility study. The board directed Munden to engage highly qualified building consultants from the design disciplines. PA management consultants recommended Ove Arup & Partners, Structural Engineers, because of their worldwide reputation. Other participants were Levett & Bailey, quantity surveyors who had practiced in Hong Kong since 1962 and J. Roger Preston & Partners, the Hong Kong office of a London mechanical and electrical engineering firm. Although the board wanted to appoint an architect "with an international reputation for outstanding quality of design" for the final building design, for the purposes of the feasibility study it was decided to engage an architect with good local knowledge and experience. As a result, the local Hong Kong architects, Palmer & Turner, were appointed for the study.

Building Consultants' Feasibility Study: January 1979 to April 1979

The feasibility study began in mid-January 1979. In mid-February the consultants produced fifty variations that showed different ways of addressing the problem of phased redevelopment and new construction on the Statue Square site. In early April 1979, two main options for the future use of the site were presented to the board: one was to retain part of the existing building and build a new tower and the other was to rebuild the entire site in phases. The time estimate for the first option was 6.5 years with a projected completion date at the end of 1986 and an estimated cost of HK$600 million (in 1979 prices). The board decided that it wanted still more investigation of the options and that the "additional creative contributions . . . from several firms of architects of international repute" should be incorporated. To keep options for the future as wide open as possible, the board directed Munden to lease any suitable alternative office accommodation "to allow both possibilities to go forward."

Selection of an Architect: June 1979 to November 1979

In response to the board's directive that international architects be invited to provide creative solutions to the bank's design problem, John Scott, from PA

Management Consultants, went to visit a recently completed multistorey bank in London, England to find out how the owner, Lloyd's of London, had selected their architect.

Scott learned that Lloyd's had followed the advice of Gordon Graham, president of the Royal Institute of British Architecture (RIBA). As president of the RIBA, Graham's role was to promote the profession of architecture and the building of fine buildings. Graham had advised Lloyd's to brief a short-listed group of architects on the client's needs and then request a submission from the firms describing the approach they would take in designing a building for the bank's consideration. The board decided to follow Lloyd's example and asked Graham to assist in preparing the terms of reference for the architects, to comment on their suitability and to help assess the aesthetic and practical aspects of each architect's submission.

In the course of the feasibility study, dozens of architectural firms were considered by the consultants, and a shortlist was developed of mainstream corporation architects with branch offices in Hong Kong. In addition to these architects, Graham recommended to Munden that Norman Foster & Associates, a small British architectural firm, be included on the bank's shortlist.

In June 1979, the board was presented with information detailing the selected architectural firms (see Exhibit III). The board formally adopted Graham's proposition of issuing a request for proposals and the shortlist of firms.

The Request for Proposals (RFP)

The Request for Proposals, "Redevelopment of 1 Queen's Road Central, Hong Kong," set out the following goals for the design solutions:

- To help the bank decide on an approach to solving the problems of whether a full-phased scheme or a south tower scheme is better
- To help the bank decide whether the best scheme is better than doing nothing
- To select and appoint an architect

The seven short-listed firms attended a joint briefing meeting on July 11, 1979. At the meeting, the architects were told that important criteria for judgment would lie in the way in which they showed an appreciation of the bank's problems and local conditions and clarified the issues on which the bank's decision depended. The bank did not request detailed proposals for structural engineering, foundations, building services, construction methods, costs, or program. The RFP stated that "the bank is determined that the building should be one of considerable architectural merit." A stringent requirement was the need to keep the bank in operation from the site throughout the redevelopment period.

The seven firms were requested to submit their proposals by October 6, 1979. The firms were told that they might be asked to do a live presentation subsequent

Exhibit III. Information presented to the board in June 1979 regarding recently constructed banks and other buildings relevant to the bank project and their height

U.S.A.:	Year	Height
Skidmore, Owings & Merrill (Chicago)		
—Chase Manhattan, New York	1960	60 stories
—Bank of America HQ, San Francisco	1969	52 stories
—Sears Tower, Chicago	1974	110 stories
—John Hancock, Chicago	1970	100 stories
Hong Kong experience		
Hugh Stubbins & Associates Inc. (Boston)		
—Citicorp Center, New York	1977	59 stories
—Federal Reserve Bank, Boston	1977	33 stories
Australia:		
Harry Seidler & Associates (Sydney)		
—MLC Centre, Sydney	1977	58 stories
—Australia Square, Sydney	1967	50 stories
—Australian Embassy, Paris	1978	
—Conzinc Riotinto HQ, Melbourne	1975	50 stories
Yuncken Freeman Pty Ltd (Melbourne)		
—BHP HQ, Melbourne		40 stories
Hong Kong experience		
Britain:		
Gollins, Melvin & Ward Partnership (London)		
—P & O Building London	1969	14 stories
—Commercial Union, London	1969	26 stories
—Banque Belge, London	1978	15? stories
—Barings Bank, London	in progress	
Hong Kong experience		
Foster & Associates (London)		
—Sainsbury Centre for Visual Arts, Norwich	3 awards	
—Willis Faber HQ, Ipswich	3 awards	
—IBM Office, Cosham	2 awards	
Hong Kong:		
Palmer & Turner		
Reserves:		
Minour Yamasaki (U.S.A.)		
World Trade Centre, New York	1974	100+ stories
Harry Rosenberg Mardall (London)		
Hong Kong experience		

to the submission. Each firm was paid HK$150,000 for their participation in the request for proposals.

Visits to Offices of Short-listed Architects

In September 1979, Munden visited the short-listed architectural firms. He met with the architects, toured buildings of their design and visited their previous clients. Munden discovered wide variances in the level of attention and interest that the different offices paid to him as a potential new client. In one office, he was left to wait in the lobby and served sandwiches for lunch. At another, he was taken for a ride in a helicopter and wined and dined. Munden returned to Hong Kong with a kaleidoscope of impressions with respect to the services provided by architectural firms and the quality of the buildings they designed.

Receipt of Architects' Submissions

On October 6, 1979, the architects' submissions were received by the bank. Each submission was unique. The submissions were assessed by Munden, Graham, David Thornburrow, a partner with the Hong Kong architectural firm Spence Robinson, and five people from PA Management and the bank. After a review of the submissions, Munden invited Norman Foster to Hong Kong for an interview on October 11. Munden told Graham of his decision and advised Sandberg.

To clarify the status of the first stage of the architectural selection process, Munden issued a circular to the directors of the bank on October 12, which stated that the design options would be placed before the board on November 13 and that in the meantime, further talks would take place with Norman Foster. Munden then invited Foster to Hong Kong to present his proposal.

Norman Foster & Associates

Norman Foster's office, a firm of twenty architects and designers, was recognized in the United Kingdom as an innovative, aggressive design firm. Foster designed buildings using up-to-date materials and technologies. He had implemented a unique design strategy for the development of innovative buildings, "developing design with industry," on small-scale projects in the United Kingdom. The technologically advanced designs prepared by his firm required the implementation of close working arrangements between the architectural designers and the manufacturers of building components. The innovative buildings resulting from the close collaboration had earned the firm several prestigious awards for buildings of architectural merit.

Developing Design with Industry

After the design of a project was initiated, Foster & Associates developed a list of qualified manufacturers with whom they could pursue their "developing design with industry" strategy. After the firm had researched companies capable of manufacturing the required components for a project and selected companies capable of doing the work, a call for tenders based on performance specifications[2] and for a detailed description of what the architect wanted the particular component to do was issued. Once a contract was agreed upon, the architect worked with the manufacturer to fine-tune the particular component. The result was the development of a kit of parts for the fabrication of a building that could be site-assembled to a greater degree than was possible in conventionally designed structures. Foster's procedure of *developing design with industry* had been exercised by his firm on projects comprising up to thirty subcontracts. Foster's goal in the design of the Hong Kong and Shanghai Bank was to design a building that could be assembled on site from premanufactured pieces and to implement a phased construction approach on a large project (Exhibit IV).

In addition to proposing innovative building designs and construction processes, Foster promoted to the owner the idea of assessing the costs of a building over the course of the building's useful life. With respect to material selection, Foster proposed to clients that a building's projected capital and operating costs should be looked at together when decisions regarding building materials were being made. Foster maintained that clients were best served by building high-quality, highly flexible, durable structures, which allowed for significant changes in use and work methods, without incurring significant costs or causing disruption to employees' work.

Preparation of a Submission to the RFP, "Redevelopment of 1 Queen's Road Central"

Foster & Associates experienced a lull in the workload of the office in the fall of 1979. A major contract to build a transportation interchange at Hammersmith in west London, which had been anticipated by the firm, had not materialized, and the remaining projects in the office were nearing completion. Foster took advantage of the downtime to prepare the competition proposal for the bank. In general, effort expended on design submissions was matched by the firm in the preparation of oral presentations. Foster practiced presentations to a high degree of polish. He even rehearsed what would be discussed during the coffee break of an actual presentation. His staff likened his presentations to theater—although highly rehearsed, the actual presentation appeared to be ad lib.

---

[2] Performance specifications were specifications written to detail the final parameters that a building component had to meet. The exact quality and details of a building element were left to the discretion of the manufacturer.

Exhibit IV. *Traditional design-then-build approach*

Historically, the traditional *design-then-build* approach has been the most prevalent form used for construction projects. In this method, the design is completed by a design firm before contractors are invited to bid on the work and before construction begins. Obviously, this requires ample time for a design to be developed and then time for the actual construction to occur. If the design requires a year and the construction two years, then the process would be as shown:

Design
| One Year | Construct |
| Two Years |

Fast-Track, or Phased Construction Approach

Owners use the *fast-track,* or *phased construction,* approach in an attempt to have the project completed in a shorter timeframe than the traditional approach would require. For instance, in the example above, the project would take three years from the start of design to the completion of construction. This may not be acceptable to an owner in terms of its needs for the project. So a method must be employed to decrease the duration for the job. In a fast-track job, the design and construction are integrated or overlapped so that the total time for the project is reduced. This is accomplished by breaking the project into specific phases and following the design of each phase with its construction. While construction is occurring for the first phase, the design is being accomplished for the second. This process continues for the entire project. Taking the previous example, the project may be divided into the following phases:

1. Site work and foundations
2. Structural steel
3. Mechanical and electrical
4. Building enclosure
5. Interior finishes

The design would begin with the first phase, site work and foundations. As soon as that phase is complete, bids are solicited for construction. While the site work and foundations are being constructed, the design is being completed for the structural steel for the building. If properly timed, the structural steel package goes out for bid such that construction can begin approximately at the end of the construction of the foundations. The process continues as shown:

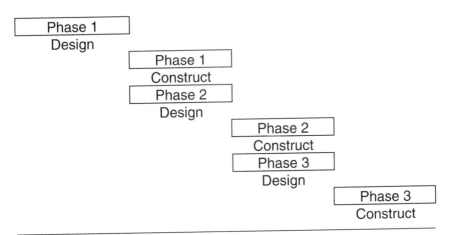

| Phase 1 |
| Design |

| Phase 1 |
| Construct |
| Phase 2 |
| Design |

| Phase 2 |
| Construct |
| Phase 3 |
| Design |

| Phase 3 |
| Construct |

Presentation to the Board

Gordon Graham, the bank's architectural adviser, attended the November 13, 1979, board meeting with Roy Munden. Although Munden had anticipated a quick hearing of his recommendation by the board, the board members discussed the submissions by the various architects and their qualifications for several hours. After the general discussion, Norman Foster was invited into the meeting. The board adopted the recommendation, put forward by Munden, and Norman Foster & Associates were appointed.

The day after the board meeting, Michael Sandberg, Roy Munden, and Norman Foster met to discuss the project. Foster's estimated completion date for the project was December 1985. In another submission to the RFP, the American firm, Skidmore, Owings and Merrill, Architects & Engineers, had identified June 1984 as its projected completion date. The bank was interested to know if Foster's schedule could be contracted. The bank was concerned with respect to schedule; cost was a lessor priority. Later that month the board advised Munden:

> In view of the prestigious and special nature of the building and having employed one of the world's leading architects it (is) not appropriate to place any particular ceiling on the price to be paid for the building.[3]

Contract Award and Project Design: November 1979 to January 1980

The award of the contract to Foster & Associates from an international field of competitors, along with the expressed mandate of the bank "to produce the best bank building in the world," had tremendous impact on the designers in Foster's office. The designers, young architectural graduates who had been handpicked by Norman Foster for their commitment to modern buildings, were eager to exercise the ideas of *popism* that they had already begun to explore in some of the firm's projects.[4]

> It was like being mercilessly overpaid," said one designer. "The award boosted the egos of everyone in the office even higher. It was like we had never left school, like writing a manifesto.

With the award of the contract, the practice suddenly became fluid. The designers recognized the tremendous opportunity to implement ideas they had previously explored on small projects. Norman Foster provided the inspiration—he put confidence into the designers and challenged them by pitting one against another in

---

[3] November 25, 1979, board paper.

[4] *Popism* was an architectural style.

design teams. The atmosphere in the office was described as highly competitive "like a training ground for pedigree race horses."[5]

The Project Team

For the preparation of the submission to the RFP, Norman Foster had chosen to work with Ove Arup & Partners, Structural Engineers. The London headquarters of Ove Arup & Partners was next door to Foster's office on Fitzroy Street. Subsequent to the award, the bank formally engaged Ove Arup's for the detailed structural design of the building, J. Roger Preston, a London-based firm for the mechanical and electrical design and Levitt &Bailey, quantity surveyors. Under the terms of the agreements, the consultants were employed directly by the bank. The architect was responsible for the direction, management and coordination of all consultants (see Exhibit V, project organization chart). Upon contract award, the design teams in the architect's and structural engineer's offices were mobilized.

Ove Arup & Partners

Ove Arup & Partners had offices in twenty-two countries and 3,000 staff. They were the largest firm of consulting engineers in the United Kingdom and had designed the structure for the Center Pompidou in Paris and the Sydney Opera House. Although the firm was renowned for the structural design of landmark buildings, the challenge of building the best bank building in the world was an uncommon request from a client. Large industrial clients of the firm usually proceeded along an organized path, according to program. The bank, a client viewed within the firm as perhaps one of the last "patrons of the arts," was focused on developing a building of high architectural quality and on keeping the bank in operation throughout the project.

At Ove Arup, from very early on, there was a sense that this project would be unique and would not follow the structured path that the firm had grown accustomed to with its traditional industrial projects. To establish a team capable of rising to the challenge, Jack Zunz, senior partner, assembled a small team of the best and brightest in the London office and located them in an office in a nearby building. There was an understanding amongst the team members that they were being offered the opportunity of their careers. The team divided itself into design area groups that mirrored the design divisions in the architect's office.

---

[5] Conversation with Tony Hackett, one of the designers in Foster's office.

Exhibit V. Project organization chart

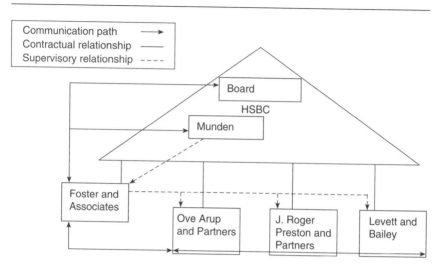

Foster's Team

The bank job was the largest, most prestigious job ever to be undertaken by Foster & Associates. The design team for the bank was broken up into three primary work areas: client relations, programming, and design. Client relations were handled by Norman Foster who flew to Hong Kong to meet with Munden and Sandberg as required. Spencer de Grey, an associate with the firm, was put in charge of developing the program for the building in conjunction with the bank's staff and was located in Hong Kong. Design was handled by a team of eight in London, led by Roy Fleetwood. Fleetwood had seven years of experience at Foster. Also on the team were Ken Shuttleworth, who had worked with Foster since 1976 and had joined the firm formally in 1977, and David Nelson and Tony Hackett, recent graduates.

The design of the project at Foster was "like fire fighting element by element."[6] Design meetings held within the office lasted between one and four hours and were intense and collaborative. The abstract concepts, which would inform the design of the building, were discussed in depth. Weaknesses in the design schemes were identified and jobs were reallocated. Team members would volunteer to take on the design of new elements of the building as new problems came up. The high energy and intensity expended during the design phase of the building were interspersed with emotionally difficult periods for the architects. The

---

[6] Conversation with Tony Hackett.

major difficulty stemmed from the highly ambiguous deliverables of the project. The only certainty was that the HSBC building would be significantly different from any other building in the world. This knowledge raised the team's excitement and commitment even higher.

Fluid Working Relationship between the Offices of the Engineer and the Architect

The conceptual design of the building necessitated a fluid relationship between Ove Arup's office and Foster—it was not unusual for a member of one team to call up another and ask for help to fix a problem or make a correction to a design. If designers at Foster were meeting about an element of the building and needed engineering input, a telephone call would be made to Ove Arup's to summon an engineer to attend the meeting. One engineer described the project environment as one in which "there was a lot of honesty." The interdependence of the members of the design team was understood—the project required them to work together, no matter what problem arose.

Managing the Project: January 22, 1980

Munden reviewed the remaining areas of work that needed to be further defined: the overall responsibility for the project, management of the construction process, and the synthesis of the Bank's spatial and functional requirements. He wondered who should take on these areas of work and what their roles and responsibilities should be.

Overall Management of the Project

The first area of work, the overall management of the project, was becoming a concern to board members of the bank. At the moment, the bank held the contracts with the individual consultants. The architect was responsible for the direction, management and co-ordination of the consultants (see Exhibit V). Norman Thompson, one of the bank's board members, believed that the bank should have a co-ordinator for the entire project. To this end, he was highly supportive of the proposal made by one of his employees, Ron Mead, project director of the Hong Kong Mass Transit Railway (MTR). Mead was credited with being responsible for the under-budget, early completion of the MTR. In mid-December 1979, Munden received a proposal from Ron Mead. In his proposal, Mead advised the bank to create a position of control over the entire project. Mead suggested that he be engaged as construction coordinator for the project. As construction coordinator, he would be responsible for transmitting the bank's instructions to the architect, running the subcontracts, and overseeing construction. This description was very close to that of a full-fledged project manager (see Exhibit VI).

Exhibit VI. Project organization proposed by Mead

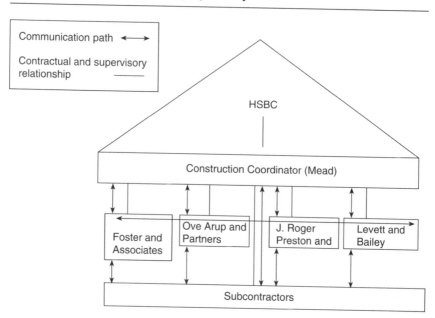

In discussing the concept with the architect, however, Munden recognized that Norman Foster was not in favor of the bank hiring Mead. As architect to the bank, Foster wanted complete authority for the project and direct access to bank personnel. Foster's concern was that a construction coordinator, as described by Mead, would impede his direct line of authority from the bank. De facto, the construction coordinator would act as a filter between the architects and the owner, thus eliminating direct access to the bank and reducing the architect's control over the project. In addition, Foster felt that there was no need for a construction coordinator to oversee the consultants. Instead, Foster suggested to Munden that the bank consider engaging a project coordinator and a management contractor.

The project coordinator envisioned by Foster was more limited in scope than the construction coordinator proposed by Mead. Foster recommended that the project coordinator be someone with experience in construction management, cost control, and project planning. Foster proposed that the project coordinator act as a focal interface between the bank and the architects. As such, the project coordinator would provide direct access to the owner and be the sole source of

Exhibit VII. Project organization proposed by Foster

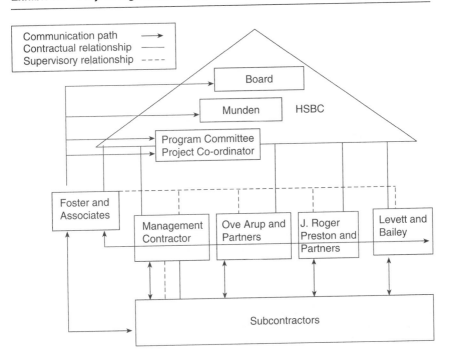

bank instruction to the architects. A second component of Foster's recommendation was his proposal for a management contractor who would participate with the consultants in the design of the building, hold and manage the subcontracts and oversee the construction process (see Exhibit VII).

Building Program

The second area of work, which needed further definition, was the establishment of a spatial program and definition of the functional interrelationships that reflected the organization of the bank.

The architects had requested input from the bank on the size of the departments to be accommodated in the new building. The bank had experienced difficulty in synthesizing this information because of the tremendous growth and changes in business practices it was undergoing. Munden had approached Sandberg about appointing a special program committee made up of representatives of the different areas within the bank to direct the architects on the bank's

Exhibit VIII.   Special committee for program development

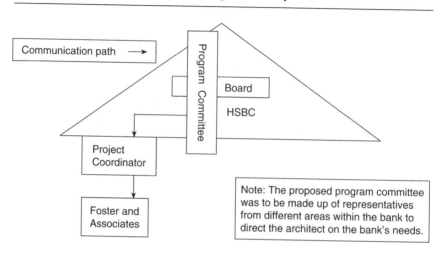

needs (see Exhibit VIII). Sandberg had rejected Munden's proposal on the basis that the architect had been contracted to provide the bank with a flexible building that would incorporate changes as they were required. With regard to the bank's special needs, Sandberg reminded Munden that the bank had selected Foster, in part, because of the understanding he had illustrated of the workings of the bank. In addition, Foster's previous clients had verified how Foster's office had worked with their staff to develop a program for the building. Sandberg advised Munden that any decisions Munden felt ill-equipped to make could be referred directly to him (see Exhibit IX).

Preparing for the Board Meeting

As Munden considered the presentation he would make at the upcoming board meeting, he reviewed the project arrangements established to date: The feasibility study was complete, the architect had been selected, the consultants were under contract and the design team was brainstorming concept proposals for the design of the building. He recognized that the organization of the project team, both internal to the bank and the external design team was as challenging as the assignment facing the architects and engineers: to build "the best bank building in the world." Munden knew that the structure set up by the relationships between the consultants would have a strong impact on innovation in the project and the final building. Strong opinions were being expressed within the bank and

Exhibit IX.   *Reporting structure for building program*

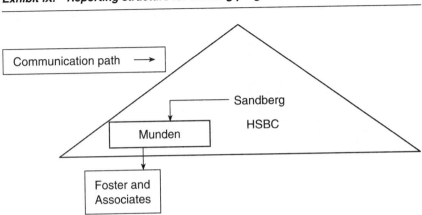

by the design team with respect to the overall management of the project. Munden considered the motivations of Sandberg, the members of the board and the consultants as he prepared his recommendation to the board for the management of the project. He knew that the outcome of the project had to be an innovative building composed of the most up-to-date systems and highest quality materials.

The Richard Ivey School of Business gratefully acknowledges the generous support of The Richard and Jean Ivey Fund in the development of this case as part of the RICHARD AND JEAN IVEY FUND ASIAN CASE SERIES.

Index